Our Food, Air and Water: How Safe Are They?

OUR FOOD, AIR and WATER: How Safe Are They?

AN EDITORIALS ON FILE BOOK

Edited by Carol C. Collins

Facts on File Publications

New York, New York • Bicester, England

Our Food, Air and Water: How Safe Are They?

Published by Facts On File, Inc.
460 Park Avenue South, New York, N.Y. 10016
© Copyright 1984 by Facts On File, Inc.

Library of Congress Cataloging in Publication Data
Main entry under title:

Our food, air, and water.

  (An Editorials on File book)
  1. Environmental health—United States—Addresses, essays, lectures.   2. Environmental health—Government policy—United States—Addresses, essays, lectures.   I. Collins, Carol Chambers.   II. Facts on File, Inc.   III. Series.
RA566.3.093 1985   363.7   84-18816
ISBN 0-87196-967-X

International Standard Book Number: 0-87196-967-X
Library of Congress Catalog Card Number: 84-18816
9 8 7 6 5 4 3 2 1
PRINTED IN THE UNITED STATES OF AMERICA

Contents

Preface

In the years following Earth Day 1970, Americans became painfully aware of the danger to their health and surroundings posed by the effluvia of an industrial society. New legislation was passed during the 'environmental decade' to regulate the pollutants belched into the air, pushed into rivers and lakes and spread across the land. With the eruption of Love Canal in 1978, public attention focused on the urgent problem of hazardous waste disposal. The chemical alphabet soup which oozed from the ground at this oddly named New York site underscored in a dramatic way the environmental consequences of the nation's previous, unrestrained technological progress. The discovery of this and other industrial waste sites soon produced a litany of chemical products and byproducts whose strange names were spoken with dread: TCE, PCBs, dioxins, EDB. The problem had literally been brought home, to the backyards of residents across the country. Public fear mounted as scientific evidence accumulated linking debilitating diseases with the foods people ate, the water they drank and the air they breathed.

At the same time, a question that had been voiced quietly a decade earlier began to be asked more frequently and more loudly—was it wise to try and restrict the proliferation in the environment of these potentially harmful substances without regard to cost? The late 1970's and early 1980's were characterized by energy shortages, high inflation and economic recessions. In this cost-conscious atmosphere, the burdens placed on industry by stringent environmental laws were seen more critically. Most Americans, after all, would be unwilling to do without the many products— from detergents and eyeglass lenses to pesticides and housing insulation— whose manufacture created these dangerous contaminants. The importance of a stable economy, industry adherents argued, should be counterbalanced against the health benefits to be gained from pollution control. This cost-benefit approach has been enthusiastically adopted by the Reagan Administration, with its anti-regulatory stance.

One problem with cost-benefit treatment of this issue, as environmentalists are quick to point out, is the difficulty of assessing the monetary value of human health, or the cost of lives curtailed. As the debate over environmental health risks continues into the mid-1980's and beyond, it is this fundamental problem that must be addressed. We have only one Earth on which to live, and must necessarily suffer the consequences of defiling it. On the other hand, if the cost of cleaning up past pollution and effectively preventing it in the future will cripple the economy, could as many lives be affected by unemployment and poverty as would have been stricken by environmentally-related disease?

Complicating this emotional issue is the uncertainty surrounding the actual health effects of low-level exposure to many toxic chemicals. While many medical researchers feel that the development of that modern scourge, cancer, can in most cases be traced to exposure to environmental carcinogens, this has yet to be proven. In May of 1984, however, the White House science office concluded that it was unsafe to assume there was a threshold level below which doses of suspected cancer-causing substances could be tolerated; its report also rejected proposals to reduce reliance on tests of laboratory animals in determining the risk of cancer in humans. Similar gaps in scientific knowledge about substances thought to cause birth defects, gene mutations or various diseases make it difficult both to regulate these substances and to assess the damage they cause.

In the following pages, newspaper editors from across the United States offer their opinions concerning many aspects of this far-reaching problem. Their words reflect the tenor of national debate about such issues as the use of pesticides, the presence of asbestos in our schools, or the growing devastation caused by acid rain. The choice of editorials is intended to include as representative a sampling of journalistic thought as possible.

September, 1984   Carol C. Collins

Part I: Food

Since the 1960's, when the dangers of agricultural pesticides became widely recognized, Americans have grown increasingly wary of the foods they consume. Not only have the natural ingredients of many foods come under suspicion, but also the ingredients introduced by man. Successive controversies over nitrites, saccharin, salt, caffeine and other food ingredients and additives have served to keep the issue of food safety very much alive. And because there is disagreement in the scientific community about many potentially harmful substances contained in the food we eat, it is an issue that is not likely to disappear soon.

Particularly frustrating is the uncertainty surrounding a subject of paramount public concern—cancer. Each time it is reported that a common food component has been shown to cause cancer in laboratory animals, it is accompanied by a disclaimer that we do not yet know how to confidently predict the consequences to humans of eating the same substance. If saccharin causes cancer when fed in massive doses to rats, does that mean it is unsafe for human consumption, even in tiny amounts? Are the effects of such consumption cumulative? The reaction of the American public to such reports depends very much on the perceived benefits of the substance under study; a known benefit can be more compelling than an unknown danger, particularly if the danger will not manifest itself for many years.

Similarly, an immediate threat provokes more opposition than a continuing, invisible one; there has been much more vocal resistance to isolated, massive sprayings of chemical pesticides (in order to control specific pests such as the "Medfly" or "gypsy moth"), than there has been to the steady rain onto food crops of pesticides that contain suspected carcinogens. In addition to pesticide residues, other contaminants entering the food chain through the environment—such as lead or PCBs—may pose a more insidious danger to human health than any yet to receive widespread attention.

1

Pesticides: Spraying Our Food with Chemicals

American farmers use an estimated 700 million pounds of pesticides each year to protect their crops from damage by insects, rodents, fungi and weeds. Designed to kill or repel these nuisances, pesticides have greatly raised crop yields and helped to hold down food prices. Yet, because of their toxicity, they are also dangerous to both humans and the environment. There are approximately 600 basic chemical compounds contained in the nearly 40,000 pesticide products on the market, and they fall into three groups. Chlorinated hydrocarbons are highly toxic and chemically stable pesticides that persist in the environment for long periods of time—these include DDT, endrin and lindane. Organic phosphates, such as malathion and parathion, are also highly toxic to humans but do not remain in the environment for very long. Carbamates have the lowest toxicity to humans.

The extensive environmental damage that could be wrought by the widespread spraying of pesticides was first made known to most Americans through Rachel Carson's 1962 bestseller, *Silent Spring*. Since the publication of that frightening book, which urged methods of pest control that "do not destroy us along with the insects," pesticide spraying in the United States has doubled. Pesticide residues have found their way into the tissues of birds and fish that eat insects or sprayed crops, and mammals, including humans, that eat the infected fowl or fish. Through a process known as biomagnification, some pesticides move up the food chain in exponential fashion, transferred in constantly growing concentrations from one predator to another. Pesticide residues are also present on many foods available at greengrocers or on supermarket shelves. In 1970, it was found that Americans carried in their bodies an average of 12 parts per million of pesticide residues, or nearly twice the levels allowed for most foods in interstate commerce.

In sufficient amounts, pesticides can cause nervous disorders or death. Among farm workers in the United States, there are about 100,000 cases of pesticide poisoning each year. Many pesticides have also been shown to cause cancer and metabolic disorders in laboratory animals. The long-term effects of low-level exposure to these toxic chemicals are less well understood. Conservationists argue, however, that even without the dangers they pose to human health, chemical pesticides are an inefficient and destructive form of insect control. Since World War II, the spraying of pesticides has increased tenfold while insect damage to crops has doubled; part of the reason for this is that many insects have developed resistances to the toxins used to destroy them. The consequences for other wildlife and even livestock in the targeted spraying areas are sometimes more severe than for the insects or weeds that were to be brought under control.

Los Angeles Times

Los Angeles, Calif., March 18, 1980

From Oregon to Washington, D.C., to the Mexican border, pesticides work faster than the government regulates them.

Last week the Environmental Protection Agency started laying out its case for a permanent ban on 2,4,5-T, a herbicide used to kill undergrowth in Oregon forests and suspected of causing miscarriages and birth defects among area residents. But the process could take years. Assistant agency administrator Steven D. Jellinek predicts the hearing, conducted like a court trial with testimony from witnesses, cross-examination and filing of legal briefs, could last 18 months. Then EPA administrator Douglas Costle must decide whether to make permanent the current emergency suspension of most uses of the pesticide. Appeals of regulatory decisions have become the rule rather than the exception, adding still more months to the process. There has to be a faster way.

But what happens when a pesticide is banned and farmers use it anyway?

Three weeks ago, some skillful scientific detective work by the Food and Drug Administration laboratory in Los Angeles identified residues of the pesticide Celathion on tomatoes and vegetables entering the United States from Mexico. The Environmental Protection Agency says Celathion has not been proved safe; it therefore may not be used on an agricultural product, except experimentally on peaches and grapes, in the United States. Imported into Mexico from Germany by an affiliate of Union Carbide Corp., it can be used only on cotton in Mexico, and cannot be sold in areas of that country that do not grow cotton. But Times reporters Laurie Becklund and Ronald Taylor discovered that Celathion was in fact being sold outside Mexico's cotton regions. Some farmers have been using it illegally for seven years.

The discoveries raise more questions than they answer:

—What, if any, long-term health hazards do consumers face?

—Why hadn't the Food and Drug Administration discovered the residues earlier?

—Why didn't Mexican pesticide authorities question the quantity of Celathion being imported, which was more than seemed reasonable if the product was being used just on cotton?

—Knowing that the U.S. government had rejected an application that would have allowed the use of Celathion on tomatoes, why did not Union Carbide Corp. keep a closer eye on its use? Since it was discovered last week that Mexican growers were using Celathion illegally, the company has decided to cancel future sales of the chemical there.

—And, finally, if laws aren't enforced, what will it take to make some people consider the consequences of their actions?   □

The Ottawa Citizen

Ottawa, Ont., March 25, 1980

Harry Parrott, it is said, is a competent dentist, and a nice guy as well. Now if Parrott the dentist was confronted with information that there were doubts about the safety of a certain anesthetic, would he continue to use it because he had a large supply in his office? Even if he were convinced in his own mind that there was no danger? Probably not. He wouldn't take a chance on harming his patients.

But Harry Parrott, as minister of environment in Ontario, applies a curious logic in his public role as protector of our flora and fauna — and of us. Last year his department temporarily banned the herbicide 2,4,5-T, a chemical sprayed on weeds on power rights-of-way and roadside ditches. Although he claims that "no environmental or health" hazards have been discovered in testing the herbicide, Ontario still plans to impose a permanent ban on 2,4,5-T for political reasons. The public doesn't trust chemicals, and the minority Ontario government doesn't want to displease the public.

Meanwhile, Parrott has a problem that underlines the public's concern with every dangerous substance. Once science produces 2,4,5-T or nuclear waste, how do we get rid of the stuff? Hazardous products are not easily dumped, as American residents in Niagara Falls, N.Y., can attest. Chemical dumping in the Love Canal came back to haunt them half a century later. Canadians in Port Hope encountered other latent environmental problems as radioactive gases began seeping into their basements from buried wastes.

Parrott is apparently determined that there will be no such concentrations of 2,4,5-T to cause trouble in the future. He proposes dumping 161,379 litres of the herbicide, now in storage, by spraying it on weeds across Ontario. He claims there is no other way to get rid of it.

This sounds too much like a drunk driver disposing of the last half bottle by drinking it before he hits the road again. If the herbicide is dangerous enough to be permanently banned, then it should not be used at all. Leave it in storage until some method of disposal can be developed, or extensive testing proves that it is not dangerous.

Roanoke Times & World-News

Roanoke, Va., November 29, 1980

Five years ago it was called an environmental disaster. Careless handling of the pesticide Kepone had poisoned scores of workers at a Hopewell manufacturing plant, causing them distressing nerve problems. Dumping of Kepone wastes in the city's sewage system had contaminated the James River.

As much as 20 tons of the pesticide were estimated to have settled on the river bottom — and in animal and aquatic life in the area. Years later, traces of Kepone still were being found in sewage lines, soil, water and creatures. "At the current rate," said a federal report in early 1978, "it will take 50 to 100 years for the James River to cleanse itself."

All ominous-sounding. But a few days ago, the state Board of Health lifted most of its five-year ban on commercial fishing in the James. This action was taken a week after newspapers in Virginia ran stories about a William & Mary college professor's frustrations at trying to persuade state officials to underwrite research into his findings that Kepone was getting into the human food chain.

Given the circumstances, the board's motives could readily look suspicious. Economic reasons certainly figured in the action. But it would be a mistake to infer cover-up or chicanery. Kepone will be around for a long while — it's one

of those insoluble chemicals, like DDT, that accumulate in the biosphere. Recent evidence, however, suggests Kepone is not nearly as deadly to humans as first feared.

Even those unfortunate people who worked directly with the stuff — some dipping ungloved hands into drums of the poison — appear to be recovering from the effects. In contrast, small doses — which is what authorities must be concerned about now — may not be overtly harmful. Research at the Medical College of Virginia indicates that people rid themselves of Kepone more quickly than do rodents. The W&M professor mentioned above, who found Kepone traces in field mice near the James, may have been headed up a blind alley.

This isn't to say that all questions about Kepone toxicity, or its possible links with cancer, have been finally answered. The environment is laced with chemicals — thousands more compounds are introduced yearly — about which we know too little. Efforts at research and control must be expanded

But pending definitive findings, all human activity can't be put on hold. Best available information indicates that it's reasonably safe to consume fish from the James. All things considered, the Board of Health action was not untimely.

The Seattle Times

Seattle, Wash., January 18, 1981

BETWEEN the fall of 1979 and the summer of 1980, 12 pregnancies in the small community of Ashford, southwest of Mount Rainier, resulted in nine miscarriages, a stillbirth, and an infant death.

Many local residents blamed herbicide spraying in the area, which is a common practice of timber companies and government agencies to keep down underbrush. There were calls for a ban on herbicides.

Now a report from the state's Department of Social and Health Services says there is no apparent link between herbicide spraying and reproductive problems in Ashford. The report is a consolidation of several studies done over the past 10 months by private physicians and state officials.

The studies, some of which have been reported previously, found "minimal" herbicide or pesticide exposure of Ashford residents in the past two years; no traces of chemicals or heavy metals in Ashford's drinking-water supply, no parasites in the women's blood samples, and no common problems among the local animal population.

Dr. Samuel Milham, who headed the state investigation, said the Ashford incidents may be "a chance time-space aggregation of unfavorable pregnancy outcomes," adding that the causes of most miscarriages are still unknown.

The D.S.H.S. report clearly will not satisfy all the residents of Ashford, some of whom contend that investigators had a predetermined notion about their ultimate findings. After such personal tragedies, we can understand these citizens' anguish and their desire to identify the cause of their loss.

But we urge them and others concerned about herbicide sprays to take a close, dispassionate look at the scientific evidence and not base their views on emotional or ideological arguments.

The herbicide in question at Ashford, 2,4-D, is the same chemical found in "weed and feed" products used two or three times a year on countless lawns throughout the country. It is also found in flea collars used on many pets. No problems have been reported from these uses.

Until a clear scientific connection between herbicides and other human health problems is proved conclusively, based on thorough studies and proven facts, we'd like to see the anti-herbicide rhetoric toned down a bit. The latest D.S.H.S. report should help accomplish that result.

THE ANN ARBOR NEWS

Ann Arbor, Mich., April 6, 1982

ONE of the landmarks of the environmental movement was the publication of the now-classic "Silent Spring" by Rachel Carson.

In that book, Miss Carson alerted the world to the danger of the chemical pesticide DDT. For that, she was vilified and called an extremist.

She documented her case well enough, however, to have DDT banned in this country. That was 10 years ago.

NOW the U.S. Fish and Wildlife Service has released a report on the results of the ban on DDT. The report contains some good news and a lesson for latter-day skeptics.

The research and wisdom and Rachel Carson have paid off. Birds which were in serious danger from the widespread appearance of DDT in food chain, soil and stream are beginning to come back.

The reason is the slow but sure breakdown and ultimate disappearance from the environment of DDT. Examples:

Bald eagles, once almost wiped out from the lower 48 states, are seen in increasing numbers along the East Coast, the New Jersey wetland areas and the Great Lakes.

The osprey has made an even faster comeback to the point where scientists foresee a return to pre-DDT levels by the end of the century.

Before the ban, people in the Great Lakes region were warned about eating fish with high DDT content; now, DDT is barely detectable. (Today, however, the warning applies to PCB).

RUNNING WATER purifies itself. That is one of those environmental givens that man has expropriated for himself and abused almost beyond recognition in some instances.

The environment *can* cleanse itself, however, and with an assist from man, if need be. The DDT case is a good example of man helping the environment to heal itself.

The lesson in all this is that clearly beneficial environmental programs and policies already *in place* should be allowed to continue. They were conceived and executed for good purposes, whether cleaner air and water, better soil conservation or whatever.

Officials such as Interior's James Watt and EPA's Anne Gorsuch should read the USFWS report and mull its implications. A better environment and an improved economy are not antithetical.

The Honolulu Advertiser

Honolulu, Ha., July 31, 1981

It is not a crisis, but there is much to be concerned about in the latest state Health Department findings about DBCP.

The pesticide was banned two years ago nationwide. It was found to cause genetic changes, cancer in laboratory animals, and infertility in males.

Hawaii, however, was exempted by the Environmental Protection Agency. The pineapple growers say DBCP is needed on some fields, particularly on Maui, to control nematodes (root worms). The industry claims a replacement pesticide, EDB, is not as effective in such areas.

OVER THE PAST few years, minute traces of DBCP have been found in wells. The levels have been so small — .01 parts per billion — that it is not thought to be a danger to those drinking water from those wells.

But the latest Health Department study has disclosed disturbing new information. DBCP, still in extremely small quantities, has turned up in central Oahu, where the pesticide has not been used for years. In addition, a Kauai well in an area where DBCP has not been used in 12 years was found to have the pesticide.

This means DBCP remains in the environment for a long time, not just a relatively few weeks as previously thought.

That, added to the recent discovery that DBCP does "percolate" through Hawaii's soil to ground water levels, means there is a strong possibility the pesticide will continue to find its way into drinking water.

This is particularly worrisome on Oahu, where ground water supplies most of the island's drinking water.

But it is an even greater concern where DBCP is still used. That it does not break down in lower soil levels as quickly as had been thought means continued use may aggravate an already existing problem.

AGRICULTURE and health department officials are to meet soon to discuss whether the state's conditional support of Hawaii's DBCP exemption should be continued.

The EPA's banning of DBCP was caused by the dangerous nature of the pesticide, and the fact it was beginning to appear in high levels in drinking water supplies on the Mainland.

One particularly hard-hit area was the San Joaquin Valley in California. In fact, that state had such a serious DBCP problem it banned the pesticide four years ago.

Things have not gotten that far along here, and officials maintain no wells need to be closed. Water from them is apparently still safe to drink.

BUT THERE ARE other questions which may not be answerable now, but need to be considered.

What are the possible long-term effects of continued exposure to low levels of DBCP, levels which are now considered "safe"? Will the continued percolation of the substance into our drinking water lead to possibly higher concentrations in the future?

A sobering fact is that much of our knowledge of DBCP is fairly recent; in the case of its longevity, literally only weeks old. There is still much to learn.

With the potential for DBCP-related problems having risen so much in the past few days, state officials would be well advised to go along with the national ban if they find the latest evidence credible.

To end DBCP use now, and perhaps reintroduce it later if it is indeed found safe, seems more prudent than to continue its use now, only to possibly discover later it may be far more dangerous than suspected.

ST. LOUIS POST-DISPATCH

The U.S. House of Representatives has evidently discovered what the chemical industry and the Reagan administration have not — that the public really cares about environmental safety. Despite heavy industry lobbying, with the administration on its side on key issues, the House overwhelmingly passed, with strengthening amendments, a two-year extension of the basic federal pesticide control law.

Among other things, the House approved a ban on toxaphene, a chemical used in cotton cultivation that is acknowledged to cause cancer in humans but that the Reagan Environmental Protection Agency had not acted against. The House also adopted an amendment erasing proposed limits on states' power to control pesticides and an amendment allowing private citizens access to federal courts to seek relief from pesticide damage. And opponents of an agricultural subcommittee version of the bill won removal of a provision that would have barred access by scientists to manufacturers' formulation data to determine the health and safety effects of new products.

Lawmakers in the House were influenced by accounts during debate of herbicide poisoning of national forests in Oregon, contamination of milk in Hawaii by a pineapple crop chemical and the tainting of water on Long Island by a potato growers' chemical. Although Congress on many issues tends to be subservient to industry lobbying, it apparently recognizes that on environmental issues, as shown by poll after poll, the public wants protection and that political futures might be jeopardized by ignoring this sentiment. Now the Senate should be acting on the same message that swayed the House.

St. Petersburg Times

St. Petersburg, Fla., August 15, 1982

Thank goodness that America's founding fathers set up a governmental system with checks and balances. If they had not been that wise, nothing could have stopped the Reagan administration from allowing industries to pollute the air, poison the water, ravage wilderness areas and jeopardize the health of the American people.

Fortunately, the members of Congress have something to say about that, too. And they did last week. Three cheers for the House members who rebuked the administration for its assault on the environment and development-at-any-cost policies. It was a week to celebrate.

Last Wednesday, the House voted down a series of amendments to the federal pesticide control act that had been backed by chemical manufacturers. One provision of the law, which was extended for two years, allows states to set controls that exceed national standards on agricultural chemicals.

THE VOTE has particular significance for Florida, considering *Network News'* recent revelations about Temik, a highly toxic pesticide. State officials are conducting independent tests to determine whether the extremely poisonous pesticide has contaminated Florida citrus or the state's ground water.

There's reason to worry. Three years ago, it was discovered that Temik had poisoned wells on Long Island, N.Y., after being used on potato fields there. Even one official with Union Carbide Agricultural Products Company, Inc., the manufacturer, said that federal regulations were not strict enough to prevent the poisoning of wells in several states.

If Temik has contaminated Florida's citrus and ground water, the state should move immediately to enact more stringent controls than the federal regulations. But if the industry-backed amendments to the federal pesticide control act had been approved, Florida wouldn't have the option of protecting its environment and the health of its residents by putting stricter controls on Temik.

Also last Wednesday, the House Energy and Commerce Committee approved tough standards for the emission of certain airborne pollutants that are suspected of causing cancer and birth defects. The amendment to the Clean Air Act won over a more lenient proposal supported by the administration and a coalition of industry groups.

Then, last Thursday, the House overwhelmingly approved a bill that would prevent the Interior Department from allowing oil and gas exploration in wilderness areas. It was another victory for the American people.

INTERIOR SECRETARY James G. Watt last year made the outrageous proposal to allow drilling in the nation's wilderness preserves. By no stretch of the imagination would drilling be compatible with the preservation of a wilderness area where man is supposed to take nothing but pictures and leave nothing but footprints. The bill now goes to the Senate. If Congress doesn't stop the administration, it will allow industry to despoil the nation's vanishing pristine lands.

The coalition between the White House and certain avaricious industries, described by Rep. Henry Waxman as "an unholy alliance held together by greed," might not be dead. But it should be gasping its last breath.

The votes in the House are the people speaking. They don't want to breathe polluted air or drink poisoned water. They don't want to exploit the wilderness. They don't want to abandon protection of the environment. It's time that the Reagan administration and industries that want to exploit the environment start listening.

The Washington Post

Washington, D.C., August 18, 1982

IN VOTING recently to extend the federal pesticide control act, the House sent a strong message to foes of the environmental movement. People from all parts of the political spectrum are worried about what uncontrolled use of chemicals can do to their health and to the world around them, and they are willing to put up with a certain amount of cost and regulation to reduce that threat.

In voting to reject a series of industry-backed amendments to the Federal Insecticide, Fungicide and Rodenticide Act (you remember FIFRA), the House also had something to say about the administration's latest modification of its views about federalism. One of the amendments would have restricted the rights of states to impose stricter pesticide controls than those set by the federal Environmental Protection Agency. EPA initially opposed the change, but the White House, under pressure from the chemical manufacturers, later dropped its opposition to the new limits.

This put the administration in the peculiar position of amending its longstanding position in favor of more control by states over their own affairs. It seems that it's all right by the administration for states to set their own standards so long as those standards are more permissive than the federal government's.

The traditionally accepted federal relationship works quite the other way—the federal government establishes a floor level of acceptable behavior for all states, but states—subject to constitutional prohibitions against impeding interstate commerce—may choose to set higher standards. That compromise has worked well in many areas, and the House rightfully declined to modify it in the case of pesticides where states have frequently been ahead of EPA in detecting health hazards.

The environmentalists scored two other victories. A House committee rejected weakening amendments to the Clean Air Act sought by industry, and the full House, by a 340 to 58 vote, approved a ban on oil and gas leasing in wilderness areas. The broad range of congressional support for these measures reinforces the finding of public opinion polls that health and the environment are not the private concerns of one region or socioeconomic group—they are of general concern.

The Des Moines Register

Des Moines, Iowa, August 30, 1982

The House recently opted for effective government regulation of pesticides when it extended the basic federal pesticide law for two years. Representative Tom Harkin (Dem., Ia.) played a key role in persuading the House that states should continue to have the ability to place stricter controls on pesticides than the federal government has.

The pesticide industry became concerned about stricter state standards when California, which uses enormous quantities of pesticides, pursued a major campaign to screen pesticides and rejected some that had been approved by the federal Environmental Protection Agency.

The California pesticide-review program was understaffed and there were bottlenecks in the system, so a large backlog of pesticide-registration cases developed. These problems largely have been cleared up in the past year.

Despite the progress made in California, the pesticide manufacturing industry pressed Congress to change the law to make it more difficult for states to impose additional controls on pesticides.

Incredibly, the Reagan administration jumped behind the industry's campaign to make it tougher for the states to protect their residents. We say "incredibly" because the Reagan administration has lectured Americans on the need to return power to the states. With the administration's less-than-aggressive enforcement of environmental laws, state review of pesticides will be more essential than ever.

The House bill should improve enforcement by permitting individuals to sue violators of the pesticide law. The measure would provide substantially more money for the federal pesticide program than the Reagan administration requested.

Pesticides are vital to agriculture. A basic purpose of the federal pesticide law is to weed out those pesticides that pose too great a risk to human health or the environment. The bill approved by the House achieves this important purpose.

THE LOUISVILLE TIMES
Louisville, Ky., February 15, 1983

While the Environmental Protection Agency fends off charges of sweetheart deals in its management of the hazardous waste superfund, controversy is building over the agency's approach to regulating dangerous farm chemicals.

When President Reagan took office, he pledged to streamline regulatory procedures, cut paperwork and relieve business of bureaucratic hassles. However enticing those promises, EPA's record so far adds to the stack of evidence that the public welfare is set back when health and safety checks are recklessly jettisoned.

Indeed, the pesticide fuss reinforces the suspicion that EPA is more interested in the convenience of influential industries and politicians than in safeguarding the public.

The new attitude was especially welcome to prosperous producers of herbicides and insecticides, who were unhappy over restrictions on the use of some popular pest killers and resent studies that delay new products.

However, the cautious approach of past administrations was justified. Most agricultural chemicals are damaging to people, and virtually everything else in nature, as well as to crop-destroying critters. Lethal dioxin, which has been carelessly spread in several Missouri cities, is the byproduct of some pesticide manufacturing processes.

Moreover, serious questions have been raised about the necessity of coating the land with poison. Although the chemical industry attributes agricultural plenty to the farmers' copious use of pesticides, many studies show, as *Washington Post* staff writer Ward Sinclair points out in a recent series of articles, that crop loss to pests is about the same now as it was 40 years ago, when the use of chemicals was less prevalent. One reason is that bugs develop immunities to sprays.

EPA's current managers seem little concerned about these issues. As Mr. Sinclair reports, the regulatory budget and staff have been cut in the name of efficiency. One disturbing result is that staff scientists are under pressure to move industry registration requests rapidly through the bureaucratic mill, even if that means skimping on evaluations of new chemical products.

This has brought charges that thoroughness is sacrificed in the rush to allow marketing of new products. A congressional committee report claims that regulation of cancer-causing substances has been relaxed.

In one notorious case, a manufacturer's evaluation of its own chemical was adopted by EPA as written. In a dispute that could parallel the toxic waste controversy, EPA administrator Anne Gorsuch is under attack in Congress for approving new rules that allow the agency to withhold safety information.

Another alarming trend described by Mr. Sinclair is the issuance of hundreds of "emergency" permits for use of banned or restricted chemicals. EPA, urged on by the state's congressmen, said South Carolina orchard operators could use DBCP, which has been restricted because of the suspicion it causes sterility. An industry-financed study claimed the substance could be used safety. Fortunately, a court order stopped its application.

Most of us would love to get government off our backs. But EPA specializes in the sort of abuse that gives deregulation a bad name.

Newsday
Long Island, N.Y., April 15, 1983

Eleven of the state's 16 pesticide inspectors are now on Long Island investigating complaints that exterminators misused two dangerous termite-killing compounds. Such misuse is only a hint of a real and serious problem — and it's not limited to Long Island or to a couple of troublesome chemicals.

In "Silent Spring," Rachel Carson argued for a broad reduction in the use of pesticides because of their cumulative and toxic biological effects. Her eloquent arguments led to the banning of DDT, one of the most effective — and most dangerous — insecticides.

But in the 21 years since "Silent Spring" was published, the use of pesticides in this country has more than doubled. And some of that increase can be ascribed to the laxity of the Environmental Protection Agency. Last year, for example, the EPA discontinued a program designed to monitor reports of pesticide misuse. A few years earlier it had shut down a hot line that received reports of problems that people were having with pesticides.

One critic of the federal government's regulation of pesticides is Dr. Robert van den Bosch, a University of California entomology professor and a consultant to the EPA, the Ford Foundation and the United Nations. He claims that, when it comes to protecting the public from the hazards of pesticides, the EPA is now "more tabby than tiger."

In his book, "The Pesticide Conspiracy," van den Bosch said: "The trouble with the EPA is that it tried to live up to its mandate. Accordingly, in the pesticide area it took aggressive action and banned such environmentally hazardous insecticides as DDT. . . This is where EPA got into trouble, for the American chemical industry wields enormous power in Washington . . . and the EPA's actions stirred the wrath of this powerful giant."

Recently, Rep. Thomas Downey (D-Amityville) complained that the EPA wasn't responding to Long Island's needs. "No federal official appears empowered to establish once and for all the seriousness of the threat — to answer the thousands of people who are terrified by the possibility that their homes have been contaminated," said Downey.

"Silent Spring" helped inspire Congress to give the EPA a strong mandate. But Rachel Carson has been dead a long time and the mandate she bequeathed seems to have dissolved. It should be restored.

The Hartford Courant
Hartford, Conn., May 17, 1983

William D. Ruckelshaus, the president's nominee to head the Environmental Protection Agency, recently said there would be no more "sweetheart deals" at the agency.

Before his nomination gets Senate approval, however, he ought to make clear that he will also try to undo some of the sweetheart deals already made. A good place to start is with last year's EPA decision to treat the pesticide permethrin as safe for human consumption.

The decision came after industry and agency officials held a series of closed-door meetings at which the government promised to allow much broader usage for permethrin. The decision could be worth $1 billion to industry.

The pesticide has been used extensively on cotton, but now will show up more generously in such human staples as meat, vegetables and eggs.

Although there is some disagreement about the possible adverse effects of the chemical, permethrin is a suspected cancer-causing agent. With the evidence already in, permethrin should be regulated as a carcinogen, but EPA policy has shifted to allow the chemical's financial benefits to outweigh its health risk.

If current thinking had prevailed in the agency in 1976, the EPA might not have banned production of toxic chemicals used widely in electrical equipment, called PCBs. The agency last week was congratulating itself for its success, based on a study showing a sharp decline in the level of PCBs found in the body tissue of Americans since 1977.

Yet EPA officials have ignored the recommendations of their top scientists, who oppose broader use of permethrin because of its possible adverse health effects. One agency scientist, M. Adrian Gross, has accused his own bosses of collusion with pesticide companies to broaden the market for the chemical.

Permethrin is only one of many pesticides implicated in a host of health problems, including cancer, birth defects and genetic mutations. They are showing up in measurable quantities in almost everything Americans eat and, by contaminating ground-water supplies, reservoirs, rivers and streams, much of what they drink.

If Mr. Ruckelshaus wants to end sweetheart dealing, he can start by backing efforts to tighten the federal law governing pesticide sales and use, to make clear to his own staff that human health cannot be bartered for greater profits and to curb the secrecy surrounding industry-agency meetings.

The Dispatch

Columbus, Ohio, November 2, 1983

The Dow Chemical Co. announced recently that it is giving up trying to sell 2,4,5-T, a once-popular weed-killer that was found to contain traces of dioxin, a toxic substance thought to cause cancer in animals. Many Americans — especially those who have heard of 2,4,5-T and dioxin — will cheer Dow's decision. Some might even conclude that the decision is the ultimate evidence that the company was wrong to begin producing 2,4,5-T in the first place, and nearly criminal in its continued production once suspicions about its safety were raised.

But a story two days after Dow's announcement gives reason to pause in this rush to judgment. That story dealt with a study conducted by Missouri health officials and doctors from the federal Centers for Disease Control in Atlanta concerning the health of residents in the dioxin-contaminated communities of Times Beach and Imperial, both in Missouri. "We have not been able to identify meaningful ill-health effects related to potential exposure to dioxin," the report concluded.

Dow surely was aware of the report's findings before it decided to terminate 2,4,5-T sales in the United States. Why, then, did Dow do what it did? Because, the company says, a $10 million effort to fight government restrictions against the chemical and to inform the public of the facts about it had exceeded the profit-return on the product. It would, in short, be bad business to attempt to continue sales here, although it is continuing sales in other countries.

Few are likely to rally behind Dow, even though the Missouri study indicates that, at best, criticism of the company was premature and could ultimately prove to be irresponsible. Dioxin is dangerous; no one disputes that. It appears, however, that the proper use of products which contain traces of it (and here we're talking about levels in the parts-per-million, parts-per-billion, and even parts-per-trillion range) poses no health hazard. The entire 2,4,5-T episode can serve as a warning that unless the public learns to view chemicals in a rational way, benefits that could result from them may never be realized.

Use of 2,4,5-T was severely restricted in 1979 by the U.S. Environmental Protection Agency after claims were made linking the use of it in Oregon to miscarriages. These claims were hotly contested by Dow and by private researchers at the University of Oregon, but the EPA decided to prohibit the weed-killer's use along rights-of-way, in forests and on pastures. It said it could be used, however, on rice paddies, sugar cane fields, range-land and some pre-harvest fruit crops.

Dow took this puzzling ban to court but a resolution of the dispute has not been achieved. Dow has now dropped the case in deciding not to pursue sales in this country.

Dow's problems were complicated by the revelations that dioxin was improperly disposed of in Times Beach and Imperial. The Centers for Disease Control say that dioxin in a concentration of one part per billion is cause for concern. Dow, while in the same ballpark, sets the concern level at 10 parts per billion. Both parties agree that the dioxin level of 300 parts per billion in Times Beach and the level of 740 parts per billion in Imperial were cause for alarm. But even with these very high levels, the recently released study of these two communities found "no meaningful ill-health effects." (Subsequent to the release of that study, the National Institute for Occupational Safety and Health challenged previously conducted research suggesting a link between dioxin and a rare form of cancer in humans. While saying that its own study "doesn't eliminate the concern and it doesn't exacerbate it," the institute stated that "the bottom line is we need larger studies to confirm or to refute these smaller ones.")

The point to be emphasized is that the scientific and health community is still learning about dioxin and its effect on the human environment. It is obvious that both the CDC and Dow have been very conservative in their estimates of the human danger threshold, and the company is not, as some critics contend, outrageously ignoring safety considerations in the rush to make a buck.

There is no need to defend Dow here. The company has a marvelous public relations and information apparatus that can do that. The point that should be stressed is that the production and proper use of chemicals can be beneficial to all of us, and that irrational fears could dissuade companies from producing substances that could go a long way toward easing suffering and enhancing the quality of life. Those wishing to halt the production of chemicals should be aware of the consequences of such an action. Those hoping to benefit from scientific or medical breakthroughs, or those wishing to find ways to feed the world's hungry, must be reasoned in their evaluation of chemicals that hold promise.

The Toronto Star

Toronto, Ont., July 20, 1983

It's hard to believe, but the deadly herbicide Agent Orange — which wreaked havoc on thousands of U.S. army troops exposed to it in Viet Nam — is still being sprayed in parts of Canada with full government approval.

The Nova Scotia government has just given the go-ahead to a pulp and paper company to spray the notorious defoliant, technically known as 2, 4-D and 2, 4, 5-T, in its forests in central Nova Scotia.

This comes a full three years after the Ontario government took action to effectively ban the use of 2,4,5-T in this province, as did Saskatchewan and British Columbia. And it's been four years since the U.S. Environmental Protection Agency banned the herbicide.

The deadliest chemical known to man — a form of dioxin known as 2,3,7,8-TCDD — can show up in the herbicide being used in Nova Scotia. It takes only 1/200th of a drop to kill a man with this compound. U.S. veterans exposed to the defoliant reported skin lesions, swollen livers, numbness and personality disorders.

Agent Orange is being used in places like Nova Scotia and New Brunswick to kill off hardwood trees and shrub, allowing softwood trees used for making pulp to grow faster. The forest industry sprayed 6,000 acres in Nova Scotia last year. Now Scott Maritimes Ltd. is set to begin spraying two additional sites later this month.

The Ontario government banned Agent Orange because it worried about the effects on workers doing the spraying, and the well-being of bystanders, nearby residents and passers-by. The risk of residues entering local streams was also judged worrisome.

Nova Scotia and other provincial governments should do like Ontario and stop declaring war on their own natural and human resources.

The Globe and Mail

Toronto, Ont., August 16, 1983

The world's longest undefended frontier? Perhaps that hoary old cliché of Canadian-American friendship should be replaced with warm references to "the world's longest defoliated frontier", in light of the recent revelation by NBC News that the border area of New Hampshire and Maine was sprayed in the 1970s with Agent Orange and other herbicides.

This was not an action calculated to win the hearts and minds of American border residents or of their Canadian neighbors. Agent Orange has been the subject of lawsuits by Vietnam veterans in the United States who blame their exposure to the defoliant when it was used in warfare for causing cancer, birth defects and other health problems. Its use in the border area reportedly continued even after it was halted in Vietnam (though not, according to Canadian officials, after its possible toxicity became known).

For residents of New Brunswick, who border on the state of Maine, unpleasant revelations about past encounters with Agent Orange are nothing new. Back in 1966, Canadian Forces Base Gagetown was having trouble clearing brush from land needed for training, so it called upon the U.S. Army for assistance. The Americans were only too happy to oblige — by spraying 250 acres of woodland with Agent Orange.

Declassified U.S. Army documents obtained in 1981 by two New Democrat MPs revealed that the U.S. Army had used the opportunity to test the defoliant for future application in Vietnam, where it was employed between 1962 and 1971 to clear underbrush and thereby deny Victor Charlie a hiding place.

Although the NBC report stated that probably no one in the border area was exposed directly to Agent Orange or Agent White (a herbicide used to kill brush that survives even Agent Orange), a spokesman for the Audubon Society told the network it was impossible to spray forest lands aerially without getting residues in the water.

Perhaps the next time a U.S. diplomat toasts his Canadian hosts, he should skip the usual effusive tributes to trans-border ties. A simple but heartfelt "To your health!" would be appropriate.

THE ATLANTA CONSTITUTION
Atlanta, Ga., February 2, 1984

The dangerous pesticide DDT — banned in this country 12 years ago — has been turning up in increasing concentrations in the tissues of fish and other wildlife in several Western states. That cannot be tolerated.

To their credit, officials in those states and the Environmental Protection Agency are already trying to locate the cause of the contamination and bring it to a halt. Urgency is justified: Runaway accumulations of DDT in the environment could bring about an ecological nightmare. The pesticide was banned in 1972 because it was found to induce cancer in laboratory animals and tended to build up in the food chain, threatening the survival of wildlife and ultimately the health of humans.

For years after the ban, the level of the contaminant in the environment declined, but now it is on the rise again. And so far, officials are not sure why. They speculate that the problem is being caused by the use of DDT imported from Mexico, where it is still legal, or — more likely — from increased use of another pesticide, dicofol, which has become popular in recent years and contains 7 percent or more DDT.

Dicofol is widely used to control insects on citrus trees, flowers, ornamental shrubs and vegetables. It was registered by the EPA before the DDT ban and apparently survived the banning on a technicality: The ban applied to all products having DDT as an active ingredient. DDT is not an active ingredient of dicofol, but an unwanted byproduct created when the chemical is manufactured.

EPA officials have begun a lengthy review of the chemical and its manufacturing process, which could lead to the banning of dicofol. Authorities must also move swiftly to punish anyone found to be importing DDT from Mexico. If this dangerous contamination is not dealt with quickly, a number of endangered species that had been making a comeback since the ban may again be threatened with extinction.

THE KANSAS CITY STAR
Kansas City, Mo., June 8, 1984

The alarming evidence continues to unfold about the toxic effects and damage to human health caused by some of the most commonplace products we use. One such is pentachlorophenol—"penta"—a versatile pesticide containing dioxin which is widely used as a wood preservative. In a series of articles in *The Kansas City Star* reporter Myron Levin, after months of examining government and court records, reading medical journals and interviewing victims' families, has provided a detailed look at this dangerous chemical.

Penta has been associated with 40 deaths and scores of illnesses, causing a variety of symptoms such as rashes, eye irritation, headaches, kidney and liver problems. Yet about 35 million pounds of it are produced yearly to treat lumber against termites and rot and for other purposes. After six years of study and public comment, the Environmental Protection Agency still has not taken action to ban or restrict usage of the product.

One difficulty is the range in human reactions to exposure. Some persons become ill in a short time around penta-treated wood, such as in a modern log cabin, while others are slow to exhibit symptoms. Manufacturers of penta say human beings can tolerate certain levels, and excrete as well as absorb it. They counsel use of safeguards in handling penta itself. Conflicting scientific claims seem to have blunted the EPA's will to act on this issue.

This is another instance of the hidden dangers in certain useful but toxic products. We may behave with caution around volatile substances, choking fumes or acids which burn on contact. But other materials may contain subtle poisons, slowly or cumulatively ingested, where the symptoms are not immediately evident and the victim may fail to realize his plight.

Sometimes it is difficult to establish cause-and-effect linkage between toxic substances and illnesses which conceivably could stem from other factors. But the medical evidence which has been and continues to be gathered on the hazards of penta justifies something more than continued study by environmental authorities. Particularly when a method exists to cut the dioxin level in this destructive pesticide.

The Miami Herald
Miami, Fla., June 30, 1984

TWELVE years ago Environmental Protection Agency (EPA) Administrator William Ruckelshaus signed an order declaring the pesticide DDT to be an "unacceptable risk to man and his environment." At that time Mr. Ruckelshaus properly banned "all uses of DDT for crop production and nonhealth purposes."

Yet five months later the EPA routinely approved the use of the pesticide dicofol. The manufacturers of dicofol listed appreciable amounts of DDT-related "impurities," but the agency was swamped with paperwork as it assumed jurisdiction over 800 pesticides and the significance was not recognized. Based on that initial registration, the agency subsequently approved nearly 50 other dicofol products for a half dozen other manufacturers. Some of those products include as much as 15 percent DDT and related compounds.

That fact alone raises serious questions about the efficacy and adequacy of pesticide regulation by the Federal Government.

More distressing, however, is the long delay between the discovery of the error in 1981 and any effort to remove from the market the offending DDT-contaminated chemicals. Indeed, under Administrator Anne Gorsuch Burford, EPA officials ruled that the ban on DDT did not apply to ingredients classified by manufacturers as contaminants or byproducts. Only after the return of Mr. Ruckelshaus did the EPA initiate the responsible steps necessary to remove dicofol from the market.

Testing for DDT has not been widespread since the 1972 Federal ban. Yet there is ample evidence that the threat of DDT remains serious and much suspicion that the widespread use of dicofol is responsible.

In California, San Joaquin Valley farmers use more than a million pounds of dicofol annually. A survey by the Natural Resources Defense Council found that the most common pesticide residue detected on state-grown produce was DDT. Dicofol was second. In South Texas, the Arroyo Colorado is so contaminated with DDT compounds that health officials warn against eating the fish. The river runs through the heart of the citrus belt, where dicofol is used to control red mites.

The situation in Florida is unknown. Citrus growers are major users of dicofol, but the state does not routinely test for DDT residues. Moreover, dicofol never has undergone the extensive testing required to introduce a new pesticide. EPA has asked for additional tests, but before undertaking the tests the manufacturer has sought assurances that DDT impurities will be allowed. To provide any such assurances would be irresponsible.

The commitment to rid the nation's environment of DDT was not lightly made. Years of research demonstrated that DDT kills wildlife and accumulates in the food chain to be deposited in the fatty tissue of humans.

The commitment to clean up the environment ought to be strengthened, not weakened. Dicofol and other pesticides contaminated by DDT and related compounds must be included in the existing ban.

The Oregonian

Portland, Ore., March 10, 1984

Recent court decisions have put the U.S. Forest Service and Bureau of Land Management in the ridiculous position of being forbidden to use the same herbicides readily available over the counter to the backyard gardener and other private landowners. Thus does underbrush of a regulatory sort choke the federal agencies' ability to kill the underbrush of a vegetative type.

The anomaly of the federal government being unable to use what the federal government has approved for others to use should prompt close scrutiny by Congress and federal environmental regulators of rules relating to pesticide and herbicide approval and application. Further, public confidence would best be restored by more intensive screening of potentially dangerous chemicals by the Environmental Protection Agency.

U.S. District Judge James M. Burns has prohibited the Forest Service and BLM from spraying any herbicides in Oregon and Washington until they complete what is called a worst-case analysis of the herbicide effects. The Council on Environmental Quality requires this analysis when exact information on adverse impacts cannot be obtained. In other words, if an agency cannot predict exactly what will happen, it must describe the worst that could happen.

The Forest Service and BLM had argued that approval of a herbicide by the Environmental Protection Agency should be deemed sufficient for its use by federal agencies without additional study. The 9th Circuit Court of Appeals — on which Burns relied — said in a recent opinion that EPA registration "is inadequate to address environmental concerns" under the National Environmental Policy Act.

If that is the case, something is drastically wrong with EPA's registration procedure. EPA, not individual federal agencies, should be the principal federal gatekeeper on chemical safety. The overwhelming use of these chemicals is by individuals and businesses not subject to the second environmental review required of federal agencies.

What matters to the environment is not whether a chemical is applied by a federal agency or private party but what it does when it is applied. If EPA cannot provide satisfactory answers to that question, it, not other agencies, should be made to do so.

The Salt Lake Tribune

Salt Lake City, Utah, February 7, 1984

For better than a decade that chemical tounge twister dichloro-diphenyl-trichloroethane (DDT) has been banned in the United States. The ban has been instrumental in increasing the populations of several endangered and threatened species of birds, including bald eagles, ospreys, brown pelicans and peregrine falcons. Now, however, there is increasing evidence that the ban is developing some weaknesses.

Ron Joseph, raptor biologist for the U.S. Fish and Wildlife Service, reported that seven known Utah wild peregrine "aeries" (nesting areas) produced only eight young birds last year. Biologists believe at least 10 young birds are needed annually to maintain a self-sustaining population. Additionally, egg shell fragments collected by biologists last summer at the nesting sites were found to be 19 percent to 22 percent thinner than normal.

In addition to thinning of the egg shells, which makes them more fragile and more susceptible age, DDT is faulted for reducing the number of eggs laid, the death of embyros and behavior changes in adult birds that causes them to neglect their chicks.

Besides the Utah evidence, increased DDT contaminations have been found in California among the fish and mussel populations of the Salina River valley, south of San Francisco. Starlings tested in New Mexico and Arizona are, reports the USFWS, showing increasing concentrations of DDT.

Monitoring studies done by the USFWS and several states showed that levels of DDT dropped in wildlife species until about five years ago, when increases turned up in migratory birds in eastern New Mexico and western Texas.

This resurgence of DDT levels in migratory birds is suspected as the reason for the troubles afflicting Utah's wild peregrine falcons. These birds feed on white-throated swifts and swallows which "winter" in Mexico and Central America and return to Utah each summer, early in the falcons' breeding cycle. Small birds make up about 90 percent of the peregrine's diet.

Contamination by DDT isn't limited to birds and fish. Nearly all Americans have accumulated some DDT in their fatty tissues as a result of the enormous quantities of the pesticide used on food crops for decades.

Mexico and many other south of the border countries have not banned DDT, for that matter the pesticide is used extensively there. Its widespread Latin American use works effectively to introduce DDT into the United States, despite the 12-year-old ban.

Until the governments of Mexico and Latin America can be persuaded to take steps to either ban or more rigidly control the use of DDT there lurks a threat to American avian wildlife and, coincidentally, to Americans themselves.

Los Angeles Times

Los Angeles, Calif., March 13, 1984

Many pesticides currently in use have never been fully tested to determine whether they can cause birth defects. A bill that has passed the California Senate and comes up today in the Assembly Health Committee would require such testing to fill a major void in the public's right to know such things.

A 1982 congressional study disclosed that 60% to 70% of pesticides in use nationally had gone on the market in the 1950s and 1960s, before the law required that they be tested for their potential of causing birth defects. The Environmental Protection Agency has been directed to fill in test gaps but, according to the study, has made virtually no progress. California law has required special attention to chronic health effects of pesticides since 1980.

In addition there is uncertainty even about some of the pesticides that were tested. Last October officials at one of the nation's largest independent testing laboratories, Industrial Bio-Test Laboratories, were convicted of fabricating key safety checks. The lab had tested at least 140 different pesticides, and the environmental agency considered only 3% of those studies to be valid.

The bill now moving through the California Legislature is sponsored by state Sen. Nicholas C. Petris (D-Oakland) at the request of the March of Dimes and the California Rural Legal Assistance Foundation. It would require a complete listing of all tests for chronic health hazards of the active ingredients of pesticides registered in California. It also would require the state Department of Food and Agriculture to ensure testing for harmful reproductive effects of any pesticides that have not been tested, with the work done either by the company that produces them or by the state with fees assessed from companies that will not do their own testing.

The provision under which the state would secure information that companies refused to supply is the principal basis for opposition by the state Department of Food and Agriculture. Hans Van Nes, assistant director for legislative services, says that the department agrees with the intent of the bill but originally found it totally unworkable. The department does not want as specific a requirement on filling the gaps as Petris advocates. But companies cannot know that there is no way to avoid the tests unless the bill is very precisely written.

California has a tradition of stronger enforcement against potentially hazardous pesticides than that of the federal government. Federal law does not prohibit state governments from setting higher standards than those in force for the rest of the country, and California should not hesitate to do so in the face of this appalling data gap.

Ethylene Dibromide (EDB): Soil, Grain, Fruit Uses Curbed by EPA

The Environmental Protection Agency in February, 1984 prohibited use of the pesticide ethylene dibromide on grain products. The controversial pesticide, commonly known as EDB, had been banned as a soil fumigant in 1983. Other uses of the chemical, determined by laboratory tests to be a potent carcinogen and cause of birth defects in animals, included fumigation of citrus and other fruit crops as well as grain-milling machinery. The 1983 ban on using EDB as a soil fumigant had been put into effect after new evidence came to light that the chemical was contaminating underground water supplies in several states, among them Florida and California. The 1984 restriction for its use on grain products followed actions by several states where detectable levels of EDB had been found in certain foodstuffs on store shelves, especially in cake, muffin and pancake mixes. Florida had begun taking contaminated products off store shelves, and Massachusetts banned the sale of 18 grain-based products. In California, high levels of EDB were discovered in imported citrus pulp and in cake and muffin mixes, and the state had requested a total ban on use of the pesticide.

In March, 1984 the EPA announced new restrictions on the use of EDB on fruit. The agency set a Sept. 1 deadline for the elimination of EDB use on imported citrus fruit (as well as imported mangoes and papayas) that was consumed in the United States, and for U.S.-grown citrus fruit sold abroad. (EDB was not currently used on domestically grown fruit sold for U.S. consumption.) It was not known whether the pesticide was carcinogenic to humans as well as to test animals.

The Orlando Sentinel
Orlando, Fla., November 15, 1983

The people of Florida are learning that there is a price for a casual attitude toward potent chemicals. And if a lesson is to be drawn from what we've seen of late, it is that we had better get serious about handling all chemicals. The state's geology and primary water source just don't leave room for mistakes.

The latest concern is ethylene dibromide — EDB — a potent chemical that has been found at higher than "safe" levels in water wells about the state. Although a state health department official says there is no reason for panic, he cautions that there is plenty of reason for concern: One more cancer-causing substance is finding its way into our bodies.

As almost everyone knows by now, Florida's own Department of Agriculture apparently is to blame for much of the EDB in our water. For years it injected megadoses of the chemical into the sandy soil of citrus groves. The target was the burrowing nematode, a prime enemy of citrus trees. Incredibly, the state had been using up to 10 times the recommended dosage long after federal authorities had urged caution in using it at all.

But what people generally don't know is that there are numerous other sources for EDB problems. Many golf courses used it to kill mole crickets. In South Florida, where the Biscayne aquifer is covered by only a few feet of sand, that has been particularly dangerous. Fortunately, federal and state bans last month ended the use of EDB as a soil pesticide.

But 90 percent of all EDB made goes into leaded gasoline and, in Florida, leaking gasoline tanks are a major source of groundwater pollution. The Legislature put some restrictions on those tanks this year but ought to look at the possible need to ban metal tanks altogether. That's not all: Ironically, EDB is used in gasoline to prevent lead buildup inside engines. When the gasoline is burned, both the lead and the EDB are pushed out the exhaust, becoming a menace to clean air.

The EDB scare emerged just as the scare over Temik, another highly toxic nematode fighter, was subsiding. The problem in both cases was the casual attitude of people who should have known better. These are not likely to be the last scares over Florida's precious water supply. Dr. Daniel Shanklin, chairman of entomology at the University of Florida, points out that as technology advances, smaller and smaller quantities of poisons will be detected and we will learn how they, too, can be the death of man.

For the person who finds that EDB has invaded his well, there is little solace in being told that the risk is "acceptable." The truth is the risk has nevertheless increased as the result of society's casual attitude toward potent chemicals.

The state finally is moving to tighten controls. It also ought to move the responsibility for regulating chemicals from the Department of Agriculture to the Department of Environmental Regulation. With Florida's growth, its dependency on underground water and the porous sand protecting that source, any chemical must be considered a threat.

THE SACRAMENTO BEE
Sacramento, Calif., October 5, 1983

The Environmental Protection Agency (EPA) has banned the use of the pesticide EDB, ethylene dibromide, and it's certainly a welcome move. But it's hard to jump up and down with enthusiasm over the protection the federal government is providing for the public.

The action, after all, came more than 10 years after the first tests indicating that EDB caused cancers, birth defects and miscarriages in laboratory animals — and more than six years after the government announced that its own scientists also found EDB to be highly toxic and almost certainly extremely dangerous to humans.

It's been, in fact, three years since the EPA promised to act, after concluding that any worker exposed throughout his or her career to the levels of EDB that were then legally allowed could be expected to die of cancer. (Fortunately, it appears that no American workers receive such continuous exposure. But the EPA scientists also found that workers exposed for only a few weeks each year had a one in 128 chance of contracting cancer, and they found that exposure to EDB is closely associated with various kidney, skin and nervous system disorders.)

For nearly two years, it seems, the EDB ban was held up by the machinations of Dr. John Todhunter, an assistant EPA administrator since forced to resign during the EPA scandal last spring. According to EPA staffers, Todhunter — at the behest of his own superiors and various industry representatives — delayed their reports by insisting on unnecessary revisions and recomputations. And when the results kept coming out the same anyway, he demanded that they find an alternative to the outright EDB ban to which all their evidence pointed.

Meanwhile, only days before the EPA ban was finally enacted, the Occupational Safety and Health Administration concluded its own study of EDB with a recommendation that exposure levels allowed in the workplace be sharply reduced — only to be prevented by the White House's Office of Management and Budget from proposing this or any other new EDB regulation.

It's going to be another several years, moreover, before much of the EPA's ban goes into effect. The ban on using EDB as a soil fumigant was issued as an emergency measure — after EDB contamination was found in water supplies in California and Florida — and thus this use of EDB will be prohibited during the period of industry appeals against the new regulation. But the ban on using EDB to fumigate stored grains, picked fruits, felled logs and agricultural machinery is only a "cancellation." It thus will be lifted as soon as the industry files its appeals and will remain unenforced for as long as it takes the courts to hear all the industry's arguments against a ban.

So what does one say about the government's response to the EDB problem? Better late than never, we suppose.

Newsday

Long Island, N.Y., October 2, 1983

Arsenic kills rats, cyanide will wipe out any bug and DDT is death on mosquitoes. The effectiveness of these poisons is not disputed. But their utility is, because they'll kill too many different living things.

Drawing the line between effectiveness in killing pests and lethal danger to others is the job of the Environmental Protection Agency. The Occupational Safety and Health Administration is supposed to make certain that workers who manufacture or handle poisons aren't harmed. At long last, both agencies have taken action to limit exposure to ethylene dibromide, or EDB.

This pesticide ingredient and gasoline additive is believed to be a potent carcinogen. Several states have moved to limit or ban it as a pesticide after it showed up in drinking water, fruit and baked goods. The federal government has spent six years studying EDB with increasingly alarming results.

The problem is its very effectiveness. The EPA's top pesticide official told a congressional subcommittee Monday: "The laboratory evidence on health effects is persuasive that EDB can induce cancer, genetic mutations and adverse reproductive effects. It is also a very effective soil and commodity fumigant, and for some of the most economically important uses no effective alternative is readily available."

Every time a product comes into question for health reasons, someone contends it is not merely the best but perhaps the only way to achieve some worthwhile purpose. Fortunately, it usually does not take long for someone else to come up with an alternative.

So far, public exposure to EDB is thought to be fairly limited. But the risk to workers who handle it is unacceptable: EPA studies have shown that the number of "excess cancer deaths" that could be expected from regular, *legal* exposure to EDB ranges from 70 to 999 for every 1,000 workers.

Last week, EPA and OSHA finally moved to ban the use of EDB as a soil fumigant for crops and to lower the allowable worker exposure limit. That's fine, as far as it goes. A total ban would be better.

SYRACUSE
HERALD·JOURNAL

Syracuse, N.Y., September 29, 1983

The ratings indicate few people saw it, but NBC's new show, "First Camera" did a devastating investigative reporting piece on a deadly pesticide Sunday night.

The pesticide, ethylene dibromide or EDB, is a killer.

Six years ago, the government described EDB as highly carcinogenic. One government risk assessment found that virtually every worker exposed to the *legal limits* of EDB throughout his career would die of cancer.

The citrus industry has lobbied heavily in favor of the chemical, which is used to kill fruit flies. But it also kills people. The NBC report showed film of a chemical worker who was accidentally exposed to EDB. He was dead within an hour and a half despite frantic efforts to save him.

Danger isn't limited to chemical and citrus industry workers. Government studies show the chemical has seeped into some Florida water supplies, contaminating at least 50 wells.

The day after the television broadcast, an Environmental Protection Agency offical said the agency will ban the pesticide within a few days. The agency is expected to impose a phased ban, probably giving the citrus industry two years to halt its use. We're not suggesting the ban is a direct result of Sunday's show but it's clear the government's six-year lag can't stand the light of public exposure.

Rep. Michael Synar, D-Okla., chairman of the House Government Operations subcommittee, is probing delays in the ban, and political and economic pressures on the Enviromental Protection Agency's decision-making process. More power to him; if this is an example of government protection, we're all in deep trouble.

It's unconscionable that a situation exposing thousands of chemical workers to exceptionally high cancer risks was allowed to exist for six years.

THE ARIZONA REPUBLIC

Phoenix, Ariz., September 30, 1983

N OW, six years after announcing that the pesticide EDB was deadly, the Environmental Protection Agency has announced it will finally impose a phased ban on the chemical.

EPA's delay was not only inexcusable, it was scandalous.

EDB [ethylene dibromide] is a pesticide used to fumigate soil in citrus fields.

Tens of thousands of workers have been exposed to the chemical.

One EPA risk study showed that almost every worker exposed to the legal EDB limit throughout his career would die of cancer.

Edwin Johnson, chief of EPA's pesticide programs, says that he and his staff formally proposed last year that it halt most uses of the pesticide.

Johnson accuses former EPA Assistant Administrator John Todhunter, who resigned under fire last spring, of demanding that he find an alternative to the ban.

Todhunter attended many meetings with Florida citrus industry members and the state congressional delegation during the uproar, which suggests at least the appearance of impropriety.

Congress has only begun to look into this latest scandal at EPA.

It should be pursued, including the role of the citrus industry and the political influence of the Florida congressional delegation in risking the health of workers.

The handling of the EDB ban only adds to the widespread perception that industries and politicians have managed to manipulate administrative decisions on the environment.

The Honolulu Advertiser

Honolulu, Ha., October 1, 1983

The federal ban on the pesticide EDB poses problems for pineapple and papaya growers, but it will come as a relief for other people in Hawaii.

For years of tests and analysis clearly indicate that the benefits to agriculture from EDB are far outweighed by potential cancer and other health hazards.

Washington reports yesterday quoted some health experts as saying this is the most carcinogenic pesticide ever used in the United States. If so, Hawaii should hardly want to be exempted from the national ban.

THAT DOES not minimize the problems for pineapple growers who use EDB in the soil to kill rootworms, or to a lesser degree for papaya shippers who use it as a spray against fruitfly larvae and now have a year to find a safer compound or alternate method.

Pineapple companies may appeal, as is their right. But even more they need to press the search for viable alternatives to EDB.

Governor Ariyoshi was also a loser in yesterday's stronger-than-expected total ban by the Environmental Protection Agency on using EDB as a soil fumigant. Despite a high level of public concern about the pesticide in Oahu drinking water, he supported the pineapple industry's bid to EPA for continued use. The governor said there was no convincing evidence that agricultural use is responsible for EDB's presence in groundwater. No doubt pineapple's economic viability and the general business climate are other concerns.

THE STATE Health Department is right in noting that the forbiddance of EDB does not answer the question of how it got into the four groundwater wells still closed in Waipahu.

EDB is commonly used as an additive to leaded gasoline, a use not being banned. There is some suspicion that a leaking underground gasoline line could be the source of the Waipahu well contamination.

So Hawaii has not heard the last of EDB, both the trouble it causes and economic problems resulting from its ban. Morever, the Health Department will still have its hands full with monitoring and other pesticide problems. In that, Hawaii's headaches are akin to those in other states.

Everything said, however, the ban on EDB amounts to a major step in a necessary direction.

The Star-Ledger
Newark, N.J., January 20, 1984

The federal ban on the widely used agricultural pesticide EDB is a precautionary move that should help allay public concern that the nation is not faced with a health emergency. While there is sharply divided opinion among leading cancer scientists and health experts over how great a threat consumers face, the suspension of the use of the pesticide is a responsible government action.

What has been definitively established by scientists at this stage is that the chemical—ethylene dibromide—has been known to cause cancer in animals used in laboratory tests. There is disagreement, however, over how potent a carcinogen EDB is, and especially so in the amounts found in food and water.

Nevertheless, it is imperative, even with the division of thinking in the scientific community, for officials to take a deliberate safe course of action. The food industry has chosen a pragamtic stand of cooperation by removing its suspected products from stores around the country.

In ordering the ban, William Ruckelshaus, head of the Environmental Protection Agency, said the suspension will "result in clearing EDB from the food pipeline in the country."

"I firmly believe that the guidelines (to the states) are fully protective of the public health," he said. The health risk to consumers, he added, "is rapidly coming down, and we think it is slight."

But even a "slight risk" represents a potential public health hazard that should be responsibly acknowledged. The environmental agency has defined safety levels of the chemical that can be used as guidelines by state governments, which should speedily implement them.

New Jersey health officials have started testing products from supermarkets, food distribution centers and farms to determine if these items have high EDB levels. Some items already have been removed from stores, and others will be taken from shelves wherever tests show they exceed safe levels.

Compliance by states would have to be on a voluntary basis, since the federal government does not have the power to regulate the amount of the pesticide in food. There is little doubt that governors would take this necessary action. But it is essential, too, for the means of long-term enforcement, that the legal impediments to direct federal intervention in dealing with this potential health hazard are removed.

This is a sufficiently serious matter affecting the public health and welfare to require a uniform national regulatory mechanism. If this needs enabling legislation, then Congress should give it a priority agenda urgency. Further, the EDB ban should remain in effect until there is solid scientific evidence that the pesticide is not a deadly cancer-causing carcinogen.

The Houston Post
Houston, Texas, January 20, 1984

Is ethylene dibromide hazardous to your health? Nobody knows for sure, but pressure is building for the federal government to take some action on the controversial chemical that has been found in a wide variety of food products.

EDB has been in use for 35 years as a pesticide and as a fumigant for grain products. It was not until the 1970s, however, that government researchers found it caused cancer and genetic defects in test animals. In 1980, the Environmental Protection Agency proposed that the use of EDB be ended in grain storage and milling. The incoming Reagan administration took no action on the proposal until last September, after William Ruckelshaus was named EPA director to put the scandal-scarred agency back in order.

Ruckelshaus banned EDB as a soil fumigant when it was found in the water supplies of some states. He also proposed that it be barred as a fumigant in grain storage and milling within 30 days and as a pesticide on fruit by next Sept. 1. But industry protests against canceling the chemical's use on grain has delayed putting the proposal into effect.

In the meantime, Texas and several other states have been testing food items ranging from wheat products to citrus fruits for EDB. The Texas Department of Agriculture has found residue of the chemical in about 100 food products in amounts varying from 1 part per billion to 400 parts per billion. A dozen states are now pressing the EPA to issue a guideline on safe levels of EDB in food. The problem is that no one yet knows how much of the chemical — if any — may be harmful to humans.

The EPA is now engaged in an emergency study to determine if a content guideline or an outright ban on EDB in wheat products may be in order. But if the agency does not act quickly, alarm over the chemical's possible threat to health could cause states to take independent regulatory steps. Florida has already banned the sale of flour and flour products that contain detectable traces of EDB.

This is a national problem and should be approached in a uniform, coordinated manner under federal direction. Otherwise we risk having a mishmash of regulations that might leave consumers in some states unprotected while causing an unholy mess in the food and agricultural industries.

FORT WORTH STAR-TELEGRAM
Fort Worth, Texas, January 7, 1984

First, based on research, the Environmental Protection Agency considers ethylene dibromide (EDB) a "potent" cancer-causing chemical, which is enough to set off alarm bells.

Second, the EPA has banned the substance from use as a soil fumigant or citrus spray. That indicates a belief that EDB is not something we should be fooling with in agriculture, where it stands a chance of getting into the food we eat.

Third, a number of food products found on grocery shelves have been discovered to contain minute but scary amounts of EDB. Because of that, Florida officials have ordered 64 such products taken off the shelves in that state.

Fourth, Texas Agriculture Commissioner Jim Hightower's department has discovered products in Texas stores with similar amounts of EDB, and Hightower says those should be removed.

Fifth, Hightower can't order that removal unless the Texas Department of Health or the EPA sets a standard, a limit, on how much EDB is safe for human consumption. The EPA can't set a standard on human consumption because the Food and Drug Administration exempted the chemical from such regulation more than 30 years ago (the EPA could, and did, ban EDB's use on soil or citrus plants — it just can't set a limit for how much a human should consume).

Sixth, Texas Health Commissioner Robert Bernstein rejects Hightower's goal of getting the foods containing EDB out of the stores because he says there is too little known about the chemical to set a "reasonable standard on human consumption."

The bottom line, given Florida's action and the EPA's concern about this chemical, is that more needs to be known and in the meantime we shouldn't be taking chances.

The argument isn't about whether EDB is harmful. The argument is about how much of it is required to actually harm — or kill — a human being.

That's the sort of thing the public needs to know *before* it innocently picks up products off the shelf and consumes those products, not *after* it eats them.

Bernstein, Hightower, the EPA — all of them are hamstrung by rules and regulations that one EPA official calls a Catch-22. And the loser would be that member of the consuming public who is harmed by eating products that contain EDB.

Bernstein says the Center for Disease Control in Atlanta is conducting a study about EDB's effects even now. If so, the state, which has a responsibility to protect the public, should find the machinery to prevent sale of those products until that study is concluded.

And if the machinery needed to cut through the rules and regulations is not present, it needs to be invented.

Where the public health is involved, and procedure stands in the way, it makes excellent sense to err on the side of caution rather than take an unnecessary chance.

The Miami Herald

Miami, Fla., January 20, 1984

JOHN BLOCK, U.S. Secretary of Agriculture, reassured neither farmers nor consumers when he belittled widespread scientific concern over the use of ethylene dibromide, or EDB.

"That's such a tiny amount, it's ridiculous," Secretary Block declared in a recent attack on Florida officials who banned the sale of products containing one part or more EDB per billion.

But Florida Agriculture Commissioner Doyle Conner and state Health Officer Dr. Stephen King issued the ban only after 10 years of studies demonstrating that EDB causes cancer, mutations, and reproductive disorders in laboratory animals. If risk to human health were ranked on a scale of one to 10, EDB probably would rate an eight, according to Edwin Johnson, director of pesticide programs for the Environmental Protection Agency (EPA).

The state's actions focused national attention on the Federal Government's failure to exercise its responsibilities. Despite prodding by responsible and concerned staff scientists, the EPA, the National Institute for Occupational Safety and Health, and the Food and Drug Administration have delayed needed regulations. In the ignorant interim, the use of EDB has increased rapidly.

It is now estimated that 40 per cent of the nation's stored wheat and 16 per cent of the stored corn has been treated with EDB. Wells in California, Florida, Georgia, and Hawaii, where EDB has been used to kill root worms, have been contaminated. The Japanese market for fresh grapefruit, which Florida growers worked so diligently to open and to build, is threatened.

The Federal Government's indefensible record in the EDB controversy cannot be explained away by Mr. Block's ridicule. His statements in San Francisco inevitably raise concern that the Reagan Administration will not yet address the issue responsibly. The statements also raise questions as to whether the Administration will support re-enactment of expiring pesticide-control laws.

The recent announcement by Florida Sen. Paula Hawkins that the EPA soon will recommend Federal tolerance levels for EDB was more encouraging. Florida alone cannot effectively deal with EDB when the marketing of foodstuffs is a nationwide industry. The EDB problem requires Federal action and protective Federal standards.

To be sure, there are economic ramifications in setting EDB standards. None is so serious, however, as to warrant the continued jeopardy to the nation's food and water supply and to the health of its people. Indeed, the Federal Government will be remiss if it does not promptly act to provide nationwide the protection that Florida officials have demonstrated is needed.

The Boston Globe

Boston, Mass., February 13, 1984

Responding to a clear threat to public health, Massachusetts moved swiftly and wisely to ban the sale of food products contaminated by EDB. The Dukakis administration should take whatever steps necessary to ensure that this ban is observed.

The state had been forced to act because the Reagan Administration has been dragging its feet for three years on this matter, with a clear tilt toward the food industry and very little indication that it is concerned about public health or safety. The food industry showed where its interests lie by rushing into court to challenge the Massachusetts ban on the narrowest of legal technicalities.

Since the industry's challenge has been upheld pending a hearing before the Supreme Judicial Court, the Dukakis administration and the legislative leadership should move at once to amend the outdated legislation which pegs state food-quality regulations to federal ones.

Whatever merit the existing legislation may have once had, it does not provide sufficient protection when the federal government is more interested in protecting food industry profits than public health and safety.

Any piece of legislation which can be as easily perverted by special-interest industry lobbyists as the existing food safety law is quite obviously one which requires amendment, if not outright repeal.

The entire incident – from the first reports that food products were contaminated through the food industry's reaction – has reinforced one of the dire lessons of modern life: People undergo great risks these days without knowing in advance what they are.

The cowboys of public relations, galloping in to rescue what remains of the reputation of their polluting, poisoning, ravaging employers, cajole the public not to be so reflexively fearful. After all, they say, mankind has always lived with risk. In support of this fact, grave actuarial possibilities are invoked.

Yes, life has its risks. It is only recently, though, that people have come to think of eating, drinking water and breathing air as being numbered among those actuarial *culs-de-sac.* Since Rachel Carson published "Silent Spring," Americans have undergone a phantomization of risk, slowly adapting to the invisible and, in extreme cases, capitulating to what in an earlier age would have certainly been paranoia but which now is only a slight fever of awareness.

If science and business have yet to find it in their self-interest to accept voluntarily a standard of proof whereby products, compounds and modes of operation must be proven safe rather than unsafe, then it remains for government to impose such a standard. To the extent that people have surrendered to government the obligation to protect them, they have the right to expect that protection.

Given the choice, any reasonable person would rather have Massachusetts public health officials setting the allowable levels of the pesticide EDB than either the Reagan Administration or the food industry.

EDB Gothic

Denver, Colo., February 26, 1984

The Morning News

Wilmington, Del., February 16, 1984

PUBLIC HEALTH officials in Delaware, we are happy to note, are avoiding hysteria on ethylene dibromide. They are adopting federal guidelines on acceptable EDB levels instead of boarding the extremist "ban-wagon" of Massachusetts and a few other states opting for instant elimination of products with *hints of traces* of EDB.

EDB is a pesticide used for more than 30 years to protect grain and fruit from insects and molds. At the same time it provides that benefit, it can also be a carcinogen — at least, animal studies so indicate. Scientists speculate that cumulative ingestion of EDB-treated foods can promote cancer. So why not steer away from EDB in our diet?

Measures taken by the Environmental Protection Agency constitute just that — a steering away, or as EPA chief William Ruckelshaus puts it, "bleeding it gradually out of the marketplace." Last fall, the EPA barred use of EDB as a soil and crop fumigant. Now the agency is setting maximum acceptable EDB levels in cereals, flour and other products. Adjustment of these limits may occur as more becomes known about the degree of risk involved and as use of the substance declines.

There are several reasons why gradualism is prudent in a situation presenting no immediate danger to the public.

● Crops need to be protected from pests, especially when stored. What assurance is there that a different protective chemical or process will be as effective or that it will not also have harmful side-effects. Irradiation is being suggested as one preservation device; experts say the food would not become radioactive. But there are doubters.

● Bruce N. Ames, a highly respected biochemist and expert in the field of carcinogenicity, points out in The Wall Street Journal that the very molds EDB is intended to destroy can cause cancer. Thus, if EDB or a "safe" substitute is not used, the molds can get us! One of the most common and most poisonous molds is aflotoxin, found in peanuts. Do we really want to increase the possible aflotoxin hazard in peanut butter? Dr. Ames says aflotoxin is "1,000 times more potent as a carcinogen in rats than EDB." By the way, Dr. Ames does eat peanut butter sandwiches.

● Scientific uncertainty about the potential cumulative impact of various substances makes decisions about what to ban and what to use extremely difficult. Mr. Ruckelshaus made that point repeatedly when in Delaware this week to commend Delaware's toxic waste disposal program.

Many of us find it hard to accept that science is uncertain. But we have to learn to live with that concept. It is not only difficult to assess the impact of one agent, EDB, on humans; it is even harder to identify all the other agents — food, air, water — that affect those same humans and may in turn interact with the EDB.

Under the circumstances, the EPA and state public health officials are setting a reasonable course on EDB.

A woman who sprayed the pesticide EDB (ethylene dibromide) near Dillon last summer as a U.S. Forest Service worker has permanent white splotches on her skin. She doesn't know what the long-term effects of her exposure will be.

The town of Frisco has said it probably will ban the use of EDB to control pine beetles, which ravage nearby forests.

The state Department of Health has recalled certain lots of grain-based products, fearing unacceptable levels of contamination.

No doubt about it: People are worried about the pesticide, and not without cause. In large doses, it gives rats cancer, although what it does in small doses to humans is less clear.

But at such times, when we labor hip-deep through warnings of doom, it's well to put the matter of man-made carcinogens into focus. Basically, the problem can be viewed in two ways. One is the view of a cadre of activists and a few government scientists, namely that industrial society is producing an ever-increasing number of carginogens that threaten to usher in an epidemic.

Cancer, in this view, is essentially a political blight.

The other, much-less-well publicized view is that espoused by most of the world's respectable epidemiologists (the people who study the causes of cancer). They will tell you that cancer is *not* predominately a disease of industrial societies, and that environmental carcinogens are not basically industrial in origin.

Cancer rates do rise as a society develops, but so do human life spans (those in the U.S. are still rising, despite all the gloomy prattle). Every Indonesian, or Pakistani, or Bolivian who dies at age 18 of an infectious disease is one fewer candidate for cancer. Not many people die of infectious diseases in America.

Yet, as Edith Efron documents in a book to be published this year by Simon and Schuster, the myth of a man-made cancer epidemic is growing. Blame it on Joseph Califano.

In 1978, the then-secretary of Health, Education and Welfare released a report claiming that up to 40 percent of all cancer deaths were occupationally related. The report was without clear authorship, nor had it been subjected to scientific publication and the inevitable critique that follows. Its views were endorsed by a few government scientists, but quietly repudiated by many of their colleagues elsewhere.

In any circumstances, Efron contends, the causes of cancer are exceedingly elusive; *all* estimates are misleading. But if occupational factors are to be assigned a figure, then 5 percent or less is probably closer to the truth. That compares with 30 percent of U.S. cancer associated with tobacco and 35 percent with dietary factors.

Without a doubt, cancer is to be feared and avoided. It ought not become a vehicle, however, for an anti-industrial ethos.

THE INDIANAPOLIS NEWS
Indianapolis, Ind., February 10, 1984

"What we are faced with . . . is not a public health emergency. It *is* an effort to eliminate a long-term, chronic and unacceptable health risk."

The words were from William Ruckelshaus, director of the Environmental Protection Agency. The "unacceptable health risk" is EDB, ethylene dibromide, a chemical used as a soil fumigant and pesticide since 1948.

Last Friday, Ruckelshaus suspended the use of EDB in grain products and issued guidelines to help states determine maximum safe levels of the pesticide in grocery products.

None of the nation's stored grain would be destroyed, he said, and no restrictions would be set on use of EDB in citrus fruit until more testing is done.

The best advice he gave, however, was "calm down."

This is not a time to panic but a time to test. Test the products on the shelves to check for traces of EDB. Test the chemical's use on citrus and vegetable crops to determine the lowest viable levels while testing for other ways to control pests. Continue to test EDB in the lab to find out more about its effects.

In Indiana, state health inspectors are going to check food products from store shelves, beginning with cereal-based baby foods. The overall risk of EDB contamination should be lower here than in some states because the chemical is seldom used by Indiana farmers and, therefore, should pose little threat to grain or water supplies.

While Ruckelshaus and his agency have been criticized for not acting sooner and doing more, we think the latest pronouncements were just right. Certainly the goal should be to reduce the use of cancer-causing agents from the food supply and eliminate them wherever possible. EDB is a potent carcinogen; it has caused cancer in laboratory animals. Yet scientists are split over the danger it poses.

EDB, though used less in recent years, has been a valuable tool. The chemicals that could be used in its place have their own flaws. Therefore, until acceptable replacements are discovered, officials should be reluctant to ban the application of EDB where it's needed most — on citrus farms.

There is time to act wisely. Ruckelshaus said the current levels of EDB discovered should pose no immediate threat to consumers. The new proposals, he said, would eliminate practically all traces of the chemical within the next several years. And that is enough.

The Philadelphia Inquirer

Philadelphia, Pa., February 1, 1984

The pesticide ethylene dibromide (EDB) has been described by the Environmental Protection Agency as the most deadly carcinogenic chemical the agency has ever tested. The EPA concluded that 999 out of 1,000 workers exposed to the legal limits of EDB during their careers would die of cancer. In the general population, three out of 1,000 people consuming normal amounts of EDB-tainted products over a lifetime will contract cancer.

There are few chemicals in use today with such an undisputed record of health risks.

The EPA is empowered under the Federal Insecticide, Fungicide and Rodenticide Act to suspend immediately use of a chemical once it is found to be carcinogenic. A provision of that act requires the federal government to reimburse the manufacturers and users of the product if such a ban is imposed. Thus, those responsible for making and using a deadly chemical product are indemnified against loss if that product proves hazardous and the federal government then decides to remove it from the market.

That provision is especially relevant in the debate over EDB regulation or lack of it. The government's failure to act on the pesticide has been purely economic. While the government ostensibly has been weighing the potential impact on the pesticide and agricultural industries of any regulatory action, those same industries are ensured reimbursement under the law if the product is banned.

Thus, the EPA has been engaged in a lengthy risk-benefit analysis in which the public has all the risks and all the benefits accrue to EDB makers and users. They can continue to sell and use it — and apply for indemnification if the EPA bans it.

The EPA has been considering an EDB ban since 1977, three years after the National Cancer Institute first recommended it. Last September, the EPA issued an emergency ban on EDB as a soil fumigant after the pesticide was found in ground water in four states. The agency, however, did not restrict use of the pesticide to fumigate grain, milling equipment or citrus fruits.

EDB now is showing up in a variety of grain-based products on grocery shelves around the country.

After six years, the EPA finally is about to deal with EDB. EPA administrator William Ruckelshaus is expected to issue EDB standards in the next day or so. The economic ramifications could be enormous. About half the nation's surplus grain stockpiles are tainted with EDB. As much as one-third of off-the-shelf grain-derived food products tested in Texas showed quantities of EDB.

The federal regulatory mechanism to remove a deadly pesticide from the marketplace is complex — and, as the EDB case has proved, open to enormous abuses. The EPA under former administrator Anne M. Burford chose to take no action. Documents introduced during a congressional hearing last year showed that a top EPA official met with citrus officials and agreed to delay any ban.

Mr. Ruckelshaus, facing mounting pressure from state officials and environmentalists, is expected to propose standards that will not entail a total recall of all EDB-contaminated products, but ones that have substantially higher levels of contamination.

That may be the only feasible solution to the EDB problem. The fact that it has taken so long to achieve is a matter for grave concern. Equally worrisome is that the law can be so easily manipulated for economic and political purposes. The pesticide and agricultural industries succeeded in blocking EPA action on EDB for years, while the public was exposed.

There are bills pending in both chambers of the Congress to overhaul pesticide regulations. Previous reform efforts have failed, in part due to the fact that the bills rested in agriculture committees heavily influenced by the pesticide industry.

The EDB crisis has made the holes in the regulatory system apparent to every American. Risks must not be borne solely by the public. Deadly chemicals must be tested stringently prior to use. Long-term effects, as well as immediate benefits, must be considered. The EPA must not be permitted to dawdle as the hazards mount.

The Dispatch

Columbus, Ohio, January 29, 1984

The Ohio Department of Agriculture is acting wisely in dealing with consumer products that have been tainted by EDB, an insecticide suspected of causing cancer.

The department has released the names of 13 products that it says contains traces of the chemical along with the amounts detected by testing. This information enables consumers in the state to identify the products and to decide for themselves whether to purchase them or not.

The department has not pulled the products from store shelves, however, because no one knows at this time what level of EDB contamination is harmful. The Ohio agency has decided to await notification from the U.S. Environmental Protection Agency on what levels are safe and which are not. Then, if the tested products exceed safe levels, the decision to pull them from shelves can be made.

This is sensible governmental supervision. It alerts consumers to a possible danger but does not overreact before sufficient information is available. And it also treats consumers with proper respect, allowing them to make up their own minds about the use of the products. We're confident the Ohio department will follow through on this and notify the state's citizens when additional information becomes available.

The Knickerbocker News

Albany, N.Y., February 15, 1984

"We're getting thousands of calls," says Howard Clark of the state Department of Agriculture and Markets. "There are a lot of people who are scared."

Their fear, not yet a panic but rapidly approaching that stage, is well founded. One reason is a pesticide known as EDB, which is found in a wide variety of foods. Tests have determined that EDB causes cancer and sterility in laboratory animals, and now consumers are worried that cake mixes, rice, flour, citrus fruits and many other diet staples might be tainted.

But are they? That's another reason for the public's anxiety. No two health officials can agree on how much EDB in food is too much. The Environmental Protection Agency, for example, has issued one set of permissible levels, but many states, including New York, find them inadequate. The result is a patchwork of standards, but no way for consumers to know which ones are necessary and which are needlessly protective.

Some objective standard is needed and soon. Experts says EDB is so pervasive in the food chain that it is impossible for consumers to avoid. But there's no similar consensus when it comes to how much of the pesticide constitutes a health danger.

At present, only random remedies exist. In New York, the state has ordered one lot of Carolina Enriched Rice, Extra Long Grain, recalled from market shelves. Meanwhile, Procter & Gamble has ordered a lot of Duncan Hines Deluxe Devil's Food Cake Mix recalled from stores in response to concerns over EDB concentrations found in the product.

Though more recalls of different products probably will follow in coming weeks, that's small comfort to millions of Americans who have been consuming varying levels of the pesticide for years.

There might be some comfort, however, if government would issue sensible and uniform guidelines for the future. Yet it seems unwilling to do so, perhaps because of the undue alarm caused in past years over such substances as saccharin and nitrates.

But that's no excuse for federal officials to keep the public wondering about EDB any longer. On Thursday, they will have a chance to end the uncertainty when they meet in Washington with representatives of New York and seven other states that believe the national EDB standards should be strengthened.

They have a responsibility to act now. A health concern, like an earthquake, sends a tremor through the populace. With the help of a Richter scale, we can accurately assess the damage done by the crumbling earth. Now we need another scale to assess the damage done to our diet.

AKRON BEACON JOURNAL
Akron, Ohio, February 2, 1984

EDB, ethylene dibromide, is only the latest cancer-causing substance to have found its way into the country's food and water supplies.

EDB has been around for 35 years, and it's used primarily as an additive in leaded gasoline. But its use as a pesticide has spurred the recent controversy, evoking that familiar sense of helplessness that has greeted consumers in the past as one toxic substance after another has been revealed.

As a pesticide, EDB has protected citrus trees from rootworm and preserved stored wheat and corn. It has played an important part in expanding world food supplies. But as residues of the chemical begin to turn up in cake mixes and cereals, consumers have every reason to worry that the hazards of ethylene dibromide may outweigh its benefits.

One thing is certain: EDB, applied in massive doses, causes cancer in laboratory animals. Scientists admit that the chemical is among the most potent carcinogens they've ever tested. States, such as Florida and California, have made this discovery on their own, and they've begun to restrict the use of EDB and remove tainted foods from grocery stores.

As yet, there are no federal standards for EDB, although the Environmental Protection Agency appears ready to act, maybe this week.

Ethylene dibromide, however, is nothing new at EPA. The agency has known for a decade that the chemical is a powerful cancer producer. But transferring that knowledge into sound policy has been difficult. The scientific community is divided over the potential threat to humans, and the food industry insists that cooking can reduce the amount of EDB in food by 70 to 80 percent.

Politics also has intruded. The New York Times reports that the citrus industry may have persuaded the EPA from taking action. It wasn't until last fall that the agency finally moved, banning EDB as a soil fumigant after traces of the chemical were found in thousands of water wells in four states.

The EPA also planned to take action against the pesticide's use in grain and fruit, but the food industry gained a delay.

EPA officials admit that EDB is exceptionally potent, and that any level of exposure entails some risk. The question is what is an acceptable risk, and the answer has proved elusive. Heavy doses of ethylene dibromide will likely produce cancer, but that is not how most of us will encounter EDB. Exposure will come in small doses over decades. As yet, effective ways of measuring that kind of exposure do not exist.

Ultimately, the EPA must err on the side of caution and restrict EDB's use. Modern technology has posed an enormous dilemma: balancing health risks against dollars and cents. The mysteries of EDB provoke helplessness, but the country is not so helpless that it cannot act to preserve its health. Uncertainty calls for caution.

Richmond Times-Dispatch
Richmond, Va., February 21, 1984

Some impressive evidence supports the conclusion that people who are concerned about the dangers of the pesticide ethylene dibromide (EDB) are worried about the wrong threat. They should be alarmed about celery, corn flakes and peanut butter, which might be more dangerous.

EDB is a fumigant that for about 40 years has been sprayed on citrus and vegetable crops to protect them from highly destructive worms. It also has been used to kill pests on stored grain and on harvested citrus to kill fruit flies. While most of the chemical is washed or cooked away, traces of it do enter the bodies of people who eat products that have been treated with it. And that's bad, for EDB has been found to cause cancer in laboratory rats.

It's bad because the ingestion of anything potentially harmful is undesirable. But it is not bad enough to justify the nation's hysterical reaction to the discovery of traces of EDB in cake mixes and other food products. The pesticide ranks very low on the list of carcinogens that threaten all who eat, drink and breathe.

Dr. John Weisburger, director of research for the American Health Foundation, is the scientist who discovered the link between EDB and cancer in rats. He has told The Wall Street Journal that the pesticide is "not a very likely cancer risk." Elsewhere he has been quoted as saying that "there is no evidence that people consuming products already on shelves will get cancer from the minute quantities of the chemicals in these foods."

Also in The Journal, Bruce N. Ames, chairman of the Department of Biochemistry at the University of California at Berkeley, has written that the amount of EDB in food is "insignificant" compared with the levels of many carcinogens that are found *naturally* in almost every meal we eat. Plants create their own pesticides to protect themselves against worms and bugs; and as a result, the "amount of nature's pesticides we are ingesting is at least 10,000 times the level of man-made pesticides."

Natural carcinogens are found in mustard, horseradish, black pepper, celery, mushrooms, parsnips and parsley. Mold carcinogens, some of which are extremely potent, are found in peanut butter, corn products, apple juice and, writes Mr. Ames, "many other foods."

This is what he says about the aflatoxin that develops naturally in peanut butter:

"Aflatoxin is about 1,000 times more potent as a carcinogen in rats than EDB. Why make a big fuss about tiny traces of EDB, when the risk from eating the average peanut butter sandwich comes out as more than eating a rare highly contaminated muffin? (The risk from eating a peanut butter sandwich is so low I don't think twice about eating one.)"

Mr. Ames emphasizes that he is not suggesting that EDB is harmless. Certainly he is not suggesting a ban on peanut butter, which is delicious and highly nourishing. The effect of his article is to show that the pesticide's threat has been absurdly inflated and that it is futile for man to attempt to bar all carcinogens from his system. A New York housewife, quoted in Newsweek, echoed this when she said in exasperation: "What can we do? Everything on the shelf is unsafe."

Indeed, life itself is unsafe. Society's only prudent course of action in dealing with a health hazard is to determine whether its contributions to the betterment of life outweigh the dangers it poses. EDB, most uses of which have been halted by the Environmental Protection Agency, seems to pass this test. But for pesticides, fruits, vegetables and grain products would be far less plentiful and far more expensive than they are. The use of pesticides should be carefully regulated, of course, but they — either EDB or effective substitutes — should continue to be available for agricultural purposes.

Los Angeles Times
Los Angeles, Calif., February 5, 1984

After a wholly unnecessary three-year delay, the Environmental Protection Agency has finally suspended the use of the carcinogenic pesticide ethylene dibromide on grain. The EPA may suspend its use on fruits in a few weeks. For the decision we have the new EPA administrator, William D. Ruckelshaus, to thank. But he should have gone even further.

Ethylene dibromide—or EDB—is a pesticide that has been used to kill weevils in stored grain, fruit-fly larvae in fruit and ground worms in the soil. Experts agree that it is probably the most carcinogenic pesticide ever registered with the government. Contact with large amounts, as during its application in the fields, can be fatal, but scientists and regulators are more concerned about long-term exposure.

There have been no conclusive studies about its effect on humans, but it has caused cancer in animals in even minute doses.

The agency also is issuing guidelines for safe levels of the pesticide that might remain in food. But guidelines are only that—indicators that states may or may not follow depending on their ability to evaluate products and enforce the standards. The federal government should set a minimum federal standard.

California has banned the use of ethylene dibromide except on some fruits. Now California health officials have taken the signal of the federal action to announce formation of an advisory group of scientists, consumers, industry and government officials to draw up a plan to eliminate EDB completely from food in no more than three years. State experts say that because even a little of the pesticide could be harmful to infants the best policy is to eliminate EDB residues completely.

The case points up the need for tougher pesticide-control laws at the federal level. California's ability to set stiffer standards than does the federal government has come under constant assault during attempts to revise the Federal Insecticide, Fungicide and Rodenticide Act. When the law is revised, a state's authority to set tougher standards than the federal government does must be made clear. As Congress goes about rewriting the law, Ruckelshaus could play a prime role if he incorporates whatever lessons have been learned in dealing with this chemical into recommendations to strengthen the nation's pesticide laws.

WORCESTER TELEGRAM.

Worcester, Mass., February 8, 1984

First, let's review some facts about EDB — ethylene dibromide.

1. It has been used for more than 40 years in this country to fumigate grain and fruit.

2. There is not a single documented case that it ever hurt a single person to eat that grain or fruit.

3. EDB is used to kill maggots and other food pests and diseases. There is no other substance that compares with it in this regard. U.S. law requires that all imported foreign fruit be fumigated with EDB. California and Florida law requires that any fruit imported from other states must be fumigated with EDB.

4. EDB, like saccharin and other substances, has been shown to cause cancer in rats. Last year the government suspended its use on grain.

5. The Environmental Protection Agency has recommended that prepared food products — mainly cake and muffin mixes — be removed from grocery shelves if they contain more than 30 parts of EDB per billion.

6. The Massachusetts Public Health Council, by a 5-2 vote, has mandated standards on EDB that are 30 times as stringent as the federal guidelines. Within a month, any prepared mix that contains more than one part of EDB per billion will be banned.

Dr. Richard Greenberg, a microbiologist and assistant director of the American Council on Science and Health, a leading educational organization, says that the federal recommendations are reasonable, even though there is no evidence that such EDB trace amounts ever hurt a single human being. He thinks it highly doubtful that any of the banned products are harmful, but concedes that he cannot prove that EDB is totally harmless.

All the excitement, he says, is over the marginal possibility that EDB in low amounts is bad for human health. There is no need for panic, he insists.

But panic is what we so often get from the anti-chemical, pro-organic food lobby, which seems to have taken over the Dukakis administration. Like the great cranberry scare several years ago, and the great saccharin scare of more recent memory, we seem about to go through a great EDB scare. It is a pity that our public officials cannot show more balanced judgment.

The Salt Lake Tribune

Salt Lake City, Utah, February 6, 1984

The wisdom and necessity of the Environmental Protection Agency decision to ban the use of ethylene dibromide (EDB) as a fumigant for stored grain and to propose limits for its residues in grocery products will probably be debated for months, yet the agency's promptness can not be faulted.

EDB causes cancer, mutations and sterility in laboratory animals. That is sufficient reason for agencies like the EPA to error on the side of caution. Already the EPA had banned the use of EDB as a soil fumigant and discovery of traces of the pesticide in corn-based and wheat-based grocery products in Florida, Texas, California and the District of Columbia was ample evidence that the possible carcinogen had been infiltrating the nation's food supply.

But the immediate ban of EDB on stored grain is no cause for panic by the general public, a fact underlined by EPA Administrator William D. Ruckelshaus when he announced his decision.

"What we are faced with is not a public health emergency," he said. " . . . Eating a grapefruit or a cupcake doesn't present a risk at all. But a lifetime of consumption, that's where the risk is."

In order to reduce to the barest minimum the risk of that "lifetime of consumption" Mr. Ruckelshaus presented the states suggested guidelines for reducing any potential hazards, among them proposing a maximum residue level of 30 parts EDB per billion parts of food for ready-to-eat products, such as cold cereals and baked goods.

Similar guidelines were presented for cake mixes and flour, along with standards for raw grains, mainly wheat, corn and oats.

In considering the ban on EDB the public will be doing itself a favor by heeding Mr. Ruckelshaus' advice to "calm down," remembering that less than an estimated 1 percent of the wheat-based products in grocery stores and warehouses and about 7 percent of the corn-based products exceed the EPA standards for EDB residues.

Additionally, the public should bear in mind that although consumers have been exposed to EDB for 30 years, primarily through grain, there has not been a documented case of human cancer traceable to the chemical; only laboratory animals have been affected.

Thus, as Mr. Ruckelshaus noted, "We are not talking about picking up something that is acutely toxic." Nevertheless, the EPA has moved with commendable quickness to reduce, probably eliminate, any potential problems that might result from a lifetime of exposure to EDB.

Minneapolis Star and Tribune

Minneapolis, Minn., February 15, 1984

The carcinogenic pesticide EDB, ethylene dibromide, simmered too long on the back burners of regulatory agencies. In banning nearly all uses of EDB early this month, the Environmental Protection Agency finally took a necessary step to protect consumer health. But that commendable step should not be the last. The federal government must set strict, enforceable standards for EDB and other pesticides already in the food pipeline.

For decades, EDB has been used to control insects in fruit and grain, to kill pests in soil before planting and to fumigate milling machinery. When the National Cancer Institute conducted research on test animals in the 1970s, it found that EDB caused cancer, birth defects, genetic disorders and sterility. EPA scientists ultimately concluded that EDB also threatened human beings; that continued exposure to EDB at levels exceeded by hundreds of products on grocery shelves would substantially increase a person's risk of cancer.

But the agency was reluctant to acknowledge the risk and slow in acting to minimize it. When the EPA first considered regulating the pesticide in 1977, food industry representatives objected. In December 1980, just before the Carter administration left office, the agency rejected industry complaints and proposed an EDB phaseout. In the long run, the agency concluded, the health risks posed by EDB would outweigh the higher costs a ban would create.

That recommendation was received coldly by the Reagan administration. EPA administrator Anne Burford dismissed evidence of EDB's health risk and instead strived to accommodate industry concerns; several of her subordinates held illegal private meetings with citrus and vegetable growers who opposed an EDB ban.

But when new data showed that EDB had contaminated groundwater and that the pesticide's cancer risk was even greater than initially estimated, the tables turned again: Burford's replacement, EPA veteran William Ruckelshaus, banned the chemical's further use for grain and soil fumigation and set guidelines for EDB levels in food.

Unfortunately, the agency's action falls short of the need. Ruckelshaus has yet to regulate EDB in citrus fruits — even though levels in some oranges have been reported to be hundreds of times as high as levels considered safe by EPA scientists. And the food-level guidelines that have been issued are only advisory: States can decide whether to ban or recall foods.

Some states, like Minnesota, are conducting thorough tests of EDB levels; others are not. Even if every state follows EPA guidelines, doubt remains that the agency's standard for "ready to eat" products is sufficient to guard against cancer risks. Some states have set stricter levels. Minnesota Sen. Dave Durenberger, chairman of the Senate subcommittee with jurisdiction over pesticides, favors mandatory federal standards three times as strict.

EPA officials admit that if EDB were a new chemical, it would not meet the standards of current environmental law. That is a telling indictment. To guard against EDB's hazards, the government must quickly ban the pesticide's use on citrus fruits and adopt mandatory food-level standards that err on the side of human health.

Roanoke Times & World-News
Roanoke, Va., March 16, 1984

THE MILLS of government grind slowly, but they don't always grind exceeding fine, either. In the aftermath of the EDB crackdown, evidence continues to emerge that federal agencies dragged their feet for years before acting on this pesticide.

EDB, or ethylene dibromide, has been used in agriculture as a pest-killer since the 1940s. In 1956 — about the time Americans became aware that some substances in food could cause cancer — the Food and Drug Administration concluded that EDB would leave no residue in foods. It granted EDB an exemption from pesticide residue standards.

But documents disclosed at recent hearings by two House Government Operations subcommittees show that: (1) Studies made as far back as 1965 indicated EDB remained at high levels in papayas, citrus fruit and whole-grain bread; (2) the U.S. Agriculture Department shared this data with FDA; (3) neither agency did anything about it, despite growing evidence in the 1970s that EDB caused cancer in laboratory animals; and (4) the data apparently was not passed on to the Environmental Protection Agency when the new agency took over pesticide regulation more than a decade ago.

The information remained hidden in the other agencies' files even after the EPA began building its case against EDB. Testimony to the congressional subcommittees indicates that USDA and FDA were more concerned about defending their former stand on the pesticide. For example, USDA signed an accord last August with a group in the fruit and vegetable industry, agreeing to pay most of the cost of compiling research data to maintain EDB's registration as a safe pesticide.

When Uncle Sam has as many agencies as he does, it's inevitable there will be turf fights over regulation. Ultimately, however, government is supposed to serve the people, not special interests. There can be no foolproof system to insure this. But Congress should seek ways to improve sharing of pertinent information among agencies, even if that means overcoming reluctance and resistance. It also should continue to encourage "whistle-blowers"; USDA once shared data on ethylene dibromide with FDA, but nothing happened.

ST. LOUIS POST-DISPATCH
St. Louis, Mo., April 2, 1984

Having acted belatedly and in a limited way to curb the use of the dangerous pesticide ethylene dibromide (EDB), officials of the Environmental Protection Agency are now saying that methyl bromide (MB), which is being used as a substitute for EDB, is, like EDB, a cancer-causing agent. As if to excuse its procrastination on EDB, the EPA said last month that it was afraid it was "merely exchanging the devil we know for the one we don't know." The "devil we don't know" has now turned up in the form of MB, which is showing up in Florida citrus fruits at levels four times greater than the EPA has deemed acceptable for EDB. A study in the Netherlands has found cancer-causing properties in MB very similar to those of EDB.

The moral of this story, it seems to us, is not that the EPA acted too quickly on EDB, but that the whole system for protecting human health against dangerous new chemicals is inadequate. Like EDB, MB has enjoyed a longstanding exemption from residue limits under the food safety laws. The process under which the EPA was enabled to act against EDB took more than a decade after the chemical was identified as a carcinogen in 1973. Must a similar period elapse before action can be taken against MB?

About 70,000 chemical compounds are already in commercial use, with about 1,000 new compounds being added each year, many of them without adequate testing. A recent study by the National Academy of Sciences noted that only about 7,000 chemicals in the current inventory have ever been tested for cancer-causing effects. With some 21 percent of the tested chemicals having been found minimally suspect, the obvious need is for stricter, congressionally mandated testing.

St. Petersburg Times
St. Petersburg, Fla., May 21, 1984

Worms so tiny that you can't see them gnaw on the roots of Florida's orange trees. To stop the nematodes from destroying this valuable crop, the State Agriculture Department used to inject the soil around citrus groves with massive amounts of the pesticide ethylene dibromide (EDB). But the agency stopped doing that last year after state officials discovered that hundreds of drinking water wells had been contaminated by the cancer-causing chemical.

Now, the problem is figuring out how to control the soil-burrowing pests without poisoning the water. Another pesticide the department was thinking about using instead of EDB may

---

The public should be assured that the environmental risks of biological engineering have been properly evaluated and that adequate safeguards are used in field tests.

---

be just as dangerous, according to experts at the University of Florida's Institute of Food and Agricultural Sciences (IFAS).

Until recently, most farmers had a nonchalant attitude about using powerful pesticides to kill bugs and other critters that attack their crops. No one knew the dangerous consequences of agriculture's heavy reliance on chemicals. Now we know better. Some of the pesticides farmers use to protect their crops also poison the water and soil.

THAT DISCOVERY has spurred researchers to search for safer ways to kill pests. Most of the emphasis has been placed on biological controls. That usually means releasing beneficial insects to prey on harmful insects. American scientists are looking around the world for natural insect enemies to import to the United States for biological control programs.

Using viruses, bacteria and manufactured organisms to kill pests is a much smaller aspect of biological control programs, but a promising one. At IFAS, researchers are exploring these methods to control nematodes in their quest for a safe alternative to EDB and other equally dangerous chemicals.

A ruling by U.S. District Court Judge John J. Sirica last week throws such genetic-engineering research into limbo. He barred what was supposed to be the first release in history of a man-made organism into the environment. The preliminary injunction prevents a University of California-Berkeley scientist from spraying a 200-foot row of potato vines with a common bacterium that has been altered in the laboratory to protect plants from frost.

The decision also prohibits the federal National Institutes of Health from approving any other releases of genetically engineered life forms until a 1983 lawsuit can be resolved. The suit charges that the experiments' environmental risks have not been properly studied.

Sirica's ruling should open a debate in Congress on how best to approach the age of biological engineering. The potential for important scientific advances is too great to abandon this exciting area of research. Researchers are trying to develop products that range from bacteria that "eat" toxic chemical spills to disease-resistant plants and microbes that enable plants to grow in arid climates.

YET, THE PUBLIC should be assured that the environmental risks have been properly evaluated and that adequate safeguards are used in field tests. The best way to do that is through a federal law governing genetic-engineering experiments. Since no such law exists, private genetic-engineering companies voluntarily have submitted their research proposals to the National Institutes of Health for safety reviews. The problem with relying on volunteerism is that some researchers might not volunteer. Congress should pass a law that would allow genetic-engineering experiments to continue with stringent controls that all researchers would have to follow.

Back in Florida, the State Legislature should reassess its decision to deny a request from IFAS that $2-million be appropriated for biological control research in next year's budget.

Pests swarm to Florida because they like its subtropical climate and all the crops grown here. The farmers need help to wean themselves from their dependency on chemicals to kill all the pests that attack their crops. Biological controls are the most promising alternative to pesticides that poison Florida's soil and water. Money spent to search for natural enemies to prey on harmful pests would be an investment in Florida's future that would return many benefits.

THE LINCOLN STAR
Lincoln, Neb., February 13, 1984

Panic is not advised. Everybody remain seated.

If you're concerned about the presence of EDB in your cake mix — throw it away; the small change you plunked down to pay for the cake mix isn't worth the cost to your mental health.

Then again, you could wait until Nebraska reveals the result of tests it's conducting to determine the extent of EDB contamination here. We'll hazard a guess that even then there won't be need for panic. EDB usage has figured relatively minor in Nebraska agriculture.

Of course, there's no excuse for its presence in our food or our groundwater. But it appears that the acute danger of EDB — ethylene dibromide, it's been registered since the late '40s for use as a pesticide — is to people who either manufacture it or work with its direct application. The National Institute for Occupational Safety and Health last year issued an alert on its possible health hazards for these people.

IN SOME CASES, EDB might have been misused. And the Environmental Protection Agency probably can take blame for some inaction since EDB's first link with cancer in lab animals was discovered in 1974. In 1980 it was ready to phase out some ag uses over a three-year period but the Reagan administration put on a brake, saying industry could not adjust in that time. Even now, the EPA is dragging its feet on banning the chemical's use on citrus fruits because of economics.

But let's not panic.

The scientist who discovered the link between EDB and cancer in laboratory rats notes that peanut butter, of all things, may be riskier because of the presence of aflatoxin B1, a toxin produced by mold that grows on peanuts. It is said to be a thousand times more powerful than EDB.

And cigarettes. EDB in the minute amounts present in foodstuffs isn't even "in the same league with cigarettes" as a cancer-causing agent, says Bruce Ames, the head of biochemistry at the University of Californa at Davis.

THE CHANCE OF contracting cancer from a slice of chocolate cake is nil. The EPA now allows 150 parts per billion in consumer products requiring cooking. The size of a part per billion is equivalent to one ounce of water in one mile of the Elkhorn River. It is not a substantial figure.

The rats that contracted cancer in the EDB experiments were force-fed 10,000 times the dose permitted in food five times a week for 30 weeks.

Also, EDB is diluted by cooking, so the amount measured in muffin mix is reduced by turning mix into muffin.

The ironic thing is that EDB replaced DBCP, a fumigant banned when workers in Colorado, California and other states became sterile. And there's always the chance EDB could be replaced with something equally or more hazardous.

The presence of pesticides is a fact of 20th century, chemically dependent life. Unfortunately if it kills off insects, it has to be toxic and caution is urged in usage.

Meanwhile, we live with EDB and much contradictory information about the consequences. It's good steps are being taken to rid our lives of chemical pests. It is not time to panic.

Rockford Register Star
Rockford, Ill., February 14, 1984

The U.S. Environmental Protection Agency was correct in taking prompt action to deal with mounting concerns that a certain pesticide might pose public health hazards.

The EPA virtually banned the use of ethylene dibromide (EDB) as a pesticide on grain and grain products after reports that traces of the suspected cancer-causing chemical had been showing up in flour, cake mixes, cookies and crackers.

Tests have shown EDB causes cancerous tumors, birth defects and sterility in laboratory animals. Though no similar threat to humans has been firmly established, the EPA was wise in not waiting for proof of such hazard before curbing the chemical's use.

Last fall, EDB was banned as a soil fumigant. Now the EPA has extended the ban to use of the chemical on grains.

The wisdom of the EPA s move against EDB is because the potential risks to the public are in long-term exposure to the chemical. By the time the effects on human health were firmly established, the number of victims could have been legion.

The Courier-Journal
Louisville, Ky., February 8, 1984

THE PLEA of Environmental Protection Agency chief William Ruckelshaus for "calm" about the ethylene dibromide situation may have been superfluous. When a problem is so resistant to a clear-cut solution, there is more inclination to be apathetic than excited. But there is plenty of need for rational concern that a huge share of the nation's food supply is contaminated with this chemical.

The bare facts are as simple as the problem is difficult. About 40 percent of the nation's stored wheat and 16 percent of its stored corn, among other products, have been treated with ethylene dibromide, or EDB, a pesticide first used as a gasoline additive but later found effective in warding off weevils and other pests. Then it was found to cause cancer, genetic disorders and other illnesses in laboratory animals whose reactions to chemicals are much like those of humans.

So both science and common sense indicate that ethylene dibromide shouldn't be in anybody's food in any quantity at all. But there is no way to say with any precision what amount of it would cause measurable harm, or to be certain that any amount at all is safe. By no coincidence, Mr. Ruckelshaus — after suspending further use of the pesticide on grain products — has declared maximum recommended tolerances for the pesticide at just the right quantities to avoid widespread destruction of existing food-grain products or disruption of markets.

Some states move quickly

The EPA levels are advisory only, because Washington lacks authority to regulate the amount of the chemical in food. States preferring to err on the side of safety already are considering much lower limits. Massachusetts, indeed, has voted immediate removal from the shelves of products containing 10 parts per billion or more of EDB, and by March 1 of products with one part per billion. The EPA recommends a limit of 30 parts per billion in ready-to-eat food, 150 parts in food requiring cooking, and 900 parts in raw grain intended for human consumption. (Heat and aeration are partly effective in removing EDB).

It's disconcerting that all this action comes so late. The EPA was warned by the National Cancer Institute in

The Evening Gazette
Worcester, Mass., February 27, 1984

New York, New Jersey, Connecticut, Maine, New Hampshire and Vermont have agreed to restrictions on ethylene dibromide that are three times as stringent as federal EPA standards.

Massachusetts is not included because Massachusetts has set its own standards. They are 10 times as strict as the northeastern standards, which are three times as strict as the federal standards. So the Massachusetts standards are 30 times as strict as federal officials think necessary.

Massachusetts stands alone.

It's not necessarily a bad thing to stand alone, especially if the principle involved makes sense. But it's foolish to stand alone merely to make a gesture. That's about what the Massachusetts EDB standards amount to.

To say that grain products on Massachusetts store shelves must not exceed one part per billion of EDB is something like demanding human perfection.

Cargill Inc., third largest producer of flour in the country, says it may have to stop shipping flour to Massachusetts if the regulations aren't changed. As goes Cargill, so may go the other flour producers.

Well, at least Massachusetts won't have any EDB in its flour. In fact, it probably won't have any flour. That's one way to solve the problem.

THE WALL STREET JOURNAL.
New York, N.Y., March 14, 1984

The policy juggernaut marches ahead on ethylene dibromide, or EDB, the pesticide linked to cancer in rats. The Environmental Protection Agency recently extended its EDB ban to fruit (it already covers grain), while the prophets of a cancer "crisis" have hit the op-ed pages, urging even stricter regulation.

We were watching this amazing performance when another cancer study crossed our desk. That study, to be published soon in the Toxicology and Applied Pharmacology journal, found that another pesticide, methyl bromide, also causes cancer in rats. Why is this news? Because methyl bromide is probably the most widely used substitute for EDB. So our regulators have in effect banned one carcinogen so that farmers and fruit growers can replace it with another. We've said before that the EPA's ban on EDB isn't justified by the public-health risk it presents. But the new methyl bromide findings, as well as a look at the few other possible substitutes, suggest the ban itself may be dangerous.

The methyl bromide study is similar to those done to test EDB. Giant gulps of the stuff were rammed down rat gullets to see if tumors developed. For methyl bromide, nearly two-thirds of the rats developed tumors after only three months. That's even faster than in the EDB tests, which took nearly a year to show a 90% tumor incidence. What's more, methyl bromide is a gas (while EDB is a liquid at room temperature), so it's more volatile and thus more dangerous to the workers who apply it. "The bottom line," says Berkeley, Calif., biologist William R. Havender, is that "one is just as dangerous as the other."

What about other substitutes? Well, for grain, there's aluminum phosphide, but that's more acutely toxic even than EDB. It's also highly flammable, and thus could cause havoc with the explosive grain dust found in storage bins and mills. But does it cause cancer? We don't know. Why? It hasn't been tested. Other chemicals offer similar risks, as do the alternative methods used to kill fruit flies in citrus. Cold storage seems to work, but it's 10 times more costly than EDB and can't be used for some 40% of the citrus crop because it damages young fruit. Irradiation also has potential, but many of its effects still aren't known, either.

The question raised by all this is: Why didn't the EPA think more about these substitutes before it pre-emptorily banned EDB? The EPA says it's because EDB was so much greater a threat that it had to be removed quickly. But in view of the methyl bromide findings, that's less than convincing.

We suspect that a more important factor is the political sensitivity of the EPA itself. Environmentalists are very good at feeding public and political fears, and in the case of EDB that was raised to a high art form. State regulators, wanting to appear cautious, joined the chorus. Once the media joined in, boxes of muffin mix were being dumped all over the country. The EPA, like any government agency, finds it hard to resist this crazy momentum.

Unfortunately, this sort of decision-by-panic rarely makes good public policy. EDB was used for an important purpose: to kill the flies, worms and other pests that destroy our food and carry diseases of their own. Like any pesticide, EDB is a poison and should be handled with care. But the scientific evidence available so far suggests that it's still needed—unless a truly safe substitute can be found.

The Hartford Courant
Hartford, Conn., March 3, 1984

The finding of the pesticide ethylene dibromide — EDB — in food and water in Connecticut seems to have aroused little panic in the state. Consumers haven't stopped buying and eating grain products, and even residents of the towns where wells have been found contaminated seem to have taken the problem in stride.

The state deserves credit for the calm by quickly adopting, with five other northeastern states, the goal of eliminating EDB from food. In most cases, officials have been forthcoming with information on the results of various tests for the pesticide. They seem to have learned that the one sure way to upset people is to create the impression that they are being kept away from information they have a right to know.

People, it seems, have also become more sophisticated in recent years and are able to put the EDB threat in the context of numerous other threats, natural and unnatural, in everyday living. As one scientist has pointed out, they would probably be taking more of a chance by driving to the supermarket to re- turn their muffins than by eating them.

There may even be a sense of fatalism involved. As long as most of the population depends for its food supply on large-scale agriculture, with its heavy dependence on pesticides, more scares are highly probable. For example, some farmers have been using a carbon tetrachloride mixture as a replacement for EDB, but animal studies have shown that carbon tetrachloride causes cancer, gene mutations and liver and kidney damage. Florida citrus growers are now using methyl bromide — for which there have also been animal studies indicating dangerous effects — as a substitute for EDB.

Will residues of carbon tetrachloride and methyl bromide be the source of future food contamination problems?

Certainly EDB is a poison that ought to be eliminated from the food supply as quickly and completely as possible. But its presence is just another reminder of far deeper problems of contamination inherent in the nation's entire system of food production and distribution.

The Washington Times
Washington, D.C., April 17, 1984

This nation has not yet learned to treat environmental concerns with common sense. The latest example is the Environmental Protection Agency's ban on EDB, the chemical that kills vermin in grain. No question that EDB causes cancer. So does saccharin. So does pepper. So do innumerable other things, natural and man-made.

But what means *causes* in the the chilling expression "causes cancer"? A bullet through the heart causes death. A cigarette in the mouth does not cause cancer. Many cigarettes a day every day may cause cancer. It depends on the person. So *cause* used with carcinogen has to do with statistical probability.

Figuring the average American consumes up to 10 micrograms of EDB a day, the EPA said this was enough to "cause" three out of every 1,000 people to get cancer. But does this statistical probability have anything to do with the real world? What *is* the real potential that EDB will give you cancer?

EPA says researchers had to feed a rat 20,000 micrograms a day every day in order to induce cancer. We got out our hand calculator and figured that, since a human weighs about 150 times as much as a rat, to risk getting cancer you would have to consume 3 million micrograms of EDB daily. This is 300,000 times the 10 micrograms the EPA estimates you consume in your normal intake of three pounds of food a day. And you would have to consume these 3 million micrograms every day of your life. To do that, you would have to eat 450 tons of food a day. That will kill you without any EDB.

The EDB ban contains other absurdities, but why belabor the point? We'd add only this reminder: When EPA took saccharin off the shelves, the reason given was it "caused" cancer. Indeed it did, but to ingest the equivalent amount of saccharin that gave laboratory animals cancer, a human being would have to drink 1,000 bottles of diet cola a day.

The alternatives to EDB, a common vermin killer, are either proven carcinogens or else have never been tested for their cancer-causing potential. All are more hazardous to workers. Without vermin killers, you can't store grain. Great news for a starving world.

We're betting the EPA ban will last no longer than the saccharin ban if the media will only do its homework instead of parroting the alarmism of environmental and health zealots, who have managed to wrap Congress — remember the Delaney amendment? —and EPA Administrator William D. Ruckelshaus around their fingers.

OKLAHOMA CITY TIMES
Oklahoma City, Okla., February 17, 1984

TESTING of wheat stored in Oklahoma has demonstrated so far that the latest chemical pesticide targeted by the environmental detectives as dangerous to human health is not much of a factor in this state. That won't keep the doomsayers quiet on the national scene, of course.

The Environmental Protection Agency has banned the use of ethylene dibromide (EDB) as a pesticide for grain because it is said to pose a cancer risk. But in Oklahoma a health department environmental epidemiologist reported the highest EDB level found in mills in this state equaled 48 parts per billion — far, far below the federal tolerance level for raw wheat of 900 ppb.

Partial test results on grain stored in eight state elevators showed none of the pesticide and fumigant present. Some EDB residue may show up when testing of grain stored on farms is completed, but apparently Oklahoma producers have been shifting from EDB to other products over the past 10 years, accounting for the generally low incidence here.

Nevertheless, officials had to go through the motions of checking it out because of the concerns raised nationally. But one wonders why all the fuss has been generated.

It seems to depend on which scientist gets the ear of someone looking for a cause to push. The more thoughtful scientists tend to put these things in perspective. Thus, a biochemist at the University of California at Berkeley has written that the trace of EDB allowed in food is insignificant compared with the level and risk of many cancer-causing agents found in every meal, most of which are natural and traditional.

In a Wall Street Journal piece adapted from his article in Science magazine, Bruce Ames notes that carcinogens come from four main sources.

One is what he calls nature's pesticides — toxic chemicals synthesized by plants in large amounts to defend against insects and other predators. And plants in the human diet are no exception, he says, adding that the "amount of nature's pesticides we are ingesting is at least 10,000 times the level of man-made pesticides."

Mold carcinogens are another source. Ames explains molds make a great variety of mutagens and carcinogens, some of which are among the most potent ever discovered. Cooking our food also makes mutagens and carcinogens, inasmuch as all burned and browned material contains them.

Finally, Ames points out, in studies on human beings high fat has been associated with colon and breast cancer and heart disease, as has a high intake of alcohol.

Ames suggests that the public health could be endangered more by banning the use of EDB as a fumigant because of the much greater cancer risk from the powerful mold carcinogens.

Detroit Free Press
Detroit, Mich., February 22, 1984

HOW DANGEROUS is the threat from EDB? Well, apparently no one's going to keel over from eating blueberry muffins, but if you drink water from a well in the Florida citrus belt, you might want to check with your local agricultural agent.

The biggest problem with EDB lies in contamination of the groundwater in citrus-growing areas, where the potential carcinogen has been used as a soil fumigant. Traces of EDB found in treated grain products are considered too minute to be an immediate threat. The possibility that it may be a long-term peril has been dealt with, to the best of anyone's knowledge, by the severe limits the federal government has placed upon its future use.

The catch, of course, is that phrase, to the best of our knowledge. The long-term questions about EDB have not been settled, no more than they have been settled for a roster of thousands of chemicals in current use. The EDB scare, like the dioxin scare before it, is a compelling argument for more research funds, not to mention more teaching of biology and chemistry and nutrition in the schools, not only to produce more scientists, but to produce more informed citizens as well.

This is not the only time political and economic decisions will have to be made before the hard scientific evidence is in. The ability to detect parts per billion — and, someday, parts per trillion — of a given substance may force society to reconsider the wisdom of its use time and again.

You might have a hard time explaining, though, why so many Americans who smoke, drink too much, eat too much or drive without seat belts also become outraged at the thought of 50 parts per 1,000,000,000 of EDB in their morning cereal, especially if the alternative is to return to the days before grain products were treated, when squiggly little things showed up regularly in flour, rice and cereal boxes.

A partial explanation is that there's a difference between risks knowingly assumed and unknowingly imposed on one. People are entitled to assurance that they aren't going to become cancer statistics or have their descendants transformed into mutants because of some chemical in their cake mix. On EDB, the federal government seems to have given that assurance. But don't think that we won't go through similar alarums and excursions again.

The Wichita
Eagle-Beacon
Wichita, Kans., February 7, 1984

Both consumers and Kansas wheat growers should feel relieved the Environmental Protection Agency has banned the future use of EDB — ethylene dibromide — as a grain storage pesticide. The agency indicated the extent of EDB pollution in foods now on grocery shelves probably is fairly small.

Health authorities in some states have ordered stores to remove cake and muffin mixes found to contain minute quantities of the elevator and mill fumigant. It was feared a state of near-hysteria over the safety of cereal foods might spread, and ultimately cost Kansas farmers millions of dollars in lost sales.

But William P. Ruckelshaus, EPA administrator, estimates only about 1 percent of the wheat-based products now in supermarkets and possibly 7 percent of those corn-based probably fail to meet EPA-recommended standards. These allow 30 parts per billion for ready-to-eat cereals, baked goods and snack foods; 150 for flour, mixes and uncooked cereals; and 900 for raw grain for human consumption. The variation in permissible levels is due to EDB residue's tendency to break down rapidly.

Use of EDB as a soil fumigant was banned last fall after the chemical was found to cause cancer in laboratory animals. The Environmental Protection Agency still is considering whether its use as a fruit fumigant should be stopped. Six states require the use of fumigation to control fruit flies, and EDB levels as high as 1,000 parts per billion have been found in citrus shipment samples from Mexico and Italy.

Obviously, the EPA and other official agencies should continue to monitor the situation closely. Criticisms of the Natural Resources Defense Council and others over the weekend should be taken seriously, and further tightening actions should be taken if the circumstances warrant.

Malathion: California's "Medfly" Spraying

Mediterranean fruit flies were found on the fruit trees of California's Santa Clara County in June, 1980. The fly, not native to the United States, deposits its eggs on a great variety of fruit, later causing the maggot-ridden fruit to rot and drop prematurely. Their appearance in California, a potential threat to the state's $14-billion produce industry, sparked one of the sharpest conflicts to date between the general populace and the agricultural industry. Farmers immediately requested aerial spraying with malathion, a chemical pesticide that is classified as an organic phosphate. One of the most widely-used ingredients in household insecticides, malathion is highly toxic to humans but does not endure long in the environment. If absorbed through the lungs or skin, it can act upon the nervous system of humans as well as insects. It was reported to be "one of the safest pesticides in use" by the California Department of Health Services, however, and its relative safety was also vouched for by the Environmental Protection Agency.

Nevertheless, residents of the heavily populated Santa Clara County vehemently opposed aerial spraying. California's Gov. Jerry Brown decided instead to try and control the infestation through the biological countermeasures of stripping affected trees and releasing sterile male flies to reduce the "Medfly" population. Despite these efforts, the flies survived the winter and emerged again in the early summer of 1981, posing a serious threat to California's crops. (The inadvertent release of more than 200,000 fertile male flies, reportedly purchased from a Peruvian supplier, apparently contributed to the spread of the pest.) Brown continued to reject aerial spraying until July, when U.S. Secretary of Agriculture John Block threatened to quarantine all California produce in an effort to keep the Medfly from spreading further. By July, about 620 square miles in three California counties were affected. Brown gave in, abandoning the more conservative approach to pest control, which by this time included extensive ground spraying. Several states, including Alabama, Florida, Mississippi, South Carolina and Texas had already instituted their own quarantines on California produce that had not been sprayed. Aerial spraying was conducted on a weekly basis over more than 1,300 square miles that summer; the spraying was resumed over a smaller area in 1982.

The Cincinnati Post

Cincinnati, Ohio, July 16, 1981

Environmental extremists are making it more and more difficult for sensible people to support them. The Mediterranean fruit fly crisis in California is the latest case in point.

On one side was the very clear and present danger of a financial catastrophe for California farmers. The state was under threat by the Agriculture Department of a total embargo of its fruits and vegetables unless it conducted aerial spraying of pesticide to control an infestation of the tiny fruit fly.

The repercussions of an embargo would have been felt by every American, since California provides nearly half the produce sold in the nation's grocery stores.

On the other side was the remote possibility that someone would be injured by the pesticide, malathion.

Predictable, hard-core environmentalists went to court to try to prevent the spraying.

Now, if malathion had been a new and untried compound, or if there was evidence that it was harmful to humans, most people would have been inclined to agree with the opponents of spraying.

But malathion appears to be about as safe as any chemical of its kind can be. It has been widely used for more than a quarter of a century and no study has ever documented any adverse effect on humans. It has the blessing of the Environmental Protection Agency, no hotbed of chemical sprayers.

The National Cancer Institute fed malathion to rats and found no evidence of cancers. That's more than can be said for saccharine.

Of course, there can never be absolute guarantee of safety in matters like this. Choices have to be made on the basis of probabilities and best available evidence. And in California, the probabilities and evidence overwhelmingly mandated aerial spraying.

Those extremists who showed themselves incapable of choosing rationally between greater and lesser evils, and especially those who tried to whip up a kind of Three Mile Island hysteria over malathion, did the cause of true environmentalism no service.

The Register

Santa Ana, Calif., July 22, 1981

Gov. Jerry Brown's footdragging and scaremongering over Malathion that began months ago has clearly led to a far more serious infestation of the dreaded Mediterranean fruit fly than imagined 10 days ago.

"The flight's lost. It is gone. The fly has established itself," frustrated state entomologist Terrance M. Allen contended Thursday. "It's so far gone that even if they throw everything at it they're not going to eradicate it."

Allen, you may remember, was given a state ag department demotion earlier this year after publicly attacking Brown's too little too late tree-stripping and sterile fly scheme. Brown was then entranced by the notion Malathion was about as deadly as atomic radiation. He has since changed his tune, of course, after realizing that mollifying the ever-fearful environmentalists just wasn't good politics with a $14 billion industry in serious jeopardy.

What makes Allen's gloomy forecast especially significant is that Allen stood alone last spring in criticizing Brown's cautious ground-level approach to the Medfly and then correctly predicted the massive hot-weather outbreak and accurately sized up how far the fly had spread.

While he pleaded and politicked for Brown's serious attention, said Allen, "All the fruit fly infestations that we found between July and September (1980) were left to breed. There was the first mistake. No one took any action on it. It was burning me up."

In reading Allen's comments in the Los Angeles *Times* it is clear the Medfly's spread occurred far earlier and to a much greater area than Brown was admitting, while the governor was panicking residents of the Santa Clara Valley with wholly irresponsible scare stories about Malathion.

It is hard to imagine now what political gain Brown saw in spreading scare and holding back on aerial spraying, but the price to California may be fearful indeed.

Now it isn't just a matter of a threatened one-year quarantine on California-grown produce, it is the prospect that the Medfly infestation is out of control.

Brown's pitiful performance under fire in this, his administration's first real crisis, cannot really be ignored. His posturing and endless public statements that took 180-degree turns overnight have even prompted warnings from the Reagan administration to "shut up."

"I think everything may be served better for a few days if the governor wouldn't make so many statements," Richard Lyng, the No. 2 man in the U.S. Department of Agriculture said Friday. "I wish he had worked more closely with his people at the (state) department of agriculture."

Lyng's comments were triggered by Brown's latest move, a request that California be given federal disaster aid because the situation is "beyond the control" of the state.

Well, Gov. Brown, who brought it to that point?

Nevada State Journal
Reno, Nev., July 17, 1981

California's assault on the medfly is a mess. Helicopters don't fly, insecticide won't spray, territory marked for attack goes unattacked, and the Medeterranian fruit fly is rushing like a Panzer blitzkrieg toward the rich commercial farms of the San Joaquin Valley.

Gov. Jerry Brown has asked the federal government for assistance, and he should get it. But more than this, the federal government should consider handling the entire entire operation itself. The spread of the fruit fly endangers not just California's agriculture, but the food resources of the entire nation. If California can't stop it, someone must. And that someone is the federal government.

Instead, White House spokesman are playing games, saying they haven't received Brown's letter asking for aid. This is absurd. If several dams broke, if fire destroyed Los Angeles, would the government wait upon a letter? Hardly. Yet this situation, in its own way, is equally serious.

The governor has made his request clear, the need is urgent, and the time for action is now.

Roanoke Times & World-News
Roanoke, Va., July 15, 1981

While he was willing to listen too much to environmentalists and not enough to farmers, California Gov. Jerry Brown wasn't standing idly by during the Mediterranean fruit fly's invasion. His administration had been fighting the infestation; one of its efforts involved shipping 3.5 billion more of the pests into the state.

That's not as strange as it sounds. This was an attempt at biological control. The 3.5 billion flies were males, hatched from eggs irradiated with radioactive cobalt to make them sterile. They were released in profusion so that they would greatly outnumber the females — making the odds 100 to 1 that a female fruit fly would mate with a sterile male.

For a time it seemed the flies had been wiped out. Then it was discovered that at least 200,000 of the males, from a Peruvian supplier, apparently were fertile — and a considerable number of them had found females to interact with. Thus the new generation of flies, now being fought with chemical spraying. The old generation — those perfidious fertile males and their cooperative mates — is gone, past punishment; fruit flies live only a few weeks.

Despite failures like these, biological control seems a basically sound tactic. When it works, it works very well. When it doesn't . . . Well, watch out for those Peruvians.

St. Louis Globe-Democrat
St. Louis, Mo., July 21, 1981

This time it appears that California Gov. Jerry Brown's eccentric, extremist behavior is going to cost residents of that state an incredible price — huge losses in their $14 billion produce industry, plus heavy costs for belated aerial sraying with Malathion.

The infestation of the Mediterranean fruit fly was discovered in Santa Clara County more than six months ago, but Brown stubbornly refused to allow aerial spraying with the most effective means of eradicating the threat, Malathion, a mildly toxic pesticide which has been used safely all over the nation for more than 30 years. It has been approved by the U.S. Department of Agriculture, the Environmental Protection Agency, and the National Cancer Institute.

Brown ignored all of this support for Malathion. He even refused to heed the advice of the head of the California Conservation Corps, B.T. Collins, who is so convinced of the safety of Malathion that he drank a glass of the stuff, diluted to spraying proportions, to prove its safety.

In the past when Brown has been on the wrong side, as he was on Proposition 13 — the initial big tax-cutting amendment in that state — he could switch after he had been proven wrong and then carry on as though he had originated the movement.

This time, however, his switching comes too late to save him from the deserved public reaction for his inexcusable delaying of the aerial spraying of Malathion.

The costs of Brown's mishandling of the Medfly crisis already are escalating out of control. Five states — Florida, Texas, Mississippi, Alabama and South Carolina — have imposed a quarantine on all California produce that could host the fast-breeding Medfly, including tomatoes, pears, plums and peppers.

This follows a limited quarantine applied by the federal government on fruit and vegetables from the three San Francisco Bay-area counties infested by the Medfly.

Monday morning nine truckloads of California produce were turned back by Florida inspectors and many other Golden State produce trucks were expected to be barred from entering the other four states while the Brown-induced crisis continues.

The spreading quarantines are coming at the height of the California's growing season as hundreds of tons of precious fruit and vegetables are being prepared for shipment.

"Nobody can remember a crisis like this, " said Steve George, chief of packing operations at George Brothers, a $7 million-a-year concern. He said the firm was building a $40,000 fumigation chamber to comply with the quarantines, which will allow fumigated fruit to cross borders.

Nobody can remember a governor of California like Jerry Brown, either. He lives in a fog as heavy as the one that covers large parts of the Medfly-infested area by night.

Sentinel Star
Orlando, Fla., July 15, 1981

CALIFORNIA'S Mediterranean fruit fly problem illustrates once again the thin thread by which a healthy agricultural industry hangs and the necessity of quick action to prevent disaster. In this case, California officials got caught in an emotional debate and dallied for almost too long.

Since the current Medfly outbreak was discovered more than a year ago, California Gov. Jerry Brown had been held captive by a small but vocal minority who feared "possible" health hazards from the aerial spraying of malathion. While the governor fiddled with half-measures, the flies spread and crop losses mounted.

Although aerial spraying of the insecticide malathion is the only known quick method of controlling the pest, the opposition — backed by the power of a governor — continued to block that action. This even though there is no documentation supporting their fears and in the face of credible evidence of the chemical's safeness.

Eventually, other warm weather states, fearing that the Medfly would spread their way, threatened a total quarantine.

According to one newspaper editor in the midst of the infested area, most people realized only last week "that we had a crisis on our hands" and that stronger measures were necessary. Indeed half the nation's fresh food hung in the balance. Finally, a federal threat to quarantine produce from the entire state forced Mr. Brown to act.

Now, even with the spraying underway, the burden is still on California to show that every effort is being made to find out where the Medfly is, contain it and keep fruits and vegetables from leaving those areas. If that isn't done, then Florida should be prepared to set up its own quarantine of California produce.

The Medfly is nothing to take lightly. It uses around 300 varieties of fruits, nuts and vegetables as hosts for its egg laying, ruining the produce. Unchecked, it will expand its home territory.

Nor is concern over chemical spraying something to be scorned. There are too many horror stories of "safe" chemicals becoming health time bombs. But malathion is one of the more common insecticides in use today, and probably is the safest weapon in man's continual battle with insects. Most homeowners have it on their shelves and the National Cancer Institute and the Environmental Protection Agency have tested it extensively and found no danger to humans.

Fighting agricultural pests always requires quick reaction if the battle is to succeed. A 12-hour delay often can mean the loss of millions of dollars. And, as our nation becomes more urbanized, those battles more and more affect non-farmers. It is just one more example of the need for effective leadership — people who can act decisively in the interest of the whole and sell that needed action to the public.

The Washington Star

Washington, D.C., July 14, 1981

The Biblical rule – as ye sow, so shall ye reap – may be applied to the California turmoil over the fruit fly. The perturbations over aerial spraying to halt the damaging infestation represent the "Ecotopian" ethic carried to the edge of absurdity: Pesticide equals peril – period. The unwillingness to make distinctions in the intricate marriage of man and nature can lead, finally, to a superstition that freezes reason.

Gov. Jerry Brown rejected the recommendations of his own technical advisers to spray from the air, using a solution containing the pesticide malathion. He preferred a "humanitarian" strategy, whatever that is supposed to mean, stripping fruit from trees by hand and using some malathion spray on the ground. The advice of agriculturists that aerial spraying was necessary if the fast-breeding Mediterranean fruit fly was to be contained detonated a fierce public outcry – one tinged with hysteria.

The public reaction cannot be dismissed as simply hysteric. After all, we have been subjected to decades of supercharged rhetoric about pesticides and herbicides – in some cases of which, from DDT to Agent Orange, there has been serious cause for concern. The obverse of that encompassing view – that there has been increasing study and control of these chemicals and that they, indeed, have contributed to unparalleled agricultural bounty – has had a poor second billing.

Governor Brown was eventually overridden, as he may have hoped he would be, by the federal government. Agriculture Secretary Block threatened a total embargo on fruits and vegetables from California, whence comes nearly 50 per cent of the nation's supply, unless products moving out of state were fumigated. The governor acquiesced, even as it was discovered that the devastating fly was extending its depredations beyond the original 97-mile area of infestation.

If the spraying begins as scheduled today (the courts are in the act at the moment), Governor Brown will be able to fan the heat eastward: the heavy hand of Washington, in contradiction of the pledge of the Republican occupant of the White House, imposed on a Democratic state executive and all that. There *is* a political element here, as there is always a political element in democratic tumult. But that is less material than the fuss over the pesticide itself.

There came a time late last week when there was little room for agriculture officials to maneuver. The federal responsibility was clear. The fruit fly has been a severe periodic problem for half a century. Malathion appears to be a prudent resort at this stage of the infestation. It has been widely tested, including two studies by the National Cancer Institute, and is not considered carcinogenic. That judgment is not definitive, to be sure, but definitive answers are not always available when action is imperative. So effective has been the anti-chemical assault, however, that scientific verdicts are widely regarded as suspect, if not dishonest.

That is not so much skepticism as irrationality. Governor Brown had an opportunity to be a voice of proportion. He tried to split the difference. That may be a very human reaction. Is it admirable leadership?

THE ATLANTA CONSTITUTION

Atlanta, Ga., July 14, 1981

In 1962, Rachel Carson's book, "Silent Spring," properly alarmed the nation to the dangers of the misuse and overuse of pesticides. In the nearly two decades since, extensive research has been conducted into the toxicity of pesticides. Certain insecticides, such as DDT, have been virtually banned from the environment. Other insecticides and herbicides are used only rarely and under the strictest controls.

Even so, pesticides are a vital element in our society. Without them, we would not be the world's foremost producer of food. We would not have one of the world's highest standards of living.

So it is with puzzlement that we witness California Gov. Jerry Brown raising objections to the use of a relatively mild pesticide, malathion, to control a serious outbreak of the destructive Mediterranean fruit fly in California's agriculture region. It appears the spraying will begin Tuesday, however the start was tied up in unnecessary litigation which had to be decided by the state's highest court Monday night.

Meanwhile, there is new evidence that the Medfly has spread.

This whole episode is not only bad news for California agribusiness. It is bad news for the Georgia consumer. As much as 65 percent of the fruit and vegetables arriving at the Atlanta Farmers Market at this time of year comes from California.

There is no evidence that aerial spraying of malathion is harmful to human beings, we are told. Florida has fought off three outbreaks of the Medfly in recent years with aerial spraying of malathion. In fact, many Floridians insist that their neighborhoods be sprayed regularly from the air with malathion to control mosquitoes.

But the Medfly outbreak in California poses a threat to the entire nation's vegetable and fruit industry if it is allowed to go unchecked. Brown and his cohorts have created an unnecessary controversy which could harm the nation. We agree with Georgia Agriculture Commissioner Tommy Irvin's call on Washington for an immediate quarantine on fruits and vegetables from the infected California area, if the spraying does not come off as scheduled Tuesday. Other Southern agriculture officials are backing Irvin's call. We hope the Reagan administration will pay heed to them.

We cannot afford to lose our own precious agriculture industry because some California officials have chosen to act irresponsibly.

The Salt Lake Tribune

Salt Lake City, Utah, July 14, 1981

To spray or not to spray, that is the question. And, if to spray, then — To flee or not to flee?

About a half million Californians around San Jose are caught up in this double dilemma all because of the tiny, but potentially catastrophic, Mediterranean fruit fly.

And then there is the politics of the situation. Gov. Edmund G. Brown Jr. first decided "not to spray" but did a turnabout and ordered spraying by helicopters using malathion when the U.S. government threatened a statewide guarantine of California crops. Gov. Brown says the Reagan administration "put a gun to my head."

But the scenario triggered an interesting shuffling of political alignments. The governor's not-to-spray decision won many of the state's environmentalists to his side, but antagonized the agricultural interests who stood to lose possibly millions because of the Mediterranean fruit fly's depredations. Gov. Brown's reversal then turned the political alignments around; the environmentalists became foes and the farmers became friends.

Of course, all the political flipping and flopping centers on whether the spraying some 120 square miles with malathion is safe. A lot of people seem to know; even if they all seem to have different ideas about the insecticide's safety, particularly as it effects the health of small children and pregnant women.

Dr. Ephraim Kahn of the California Health Services Department says a naked infant could roll around on the ground after an application of malathion and suffer no harm.

However, San Jose physician Dr. Alan Levin is advising more than 200 of his patients — including pregnant women, nursing mothers and people with chronic diseases — to leave the area during spraying operations.

And many people seem to be thinking the same way as Dr. Levin — they are preparing to leave. And this makes Jerry Scribner, the spray program director, upset.

"I think its ridiculous," he said. "People don't leave because of smog that eats away the sides of buildings. They don't leave when their neighbors spray a whole bottle of malathion. They're just not looking at the realities."

And the courts are not much help in settling the issue. A Superior Court judge said he wouldn't approve the spray plan until he was certain it was safe, however, he didn't do anything to stop it. So, plans for spraying went right ahead.

Such combination of controversy, confusion and conflict is a very infrequent event. It is hard to realize it was caused by something as tiny as the Mediterranean fruit fly.

Chicago Tribune
Chicago, Ill., July 13, 1981

Because 200,000 male Mediterranean fruit flies had not been sterilized by radiation before they were released, Gov. Edmund G. Brown Jr. of California was put in a painful dilemma. And Americans may have to pay more for fruit.

In an area between San Francisco and San Jose infested by fruit flies, a $22-million campaign of stripping trees, spraying, and biological controls was thought to have succeeded. But then there was a fresh outbreak, traced back to the accidental release of unsterilized male flies. With 620 square miles under quarantine and a wider quarantine threatening, agricultural interests were willing to do almost anything. But Gov. Brown resisted aerial spraying with malathion, widely believed to have toxic effects on human beings as well as insects. Pressure on Gov. Brown became irresistible when the U.S. Department of Agriculture said that in the absence of aerial spraying it would impose a statewide quarantine on 200 types of fruits and vegetables. This, Gov. Brown said, "leaves California with no alternative. We are being forced into aerial spraying."

This awkward conflict between the need to spray and the need not to could never have arisen were our species not as clever as it is. It takes impressive ingenuity to develop contemporary methods of growing food and of controlling pests by biological and chemical means.

Through sad experience of damaging side-effects from chemical pesticides, biological controls have come to the fore. Sterilizing male fruit flies by radiation and then releasing them in large numbers was a brilliant idea. California has released 3.5 billion fruit flies, happily anticipating that eggs from the mating of a sterile male with a fertile female would never grow into destructive larvae or a new generation of fruit flies.

This kind of pest control is far more elegant than spraying toxic chemicals on the landscape, with undesirable effects on many creatures, including human beings. But somebody blundered, and 200,000 of those presumably sterile male fruit flies were not sterile. Result: a tough conversation between Secretary of Agriculture John Block and Gov. Brown, causing Gov. Brown to move from one horn of his dilemma to the other one.

It will be hard indeed to exterminate the basic villian in California's hard choice about aerial spraying — human error.

The News and Courier
Charleston, S.C., July 22, 1981

Amid the confusion, false starts and missteps attendant attempts to control the California fruit fly infestation, nothing is more typical of misdirected efforts than the criticism of and challenges to quarantines imposed by five Southern states, including South Carolina.

Agriculture Secretary John Block has taken the five states to task for imposing quarantines stricter than the federal quarantine on fruits and vegetables. Mr. Block has declared most of California "free and clear" of fruit flies, the implication being that the Southern states overreacted. At the same time, a coalition of California growers asked a federal judge to lift the Texas quarantine, arguing it was unfairly imposed.

There is nothing unconstitutional, however, about the Southern states' quarantines on produce shipped from California. The quarantines were imposed not to restrict interstate trade but to protect the crops and the means of livelihood of farmers and fruit growers in South Carolina, Texas, Florida, Mississippi and Alabama. Surely the growers in these states, and consumers as well, are entitled to reasonable measures to protect their interests. If the fruit fly problem did not pose a threat in California, then the big flail out there has been over nothing — and Mr. Block himself was talking through his hat earlier when he threatened a blanket quarantine unless California agreed to aerial spraying.

Instead of drawing criticism, quarantine inspections by the Southern states should be regarded as back-up efforts to control an insect infestation that could prove exceedingly costly should it get out of hand. Secretary Block should recognize that point of view, as should the federal judge asked to overrule the considered judgment of agricultural officials in Texas.

THE MILWAUKEE JOURNAL
Milwaukee, Wisc., July 14, 1981

As a budget-weary Wisconsin resident who can't abide another round of price rises at the supermarket, you should favor the aerial spraying of pesticides on fruit fly-infested areas of California, where much of the nation's fruit is grown.

Or should you?

The spraying of malathion over 575,000 persons in three California counties raises questions we don't think have been properly addressed. Is malathion truly safe? Evidence from tests on laboratory animals shows that it is but, as with so many chemicals, the long-term effects cannot be determined without trying it. If the application eradicates the "bad" bugs, what will it do to the "good" bugs, including those that pollinated the fruit trees in the first place? Was the indiscriminate aerial spraying — surely a last-resort tactic — truly called for at this time? Could the pests have been contained by ground spraying and by destroying infested fruit? Was there another way?

If these questions sound familiar, they are. They were the issues of the great DDT debate a decade ago. They surround the widespread use of any pesticide or herbicide. The DDT war, which led to DDT's being banned, followed years of pesticide abuse and was waged bitterly. It politicized the old environmental movement as no other issue did. Those favoring the ban were cast as favoring malaria and human starvation; those who opposed the ban were drawn as greedy types willing to risk environmental catastrophe for short-term gain.

Well, we think we hear in the California situation the opening salvo of a DDT-like war. We hope not. Threatening to quarantine fruit from throughout California to force the aerial spraying of a limited area is an imperious act by the US Department of Agriculture, unless it can prove a true national emergency. However, to characterize the measure as *solely* the expression of the administration's anti-environmental stance, as California's Gov. Jerry Brown has done, sounds like a candidate making political hay.

We would prefer to hear scientific voices, rather than political ones, describing the extent of the fruit fly threat and the alternatives for controlling it. What we don't need is a reopening of old environmental wars and a return to the rotten hyperbole of the DDT debate.

Oregon Journal
Portland, Ore., July 14, 1981

Can this be the same California that used to stop us at its borders and remove half a sack of sweet cherries and two squashed plums from the picnic basket?

Yes, indeed. And annoying as the search-and-destroy mission was, it was a sincere attempt to prevent the migration of dangerous pests into an environment all too hospitable to alien insects.

How is it that the migrants — animal, bird, insect and disease — are so much worse in the new territory than they were at home? Starlings are a blight in America, whereas in Europe they are far less aggressive and numerous.

Water hyacinths released in Florida's canals and ditches threaten to choke natural swamps and waterways with their rafts of solid vegetation.

Russian thistle drifted across the Western dry plains like an oil slick spreading across water. A sort of milfoil was introduced into Washington waters, where it promptly got out of hand and required strong treatment to contain it. It remains a threat.

Some migrants are welcome in their new territories, like California house finches and Anna's hummingbirds, both small birds which have come north to Oregon to make it a permanent home.

When a migrant from another environment, be it microbe or mammal, moves to a fresh place, the local population will not be resistant to it, will not recognize it as prey, and the natural defenses built up over years in its place of origin will not be operating.

Dutch elm disease, long the scourge of the beautiful elms in Eastern towns, has showed up in Oregon. The city of Portland takes stringent measures to control it and to remove trees stricken with the disease in such a manner as not to spread it to healthy trees.

In such a crisis as California's, spraying to control the fruit fly is the only method large enough and speedy enough to have a hope of success. Sterile males have been tried and were not numerous enough (or sterile enough) to interrupt the fearful reproductive cycle of the pest.

Sprays should be held in reserve for such times, instead of being used as ho-hum routine. Insects can build up immunity to the chemicals, and the fighters are left with nothing that will work.

The Kansas City Times

Kansas City, Mo., July 17, 1981

Any adult who stands before an assembly of 900 people and drinks a glass of pesticide either needs a long rest or counseling. Maybe both.

To persuade his employees they needn't fear to work on California's fruit trees sprayed with malathion, B.T. Collins, California Conservation Corps director, swallowed the stuff, diluted to the strength being used now in the war against the fruit flies. It's reminiscent of Tom Sawyer inching across the board fence to impress Becky Thatcher and probably about as effective.

Certainly it's attention-getting, this dramatic display of confidence in the safety of the chemical. With the conflicting opinions of experts on the matter swirling about, he's far more convinced than we are.

Beyond that, Collins demonstrated an irresponsible attitude toward chemicals and, by doing so, delivered a dangerous message not only to his employees but to the broader public.

Even a chemical that is safe when used for the intended purpose can be lethal when misused, and no one should be encouraged to play around with silly experiments or macho demonstrations.

It is, finally, a horrible example to set for children. Enough danger — and permanently damaged little bodies — exists because curious minds and exploring hands find medicine, household cleansers and other toxic substances. It takes constant vigilance to keep them out of kids' mouths. Collins' performance throws a chill over the whole campaign for preventive health care.

The Wichita
Eagle-Beacon

Wichita, Kans., August 13, 1981

The question of which citrus-producing state, California or Florida, produces the best orange never may be resolved. But there shouldn't be any doubt about which of the two has better handled the potentially devastating Mediterranean fruit fly.

Florida, faced originally with the presence of three medflies, wasted no time in launching a pragmatic, all-out aerial spraying attack on the insects, rather than jeopardize its multi-billion-dollar fruit crop.

California, meanwhile, has had to quarantine at least three counties after putting off a similar spraying program its farmers had urged be started before the year-old problem got out of hand. Delays instigated by Gov. Jerry Brown, while he rightfully considered the environmental and human health risks of spraying the insecticide, malathion, from the air allowed the problem to get progressively worse.

It also prompted several states to initiate bans on the importation of California fruit, and prompted a federal quarantine threat before the governor finally relented. No one has come right out and said it, but the implication is strong that Florida's fledgling medfly problem may, in fact, have been imported from California because of the delay in taking action on the West Coast. Florida, however, appears to have a much better chance of successfully battling the pest.

The lesson to be learned here is that any agriculturally oriented state must be prepared, in advance, to deal with any predictable eventuality threatening its crops. It then must be ready to follow through with the appropriate decisive action when the time comes.

The Dallas Morning News

Dallas, Texas, July 18, 1981

THE present question isn't whether California will blitz its Mediterranean fruit flies from the air. The question is whether, at this late date, the blitz will stop the spread of the pest in time.

A Texas A&M entomologist, Frank Gilstrap, puts at dead-even the aerial-spraying program's chances for success. He darkly warns of the flies' imminent descent upon the rest of the country.

Likewise a spokesman for Texas citrus growers, Mike Wallace, claims, "The program they're using to get rid of the flies cannot succeed. That's the tragedy of the whole thing. It can't work."

Time will tell. The spraying program has been weighed down with mechanical problems, such as clogged helicopter pumps. By the third day of spraying, only half the job projected for that stage had actually been accomplished. However, a U.S. Agriculture Department official is optimistic. "I think the Medfly project is on track," says deputy director Richard Lyng. "We are going to eradicate this pest in relatively short order."

Let's hope Lyng is right. Having eaten gaping holes in California Gov. Jerry Brown's political prospects, the medfly, if uncontrolled, could devastate American fruit crops, to say nothing of American budgets.

That bridge is better crossed when we come to it. For now it is enough to wonder whether a lesson is being learned. The lesson is a familiar one, grounded in folk philosophy. One expression of it is, "A stitch in time saves nine." Another: "Strike while the iron is hot."

Meaning, why couldn't California have done what so plainly needed doing at the outset — spray demonstrably safe malathion from the air? We'll tell you why. Because trendy environmentalists, from California's governor on down, are inveterate foot-draggers, when it comes to environmental problems.

It isn't that the trendy ones love fruit flies, however lovingly crafted by nature. What they fear is man-made pesticides. Pronounced perfectly safe by the Environmental Protection Agency, malathion nevertheless strikes trepidation into trendy hearts. For instance there's the Palo Alto, Calif., man who said he'd keep his family inside during the spraying, because "It's like in a gas chamber."

Maybe, if you're a fruit fly. But most of us aren't; thus we have an interest in the rapid stamping out of this plague.

If you love the environment surely you want to protect it even from natural creatures that would destroy it. Nature's fruit flies nibbling nature's avocados is an issue that has nothing to do with belching smoke stacks. Why deal gingerly with the depredator at the expense of his victim?

Jerry Brown's anguish over spraying people along with fruit flies may inflict a high cost not only on Californians, but on consumers and farmers elsewhere. Brown is one of those environmentalists who, for every reason to act, can think of ten reasons not to.

It is difficult to see how this dangerous delay is going to help him become U.S. senator next year. Fruit flies don't vote. People who prefer action to stalling, do.

Democrat Chronicle
Rochester, N.Y., August 10, 1981

INSECT pests are now so prevalent in this country that agriculture is forced to rely more and more on both aerial and ground spraying.

But how much spraying is enough? At what point do workers and consumers become endangered? These are some of the questions being debated in California in the wake of the spraying against the Mediterranean fruit fly.

Environmental and health groups, labor unions and other critics say that pesticides are now posing long-term risks of cancer, birth defects, sterility and genetic mutations.

Chemical companies and the farmers respond by saying that health risks are not unreasonable when the spraying is done properly and that large-scale agriculture could not survive without it.

Like so many issues these days, the battle will probably be decided in the courts. California's increasingly strict regulation of the role and use of pesticides is now under challenge in federal court, and in Congress as well.

The difficulty, of course, is that the long-term effects of pesticides can't be decided in the short term. Which is all the more reason for caution.

As Richard L. Worsnop recently wrote in Editorial Research Reports, experience has shown that indiscriminate spraying to eradicate insect or plant pests ultimately is self-defeating. "Harmful insects become resistant to pesticides, and thus more difficult than ever to control, while pesticide residues contaminate lakes, streams and underground reservoirs used for drinking water."

The preferred method now is biological, as in the release of parasites, predators, microbial agents and other natural enemies of destructive insects.

Applying the right tools in the right circumstances is an important part of the answer.

The Times-Picayune
The States-Item
New Orleans, La., July 16, 1981

The controversy over the aerial spraying of the pesticide malathion to eradicate Mediterranean fruit fly infestations in California centers on environmental and public health concerns. Those concerns, it seems to us, have been exaggerated. For that, Gov. Edmund G. Brown Jr., who initially opposed the aerial application of the pesticide in favor of spreading it on the ground, must shoulder some responsibility.

There are some pesticides, such as the largely discredited Mirex, which was aerially sprayed all over the Southeast for more than a decade, that are best applied on the ground — if at all — to minimize and localize potential environmental and health hazards. But malathion is one of the most widely accepted pesticides around. Many gardeners use it, and it is routinely broadcast aerially by the mosquito control agencies in Louisiana.

Aerial spraying of pesticides can be a tricky business when the more controversial pesticides are involved. When the chemicals drift where they are not wanted, lawsuits such as one filed by Plaquemines Parish citrus growers who claimed the aerial application of 2-4D was damaging their crops, can result.

But the extent of the threat to the California fruit crop, when balanced against the unsubstantiated health threat, would seem to more than justify the aerial use of malathion in this case. The Mediterranean fruit fly is said to be a fast breeder that, if not stopped, could endanger the entire California fruit and vegetable crops, which, according to reports, are half of the nation's. Even worse, the infestation could spread to other states and endanger their crops. In that respect, Louisiana Agiculture Commissioner Bob Odom is justified in ordering inspections of fruit and vegetables trucked in from California.

The possibility that the reproductive cycle of an exotic little fly in distant California could drive up fruit and vegetable prices across the nation tells much about the state of American agriculture and the risk of putting too many eggs in too few baskets. Prime farm land is disappearing to other uses all over the nation, including in Louisiana and, even more to the point, California and Florida, another state on which much of the nation depends for fruit and other produce.

Commissioner Odom notes that 80 percent of the produce sold in Louisiana comes from out of state. Why is that, with all the fertile agricultural land remaining in the state? Where are the agricultural and marketing policies, not only in Louisiana but in many other states, to encourage local production, marketing and consumption of farm products, particularly in this era of high energy costs when shipping goods long distances adds substantially to their price?

Why should Louisiana consumers of fruits and vegetables be potentially at the mercy of the Mediterranean fruit fly, not to mention the complexities of California politics?

THE RICHMOND NEWS LEADER
Richmond, Va., July 16, 1981

So. Helicopters have sprayed the pesticide malathion over large areas of the Santa Clara Valley, and none of the dire consequences predicted has occurred. Californians did not collapse in the streets, and the skies did not fall. The Henny Pennies who forecast doom and disaster thus have been proved hysterics who could use some est or encounter sessions to calm their anxieties.

Yet the airborne battle to subdue the Mediterranean fruit fly probably comes too late to save California's $14 billion farm industry. For the past year, while the medflies have ravaged acre after acre of produce, Governor Jerry Brown persisted in a low-key state effort to eradicate the pest. While the medflies proliferated, California spent $23 million in ground spraying and in releasing sterile medflies. Now the medfly has been found outside the area targeted for aerial spraying, and it soon may spread to other lush farmland in the state.

As a consequence of Brown's too-little-too-late policies, his state's farm industry may suffer horrendous losses. Now that the medfly has been discovered elsewhere in California, 11 Southern states have slapped a quarantine on all unfumigated produce shipped from California. Virginia will monitor California produce. As California provides more than 50 per cent of the nation's produce, consumers soon will face soaring prices for shorter supplies. Market disruptions may reach devastating proportions.

This unfortunate situation could have been prevented if Jerry Brown's political ambitions for election to the Senate had not transcended his duty to act as a responsible elected official. No doubt Brown still would be vacillating if the federal government had not forced his hand by threatening a national quarantine of California produce. Brown has made a political career of catering to the shrill wackos California produces in great abundance, to the detriment of his state's more responsible residents.

The campaign against aerial spraying was both misleading and vacuous. Malathion has an unpleasant odor, and it is mildly toxic, but thousands of gardeners routinely use it to control a variety of pests. Used properly, it is a safe and highly effective means of protecting fruit and vegetables. The National Cancer Institute, the Center for Disease Control, and even the federal Environmental Protection Agency absolve malathion of any serious threat to human health. It is not DDT or Agent Orange, despite its critics' overblown contentions to the contrary.

In time, aerial spraying may subdue the medfly, but that spraying now must be more extensive and of longer duration because of the delay in initiating it. Brown, never one to accept defeat with grace, still issues wild statements to alarm his constituents: "We're going to face lawsuits from women who have miscarriages, if there are any defective babies born, if cars are wrecked, if goldfish are killed. Who knows what's going to happen?" Thanks to his lollygagging, many people know: more medflies in more areas and less marketable produce. An advocate of the silly "less is more" theory, Brown thus will find that his state's farmers will have more of less, and they are not going to be happy about his efforts to win election to the Senate.

The Miami Herald
Miami, Fla., July 25, 1981

FOR A year, California Gov. Jerry Brown played hero to the hysterics who were blocking the aerial spraying that would have confined the Mediterranean fruit fly outbreak to a minor nuisance. When the pests got out of hand, he was forced to accept widespread spraying and was hit with quarantines.

Now, Mr. Brown is crying foul because the Federal Government has refused California the "disaster area" designation that would absorb the financial losses. The Administration's refusal, says this Democratic U.S. Senate hopeful, is politically motivated.

How about a compromise? Just declare a disaster in one small section of Sacramento.

The Oregonian
Portland, Ore., July 15, 1981

The public hysteria that bursts out every time there is a chemical accident in the environment is due to a massive lack of confidence in public officials, independent experts, the media and other varieties of authority that society once believed it could rely on.

The chemical war being waged in California against the Mediterreanan fruit fly, an insect capable of devastating fruit crops in that state and others, including Oregon, if not contained, has brought protesters into the streets and before the courts, violently battling a plan to attack the insects with aerial sprays.

The spray being used is malathion, a chemical pesticide common on the shelves of most backyard gardeners. It has been used for decades, and its effects on human beings is well understood. The British, who are careful with toxic substances, even sprinkle it on their children to combat head lice. But the fact that the California health department has said it offers "no significant health risks at all" has not reduced the anxiety level in Santa Clara County.

Fear of pesticides began with DDT, a chemical which almost alone triggered the environmental movement when it was one of the substances attacked by the late Rachel Carson in her book "Silent Spring" in 1962. That DDT took the blame for other environmental pollutants, or that it had saved hundreds of thousands of lives in malarial and yellow fever areas, did not save it. It had been profusely overused and carelessly and needlessly sprayed. Worse, resistant insect strains were developing that it could not touch. Better insecticides would be needed for many pernicious bugs.

DDT eventually was swatted down in this country. William Ruckelshaus, administrator of the Environmental Protection Agency, ordered a virtual total ban on its use in 1972, although it continued to be used abroad.

But DDT, whatever its merits or problems, had became a national symbol, a rallying cry, bringing out protests against all pesticides, chemicals and, later, nuclear energy, regardless of their values or dangers.

The public, often misled by government officials and confused by conflicting reports from so-called medical and health experts, has been bombarded by sensational stories. It is often upset by warnings of known and unknown cancer-causing pollutants in the environment. Under the circumstances, it is not surprising it does not more often take to the streets.

WORCESTER TELEGRAM.
Worcester, Mass., August 8, 1981

Three dead fruit flies found in an insect trap wouldn't seem to be worth much fuss. But these were found in Florida and identified as Mediterranean fruit flies, which have staged a costly invasion of California.

So Florida agriculture officials are alarmed, with good reason. As in California, the economy there depends on citrus, fruit and vegetable crops that are marketable, not infested with medflies. As in California, produce moved in or out of Florida must be inspected to be sure there is no infestation. And as in California, Florida has its advocates and its opposition to aerial spraying if and when there is an actual infestation.

At the moment, there is a possibility the three fruit flies found in a trap near Tampa were planted there as a hoax. However, it's no laughing matter if they arrived with an advance party, maybe planning to invade Florida as they did California.

Actually Florida can probably withstand an assault of medflies because it already has in place fumigation equipment and procedures that were not readily available in California. Still it would be a double whammy to have the infestation in both California and Florida crops at the same time. If that happens, prices of fruits and vegetables will soar.

Shoppers in the supermarket had better hope that the three medflies in Tampa were loners, a freak occurence or even a hoax.

AKRON BEACON JOURNAL
Akron, Ohio, July 15, 1981

NO CHEMICAL that will kill bugs can be altogether harmless to other creatures, including human beings. But the stuff that has caused all the uproar in California comes about as close to it as anything currently available. This makes the protests look like 5 percent reason and 95 percent emotion.

A dead medfly is examined on the point of a finger.

Reluctantly, after other approaches proved to be too slow or otherwise inadequate and the state was threatened with a federal quarantine on its $14 billion fruit and produce crop, California has opened an aerial attack on the Mediterranean fruit fly — a pest that infests and destroys fruit and vegetables.

The weapon being used is malathion, a chemical lethal to virtually all insects but described by its principal maker, American Cyanamid, as having no known harmful effects on larger creatures unless ingested in very large amounts.

Malathion breaks down into harmless substances within two or three days of exposure to outdoor air, says a Cyanamid researcher, and it would take about two ounces of the chemical to cause serious harm to a human adult. That much common table salt, taken all at once, would also give an adult severe problems.

Malathion is being sprayed from low-flying helicopters in California at a rate of three ounces to the acre. This is enough to kill the flies and other insects, but the few millionths of an ounce to which this spraying rate might subject any person entails no known danger to health.

A good many small towns spray malathion from the air to keep down their mosquito populations. Fruit and produce growers, including many in Ohio, use it as one of the least toxic insect control chemicals available.

Labeling laws, which require explicit and prominent danger warnings on many other pesticides, allow marketing of malathion with no warning except advice to keep it out of reach of children — though people who use it are advised by health authorities to wash their hands afterward.

But in spite of these things and the fact that the chemical has been used for decades without any record of human harm, pickets are demonstrating against aerial spraying of malathion over the 120 square miles of the Santa Clara Valley around San Jose where medfly infestation has been detected.

And the mayors of several of the towns in the area went all the way to the California Supreme Court in a futile effort to prevent the spraying. They even tried in vain to get a hearing from the U. S. Supreme Court.

This is emotion rampant. It is conceivable that there are still unknown risks in the use of malathion — as there may be with virtually anything people use. But they appear to be exceedingly small and remote, while the danger from the bugs is clear, immediate and pressing.

It is good that people be intelligently concerned about health risks that have long had too little attention. But fears that have no known rational basis should not be allowed to keep us from doing what has to be done.

SAN JOSE NEWS

San Jose, Calif., July 22, 1981

ROUND 1 in the aerial medfly war is over. As is so often the case, the event itself was a lot less traumatic than the anticipation.

Anxious Santa Clara Valley residents who awoke the morning after the spraying half-expecting to see birds dropping out of the sky and a toxic fog covering the landscape were agreeably surprised to find nothing worse than a scattering of pinhead-sized, yellowish droplets on foliage and cars.

Despite all the advance hysteria, there was little panic and no violence. The protesters were few in number and generally well-behaved. Although there were times when it seemed to be acting out a script by Monty Python's Flying Circus, the medfly spray program eventually did what it was supposed to do. There were no reported pesticide spills, no helicopter accidents and no known cases, so far, of malathion-related illness.

At least five more rounds of spraying are scheduled, the first of them to begin Thursday, and it will be weeks or months before anyone knows whether the medfly has been knocked out. Even at this early stage, however, there are some important observations to be made.

In the first place, it was painfully obvious from Monday's meeting between Gov. Jerry Brown, Medfly Project Director Jerry Scribner and local police chiefs that the state still doesn't have its act together on enforcement of the order to remove produce that might harbor medfly larvae. The governor should make it clear that *everyone* in the quarantine area is required to strip fruits and vegetables. If local agencies don't have the manpower to enforce the edict, the state should provide it.

Second, now that it's had some experience, the Medfly Project should be able to give residents more reliable information about the times and locations of spray applications. Incidents like Tuesday's in which Los Altos Hills residents, including a group of schoolchildren, were accidentally exposed to malathion because of confusion over the spraying schedule are inexcusable and must not be repeated.

Third, the Bay Area medfly war creates a unique opportunity to study the short- and long-term health effects of exposing a large urban population to low dosages of malathion. Such a research opportunity was missed when malathion was sprayed to control the medfly in Florida in the early '50s; it should not be missed now.

The Governor's Medical Advisory Committee, a group of local physicians and other health professionals created to monitor the health effects of the spraying, advise government and answer citizens' questions, already has taken important initiatives. Its phone hot line has handled more than 8,000 calls to date, including a dozen complaints of illness, which turned out to be unrelated to spraying. It also is keeping track of ambulance runs, emergency room activity and physicians' reports of poisoning; none of these appears to indicate any malathion-related health problems.

More ambitiously, the committee is doing a random telephone survey comparing people in an area scheduled for spraying with a similar group in a non-spray area. Persons in each group were questioned in detail about their health status before the spraying; follow-up questioning will be done immediately after spraying and again six weeks later. The panel also has commissioned a door-to-door survey which will question people before and after spraying.

Studying possible long-term health consequences, such as cancer and birth defects, is more difficult and more expensive.

Medical authorities believe that if malathion is a carcinogen at all, it's a very weak one — capable of producing, say, just one "extra" case of cancer in a population of several million. Such a low-level effect would be hard to spot through any epidemiological study. Moreover, it would be virtually impossible to distinguish the effect of malathion from the effects of other known or possible carcinogens like smoking, auto fumes or industrial chemicals.

With all the difficulties, though, we think it would be worthwhile to monitor cancer cases in the Bay Area for several decades and compare incidence of the disease among sprayed vs. non-sprayed populations. Such a study would at least uncover any gross aberrations that might be related to the pesticide.

A big obstacle to such a study is the lack of a central tumor registry for Santa Clara County. The Resource for Cancer Epidemiology in Oakland, an agency of the state Health Department, tracks cancer incidence in several Bay Area counties, but not this one. The malathion spraying is an excellent reason for extending the monitoring.

Malathion's effects on the unborn are easier to study because any problems would show up in a year or less. A pending bill by state Sen. Diane Watson, D-Los Angeles, would create a birth defect monitoring system in Alameda and Contra Costa counties. The health advisory panel recommends expanding the system to cover Santa Clara County, as well as setting up a study comparing the outcomes of 1,000 or more pregnancies in sprayed and non-sprayed areas. We hope the Legislature will vote the needed funds for these projects as quickly as possible so that the pregnancies can be followed from the early stages.

One final, and perhaps most important, point: The medfly battle in Santa Clara County was seen from the beginning as a contest between proponents of integrated pest management, which emphasizes non-chemical methods and relies on toxic pesticides as a last resort, and the traditionalists who believe in chemical bombardment. The zealots in the pesticide camp no doubt will cite the outcome as proof that non-chemical controls can't work and that all-out chemical war is our only chance against marauding bugs.

In fact, the medfly episode doesn't prove anything of the kind. Integrated pest management didn't fail against the medfly — it was never really tried.

The campaign to control the fly through fruit stripping, ground spraying and release of sterile males got under way too late and was crippled by inefficiency, poor judgment and lack of leadership from the start. Significantly, a medfly infestation that was discovered in the Los Angeles area about the same time as the Santa Clara County outbreak was quickly eradicated without aerial spraying. The experience here doesn't prove non-chemical controls won't work; it proves that clumsy, ineffective non-chemical controls won't work. If the bungling of the medfly control program ends up discrediting the whole concept of integrated pest management and convinces governments and farmers that they have to push the spray button as soon as a bug rears its head, it would be perhaps the saddest result of this whole sad business.

The Honolulu Advertiser
Honolulu, Ha., August 20, 1981

As if California's embattled agriculture industry did not already have enough headaches, the threatened boycott of that state's produce by Japan could have dealt another serious blow. Japan is one of the largest customers of California fresh fruit.

The quick compromise reached between State Department and Japanese officials to allow continued shipments from areas not infested by fruit flies thus reduces the fear that much of California's $118.5 million foreign market might be lost. Last weekend, for example, South Korea and Taiwan began accepting only treated fruit.

IDEALLY, of course, if all produce leaving the state were fumigated there would be reduced concern in other states and overseas. But there are not enough facilities in California to handle a job of such magnitude.

So if Japan had not softened its position, other nations (and possibly other states) might have followed suit. Any continued total Japanese boycott — and they import almost all of their foreign oranges and lemons from California — would have sent a powerful signal.

But if California growers feel relieved about Japan's decision, they still know their battle against the Mediterranean fruit fly is far from over. The area of probable infestation continues to grow, and it now includes the San Joaquin Valley, the heart of the state's $14 billion agriculture industry.

Because of the damage fruit flies can inflict — Hawaii knows this first-hand from dealing with its "trifly" problem — importers of California fresh produce cannot be faulted for being extra cautious. It is possible to control the pest, but as Californians know from past experience the cost is high.

As long as a fruit fly problem exists, that state's growers can expect worry by others.

IN THIS DISMAL situation, then, the good news is that major importers such as Japan are not overreacting to the real danger.

The bad news is that it may take many months before the situation returns to normal. One of California's last fruit fly infestations took almost a year to bring under control, and the area affected was much smaller than the acreage under fruit-fly siege now.

RAPID CITY *JOURNAL*—
Rapid City, S.D., August 23, 1981

When the last chapter unfolds, the story of the Mediterranean fruit fly infestation in California may become a classic in in the annals of environmental zaniness.

The mist had hardly settled over the fruit orchards when the first of the lawsuits appeared.

One plaintiff claimed the spraying of pesticides to stop the Medfly infestation upset his dog's stomach and caused the 17-year-old poodle to vomit. He asked the state to pay $45 for the visit to the veterinarian. A woman asked $1 million for herself and her two daughters for alleged health damage even after a state conservation corps official downed a glassful of the malathion liquid to demonstrate its safety.

All the elements for extremism were there. Exotic, alien bugs infested the world's most productive agricultural area. Gov. Jerry Brown refused to heed the advice of the experts and instead of ordering spray to kill the bugs, he ordered out the National Guard to engage the bugs in hand-to-hand combat. One environmental group even went so far as to argue that the decision to go to aerial spraying was a cynical effort that would poison people, not the Medfly.

When the belated aerial spraying began, one of the spray helicopters was sprayed with gunfire.

Whatever the final outcome of the Medfly operation, the public is the loser.

At a time when the United States, for moral and economic reasons, should be seeking ways to increase farm production, environmentalists are battling against the use of a much-needed component of agricultural technology — farm pesticides.

Agricultural bug and weed killers are vital to the continued well-being of American people. Hysterical ranting every time the subject of pesticides moves into the policy arena is counter-productive.

Certainly there is a government responsibility to insure that pesticides are safe and used properly. But if all pesticides are eliminated arbitrarily, it is estimated there would be a reduction of 30 to 40 percent in the availability, not to mention the quality, of America's food supply.

In the next 35 years, the world will need to produce as much food as has been produced in all of history. Pesticides are among the proven scientific tools needed to do the job.

The Morning News
Wilmington, Del., August 19, 1981

"The worst thing that can happen now is for a bunch of little politicians to run around the state seeking their own narrow gain." That noble sentiment would do any responsible public official proud, unless he had been guilty of the same thing himself just a few weeks ago. And so he had.

The orator of the fine words was California's Gov. Edmund Brown Jr., who was seeking his own narrow gain a month ago by denouncing calls for pesticide control against the Mediterranean fruit fly with the irresponsible sloganeering equivalent of "better Med than dead."

Now the governor is forced to defend his conduct during a month that has seen the destructive pest spread from three quarantined counties near San Francisco to the state's fruit-basket, the San Joaquin Valley, seriously threatening a $14-billion-a-year industry. He dismisses his critics as people who are busy seeking their next elective office and says he's too busy being governor to waste time arguing with them.

It hadn't escaped many observers last month when Gov. Brown was outrageously fanning his misinformed constituents' fears of poisoning that he too is seeking another elective office, U.S. senator from California. It was only after the federal government threatened to embargo all fruit from the state that Gov. Brown relented and allowed the start of aerial spraying with malathion.

Gov. Brown is understandably impatient with those who point out that the aerial spraying eventually begun could have been considerably more effective if it had begun earlier. If he had exercised the sort of responsibility expected of a governor instead of playing to misinformation and fear, he might have destroyed the Mediter-

ranean fruit fly instead of his political credibility.

This is not the first time California has had to resort to aerial spraying with malathion against the Mediterranean fruit fly, and Gov. Brown knows it. He also knows that malathion is notable among synthetic organic phosphate insecticides for having a short-lived residue and being safe enough to handle without special precautions. Had he attempted to explain clearly to fearful and uninformed constituents the choice between intelligent use of aerial spray and economic disaster for the state's farm industry and consumers of fruit in general, he might have a little more credibility now.

Instead, what he has done from the beginning about the fruit fly comes down to too little, too late. The worst thing that can happen now, to paraphrase the governor, is for that to be forgotten by Californians and any other Americans to whom he might appeal for support.

THE PLAIN DEALER
Cleveland, Ohio, August 26, 1981

After the initial "to spray or not to spray" hysteria that gripped California, where Gov. Edmund G. Brown Jr. staved off the malathion crowd almost single-handedly for weeks and in the process lost a significant amount of his popularity with the average voter, the war of the flies seems to have settled down to skirmishing and containment, both long-term efforts.

Japan's decision not to impose a blanket ban on all California fruit because of the medfly's presence in four quarantined counties was welcome news last week; would that such restraint prevailed in the United States, too — Georgia, for one, has excluded from its borders all California fruit not fumigated or held in infestation-killing cold storage, a ban far in excess of

the federal quarantine. California is fighting Georgia's action in the Supreme Court.

The coming of fall may help the anti-medfly effort since the insect can't survive and reproduce in cold weather. But only systematic, thorough eradication efforts will keep California's fruit industry healthy in the meantime.

All that can be done is being done, or so it appears. What more is needed? Calm from the purchasers of California's produce, patience and cooperation from both landowners within the medfly-infested areas and travelers inconvenienced by the vehicle searches necessary to ensure that infected fruit doesn't leave the area in a camper's fruit basket.

THE SACRAMENTO BEE
Sacramento, Calif., September, 15, 1981

We'll gratefully accept even the cautious optimism of state and federal agricultural officials who now believe they're winning the Medfly fight.

The caution is amply justified; there's been optimism before, most notably early last June, just before the infestation — until then contained in small parts of Santa Clara and Alameda counties — really begun to erupt. What's most important now is not that only one fly has been found in more than a week or that the aerial spraying program appears to be well-organized and running smoothly, but that the pest now seems to be behaving more or less as expected — is not without the ability to survive, but seems less able to surprise and confuse and frighten. Which is to say that the most important element — the sense of control — seems to have been re-established.

The fight, of course, is a long way from being over. Even in the unlikely event that no new flies are found in the coming days, spraying will have to continue for months. Nor is there yet any certainty about how the pest spread or why such costly errors were made in June. What does seem clear is the course that has to be followed now, both in the air and on the ground, to control the pest and to restore confidence in California's produce. What's also clear is that an enormous amount of damage has already been done to California agriculture, both by the pest and by the confusion and fear it caused.

Undoubtedly there are lessons to be learned: biological lessons, political lessons, administrative lessons. As the spraying continues, it therefore would be altogether appropriate if the state created an impartial body — a body immune to the political positions and controversies attending this issue — to study the causes of this near-catastrophe and to establish more reliable ways to deal with such emergencies in the future. Perhaps the most appropriate locus of such a study would be in the University of California, though there may well be other places in which it could be conducted.

At this point, state officials still blame those fertile Peruvian Medflies that were supposed to be sterile for the fact that the pest wasn't controlled in June; the U.S. Department of Agriculture argues it wasn't the Peruvian flies, and that if aerial spraying had started sooner there would have been no problem. That controversy may never be fully resolved, but it would certainly be useful to make the effort to understand what happened and what, if anything, can be done to prevent similar occurrences in the future.

California has led the nation both in the management and protection of its agricultural enterprise and in protecting the environment and public health from the hazards of agricultural chemicals. In the Medfly situation, something obviously went wrong; if that something can be prevented in the future, now is the time to find out. A $14 billion industry is too valuable to treat as casually — and then as hysterically — as this situation was.

Los Angeles Times
Los Angeles, Calif., September 21, 1981

Controversial new questions are being raised about the possible side effects of the pesticide malathion, a mild organophosphate being used throughout the state to help eradicate the Mediterranean fruit fly. The compound has been studied before, but it should be again, to answer any doubts that remain.

Times Medical Writer Harry Nelson reported last week that a prominent scientist who has specialized in cancer research for a U.S. government laboratory has prepared a report concluding that malathion causes cancer in laboratory animals. The report, by Dr. Melvin D. Reuber, contradicts the preponderance of scientific evidence on malathion, which indicates that it has no harmful effect on humans, and the opinion of the National Cancer Institute, which has concluded that the pesticide does not cause cancer

Coincidentally, a team of Harvard University neurophysiologists are complaining that their proposal to conduct tests on persons exposed to malathion has been turned down by the state Department of Health Services.

Reuber's report is clouded in controversy because of the way it became public knowledge. Rather than publishing his conclusions in a scientific journal, Reuber bypassed normal scientific and government channels and wrote a letter this past summer to California agricultural officials. At the time, officials here were weighing the possibility of aerially spraying malathion over urban neighborhoods where the Medfly infestation began. Reuber was severely criticized by his superiors for his actions, and resigned from his job.

The impasse between the Harvard researchers and the state also centers on scientific questions. California health specialists doubt that the complex research the Harvard team wants to conduct, which would study the effects of malathion on the human brain, can be effectively and ethically carried out.

Malathion is one of the most widely studied pesticides in general use. And in the current campaign against the Medfly, it is being used in amounts so small that, even if it were shown to be harmful, the likelihood of its causing any problems is minimal. Nevertheless, some effort should be made to resolve these two scientific standoffs.

The federal government should ask the National Academy of Sciences to appoint a panel of objective pathologists to review Reuber's findings and those of previous malathion researchers. This has been done before to resolve controversies about other substances suspected of causing cancer, like nitrites in meat products.

In the meantime, state officials should continue discussing the Harvard study with representatives of that university to determine if ways can be found to carry it out.

Our rationale for this position is not scientific, but political. The more effort that is made to answer the questions about malathion, the less likelihood there will be that people will react irrationally to its use, which will have to continue for several more months as the Medfly eradication effort goes on.

The Medfly became a major issue in California only when serious public opposition arose in the Santa Clara Valley to its use in aerial spraying over heavily urbanized neighborhoods. One reason for that opposition was that proponents of spraying did not make enough of an effort to address the understandable public concerns about malathion's effects. The same mistake should not be made twice.

EVENING EXPRESS
Portland, Maine, August 20, 1981

The Mediterranean fruit fly is teaching California Gov. Edmund G. Brown Jr. a painful political lesson: Small is not always beautiful.

And the medfly is teaching us all that state governments cannot be counted on to deal promptly and effectively with rapidly spreading infestations. The federal Department of Agriculture might do better.

It's too early to say that the USDA should have clearly controlled the medfly battle from the beginning. But it's not too early to think about it.

Already medflies have turned up in citrus-rich Florida and Arizona. And their impact could reach into every state where eventual shortages will create higher fruit and vegetable prices. Internationally, the U.S. balance of payments stands to be adversely affected by Japan's limited, temporary embargo on California fruit.

Fresh produce is a mainstay of this country's interstate—and international—commerce. As such, it would seem to be a proper focus for federal concern.

There are those in California who blame politics for delays in medfly spraying that may have encouraged its spread into the lush San Joaquin Valley, centerpiece of that state's $14 billion-a-year agricultural industry.

Direct and immediate federal control of the medfly battle might have made a difference. And it could have averted the kind of private, uncontrolled spraying that farmers undertook this week.

When it comes to protecting the nation's food, big may be better after all.

Paraquat:
U.S. Sprays Marijuana Fields

Paraquat is a toxic herbicide usually sprayed from ground level. In 1983, the U.S. Drug Enforcement Agency used helicopters to spray the herbicide on several patches of marijuana plants in Georgia. The highly publicized federal action drew widespread criticism, most of it concerning risks to the area's residents, who were notified of the spraying only 15 minutes before it began. To protect against the possibility that the sprayed plants would reach marijuana smokers, the plants were pulled up after they had been sprayed. (One argument advanced against this use of paraquat was that the uprooting of the plants alone would have been sufficient to kill them.) The federal government said it was going to use films made of the spraying operation to try to convince authorities in Colombia to use paraquat to destroy marijuana crops in that country, the source of much of the marijuana used in the United States. In recent years, however, commercial production of marijuana had also become a flourishing agricultural enterprise in California and some other domestic areas. The DEA spraying was the first of its kind undertaken by federal officials.

FORT WORTH STAR-TELEGRAM
Fort Worth, Texas, May 13, 1981

Those who would wage war on drug trafficking in the United States are recommending the use of paraquat, the controversial herbicide that destroyed Mexico's once-thriving marijuana industry, on marijuana fields in the United States.

It is a recommendation that deserves serious consideration for a number of reasons.

Use of the chemical on domestic marijuana fields could save drug enforcement agencies, and therefore the taxpayers, a considerable amount of money.

Los Angeles Police Chief Daryl Gates made a convincing plea for use of paraquat:

"Why should the people of California continue to spend so much money on marijuana eradication — paying for armies of drug enforcement agents to go in and harvest these crops — when paraquat could do it quickly and easily?"

The possibility of domestic use of the spray has even become something of an international issue, linked to diplomatic efforts on the part of this country to persuade foreign nations to use the herbicide on their drug crops.

Federal officials met with state and local agents from throughout the country recently in Nashville, Tenn., to study ways of fighting the burgeoning marijuana traffic and acknowledged that the United States is under pressure from abroad to use paraquat on domestic fields.

Colombia, which supplies about 70 percent of the U.S. marijuana market, is one of the countries applying the pressure. Colombians, afraid that unilateral use of paraquat on their part might help American growers, are reluctant to begin eradication procedures if this country doesn't do likewise.

Paraquat spraying is just one of several strategies being studied by law enforcement agents in a stepped-up assault on drug trafficking. Others include:

■ Use of military forces in surveillance of suspected smuggling routes.

■ Expanded utilization of property seizures, such as aircraft, boats and automobiles, plus seizure of property purchased with drug trade profits, such as houses, vacation property and race horses.

The most frequently voiced objection to the use of paraquat is that marijuana sprayed with it but salvaged before it is destroyed can have a toxic effect if smoked.

Since marijuana smoking is supposed to be against the law anyway, that appears to be a small risk indeed when compared to the substantial savings made possible by the use of paraquat and the proven effectiveness of the herbicide in destroying marijuana crops.

DAYTON DAILY NEWS
Dayton, Ohio, May 18, 1981

The U.S. government apparently is going to resume poisoning Americans who smoke marijuana.

The Senate Foreign Relations Committee has cleared away a ban against helping pot-growing nations spray their marijuana crops with paraquat, a herbicide. The Reagan administration reportedly plans to press Colombia and Jamaica, where most of the marijuana for the United States is grown, to begin paraquat spraying programs.

The original program was scrapped when the Department of Health, Education and Welfare reported that paraquat could cause lung fibrosis in persons smoking marijuana tainted with it. Later studies pooh-pooh that worry, but none has ruled out the danger altogether.

Why risk damaging the health of millions, when the past paraquat program proved to be a bust anyway? It did not reduce marijuana use. Mexican marijuana farmers whose fields had been sprayed harvested their crop fast and dumped it on the market. With Mexican pot discredited as a result, cultivation jumped to Colombia and Jamaica, whose growing conditions produce plants higher in the intoxicant THC.

Legalization would allow better public controls than are possible through the doomed efforts to keep marijuana off shore. Distribution could be regulated in ways that at least would reduce the serious problems with the drug in many schools, and THC content could be held to more or less safe levels.

Marijuana can be a destructive drug, especially at some high potentcy levels and with dependency-prone personalities. But decades of studies have failed to show that it is especially harmful in its typical patterns of use, and clearly its use is not going to end.

The controls on the booze trade are less than perfect, but they work a far sight better than Prohibition's attempt to suppress alcohol.

ALBUQUERQUE JOURNAL
Albuquerque, N.M., July 15, 1981

The Drug Enforcement Administration exhibits an acute disconnection with reality when it states it has not ruled out the use of the controversial herbicide Paraquat on marijuana fields within the United States.

Federal law currently prohibits any federal agency from using the herbicide in the United States, but DEA Public Affairs Director Robert Feldkamp said states were not bound by the same stricture.

In the environment-conscious 1980s it is highly unlikely that any state's citizens would tolerate the spraying of such a potentially human-injuring chemical within its borders.

Most citizens side with authorities in the war on illegal drugs. But most would draw the line at the possible poisoning of illicit drug users.

The State
Columbia, S.C., August 20, 1983

GOV. DICK Riley and Agriculture Commissioner Les Tindal are reasonably alarmed at the U.S. Drug Enforcement Agency's plan to use aerial sprays of a toxic herbicide on marijuana fields DEA agents may find in South Carolina.

A federal judge in Georgia has granted a temporary injunction against DEA's activities there. And now Governor Riley has wisely asked DEA to talk with him before deciding to spray in South Carolina. DEA says it will do so.

The federal agency is beginning a 40-state marijuana eradication program, according to DEA spokesmen. It plans to use the herbicide called paraquat, a highly toxic chemical which is used in small direct applications by farmers against weeds.

Aerial spraying raises real apprehension, however, and state officials should satisfy themselves that there is little or no risk to the environment or to humans. Mr. Tindal, himself a farmer, acknowledged the controlled use of paraquat to kill weed stubble in crops such as sugar cane, soybeans and sunflowers, but he had doubts about aerial spraying.

"I am not at all satisfied," the agriculture commissioner said, "that aerial application of this chemical in large areas is safe. Paraquat even in diluted form is potentially hazardous to humans and wildlife. Not only could it destroy wildlife, but it could also destroy human life."

Frank Monastero, chief of DEA's enforcement operations, says to the contrary that paraquat has been sprayed from aircraft in Mexico since 1975 and has been effective in destroying marijuana crops.

Paraquat, indeed, made headlines in 1978 when the Department of Health, Education and Welfare warned that an irreversible lung condition, pulmonary fibrosis, could result from smoking marijuana sprayed with paraquat. HEW reported that about 60 percent of the marijuana coming into the United States was from Mexico; therefore paraquat posed a serious risk to marijuana smokers.

Scientists at the Center for Disease Control in Atlanta subsequently found that marijuana contaminated by paraquat was not as prevalent as reported by HEW. But a spokesman for the National Institute on Drug Abuse said the possibility of cumulative effects from long-term smoking of contaminated marijuana could not be excluded. Mr. Monastero's comment: "There's never been a proven case that it has affected anyone."

Last year, Florida state officials were taken to court by the National Organization for the Reform of Marijuana Laws (NORML), which objected to their plan to spray paraquat. Gov. Robert Graham denied there would be any health risks and said the state would police the sprayed fields so that no contaminated marijuana could be taken away.

Governor Graham also said the paraquat spraying would encourage Latin American countries to also use the herbicide treatment to reduce the amount of marijuana coming to the United States. That view seems to figure in the paraquat campaign launched by DEA — that the United States will serve as an example.

We do not know what to believe in this controversy, but, if there is any doubt of paraquat's dangers in aerial spraying, then it should not be used. Our state officials should insist on being convinced before the DEA's operation begins in South Carolina.

ST. LOUIS POST-DISPATCH
St. Louis, Mo., August 17, 1983

Granted, it takes a certain audacity to plant or harvest plots of illegal marijuana on U.S. government property. But that has been matched by the audacity of the White House Drug Abuse Policy Office, which has come up with a terrific "solution" for the problem: Spray the weed with the poisonous herbicide paraquat, which will, in time, either kill the illicit plants or kill the people who smoke it.

Not everybody agrees with this exquisitely simple approach. After all, it was tried in Mexico in 1979 (with assistance from the United States) only to be discontinued after Public Health Service doctors pointed out that individuals who smoked paraquat-poisoned pot could suffer irreversible lung damage and death. Fortunately, a federal judge in Atlanta has ordered a temporary halt to the spraying in response to a suit filed by a citizens group.

But the White House drug fighters say not to worry. Marijuana plants sprayed with paraquat are rendered useless within 72 hours, they insist. And if that is so, the chances that poisoned pot will be harvested are diminished. But they are not eliminated for those who still might unwittingly use the government-poisoned pot.

Moreover, if federal authorities can find and identify illicit fields of pot, they can use other means to destroy them without dousing them with poison and imposing what amounts to a federal death penalty on those committing what in most jurisdictions is a minor crime. What could be more perverse than a policy that threatens the lives and health of pot smokers, but leaves those who grow and profit from the marijuana trade unscathed?

The Chattanooga Times
Chattanooga, Tenn., August 18, 1983

Tennessee has now become the pawn in a game being played by the federal Drug Enforcement Administration. The object of the DEA's paraquat campaign is not so much the eradication of domestically grown marijuana as it is to convince officials in Peru, Columbia and Bolivia that they should poison the large marijuana crops in their countries. South American officials have resisted DEA's urgings to use paraquat by pointing out that the highly toxic herbicide is not employed in the United States. So the DEA has chosen to neutralize that argument, at the risk of the environment and the public health.

Gov. Lamar Alexander wisely rejected the DEA request to spray paraquat on private or state-owned land in Tennessee but said the federal authorities make their own decisions concerning federally-owned property. So DEA agents are searching for marijuana in the Cherokee National Forest and may extend their efforts into the Great Smoky Mountains National Park.

Paraquat is used in very diluted form as a herbicide in normal agriculture, but it can be lethal. Just last year a Florida man died after accidentally inhaling the chemical while spraying weeds, and doctors from Malaysia reported 27 deaths from a single accidental paraquat poisoning. Use of the chemical on Mexican marijuana fields was reportedly discontinued by the federal government in 1979 after Public Health Service doctors stated the herbicide could seriously injure smokers by causing irreversible fibrosis in their lungs.

Because of the serious public health and environmental risk involved with using paraquat, DEA officials have stated that aerial spraying will not begin until the target area is checked to make sure there are no people present and that no streams or wildlife are nearby. Little comfort can be taken from those statements. How can reliable assurances be given as to the exact location of wildlife in a national forest or a national park? What is to keep wildlife from eating contaminated vegetation after the spraying? And is a deployment of guards to be made to ensure that human visitors to the federal reserves will not stray into the target area? We are also assured by DEA that the poisoned marijuana will not reach the market because the crops will be guarded for three days after spraying.

State drug enforcement officials have said Tennessee marijuana growers are planting more and smaller patches of the illicit weed rather than large fields, in hopes of avoiding detection from the air. So each paraquat spraying will likely target a very small area. In each case the officials supposedly will go to great lengths to guard against intrusion by wildlife and humans. Then guards will be posted at each patch to watch it for three days. Is the public to believe this is the most effective means of eliminating marijuana growing? No. With considerably less effort and expense the plants can be pulled up and burned as the state is doing.

The DEA approach makes no sense — except as a bargaining ploy aimed at the countries of South America. The greater public interest in protecting the environment and human health demands that use of the posion be stopped.

The Boston Herald

Boston, Mass., September 26, 1983

SHOULD you need a graphic example of doing the right thing the wrong way, you'd be hard put to find a better one than the way the Drug Enforcement Administration wants to kill pot plants on federal land and wilderness areas.

It proposes to spray them with paraquat, one of the deadliest herbicides in use today. Contact with it can irritate lungs and burn skin, and ingesting as little as a teaspoon of it can kill an adult.

For that reason environmentalists have been resorting to both protests and legal delays, and by and large have been successful in making it difficult for the DEA to get the program off the ground.

Perhaps, as laudable as the DEA's aim is, the public has become so conscious of chemical hazards in recent years to be alarmed at the use of any herbicide. That being so, it might be possible to burn marijuana patches in remote federal areas without getting half the county high.

Other than that, though, the DEA ought to be able to use a less toxic defoliant that would kill the plants without harming birds, other wildlife, or human beings.

With paraquat, the cure might do more harm than the pot.

THE ATLANTA CONSTITUTION

Atlanta, Ga., August 12, 1983

The U.S. Drug Enforcement Administration is trying to decide whether to use the controversial — and extremely toxic — herbicide paraquat to destroy illegal marijuana crops on government-owned lands in Georgia. It should not.

Granted, paraquat is an effective, quick-acting and cheap tool in the fight against illicit drugs: It destroys marijuana plants within three days of application. But the herbicide is so toxic that when it is ingested by humans, it can cause serious disability or death within a few hours. It attacks the lungs and the damage it does is irreversible.

The concern is not that some pot farmers will lose their cash crops, but that some of the lethal herbicide will find its way to the unsuspecting user — maybe a high-school kid just experimenting with the drug — and result in a greater danger than any potential harm caused by the occasional smoking of the drug itself.

Law-enforcement officials say they try to guard fields after spraying them to make sure the crop cannot be harvested during the three-day period it takes for the herbicide to destroy the plants. But that coverage is spotty at best.

In Mexico, where paraquat has been used on marijuana crops for years, the fields are supposedly guarded after they are sprayed, yet paraquat-laden marijuana turns up frequently in this country. The U.S. Centers for Disease Control estimated that between 1975 and 1979, more than 9,000 Americans were exposed, annually, to paraquat which had been sprayed on Mexican marijuana.

Congress suspended its support for the Mexican spraying program in '79, after health questions were raised, but the federal government now wants to resume spraying and extend it to other nations, including Colombia and Jamaica.

The use of paraquat by the federal government on fields in Georgia and other Southeastern states appears to be mainly a kind of macho effort to persuade Colombia — where an estimated 90 percent of the marijuana consumed in the United States is grown — to adopt the practice.

No one questions the right of the government to control illegal substances, but the tools used for that control should not pose as much — or more — of a potential health hazard than the substances being controlled.

The San Diego Union

San Diego, Calif., August 24, 1983

There were no headlines when 10 million acres of American farmland were sprayed with the weed-killer paraquat last year. So why the fuss now when federal agents are destroying illegal marijuana crops by aerial spraying of the herbicide?

The principal reason is a fraudulent claim advanced by organizations of marijuana users. In contending that paraquat spraying constitutes a health hazard, they have ignored overwhelming evidence to the contrary. Nevertheless, the marijuana lobby continues to predict that portions of the contaminated crop will find its way onto the market and severely damage the lungs of those who use it.

Considering the precautions taken by the U.S. Drug Enforcement Administration (DEA), this fear is groundless. Marijuana plants die within three days after spraying and guards posted at each spraying site ensure the plants are not harvested. Moreover, the guards remain on duty until the plants are burned by DEA agents.

This is not the first time the marijuana lobby has played on health fears in an effort to block paraquat spraying. The United States stopped its paraquat spraying subsidy to Mexico in 1978 after complaints that poisoned pot was crossing the border.

But the ban on subsidizing paraquat applications in other nations was lifted in 1981 when a Congressional study found the spraying was not harmful. Indeed, the Center for Disease Control was unable to confirm a single case of lung damage due to paraquat.

We would have expected California Attorney General John Van de Kamp to support federal efforts to combat the illicit marijuana trade. We were suprised, therefore, when Mr. Van de Kamp announced last week that he would oppose spraying in California.

Perhaps Mr. Van de Kamp was just trying to score some political points with marijuana users. In any event, his opinion is moot. DEA officials told us they currently have no intention of spraying in California. Most of the state's marijuana farms are too small and too well camouflaged for aerial spraying to be effective.

The DEA spraying program that began in Georgia has moved into Kentucky. Spraying is expected to begin soon in other states. Meanwhile, American officials are negotiating for paraquat spraying in Colombia where it is urgently needed. Colombia supplies an estimated 50 percent of the marijuana sold in the United States.

In destroying marijuana plants before they are harvested, the paraquat spraying program is a highly valuable law enforcement tool. And, if marijuana users stop purchasing the illegal drug for fear of health hazards, so much the better.

The Courier-Journal

Louisville, Ky., August 18, 1983

TO THE AVERAGE citizen, the word Paraquat is rarely heard except in relation to marijuana. But as an article in June's *Science Digest* observes, killing marijuana is perhaps the least important use of this effective — and deadly — chemical.

Paraquat, says the article, is used to control weeds on more than 10 million acres of U.S. cropland. It's called the world's most effective herbicide, and it's expected to become increasingly important as farmers go to "no-till" techniques that mean lower costs and greatly reduced soil erosion.

But it's a two-edged sword. Paraquat can kill if it's swallowed, inhaled or spilled on the skin. Horrifying death rates have been recorded in Third World countries where Paraquat sometimes is used without the care normally exercised in this country.

What Paraquat does in plants is to cause individual cells to collapse like deflating balloons. In humans, the article says, it becomes concentrated in the lungs, which rapidly degenerate.

Even Chevron Chemical, which holds the American license to manufacture Paraquat, uses strong language in opposing its use to spray a substance that may be smoked by humans. In a letter to the Drug Enforcement Administration last year, Chevron declared: "Terrifying people (by poisoning marijuana) in order to modify their social behavior" is not an acceptable use of the product. Nothing, the letter added, is achieved with Paraquat that can't be accomplished with a hoe. "The hoe, in this instance, would be the preferred means," DEA was told.

The TENNESSEAN
Nashville, Tenn., August 18, 1983

THE federal Drug Enforcement Administration is using steamroller tactics in its program to spray marijuana crops with the controversial herbicide paraquat.

The DEA was forbidden by the federal courts to spray the dangerous chemical in north Georgia this week. Then the agency simply moved out of that court's jurisdiction into Tennessee and announced it would spray on federal property in this state.

Regrettably, Gov. Lamar Alexander said he would not protest the federal plan, although it was clear that state officials were not pleased with the idea. The state did reject a DEA offer to spray with paraquat on state property, and that is commendable.

The DEA's program calls for spraying in the Cherokee National Forest where thousands of vacationers visit at this time of year. Marijuana crops in Tennessee should be eradicated. But there are better, safer, and apparently cheaper ways to do it than by spraying a poisonous chemical over large areas and possibly endangering the health of unsuspecting citizens. The state has a safer way. Its agents simply pull up the marijuana plants and burn them. TBI Director Arzo Carson says this method is satisfactory and that Tennessee has no need for paraquat.

By contrast, the DEA sprays paraquat and kills the marijuana. Then it has to post guards around the plot for up to three days to keep growers from harvesting and selling the contaminated marijuana.

The DEA says the cost of guarding the fields is less than the cost of pulling up the plants. This may be true if the DEA comes across a 50-acre field of marijuana spread out in the Cherokee National Forest. But what if it discovers a crop no larger than a bean patch in a backyard garden? How will the DEA justify the cost of guarding that for three days, against the cost of pulling the plants and burning them?

State officials quoted the DEA as saying the spraying would start as soon as agents found a "suitable place," that is, a patch where marijuana is growing. The DEA sounds like it is planning a military campaign in the jungle to test its equipment rather than a sensible program to destroy marijuana. This seems typical of the tendency of some federal agencies to want to use a sledge hammer to kill a fly.

There is a great deal of disagreement about the degree of danger of paraquat. Some contend it can be used safely, which may be true. But it could also cause serious health problems for some if its use should become widespread.

In any event, millions believe it is dangerous and don't want it sprayed in areas which they will be visiting. Few would want to venture into a national forest to camp or fish where paraquat has been sprayed. There is just no need to take the chance with paraquat when other effective means are available for disposing of marijuana.

Newsday
Long Island, N.Y., September 5, 1983

One day last month, federal drug enforcement agents descended on the Chattahoochee National Forest in Georgia and proceeded to spray 70 marijuana plants with the toxic herbicide paraquat. That done, they pulled up the plants — leading local residents to wonder why they had sprayed them in the first place.

One reason was that the Reagan administration wanted to make a film of the operation, to show government officials in Colombia and elsewhere. Demonstrating that the United States has no compunction about spraying domestic marijuana with paraquat might persuade other governments to use more of it.

Current plans by the Drug Enforcement Agency call for spraying marijuana growing on federal land in about 40 states. But residents and officials in the targeted states have been quick to organize protests, and the government has agreed to a brief postponement.

Paraquat can cause severe lung damage and even death. There was widespread concern five years ago when it was revealed that the United States was sponsoring spraying operations in Mexico. Although the plants wilt and die within a few days of spraying, they can be processed if harvested quickly enough. Laboratory tests are required before it's possible to tell whether a particular batch is contaminated.

The protests now are coming mostly from residents worried about danger to people, other vegetation, wildlife and water supplies in sprayed areas. The DEA has refused to provide advance notice, saying that doing so could endanger its agents and make it possible for the growers to harvest their crops. That may be true, but it doesn't make the spraying less objectionable.

While growing marijuana is a federal crime, its use has been decriminalized in many states and millions of Americans smoke it. There's no consensus as to its potential health hazard. In contrast, the adverse effects of paraquat are well known, and people — whether they're marijuana smokers or not — have reason to be concerned about them. If the administration is determined to eradicate marijuana on federal lands, it ought to find a safer way.

The Dispatch
Columbus, Ohio, August 18, 1983

The federal effort to halt illegal drug use must be a vigorous one if the public can ever hope to be protected from the direct and indirect harm caused by drug abuse. To succeed, it must have the support of law-abiding citizens across the land, and it must use every means at its disposal to combat the drug problem.

But federal agents must also seek to engender good will in the way they go about their work. The Drug Enforcement Administration missed a recent opportunity to generate public support when it launched a surprise aerial spraying drive, using the herbicide paraquat, against suspected marijuana fields in White County, Ga. County residents were justifiably alarmed over the spraying and soon obtained a court order prohibiting further spraying until the effects of the federal action could be determined.

The spraying was one of the first times that the federal government used the chemical against illegal marijuana plants. Paraquat can cause severe respiratory ailments and death if ingested by humans. The paraquat-tainted marijuana is unsuitable for use.

The White County residents fear that some of the chemical could travel to legitimate food plants and to water supplies and they charge that the spraying amounts to an indiscriminate application of the chemical and poses a threat to the county's residents and food chain.

Their concerns are legitimate, and while the fight against illegal drug use should be a vigorous one it should not — cannot — be allowed to endanger the health of residents who happen to live near the sites where the banned substances are produced. Any spraying that is done should be conducted with the full knowledge and assistance of local officials to ensure that area residents are not endangered.

And while the federal agents have every right to go after marijuana, they would be better off concentrating their efforts against the widespread and growing use of a more dangerous substance: cocaine. "Coke" is a mind-warping, life-threatening drug that poses a far greater threat to users than does marijuana.

The DEA should rethink its tactics and reconsider its objectives if its campaign is to have the public support it needs to succeed. The spectre of government planes swooping down and spewing out dangerous herbicides is no way to engender good will.

Polychlorinated Biphenyls (PCBs): Contamination Through the Food Chain

Polychlorinated biphenyls (PCBs), produced in the United States since 1929, were banned from manufacturing processes fifty years later, in 1979. They are synthetic organic chemicals which do not exist in nature, and are extremely stable, resisting decomposition for decades. Their stability has made them useful to industry, particularly as insulating fluids in electrical equipment, such as transformers used by utility companies. (A federal deadline of 1988 has been set for the elimination of their use in electrical equipment.) PCBs have also been used in the manufacture of adhesives, sealants, printing inks, waxes and other products.

The very stability that has made PCBs an industrial asset has increased their threat to the environment—once dispersed, they do not degrade but persist in the environment for years. It was estimated by the Environmental Protection Agency in 1980 that 440 million pounds of PCBs were present in the environment or in landfills. (About one and a half billion pounds of PCBs had been produced before regulatory action was taken in the 1970's to limit the uses and disposal of these industrial chemicals.) PCBs have been shown to cause reproductive defects and death in fish, birds and mammals, and are also suspected to be carcinogenic. In humans, they have caused severe skin and eye irritations, and have been linked to reproductive disorders, kidney damage and liver ailments. There is also evidence of a correlation between high levels of PCBs with a lowered sperm count in males.

Because of their spread through the food chain from fish and birds to mammals and man, PCBs are now present in the flesh of nearly all humans, and have been found in the tissues of such far-flung creatures as Arctic polar bears and organisms living in the depths of the Atlantic Ocean. It is suspected by some scientists that they may now be present to some extent in all living creatures. A major source of contamination in humans is the consumption of fish from lakes and rivers into which the PCBs have been dumped; through a process known as "biomagnification," the fish harbor far larger concentrations of these and other chemicals in their bodies than are present in the water surrounding them. PCBs have also been found to concentrate somewhat in mother's milk; in studies of breast-feeding mothers, the level of PCBs found in many milk samples far exceeded the federal limit for cow's milk set by the Food and Drug Administration. Like other chlorinated hydrocarbons, the PCBs are stored in body fat; in a nursing mother, the stored fat is drawn upon for the manufacture of milk.

ST. LOUIS POST-DISPATCH
St. Louis, Mo., April 21, 1980

Food contamination by PCBs (polychlorinated biphenyls) has reached "rather alarming proportions," said Deputy Assistant Agriculture Secretary Sydney Butler the other day and because of that, the Agriculture Department is proposing that meat, poultry and egg processing plants get rid of PCBs in old equipment. That is all to the good. PCBs are a toxic chemical used widely in electrical equipment which can cause birth defects and possibly cancer.

Although manufacture and distribution of PCBs was banned last year by the government, it continued to allow their use in totally enclosed systems such as transformers. The argument for retaining them in those circumstances was that leakage was unlikely. But an incident last summer disproved that notion: PCBs leaked from a transformer in a Montana slaughterhouse into animal feed produced in the plant, which subsequently contaminated meat and poultry products in 19 states. Moreover a recent government report showed that of 822 PCB transformers in the Washington D.C. area, 494 were leaking or showing signs of having leaked.

Coupled with some Agriculture Department plans to improve its rather unwieldy food monitoring system, the removal of PCBs ought to reduce the chances of food contamination. Food processors stand to gain, too: Although PCB removal will cost the industry $20 to $40 million, that cost is relatively small when compared to the expense of contamination accidents. The Montana episode, for instance, ran up a bill of about $10 million.

Detroit Free Press
Detroit, Mich., May 28, 1981

THE INDUSTRIAL chemicals known as PCBs have been found in polar bears, mother's milk, the fish in the Great Lakes, eagles in Sweden and snow falling on Milwaukee — just about every place scientists have cared to look. The worldwide contamination occurred in the roughly 50 years since commercial production of PCBs began in 1929; in the U.S., some 750 million pounds of PCBs — half of all that was ever manufactured here — has already found its way into the environment through leaks, spills and dumping into lakes, rivers and landfills.

The rest remains in use or in temporary storage while the search for safe disposal methods continues. In Michigan, the auto industry, utilities and other large users of coolants and insulating liquids containing PCBs have been forced to store tens of thousands of gallons of wastes since dumping of PCBs was banned in the 1970s. The only alternative has been to pay the considerable cost of shipping the wastes to incinerators outside the state.

So the EPA's initial approval of an apparently safe and economical way of breaking down PCBs should give rise to some cautious optimism. This is not the first time a breakthrough on PCB disposal has been announced. But if this particular method, developed by the Sunohio Corp., lives up to its promises, it will relieve industry of a major headache and ease a threat to the environment and to human health.

PCBs are suspected carcinogens and have been linked to reproductive failure in birds and fish. Although the manufacture and dumping of PCBs ended in the 1970s, warnings against the consumption of certain kinds of Great Lakes fish are still in effect. The Sunohio process will not erase the PCBs that have already escaped into the environment; only time and perhaps new discoveries will do that. But by solving the industrial disposal problem, the Sunohio process and others under development may forestall the further release and accumulation of PCBs in the air, land and water.

The PCB story has been a classic case of chemical contamination, in which an extremely useful compound was widely adopted for a variety of manufacturing activities and products, dumped indiscriminately — and then discovered to be long-lasting, accumulative and dangerous in the environment. But we know now that industry can be encouraged, through regulation, tax incentives or whatever means, to recycle and to plan responsibly for disposal of hazardous wastes. People can be made aware not only of the perils to health and the environment but of how to deal with them safely and economically. The lesson in the PCB story isn't that we can never use chemicals for human benefit and economic progress — only that we have to use them more carefully.

The Pittsburgh Press
Pittsburgh, Pa., June 3, 1981

Many Americans no longer believe that the solution to the problems caused by technology is more technology.

Yet short of repealing the Industrial Revolution and halting all progress, it's the only choice there is.

Consider the case of the PCBs, a family of chemical compounds used since 1929 in electrical transformers and other applications.

PCBs were banned in 1977 — as suspected causers of birth defects, cancer and liver problems. But they remain as one of the most threatening environmental contaminants.

These toxic compounds apparently persist forever, and hundreds of thousands of pounds of them are already contaminating rivers, lakes and ground waters.

So — how do we get rid of the vast quantities of PCBs produced over half a century?

The answer obviously must come from technology.

Indeed, thanks to a new process developed by a subsidiary of Sun Oil Co., the ecological time bomb of the PCBs should be defused to a large extent. This process breaks the PCBs down into relatively harmless salts and an inert plastic-like compound that can be safely buried in landfills.

Unfortunately, this process does nothing about the PCBs already loose in the environment. But perhaps another process will.

Human ingenuity gets us into these messes, and only more human ingenuity can get us out of them.

The Register
Santa Ana, Calif., March 19, 1981

Trendy causes have run their course when a big-name politician gets off the bandwagon.

New York Gov. Hugh Carey, apparently believing that the PCBs scare is no longer politically in, is offering to drink a glass of PCBs to demonstrate his belief that fears about the chemical are overblown. Carey's offer followed the mass exit of a state building in Binghamton March 5 following an electrical transformer fire.

"I'm trying to draw a line between a reasonable risk and an unnecessary degree of precaution," commented Carey. "Life is a risk."

Ruffled by the remarks were members of the state's 215,000-member Civil Service Employees Association, which asked its members to chip in a penny each to send the governor to Binghamton to carry out his offer to drink a glass of PCBs — a chemical considered toxic and cancercausing.

So far, the union hasn't raised the money and Carey hasn't gone to Binghamton.

In light of numerous news stories dealing with the alleged deadlines of PCBs, an aside to Carey's offer was a UPI story from Alfred, N.Y. in which an Alfred University professor conceded that a glass of PCBs probably wouldn't hurt the governor.

"No way would I be drinking it," said Prof. Richard Sands. "On the other hand, if he (Carey) wanted to do it one time, I don't think there's any danger to it."

You can't help but wonder where the truth lies in today's world of environmental scare stories. Perhaps, in fact, there is much substance to the remarks of former Washington governor Dixy Lee Ray who told an energy conference in Laguna Hills Sunday:

"There is a small band of activists determined to destroy the corporate structure that produces energy, especially electricity in this country."

We know of no civilization past or present bombarded with more environmental scare stories than the United States in recent years. We all know the result — delays, cancellation and regulation to the point of work halts on anything smacking of progress in meeting the needs of today. How much of it is "anti-establishment" is anyone's guess. How much of each scare is valid is another guess. If Gov. Carey gets to Binghamton, maybe we'll find out.

THE INDIANAPOLIS NEWS
Indianapolis, Ind., June 2, 1981

It's as if an environmental time bomb were about to be diffused.

Polychlorinated biphenyls, better known as PCBs, have been among the most dreaded of industrial wastes because they are so toxic and so prevalent. Used for almost a half century as a fire retardent in electrical transformers before studies showed PCBs caused birth defects, tumors or other health problems, careless disposal techniques of the nonbiodegradable chemical have already polluted the environment throughout the country. As a result, 9 out of every 10 Americans carry some PCBs in their bodies.

U.S. production of PCBs was ordered ended in 1977, but the threat did not stop with the ban. More than 750 million pounds of PCBs are still in industrial use. About 20 million pounds more are in storage. Until recently, the only government-approved methods of disposing of PCBs were incineration at high temperatures and storage at a limited number of landfills.

But now, the Environmental Protection Agency has given limited approval to the use of a new process that promises to remove much of the PCB threat. Dubbed PCBX by its creator, Sunohio, the process breaks down PCB molecules in contaminated electrical transformer oil and produces a salt that can be disposed of harmlessly and oils that are reusable. The process is done in a trailer which is brought to sites where the contaminant is produced. The dangers, then, of transporting the deadly substance for disposal are eliminated.

Chemical disposal processes ordinarily may not make exciting news. But the value of this discovery must not be overlooked. When man began using PCBs he could not foresee the danger in them, a danger that has threatened the health of people in Indianapolis, Bloomington and elsewhere.

The discovery is significant in itself, but it also gives promise to finding similar solutions to problems cited by alarmists as permanent plagues upon an industrial society.

The Hartford Courant
Hartford, Conn., May 29, 1981

A two-year clinical study of fishermen who have eaten trout from the polluted Housatonic River has produced predictable findings.

People who have eaten the fish have more of the toxic PCBs (polychlorinated biphenyls) in their blood than people who have not eaten the trout. However, according to the state health department, there was no evidence that those who had the higher level of PCBs suffered any physical damage as a result.

The problem, according to health and environmental officials, is that scientists cannot yet predict the long-term effects of frequent exposure to low doses of the PCBs. The officials repeated their advice that fish taken from the Housatonic should not be eaten.

PCBs, found in the river 10 years ago, have been shown to cause cancer and birth defects in laboratory animals.

While state officials continue to issue warnings against eating the trout, they have resumed stocking the river with the fish. The program attracts fishermen to the river, and some of them have acknowledged they eat the trout they catch, even though they know about the warning.

The trout stocking program, which had been suspended for two years because of the PCB hazard, is a step backwards for Connecticut environmental policies. Further, state environmental officials have delayed dredging the PCBs from the river. They say that even if the PCBs were removed, they wouldn't know how to dispose of the hazardous waste.

In the meantime, New York State is well along on plans to dredge tons of PCBs from the depths of the upper Hudson River.

After months of planning and discussions with federal environmental officials, New York recently received tentative approval to dredge 140,000 pounds of the chemical contaminant from the river. The state is to receive $20 million from Congress for the dredging and for a landfill site lined to a depth of 50 feet with clay.

Federal and state environmental experts believe the clay will prevent the PCBs from leaching into other areas.

Connecticut should follow New York's lead in setting up a dredging and disposal program.

In the meantime, this state can take a step forward in dealing with the PCB problem. The trout stocking program must be stopped, until more is known about the effects of this dangerous substance.

THE TENNESSEAN
Nashville, Tenn., June 3, 1981

THE first word in dealing with hazardous wastes is "caution"— even when the news is good.

Still, the reports that a new process to neutralize polychlorinated biphenyls (PCBs) has been developed is good news of the first order. The U.S. has banned the manufacture of PCBs since 1977 because of laboratory tests showing them to be carcinogenic. A fire-resistant insulation in transformers and other electrical equipment, 750 million pounds of PCBs are still in use in this country today. It is estimated that another 20 million pounds are stored, awaiting disposal.

That indicates the magnitude of the waste problem caused by PCBs. Their properties of toxicity and slow decomposition increase the magnitude.

The new process — developed by an Ohio company — promises help. It is a chemical process that transforms the PCB compounds into harmless salts and a non-toxic clay residue. The treatment even allows the oil in transformers to be reused, and the equipment to do this can be taken to the transformers, thus eliminating transportation of the toxic PCBs.

If this sounds too good to be true, that remains to be seen and that is why "caution" is important. The Environmental Protection Agency knows this and has given limited approval to the use of the process.

Two places where the process will be implemented on a limited basis are the federal facilities at Oak Ridge and TVA which has PCB transformers throughout its seven-state service area. "Every utility is faced with this problem. If the method works as well as they say, it could be the solution," said a TVA spokesman.

That is the hope of all concerned with the public safety.

ALBUQUERQUE JOURNAL
Albuquerque, N.M., July 7, 1981

A New Mexico chemical laboratory has shown once again the state's key role in science and technology.

Science has been both blessed and jinxed by strides it has made in the last few decades. On the one hand, its advances have helped people live more comfortably and longer. But it has also helped create some problems with which it wasn't yet ready to deal.

When scientists put the chemicals together to make polychlorinated biphenyls, or PCBs, they had a substance that would benefit the whole industry involved in power generation. But the substance is both toxic and pervasive.

AnaChem Inc., a chemical firm from Albuquerque, has found a method that is much more accurate at detecting the presence of PCBs than any other known. Coupled with last spring's discovery by an Ohio company of a way to neutralize PCBs, it appears science has licked another tricky problem.

Credit goes to both companies for using advances in technology, and good old American ingenuity, to conquer a problem that, last year at this time, looked impossible.

The Houston Post
Houston, Texas, June 15, 1981

So much attention has been focused on the hazards, real and potential, of toxic chemicals that it is encouraging when the technology that spawned these substances also comes up with antidotes to combat them. The Environmental Protection Agency has tentatively approved a process described by an EPA official as "a significant step forward" in reducing the threat of PCBs to public health and the environment. PCBs are polychlorinated biphenyls, a family of chemicals widely used in high-temperature coolants and lubricants until their production was banned by Congress in 1977 after they were suspected of causing cancer, birth defects and liver ailments.

The new process reportedly will not destroy PCBs already loose in the water and soil, nor is it yet effective on some electrical transformers that contain high concentrations of the compound. But government officials say it could be useful in disposing of more than 750 million pounds of PCBs in storage or still in use in electrical equipment and elsewhere in industry. Until now, the only safe ways to dispose of these potentially hazardous chemicals were by burning in one of two federally approved incinerators or burial in one of eight government-approved landfills.

The new process, developed by Sunohio, a subsidiary of the Sun Oil Company, has the added advantage of being portable. Instead of an industry having to transport PCBs to a disposal site, with the attendant risk of accident, a truck can bring disposal equipment to a plant. The government says the process reduces PCBs to salts and plastic-like material that can be disposed of safely. Sunohio is also experimenting with the process as a possible neutralizer of other potentially hazardous chlorinated compounds, such as pesticides.

The danger of PCBs were recognized only after their long and widespread use. Though safeguards against hazardous chemicals have been improved, no system of screening and testing can provide absolute protection of the public health and the environment. Thus, development of processes like the PCB antidote should be encouraged to counter both existing chemical poisons and to prevent new ones from becoming a major threat to our complex industrialized society.

The Boston Herald American
Boston, Mass., December 27, 1981

Alchemists tried to turn base metal into gold. Their modern counterparts may be on the brink of unlocking an even more useful secret — how to turn toxic wastes into harmless chemicals.

If they are successful we won't need hazardous waste dumps. The wastes will no longer be hazardous.

It sounds almost too good to be true—and yet it is quite logical. Those dangerous man-made substances are composed of chemical elements that are individually quite harmless. So why not take them apart when we are finsihed with them, rather than burying them in dumps to remain a threat for decades?

Cancer causing PCB, widely used in electrical transformers for 50 years and now a hazardous waste problem, has been reduced to ordinary table salt by a process developed at the Franklin Institute of Research Laboratories, Inc., in Philadelphia. Modified sodium salts, when mixed with PCB and heated, dismantled the deadly compound.

Land contaminated with Agent Orange can be cleaned up by a man-made bacteria that eats the main component of the controversial herbicide. The University of Illinois has applied for a patent on the laboratory-created microbe, which doesn't exist in nature.

Carbon tetrachloride, a widely used cleaning solvent that can cause liver damage and possibly cancer, has been reduced to the equivalent of baking soda by a process developed at the University of California.

The process works on a whole family of petroleum-based compounds that have been among the most difficult to detoxify. Using only air and electricity it produces a superoxide that shatters the chemical bonds holding the compounds together.

Even the seemingly insurmountable problem of nuclear wastes might be solved by a method developed by other University of California researchers. With a new chemical called Licam-C, they are able to remove radioactive plutonium from the living issue of laboratory mice. They believe the technique is applicable to the mounting accumulation of wastes from nuclear power plants.

Even if these particular experiments don't produce the ultimate solution, neutralizing toxic wastes is a more sensible avenue to explore than burying them on land or at sea—and perhaps boobytrapping some future generation.

THE   SUN
Baltimore, Md., September 24, 1982

The Environmental Protection Agency, in its admirable efforts to rectify chemical mistakes of past decades, has taken a step in the right direction—but has not gone far enough—in restricting the use of PCBs in electrical equipment.

One problem with the EPA final ruling in this matter is that certain electrial equipment currently insulated with PCB materials can remain in operation for the rest of its useful life. That would seem to be an economical way to handle the problem from the point of view of utilities and other enterprises, but it is hardly reassuring to people concerned about the potential health hazards of these toxic substances.

PCBs, or polychlorinated biphenyls, were developed in 1929 and they have proven to be a highly stable, fire-resistant insulating material for transformers, capacitors and electromagnets. But, more recently, they have also been found to cause cancer in laboratory animals. True, laboratory animals are not human beings, but these kinds of scientific tests certainly should raise a red flag for health officials and envionmental regulators.

Millions of gallons of these organic compounds are still in use and they do not deteriorate readily under normal environmental conditions. Further, small amounts of PCBs have entered the food chain through fish, and have even been detected in the air over the Antarctic.

The EPA decided in its final regulations issued in July that PCBs in transformers or electromagnets that "pose an exposure risk" to food or animal feed shall be prohibited after October 1, 1985, and that PCB capacitors near food or feed, or outdoors where the public could be exposed to them (on telephone poles, for example), must be phased out by October 1, 1988. Capacitors in restricted areas away from the public can be used for their remaining useful life.

The Baltimore Gas & Electric Company estimates it will cost about $8 million to replace its PCB capacitors, and the company is concerned about the availability of substitute materials which will be in great demand as industries attempt to comply with the EPA ruling.

Dr. Ellen K. Silbergeld, a toxicologist with the Environmental Defense Fund, has charged that the new EPA rules "were drawn up to meet the wishes of the electrical industry and at the expense of public health." That charge is a bit too severe, but the EPA's new restrictions in its PCB ruling does suggest a rather casual approach to helping rid the environment of a seriously suspect threat to public health.

The Idaho STATESMAN
Boise, Idaho, September 27, 1981

Neighborhood residents should have been notified after dangerous chemicals contaminated the ground near a power pole in west Boise. Idaho Power Co. and the Environmental Protection Agency had their reasons for not advising residents that a 15-foot radius of earth around the power pole was contaminated with PCBs (polychlorinated biphenyls), which are known to cause cancer. In our view, the reasons are insufficient.

Idaho Power officials thought that the chemicals, once used in capacitors, insulators and home-heating equipment, had been cleaned up adequately. Nothing in EPA regulations governing spills required the company to give neighboring property owners notification. Back in September 1979, when the chemicals were released into the environment by an exploding capacitor, Idaho Power had little experience in dealing with spills of PCBs, the regulation of which had been in effect for only a matter of months.

No matter what the company's reasons for its silence, however, people have a right to know of contamination around them. That rule also should have been observed by the EPA.

EPA officials say the contaminated earth wasn't hazardous. An adult would have to eat a pound of PCBs to ingest an acute dose at one time, they say. Obviously, no one is going to eat a pound of dirt.

An irrigation ditch runs through the contaminated area, but the EPA officials say PCBs aren't taken up by plants, eliminating the danger that vegetables grown in gardens irrigated with water from the ditch would pose a hazard. The officials add that since PCBs are heavier than water, the chemicals would settle out of the water in the irrigation ditch before livestock drank it.

Still, notification was necessary. PCBs, ever-present in the soil and water in small amounts because of past pollution, accumulate in the fatty tissues of the body. They are so pervasive that legal limits have been set for their presence in fish, shellfish, poultry, milk and other dairy products. Congress' intent in banning the manufacture and use of PCBs, which remain stable for long periods, was to reduce their presence in the environment over time.

The power pole is adjacent to a children's playground and a Mormon Church school. A child may have played in the contaminated dirt near the power pole and eaten dirt, as small chidren will do. Perhaps that child wouldn't have played near the pole if that child's mother had been advised of the contamination. We don't like the thought that that might have happened.

The EPA's record on the notification point is even worse than Idaho Power's. The EPA took no action to notify either Idaho Power or neighborhood residents when tests conducted months after the spill determined concentrations far in excess of the standard of 50 parts per million. One test showed results of 73,000 parts per million. Don Bliss, the EPA's regional public relations officer, says the probable lack of a hazard "doesn't mitigate our stupidity" in not notifying neighborhood residents. His comment is apt.

It's unclear whether a real threat ever existed. Idaho Power again is having samples of soil tested after having received a formal complaint from the EPA on Aug. 18. The company's tests never have showed concentrations as high as those found by the EPA.

Even if it's determined that no threat existed, neighborhood residents still deserved to be notified. Throughout the incident, they have had no choice but to rely on the company and the regulatory agency to give them the complete story.

EVENING EXPRESS
Portland, Maine, June 11, 1983

The state Board of Environmental Protection has wisely adopted tough new regulations governing the storage and disposal of polychlorinated biphenyls—PCBs. Addition of the suspected cancer-causing material to the state's list of hazardous wastes will give authorities the ability to effectively monitor and control PCB handlers in Maine.

State Sen. Judy Kany, co-chairwoman of the Legislature's Energy and Natural Resources Committee, says adoption of the new rules by the BEP makes it unnecessary to consider a bill that would do essentially the same thing by statute.

The state Department of Environmental Protection says that with the exception of regulations governing hazardous waste spills, the only rules on PCBs until the BEP acted were federal rules enforced by the U.S. Environmental Protection Agency. But enforcement has been so lax that the federal PCB regulations have been ineffective here.

The new rules establish a reporting system for PCB handlers so that the state can keep track of the material. And a public hearing process is set up to give citizens an opportunity to comment on potential disposal sites.

The state rules, which become effective Wednesday, are more stringent than EPA's and will be strictly enforced, the DEP says.

That's good. PCBs can present a serious public health hazard if not carefully monitored and controlled. And if the federal government does not provide adequate regulation of dangerous substances, the state must assume that responsibility.

The Providence Journal
Providence, R.I., September 6, 1983

In 1981, a state office building in Binghamton, N. Y., was contaminated by PCBs as the result of a fire in an electrical transformer. Last May, the same thing happened to a high-rise office building in San Francisco. The Binghamton building is still closed. The other one was closed for a long time. Both required extensive decontamination work — and officials aren't sure yet how thorough such cleaning has to be.

What is sure is that there are going to be other incidents just like these in this country and abroad. And they pose a much greater potential hazard to the public than some of the more publicized poisons like Agent Orange or dioxin.

Why will there be more incidents, especially since the use of PCBs in transformers has been banned since 1977? PCBs are a kind of oil that is used as insulation to contain the heat that transformers build up. (These are the big transformers that are frequently placed in underground vaults outside modern buildings or under city streets). They are efficient insulators. But spread by the smoke of an electrical fire, they will coat all surfaces in a building; and until the contamination has been cleaned up, human beings can be affected by contact with it.

Before the ban went into effect, 1.8 million capacitors and 100,000 transformers with PCBs were installed in this country. The likelihood of the poison being released increases as the electrical equipment ages. Thus, there's a herculean job ahead to replace all of it as rapidly as power companies can manage to do it. Expensive, too.

But the job has to be done. Fire followed by contamination "can happen in any building with PCB transformers," says one authority. "They are public health time bombs, medical time bombs, waiting to go off."

The danger can be over-estimated. But the people who have been exposed to PCBs don't think it a minor matter. And neither do the occupants of buildings that have been contaminated. When they can't get back into their offices for weeks or even longer, the loss and disruption can be substantial.

It is not a problem that the public can do much about. But public and power-company officials ought to be laying plans for systematic and timely replacement of the transformers that pose the hazard. Other insulation material is available; and the PCBs had better be replaced before new incidents occur, with the potentiality for worse injury and damage than has occurred thus far.

The Wichita Eagle-Beacon
Wichita, Kans., August 24, 1983

Emergency services and private enterprise worked very well together Monday evening, and with commendable speed, after a capacitor exploded on an electric service utility pole, spraying insulating oil that contained PCB. The acronym stands for polychronated biphenyls, a class of industrial chemicals now known to be a potential cause of many serious health problems, including cancer.

Some of the oil spilled on a passing van and a bit splashed into the eye of a 10-year-old boy riding in the vehicle. Fortunately, fire department and Emergency Medical Service units and personnel from the health department and the new city-county hazardous chemical spill unit all arrived quickly at the scene, near 22nd Street and Hillside, and police quickly rerouted traffic around the area.

What might be called decontamination first aid proved to be possible with soap and water, and the boy whose eye had been sprayed was rushed to a hospital for further treatment.

Television and radio stations cooperated by airing an appeal for anyone whose car might have become contaminated while driving through the area to return for cleansing or to call in for advice.

Cleanup of the street and nearby property — even to the extent of physical removal of contaminated soil — was handled by workers from KG&E, which owns the capacitor. Because of its good insulating qualities, PCB was used widely until a few years ago in electric transformers and capacitors. Thousands of such units still are in use, although KG&E currently has a replacement program it hopes to have completed by 1988.

Meanwhile, the all-important rule is that if a spill does occur, removal of the oil must be immediate. Extended exposure to the effects of even low-grade PCB products can be quite harmful. The reaction to Monday's spill was a textbook example of such a response.

The Courier-Journal
Louisville, Ky., May 18, 1983

IT'S ENCOURAGING to hear that the fraction of Americans with "high" PCB levels in their systems has declined from 9.7 to 1 percent since federal law stopped manufacture of this family of toxic chemicals in 1977. But there's gloomy news, too, as in the report from Bloomington, Indiana, that PCBs in such places as still-undetected dumps will continue to be a serious health problem for a long time.

Virtually all Americans carry at least a bit of PCB in their body tissues. It's a molasses-like liquid once widely used for such purposes as insulators in electrical equipment, and it has been linked by scientific evidence to severe forms of acne, eye and liver diseases, stillbirths and cancer. Nobody knows precisely what harm results from low-level accumulations in the human system. But, as in the case of some other chemicals, there is a distinct possibility that long-term exposure increases the likelihood of cancer. It is certain that the stuff is toxic and has no place in the human system in any amount.

So the marked reduction reported last week by the Environmental Protection Agency is gratifying. But there still are enormous amounts of the stuff in circulation — either in storage, in electrical equipment still in use, in dumps, and in areas, like Bloomington, that have pockets of pollution because of heavy use of PCBs in the past. The Westinghouse plant at Bloomington once used large amounts of PCBs in manufacturing electrical equipment.

Westinghouse now is a defendant in several lawsuits seeking a cleanup of PCBs. The Monroe County Health Department reported the discovery of three new apparently contaminated sites this month, and expects to find more as time goes on.

Decades will pass before the full story of the effects of PCBs — and many other pervasive toxic chemicals — will be known. But the environmental policies of the 1970s are clearly bringing results. So are gains in scientific knowledge that make it more possible to foresee such dangers and cope with them. The thought that toxic chemicals are now part of everyone's makeup isn't pleasant. But on balance, the news is good.

The Dispatch
Columbus, Ohio, May 17, 1983

THE U.S. Environmental Protection Agency was tooting its own horn the other day and it appears the self-praise — coming after months of harsh outside criticism — is deserved.

The agency released a report which said that the levels of PCBs and DDT in humans have declined dramatically since the mid-1970s. The two chemicals have been found responsible for a variety of illnesses in laboratory animals and strict controls were placed on their production and use. PCB was an industrial processing agent often discharged as a pollutant after use. DDT was most commonly used as a pesticide before its use was banned in the 1970s.

The agency's report showed that the number of Americans with high levels of PCB had declined from 9.7 percent in 1977 to 1 percent in 1981. The average DDT level in humans, which was 8 parts per million in 1971, dropped to between 3 and 4 parts per million in 1981.

The agency can rightly take credit for the improvement and it should be encouraged to keep keep up the good work.

Anchorage Daily News
Anchorage, Alas., June 16, 1984

Reports that a local road-oiling contractor recently found unacceptable levels of PCBs in his storage tanks have caused a minor panic among people living near oil-covered Anchorage roads and streets. Their fears — that dangerous PCBs have been spread on the roads — have not, in any way, been confirmed. But those fears have not been refuted either. What should grow from the intervening uncertainty is a determination to craft and support an effective web of regulatory controls over use and disposal of hazardous substances in our community.

The current case is a fine example of how innocent people can be victimized in the absence of an appropri-ate regulatory web and effective enforcement of its requirements.

One victim is Albert Bendle, a road-oiler who by most accounts is a responsible operator conscientiously following the rules — and who learned of the PCB contamination only after carrying out required new tests. Mr. Bendle surely bears some responsibility for the substances he spreads on public streets and roads; that's his business, and he carries a professional obligation to operate it safely. But without a required mechanism for testing and safeguarding the safety of waste oil stocks gathered from all over town, it is difficult to know what contamination may exist. Now Mr. Bendle may be forced into bankruptcy because oil he bought and assumed to be safe turns out to be contaminated and expensive to clean up.

The more important potential victims are people whose streets and roads may have been contaminated — however inadvertently — in years past. New tests required by the state Department of Environmental Conservation uncovered Mr. Bendle's contaminated oil and presumably will prevent the use of other PCB-laden oil by other road-oilers this year. But because the source of the PCBs in Mr. Bendle's batch has not been determined, there is no way yet to rule out the chance that contaminated oil has been spread on Anchorage-area roads in the past. Some roads have been sampled; test results are due any day.

If the results show hazardous contamination of the roads, Anchorage will have a major headache on its hands. But more likely, there will be only a debate about what rules and regulations are needed to protect the public safety.

The DEC deserves commendation for its foresight and responsibility by adopting emergency rules this spring in response to national concern about road-oiling operations. The road-oilers themselves must be encouraged to follow the rules in good faith, and their suppliers must be prevented from mixing hazardous substances with the waste oils they sell.

Most of all, the public deserves assurance that PCBs will be handled safely and disposed of responsibly in Anchorage — and, indeed, around Alaska. Both state and local government are working to develop appropriate rules for handling PCBs and other hazardous substances. They surely deserve Alaskans' guidance and support.

The Star-Ledger
Newark, N.J., July 30, 1983

The American experience with toxic materials and wastes is permeated with depressing findings of contamination caused by exposure of people to these elements. There is a chilling documentation in the periodic disclosures of regions and communities where deadly toxic levels required evacuation of residents because of serious health threats.

Many parts of the country have been inflicted with these dark incidents, a dreary continuity of widespread contamination that revealed decades of environmental dereliction. In short, virtually all of the official and scientific evaluations regarding hazards to health caused by pollution excesses have been disturbing and bleak.

But there may be a major break in this otherwise dismal trend, a reassuring conclusion reached in a federal study on one of the most dangerous contaminants—polychlorinated biphenyls (PCBs). The number of persons with high levels of PCBs in their bodies decreased significantly between 1977 and 1981.

The survey by the Environmental Protection Agency showed a continued decrease of PCBs in the subjects examined. But this deadly substance is so pervasive (it is present in the environment) that practically everyone has accumulated some of these chemicals. More people will be exposed, but at lower levels resulting from enforcement of the Toxic Substances Control Act (TSCA).

Under that statute, the manufacture of PCBs, used in the manufacture of electric components, was banned in 1976. The dramatic improvement in the smaller levels of this contaminant is a highly encouraging sign that environmental regulations are effective in finally abating this serious pollution problem.

A New Jersey congressman, Rep. James Florio (D-1st Dist.), a major figure in drafting hazardous waste legislation, singled out the enforcement factor in reversing the effects of PCB contamination. Improper enforcement of statutes covering hazardous waste disposal, he charged, is responsible for the spread of low-level exposure from this source.

"If these laws worked as they were supposed to, there wouldn't be a problem," Congressman Florio said. With a new, vigorous EPA leadership under William Ruckelshaus, there is a strong commitment to make the "laws work," which obviously was not the case with the agency's deposed administration.

Without proper—let alone strong—enforcement, reversing the effects of the wanton ravages of the environment would be futile. The governing protection statutes provide the needed tools. It is imperative that they don't again fall into disuse, subverted by political favoritism and sweetheart deals.

THE KANSAS CITY STAR
Kansas City, Mo., May 15, 1983

After all the recent chaos and sorry evidence of nonperformance at the Environmental Protection Agency, it is good to encounter a long-term success story at this troubled organization. Continued EPA monitoring of the toxic chemical PCB (for polychlorinated biphenyls) now shows a sharp decline in the percentage of the population with relatively high PCB levels in their body tissues. Studying tissue samples from hospitals in 48 states, EPA watched the percentage of those containing more than 3 parts per million rise from 4 percent in 1972 to 9.7 percent in 1977.

In 1977, however, a congressional ban on further production of the substance—used as an insulator or fire retardant in electrical equipment—took effect. And the percentage then declined steadily to only 1 percent in the latest survey in 1981. Experiments on laboratory animals in the 1970s linked PCB to reproductive failures, gastric disorders, skin lesions and tumors, although the effects on human health were less clear. But since PCBs do not readily break down in the environment, and accumulate in living organisms, such as the Great Lakes fish, EPA scientists were alarmed by the rising incidence shown by the tissue sampling.

The ban on production yielded prompt and continuing results. But of the 1.25 billion pounds of PCBs produced since 1929 perhaps half are still around in storage or in large electrical equipment. The EPA's first set of rules to phase out existing uses of the substance were rejected in court as too lax, but new standards now would require the replacement in the next six years of 1.5 million transformers and capacitors.

The EPA survey has also shown declines in the percentage of human tissues showing high levels of DDT, a pesticide banned in 1972, and less dramatic drops in two other pesticides prohibited in the mid-1970s. The revelations and implications are still unfolding of how modern man, with his miracle chemicals, has inadvertently threatened to poison much of the human race. This latest EPA report is heartening evidence that these dangers now are being recognized and the counter-measures that have been taken are producing results.

Food Irradiation: Gamma Rays Instead of Chemicals

In the wake of the nationwide EDB scare in 1983, an alternative technology for controlling insect damage to food began to receive much attention. (See pp. 10-21). Food irradiation, already in use in many countries, employs very low levels of gamma rays to kill insects, bacteria and molds on fresh produce, thus checking the spread of such pests as the Mediterranean fruit fly. In slightly higher dosages, irradiation could greatly extend the shelf life of many foods, and its use on meats and canned foods could drastically reduce the need for chemical preservatives such as salt and nitrites. Widespread use of the technique in the United States has been hindered, however, by its classification in federal regulations as a food additive. Because large amounts of radiation can cause cancer in humans, irradiation of most foods had been prohibited by the 1958 Delaney clause, which bars the use of any food additive that "induces" cancer in humans or animals. (Named after its sponsor, Rep. James Delaney of New York, the flat prohibition was passed in large measure because of the limits of scientific knowledge about the nature of the causes of cancer, and about whether there are "safe" limits of carcinogens in foods.) Irradiation is already used commercially in the U.S. to protect wheat and wheat flour from insects, and to retard the sprouting of potatoes, and in 1983 its use was extended to spices and dried vegetable seasonings such as garlic powder. The Food and Drug Administration proposed in February, 1984 that irradiation also be permitted for other fruits and vegetables.

Unlike chemical treatments, irradiation does not leave any residue in food, although it causes chemical changes similar to those caused by cooking or canning foods; it can, for example, tenderize tough meat. (Scientists compare what occurs during microwave cooking to the process of passing low levels of ionizing radiation through foods.) The irradiation of food is expensive, and would probably raise the cost of treated foods somewhat, but its proponents argue that it is safer both for consumers and especially for farm workers who are currently exposed to the toxic chemicals contained in pesticides.

The Hartford Courant
Hartford, Conn., February 23, 1984

The furor over the pesticide EDB — ethylene dibromide — found in fruit and grain products has the Food and Drug Administration looking for a safe alternative for eliminating pests in foods.

It does not seem, however, that the FDA's proposal for wider use of radiation on food is going to relax the concern of many consumers.

Scientists agree that the use of radiation at the levels proposed by the FDA would not make food radioactive — no glowing oranges. Nor is there any evidence that it would change anything in the food to pose a health threat. In fact, the process is already used — to a very limited extent in the United States — in about 20 countries. The process, which involves bombarding the food with gamma rays or high-energy electrons, is an effective way to kill insects and to retard spoilage of food.

Yet the process raises enough questions to justify the use of some kind of labeling, so that consumers know that the food they are getting has been so treated. Secretary of Health and Human Services Margaret M. Heckler, who announced the irradiation proposal, dropped a recommendation from the FDA staff that labeling be required.

The lack of evidence of adverse health effects from food irradiation does not preclude the possibility that it could have some long-term impact as yet undiscovered. Some scientists say that to fully determine the health effects on humans would require a study spanning 20 or 30 years — a study not likely to be undertaken and which, at any rate, would be of no help for this generation.

The government is making a presumption of long-term safety that many consumers might not want to make. Labeling would give them that option.

Further, the FDA acknowledges that use of radiation can affect the taste, color, smell or texture of food "in ways that could be important to consumers." California and Florida officials, including the California Citrus Council, claim that irradiation at levels needed to kill fruit flies in fresh citrus causes blemishes in the peel and makes the pulp mushy.

It is true that a label might scare off some consumers and therefore hurt industry. But the primary responsibility of the FDA is to the public, not to the food industry.

Besides, a non-provocative symbol for irradiated food can be developed. The United States might even adopt the one already in use in The Netherlands — a flower in the middle of a partly segmented circle (reproduced in this column).

The Honolulu Advertiser
Honolulu, Ha., November 18, 1983

Senator Spark Matsunaga calls it "papaya processed with ionizing energy from the Pacific paradise."

That's just a palatable way of describing the potential use in Hawaii's papaya industry of irradiation, one of the hottest, indeed most controversial, topics of discussion among food growers, manufacturers, scientists and government officials in the world today. Some of these experts have been attending a conference this week in Hawaii on irradiation disinfestation of food and agricultural products.

WHAT EXCITES these people are the anticipated commercial applications of irradiation food processing, which is soon expected to gain federal approval as a quarantine treatment.

In essence, the irradiation technique kills bacteria and prolongs the shelf life of food by exposing it to gamma rays.

The technique goes back more than a quarter-century but it remains, with minor exceptions, illegal in this country. More than 20 nations, including Japan, the Soviet Union, Canada and Great Britain, use food irradiation commercially or experimentally.

Locally, irradiation has been hailed by some as a potential savior of the papaya industry, which is looking for a way to control pests on exported fruit in the wake of the federal ban on the pesticide EDB.

Two companies this week announced a joint project to build and operate two irradiation processing plants in Hawaii. But so far papaya growers have made no commitment about using the proposed facilities and they're exploring other methods of disinfecting fruit flies.

However, the plants could be used to process seafood products and as a way to prolong the shelf life of such other Island agricultural exports as avocados, guava, mangoes, orchids and even coffee.

Globally, too, the process has considerable appeal, particularly as a method to preserve food, be it Kansas beef or rice in India. Experts contend about a quarter of the world's food supply spoils annually.

THE IRRADIATION technique is not without its detractors. For while some say chemicals in current use pose greater problems, there is concern that irradiation would result in unforeseen chemical changes in food.

More of a problem, however, is the public's reaction to eating radiated food. Just as teeth don't become radioactive after dental X-rays, radiated food does not become radioactive. Proponents also stress that the process has little effect on taste or on food's nutritional content.

Despite such reassurances, convincing the American public of irradiation's safety may loom as a major public relations exercise for the food industry.

Having crossed that hurdle, irradiation is likely to become an accepted quarantine treatment by the end of the '80s, opening a variety of opportunities for American food growers and manufacturers in U.S. and world markets.

The Oregonian
Portland, Ore., March 9, 1984

Pesticides and preservatives that are proving health menaces in many foods would not be used now if the federal government had not long delayed the use of irradiation as a method of sterilizing foods in order to kill harmful bacteria and bugs.

A chief cause for the delay has been a failure of leadership, permitting irrational public fears to grow. These are founded on the total misconception that the irradiation of foods with radioactive materials, such as cobalt-60 or cesium-137, will produce radioactivity in the food.

This ignorance is not shared by 28 other nations, including Japan and Canada, which permit food irradiations. A joint study in 1977 by the Food and Agriculture Organization, the World Health Organization and the International Atomic Energy Commission gave irradiation of food virtually unconditional approval.

The use of harmful chemicals, such as the recently banned ethylene dibromide in grains and nitrite preservatives in meats, could be eliminated from U.S. food processing.

Irradiation will dramatically increase the shelf life of foods, while destroying harmful contaminations, such as larvae, salmonellas and organisms that cause botulism. Astronauts have long consumed irradiated foods, while the Defense Department, which has done years of research on the process, has advocated irradiating meals for troops as a safe way to make food taste fresh despite long storage periods.

Congress caused the U.S. restrictions when in 1958 it enacted a law classifying irradiation as an additive instead of a preservation method. This law requires all irradiated foods be labeled, and it forces clumsy testing methods, making it difficult for processors to comply with the law. New legislation before Congress would change some of these provisions, freeing the Food and Drug Administration to move forward.

Rep. Sid Morrison, R-Wash., backs the new legislation in an effort to help apple growers by lowering their worries over losing the use of EDB. Also, one of the chief materials used to irradiate food is cesium-137. The world's biggest supply is stockpiled at Hanford, Wash. Thus, its greater use could provide the state with a new industry, Morrison believes.

Not all irradiated foods taste better. There are debates about how much irradiation should be allowed. But these are not new problems, and international rules are in effect. Studies show that present low levels can be raised as much as five times without producing any harmful compounds in the irradiated food.

The U.S. food industry will suffer growing export and import penalties, raising food costs and lowering trading activity, if Congress fails to ease some of the restrictions now handcuffing the processing of food.

The Courier-Journal
Louisville, Ky., February 24, 1984

IRRADIATING fruits and vegetables to kill insects and reduce spoilage sounds like a wise alternative to the noxious pesticides that so often have proven harmful to human health. Given a few safeguards, there appears to be no reason proposed Food and Drug Administration regulations approving irradiation shouldn't go into effect.

The proposed rules were issued last week, even as the nation's grain industry was reeling under new restrictions to eliminate the pesticide ethylene dibromide, or EDB, a potent cancer-causing agent. The Environmental Protection Agency ordered an immediate halt to use of EDB and set maximum recommended tolerances; various states have imposed stringent restrictions of their own. Massachusetts, for instance, has banned food products containing 10 parts per billion or more of the pesticide, and on March 1 that will be lowered to 1 part per billion.

Irradiation appears to be both safe and effective, and studies show it has little or no effect on nutritive values.

The typical method used is to load produce into metal containers and put them on conveyor belts; the boxes are exposed to either cobalt-60 or cesium-137 in special rooms.

In announcing the proposed rules, Margaret Heckler, secretary of the Department of Health and Human Services, said 30 years of research had proven the process safe. Irradiation has been used in the U.S. for years on spices and to inhibit sprouting of potatoes; though it has been approved to kill insects in wheat and flour, that use never caught on because treating with pesticides was cheaper. In addition, some 28 other countries already irradiate some of their food.

Some scientists disagree with its safety, however, and some concerns must be addressed. Companies doing irradiation must be strictly supervised, to protect the health of workers. Monitoring exposure to even low levels of radiation is vital. This could become even more important, and more difficult, if crops are irradiated while still in the field. In addition, the process would produce a certain amount of low-level radioactive waste that must be disposed of in proper fashion.

The proposed regulations do not require that irradiated produce be labeled, though this might be a good idea. Research indicates that radiation might noticeably affect some foods, which is certainly of concern to consumers. In addition, there always will be those Americans who want to avoid anything to do with radiation, and labeling would allow them to do so.

ST. LOUIS POST-DISPATCH
St. Louis, Mo., February 28, 1984

In proposing to allow the widespread use of radiation to kill insects in fruits and vegetables and to prevent the spoiling of food, the Food and Drug Administration would replace one bad idea with another. Secretary of Health and Human Services Margaret Heckler said radiation could substitute for the chemical pesticide, ethylene dibromide (EDB), which the Environmental Protection Agency has banned as unsafe for grain products. A ban of EDB for fruits is also being considered.

Although Mrs. Heckler and FDA officials said the proposed levels of radiation — 100 kilorads as a pesticide in food and as much as 3,000 kilorads against bacterial contamination and insects in spices — would be safe, that assurance was promptly contradicted by a recognized expert on radiation.

John W. Gofman, professor emeritus of biophysics and medical physics at the University of California at Berkeley, said: "To really be able to say whether this technique has serious adverse effects on humans would require epidemiological studies of 20 to 30 years on 100,000 or more subjects. That study has not been done." Dr. Gofman added that if consumers were not informed about the use of radiation on specific products, that would be "a fraud upon the public."

If the government allows the use of radiation on food, it would be following the same procedure that has been proved to be disastrous in the past — that is, one pesticide after another has been allowed on the market without adequate testing for safety, only to be found unsafe many years later and banned after untold harm was done. That was the story of DDT, 2,4,5-T, EDB and many others. By now the government should have learned to say No to the would-be marketers of poison.

The Pittsburgh PRESS
Pittsburgh, Pa., March 3, 1984

The U.S. Food and Drug Administration's banning of the agricultural pesticide EDB as a suspected cause of cancer has ended one controversy, only to open the way for another.

As a replacement for EDB and other chemical pesticides, the agency has proposed allowing widespread use of irradiation to kill harmful insects in fruits and vegetables, as well as to prevent spoilage.

Treatment of food with small amounts of radiation has been permitted since the 1960s for such purposes as inhibiting the growth of sprouts on potatoes. But the technique has never been used on a large scale because of the availability of cheaper chemicals, like EDB.

Some scientists, however, say that not enough is known about the long-term effects of irradiated foods on human health. Others suggest possible dangers to food workers exposed to radioactive materials.

Still another issue is what kind of indentification labels, if any, should be put on irradiated foods to inform consumers.

Critics and supporters alike have until April 16 during a public comment period to make their views known to the FDA before the proposal goes into effect. The objections certainly deserve a thorough hearing.

At this writing, one hopes that the informed scientific verdict comes down on the side of irradiation because the technique holds out so much promise, not only of ridding the environment of harmful chemicals but of greatly extending the world food supply.

THE DENVER POST
Denver, Colo., March 11, 1984

THE NATION'S near panic over ethylene dibromide, or EDB, is abating slowly. A recent Colorado health warning about a shipment of oatmeal got few headlines compared to the cancer scare a weeks ago.

But a blessing in disguise has emerged: federal substitution of nuclear irradiation for EDB. Involved is the simple realization that some sort of disinfecting is needed if grains and vegetables are to reach U.S. tables free of insects and other pests.

Some 36 other countries, including Canada and Japan, irradiate food products. But not Uncle Sam. We pioneered the technique but have allowed vague public fears to immobilize policy.

Irradiation is logical. U.S. Food and Drug Administration scientists discovered years ago that low-level gamma rays will kill fruit fly larvae in a grapefruit without resi-due. Increase the dosage enough to kill germs and other micro-organisms, and perishable foods are given a shelf life of many months.

Critics argue that radiation alters food consistency slightly. Scientists agree but add that so do many food processes, including microwaving. The key point is that a product's value as food would be destroyed long before a damaging radiation level is reached.

Besides killing insects, radiation has other applications. U.S. pork, for example, is seldom served rare because of fears of trichinosis. Europeans forbid importation of U.S. pork for the same reason. Irradiation could do wonders for pork sales. Benefits to Third World nations, which lack refrigeration, could be immense. For all these reasons we hope the FDA — having authorized low-dose radiation for insects — develops this valuable option to its full potential.

Houston Chronicle
Houston, Texas, February 21, 1984

It's hard enough to avoid unhealthy pursuits. Now it seems we must be on guard against "healthy" habits, as well.

Saccharine was once a great alternative for dieters who craved sweets — until it was linked to cancer. Diligent city joggers have been warned that the pollutants they inhale may offset exercise benefits. Even tried-and-true basics can sabotage plans to live right: Recently, EDB dangers have cast a pall on resolutions to eat healthful fresh fruit and vegetables.

This trend has not gone unnoticed. More than 100 labor, environmental and consumer groups have launched a campaign to reduce public exposure to toxic chemicals. They call the chemical threat "America's No. 1 hidden health problem."

But the federal government has a plan of its own: using gamma ray irradiation, rather than chemicals, to disinfect and preserve food. Critics of the plan suggest that the process may indeed leave harmful residues and point out dangers of transporting the nuclear materials. A Texas A&M entomologist says irradiation is a safe but more expensive alternative to chemicals.

Ironically, one of the most controversial aspects of the proposal is whether or not to label the food as irradiated. Retailers fear it could cause negative reactions to the produce.

A strange choice is being posed: On the one hand, chemicals which are suspect if not proven dangerous. On the other, radiation, that bug-a-boo of environmentalists, which may well be the safest approach.

This business of staying healthy is getting increasingly complicated.

DESERET NEWS
Salt Lake City, Utah, February 19, 1984

Faced with growing problems involving chemical pesticides, the U.S. is reaching back a quarter century for another way to protect food — exposing farm products to atomic radiation.

The Food and Drug Administration, after a three-year study, this week published in the Federal Register the guidelines for irradiating fruits and vegetables.

There will be a 60-day wait for public comment and the rules could take effect by the end of 1984.

Proposals to allow the use of radiation on meats, fish and poultry are planned for the future, an FDA spokesman said.

Such a step may cause an outcry among environmentalists and others for whom "radiation" is a fighting word. But before they start complaining, let's pause long enough to look at irradiation calmly and clearly.

Under the proposed rules, produce would be placed in containers and moved into concrete-lined rooms where it would be exposed to cobalt 60 or cesium 137, both powerful radioactive isotopes.

The radiation kills pests, bacteria, and food-transmitted diseases like trichina and salmonella. That's how the process dramatically extends the shelf-life of many fresh foods that otherwise would have to be frozen.

The irradiating process does not make the food radioactive any more than a dental X-ray makes a person radioactive.

Irradiation of food has been known for a long time and has even been used commercially in other countries.

But the use of pesticides, wider use of frozen foods, and public fear of anything with the word "radiation" associated with it, has kept the irradiation technique from being applied commercially in the U.S.

Some critics, while admitting that irradiated food is not radioactive in any way, say the technique would create hazards for workers at irradiating plants, pose possible dangers to nearby residents, and add to the problem of what to do with nuclear waste.

However, something has to be done to protect the nation's food supply if more and more chemicals are ruled out. Irradiation should be given a close second look.

The Philadelphia Inquirer
Philadelphia, Pa., April 30, 1984

When proposed, it seemed to be an exceedingly promising method of preserving perishable foods by reducing or destroying micro-organisms that caused spoilage, and thus increasing products' shelf life. What was it? Ethylene dibromide (EDB).

Federal authorities ignored early warnings that residues of EDB, a cancer-causing substance, remained in the treated products and posed a health risk. Not until this year did concern about the effects of EDB finally prompt the federal government to ban its use.

The Food and Drug Administration now is considering approval of what seems to be another promising method of preserving foods. The process destroys micro-organisms that cause spoilage, vastly increasing shelf life. Its name? Irradiation.

Food irradiation, used abroad since the 1960s, holds great promise. Foods are subjected to massive amounts of gamma radiation that destroys bacteria. This sterilization extends the usable life of produce, meats and other foodstuffs.

Some things are known about irra-diation. It does not make the food radioactive. On that there is no disagreement. On other things, however, there are far more questions than answers.

Does irradiation destroy beneficial micro-organisms? Does it change the chemical structure of the product, creating a new variety of cancer-causing substances? Will some bacteria become resistant to radiation?

In the wake of the EDB ban, the FDA and its parent, the Department of Health and Human Services, appear to be racing to institute long-delayed standards on food irradiation, despite evidence that further studies are necessary. In response to those concerns, HHS Secretary Margaret M. Heckler recently extended a public comment period on proposed federal regulations on the process, but only until mid-May — 30 days.

Among the issues yet unresolved is whether irradiated food must be clearly labeled so that consumers can know what they are buying.

There is no vital interest to be served by rushing through inadequate standards. There are, however, many potentially troublesome consequences.

Part II: Air

While we can to some extent choose what foods to eat, we have less choice about the air we breathe. Many respiratory illnesses are induced or aggravated by a host of airborne pollutants from automobiles, factories and power plants. Indoor pollutants, from such sources as cigarettes, gas stoves, building materials and carpeting, add to the harmful substances drawn into our lungs daily; this problem has increased in recent years as more homes and buildings, in an effort at energy conservation, have been provided with air-tight insulation. Some industrial pollutants, such as arsenic and asbestos, have been shown to cause cancer. Lead, one of the most dangerous, has been implicated in mental disorders as well as kidney disease.

Coal-fired power plants are among the worst industrial polluters, emitting large amounts of sulfur dioxide, nitrogen oxides, particulates and carbon dioxide. In addition to the respiratory illnesses linked to sulfure dioxide, these plants have been implicated in long-range environmental problems less directly associated with human health. Foremost among these is the phenomenon of acid rain, widely held responsible for the elimination of life from many lakes and extensive damage to forests, particularly in the Northeast. Scientists theorize that another result of the continued burning of fossil fuels may be an eventual warming of the Earth, referred to as the "greenhouse effect," with potentially drastic consequences for agriculture as the average global temperature rises.

The National Commission on Air Quality, appointed by President Carter and reporting in March, 1981, found that in the decade since passage of the Clean Air Act, the nation's air had become "measurably better," and that the requirements of the law had not significantly impeded energy development or inhibited economic growth. Under the Reagan Administration, however, the law's enforcement as well as its future are in doubt, as federal officials lend a sympathetic ear to industrial adherents lobbying for less stringent requirements and delayed deadlines for pollution control.

Asbestos:
A Workers Provide Early Warning

Asbestos is a mineral composed of non-flammable, needle-like fibers that are small enough to be inhaled. Over the past 15 to 20 years, it has been established that even slight exposure to asbestos can result in the development of several serious diseases: asbestosis, a chronic disease of the lungs similar to emphysema; mesothelioma, a rare but fatal cancer of the chest or abdomen lining, and lung and gastrointestinal cancer. Asbestos has been widely used as insulation and in construction materials because of its tensile strength, flexibility and resistance to heat and corrosive chemicals. Although asbestosis is usually thought of as an occupational illness, the effects of exposure to this "magic mineral" are not limited to those who work in asbestos mines, factories or with products containing asbestos. The fine slivers of asbestos, which become airborne whenever an asbestos product is damaged, have embedded themselves in the lungs and digestive systems of most Americans, particularly in urban areas. Asbestos is released whenever a building on which it has been used as acoustical material, fireproofing or insulation is demolished; it has been widely used as a spray-on wall covering in schools. The mineral has also been used to line automobile brakes and clutch plates, filter beers and wines, and as a component in cement, shingles, roads and driveways. It is a good example of a material whose danger to society was first evidenced by the high incidence of disease among workers exposed to it. Many of those workers and their families have now filed suit against Manville Corp., the leading producer of asbestos in the Western world. In 1982, the giant company, with a healthy net worth of $1.1 billion, filed for reorganization under Chapter 11 of the Federal Bankruptcy Act. Lawyers for Manville's insurers and for victims of asbestos-related diseases claimed the company knew of the health hazards to its workers as long ago as the 1930's. When an outside study contracted by Manville concluded that the company could eventually face as many as 52,000 lawsuits at a cost of about $2 billion, the corporation made the decision to declare bankruptcy. By filing for reorganization, Manville achieved an immediate freeze on all legal action against the company.

The Houston Post
Houston, Texas, March 8, 1982

The Labor Department estimates that over the next two decades between 8,500 and 12,000 workers will die each year from asbestos-related cancer. Such statistics, coupled with the mounting medical, legal and social costs of asbestos pollution, are putting increasing pressure on Congress to deal with this grave national health issue.

Legislation has already been proposed to create a federal compensation program for victims of asbestosis. The idea is supported by the asbestos industry, which faces 12,000 damage suits and expects more to be filed. Some manufacturers fear that awards could total billions, forcing them into bankruptcy. The Asbestos Compensation Coalition, a group of nine firms that is seeking a federal program to provide quick out-of-court settlements of claims, warns that if the industry goes under there will be no one to sue.

Congress will have to decide if the government should establish a compensation program for disabled asbestos workers and possibly those in other occupations involving hazardous materials. If it approves such a program, it will have to determine whether the cost should be borne entirely by industry or whether the government — the taxpayers — should assume part of it. The taxpayers are already providing benefits totaling $3 billion a year to asbestos victims and their survivors through such programs as Social Security, Medicare, Medicaid and welfare.

The health risks posed by asbestos are an example of the kind of dilemma that increasingly plagues industrial societies. Though the mineral was most commonly used in insulation and fire-proofing material until it was proven to be a carcinogen nine years ago, it has some 3,000 industrial applications. It is thus widely distributed throughout the environment. But its fine fibers can lodge in the lungs and abdomen, causing cancer. Not only asbestos workers, but their families and others exposed to the fibrous material may be susceptible to its hazards. Furthermore, asbestos-related health problems usually don't show up until years after initial exposure. That means victims usually do not benefit from existing workmen's compensation programs, which require that a claim be filed shortly after disability occurs.

Some members of Congress believe that the public, which has benefited from products containing asbestos, should bear part of the cost of caring for victims of the mineral. Sen. Gary Hart, D-Colo., contends that World War II shipyard workers suffering from asbestos poisoning should have federal compensation since the government ordered the fire-resistant material's use in the ships.

At the same time, however, there has been congressional criticism of industry lobbying to lower federal standards on asbestos in the workplace. The Environmental Protection Agency has already delayed or dropped half a dozen regulations and education programs to control asbestos pollution. Lawmakers contend that if the industry wants a federal compensation program to relieve it of part of the burden of aiding disabled workers, it should accept strict standards on the use of asbestos.

New studies, in fact, indicate that the present federal standard of two asbestos fibers per cubic centimeter in the workplace may not be strict enough. Whatever Congress decides about compensating those afflicted with asbestos-related ailments, it would be callously shortsighted to relax existing standards without solid evidence that it would not expose the public to increased health hazards and boost the cost of any future compensation program for asbestos victims.

The Des Moines Register
Des Moines, Iowa, June 16, 1982

Under rules announced by the Environmental Protection Agency, officials of public and private elementary and secondary schools must identify hazardous asbestos in their buildings and keep records. If any is found, employees must be told where it is and how they can avoid exposure to it. Parents or parent-teacher organizations must be notified. But that's all.

School authorities need not remove one square inch of the potentially cancer-causing material. If they can not competently identify it, they need not get expert help. If they decide to remove it, they are not required to learn how best to proceed. If asked, the EPA will provide advice and assistance.

The use of friable asbestos — asbestos that easily can be broken apart or powdered when dry — has been forbidden in schools built since 1973, but it was used extensively during the previous 30 years, before its carcinogenic qualities were known.

The Carter administration had a voluntary program to call attention to asbestos in schools. The Reagan administration, to its credit, realized that this was insufficient. But it didn't go far enough. Instead of ordering elimination of the hazard, it decided merely to make the awareness program mandatory.

Because the rules are so weak, their key provision probably is parental notification. Parents at least can prod reluctant school administrators to get rid of the stuff, or they can switch schools.

The notification provision is weakened, however, by the fact that all parents do not have to be notified. A principal *either* may tell parents or tell the parents' organization. Moreover, the rule applies only to parents of children in school during the 1982-83 school year. How are parents of children enrolling later to know if there is a health hazard?

EPA estimates show that, in the "vast majority" of schools, friable asbestos poses no "current health hazard," according to Susan Vogt, special assistant to the assistant administrator for pesticides and toxic substances. That's not much comfort. What about the other schools? What if the estimates are off? What about materials that could become hazardous?

If the EPA is serious about the asbestos hazard, it will set a deadline for removing all materials containing friable asbestos that pose a potential threat to building users. It will ensure that all buildings are thoroughly searched to identify the stuff. It will provide technical assistance, and money, if needed. And it will check to see that the work is done properly.

DESERET NEWS
Salt Lake City, Utah, March 1, 1982

More than 12,000 lawsuits have been filed against the asbestos industry by victims of asbestos poisoning or their families. Some companies could be bankrupted by damage awards and want the federal government to come to the rescue.

The problem won't go away either. An estimated 10,000 people are expected to die each year between now and the end of the century because of asbestos-related illness. Damage claims could run into the billions of dollars.

While appealing for federal help on one hand, the asbestos industry wants government restrictions on asbestos pollution relaxed on the other. Neither request seems like a good idea.

Asbestos is a fibrous material and for many years was used freely as a fireproofing agent in addition to hundreds of other applications in construction and heavy industry. More than 13 million workers have been exposed to it in some degree.

Industrial workers aren't the only ones at risk. The fibers can be dangerous when released from walls or insulation during home or commercial renovation or when paint crumbles. Family members also can be exposed to the fibers carried home on the clothing or hair of workers at places where asbestos is used.

When inhaled, the invisible fibers lodge in the lungs and stay there, causing a buildup of scar tissue or leading to cancer of the lung, the lung lining or abdomen. The diseases usually appear 10 to 40 years after sufficient exposure.

"You have potentially millions of victims," said Rep. George Miller, D-Calif., who has conducted hearings into the asbestos problem.

Asbestos companies would like a federally-funded workers compensation program allowing quick out-of-court settlements to save court costs and lawyers' fees.

Some Congressmen like the compensation idea, but with a difference — the asbestos industry should put up the money. That seems reasonable. Ideally, the federal government should stay as far away as possible from propping up problem-plagued private industries.

And the environmental rules shouldn't be relaxed either. It wa failure to take extreme caution wit asbestos that brought about the heal problem in the first place.

Salt Lake City, Utah, August 30-31, 1982

Six months ago this page warned that some asbestos companies could soon go bankrupt because of the avalanche of damage suits being filed against them by victims of asbestos poisoning.

A few days ago, the world's leading producer of asbestos, which is also one of the most sued firms in the world, filed for bankruptcy.

But because the firm, Manville Corp., is far from broke, there's room for suspecting it of trying to hide behind the bankruptcy laws and prod Congress into bailing out the entire asbestos industry.

A reasonable case can be made for providing some help — but not the wholesale rescue effort that Manville and some other firms want.

The situation arises because of a growing number of asbestos poisoning cases and a growing number of law suits. Exposure to asbestos can cause chronic lung disease and several forms of cancer. But the medical problems aren't always manifested until 20 to 40 years after exposure. Consequently, even though safety regulations were tightened in the mid-1970's, the number of victims is expected to keep increasing for years to come.

So is the number of court cases. Because workers' compensation programs were not designed to handle ailments and injuries that don't show up for decades, the victims have turned to the courts.

The Manville Corp. alone has so far been named a defendant in 11,000 asbestos health lawsuits, and expects 32,000 or more new suits that eventually could cost the firm a total of $2 billion. Other asbestos manufacturers, many with only a few hundred employees, typically have 10,000 or more cases pending against them. The suits are said to claim more time and attention of top management than does current business.

Meanwhile, many of these firms have exhausted their primary insurance coverage and, because of their immense but unpredictable liability, it is extremely difficult to raise capital for their current needs.

But from the standpoint of its balance sheet, The Wall Street Journal reports, the Manville Corp. is considered the healthiest giant corporation ever to file under Chapter 11 of the federal bankruptcy code. The filing automatically stops further proceedings in pending lawsuits and the commencement of new suits against Manville.

To some extent, the asbestos firms are morally, if not legally, blameless because the industry wasn't entirely aware of the extent and seriousness of asbestos poisoning until comparatively recently.

Part of Manville's problems, however, result not from ordinary damages it must pay but from punitive damages — the kind not covered by insurance but which must come directly from a company's pockets. Two years ago the California Supreme Court ruled that plaintiffs were entitled to punitive or other damages because Manville concealed its knowledge of asbestos hazards from workers.

Yet Manville has been trying to get Congress to pass a bill that would require the federal government — meaning the taxpayers — to pay almost all of the cost of compensating asbestos victims. This would put the government in the position of bailing out not just individual firms but an entire industry. And the taxpayers would be penalized for what to some extent is corporate mismanagement.

The idea that the asbestos industry should get off scot-free is indefensible. No one wants to see an entire industry go bankrupt. But asbestos firms should put up at least a sizeable share of the money needed to save them. What's needed from Congress is just a cushion, not a wholesale bailout.

San Francisco Chronicle
San Francisco, Calif., August 30, 1982

ALTHOUGH IT IS financially healthy, the Manville Corp. has filed for the protection of the bankruptcy court before it is overwhelmed by some 16,500 current asbestos health damage suits and before still thousands more are filed. The action brings sharp attention upon a legal, economic and social failure of catastrophic scope.

The bankruptcy filing brings to a halt action on suits already filed. Attorneys for the plaintiffs, who are former employees of Manville or of firms which used its products, are predictably outraged. They see a giant corporation fleeing from its responsibilities, evading the just claims of seriously damaged employees, using a legal evasion at a time when the company is enjoying good business. It is easy to sympathize with these views and even easier to share indignation their clients must feel.

But this is not the most important issue in this controversy. The paramount issue is to ensure some satisfaction, some compensation and medical care for these victims of a medical disaster which was decades in evolution. And it is, perhaps, time to acknowledge that adversary duels in the courtroom may not always be the most effective means for solving a problem of such sweeping dimensions. If Manville is brought to insolvency, the asbestos victims will receive nothing.

THE MANVILLE CORP. and its asbestos competitors have acknowledged that certain exposures to asbestos fibers harm human lungs and can cripple and kill people. The point that the company made with its bankruptcy filing, and made with explosive force, is that it cannot continue to operate with predictability or profitability if it is being confronted with 500 new damage suits monthly which threaten unknown but massive future liability. The liability could easily reach the billions. And, the Manville Corp. alleges, it is involved itself now in suing its insurance companies because all but one have refused to honor claims.

Congress, it seems to us, must now act to adopt some form of asbestos claims commission, as the industry has been urging. Federal financing will, no doubt, be involved and legitimately so. Thousands of the damage suits have come from shipyard workers who were employed during World War II in outfitting cargo ships and warships with fire-resistant asbestos, which was in wide wartime use. Senator Gary Hart, D-Colo., is the author of a pending bill that would require the federal government to contribute directly to meeting the asbestos industry's compensatory obligation. He justifies this on the basis that thousands of afflicted victims of asbestosis incurred their disease in the shipyards. Congress should face this responsibility. The industry is also urging the establishment of a claims procedure administered by medical panels from which all successful claimants would receive adequate and prompt compensation. That, too, would be reasonable and equitable. As an industry spokesman said, in the wake of the Manville bankruptcy filing, "Putting companies out of business is not a good way of obtaining compensation."

The Wichita
Eagle-Beacon
Wichita, Kans., September 3, 1982

The filing of bankruptcy by the Manville Corp. is evidence of serious problems with U.S. bankruptcy laws. To provide financially sound business entities the ability to be insulated from pending lawsuits in such a manner was not the intent of Congress in revising the nation's bankruptcy laws in 1978. In the case in question, a disservice also is done to the thousands of people who have had their health permanently damaged.

In the past 40 years, as many as 9 million American workers have been exposed to asbestos. The resulting cases of asbestosis, a sometimes-fatal lung disease, have generated more than 16,000 lawsuits against the Manville Corp., the largest single producer of asbestos in the world. It's estimated the company ultimately could face as many as 52,000 suits totaling more than $2 billion. On Aug. 27, attorneys for Manville filed under Chapter 11 of the Federal Bankruptcy Act, which suspended all pending damage suits and the payment of any of the company's debt until a plan for reorganization can be approved by the court.

With a net worth of $1.1 billion and total assets of nearly $2 billion, declaring bankruptcy hardly seems a justifiable way out. Admittedly, insurance companies have refused to honor Manville claims in an ongoing dispute over the extent of their policy coverage. There even may be some validity to the argument that the federal government has some responsibility in the matter, since a significant number of people came into contact with asbestos while working in government shipyards during World War II.

The overriding fact, however, is that Manville has been found legally responsible for physical impairment resulting from contact with asbestos. When Congress addresses the matter after the Labor Day recess, bankruptcy regulations should be revised to prevent similar actions by other companies, and to ensure that the people suffering from the effects of asbestos won't be ignored by the responsible parties.

THE BLADE
Toledo, Ohio, September 3, 1982

THE Manville Corporation's surprise decision to file for bankruptcy raises many unsettling questions. The basic one among them, of course, and the crux of this entire problem, is this: Just how much culpability does a company have for damaging effects on its employees or others who were exposed to products that, so far as was known at the time, were not dangerous?

Manville, a diversified mining and manufacturing firm, filed for protection under federal bankruptcy laws while still in good financial health. The firm, however, is being overwhelmed by lawsuits related to the health effects of asbestos.

About 16,500 lawsuits already had been filed against Manville, the nation's largest manufacturer of asbestos, a heat-resistant mineral used in pipe insulation and many other products. And Manville attorneys anticipate that more than 30,000 additional suits will be filed.

That would mean a financial burden estimated by Manville attorneys at about $2 billion, far in excess of what the firm could withstand without outside assistance. So Manville decided upon self-preservation. Bankruptcy will halt existing lawsuits and prevent future filings while the firm formulates a plan to handle its liabilities. It will continue in operation as a major industrial concern and employer of 25,000.

The largest group of asbestos-related suits so far has involved World War II shipyard workers exposed to the mineral while insulating steam pipes and boilers of ships being built or refitted. The workers claim that occupational exposure to asbestos caused lung cancer and other diseases that appeared 30 to 40 years after initial exposure.

Indeed, that seems to be the potential consequence of excessive occupational exposure to this valuable material, which is used to make more than 3,000 different products. Toxicologists worry that other industrial and commercial products that help to make modern life possible may fall into essentially the same category as asbestos. They are seemingly beneficial products that decades from now also may be shown by more sophisticated tests to be harmful.

Asbestos, for example, has been used since antiquity. It was not until 1964, Manville maintains, that scientists became certain of its link with certain respiratory disorders. The Federal Government did not begin regulating asbestos as a potential workplace health hazard until the late 1960s.

Who should bear financial responsibility, the firm that in good faith markets a new product without any notion of what health effects loom 30 or 40 years hence, or society, which collectively benefits from new products?

The question is valid enough to warrant consideration in Congress on Manville's request for a federal compensation program for asbestos victims. Many of these, after all, did work in government shipyards. And there are other government-compensation programs for victims of occupational illness, such as that for coal miners who develop black lung disease.

One certainty emerges from this complex situation: The current remedy available to firms such as Manville is unsatisfactory. For present and former employees suffering with legitimate occupational illness, it is doubly so. Bankruptcies can deprive them of legitimate compensation. And the need to resort to the court system in the first place means paying stiff fees to the army of attorneys all too eager to capitalize on human misfortune.

Product-liability attorneys, in fact, are having a field day in reaping huge contingency fees in what has become known as "toxic torts" law. Torts are wrongful acts, and lawyers argue that persons injured from the toxic effects of products deserve compensation as a result.

But as Manville's chief executive John McKinney maintained, much of the asbestos-related compensation so far has gone not to victims, but to their attorneys. The Manville case should provide a stimulus for improving ways to compensate victims of product injury without enriching their attorneys.

Rocky Mountain News
A Scripps-Howard Newspaper   Reg. U.S. Pat. Off.   Colorado's First Newspaper—Founded in 1859

Denver, Colo., September 1, 1982

NO matter what, the Denver-based Manville Corp. is going to have to pay a bundle to victims of asbestos-related diseases. No one, least of all the company, doubts that.

What is in doubt, especially in the wake of Manville's filing for bankruptcy last week, is whether liability will be shared, and whether a reasonably uniform system of compensation will prevail. At present neither is the case. Instead, compensation is determined in court, and therefore hostage to the fluctuating sympathy of juries and inflated by lawyers' fees. What's more, asbestos manufacturers are entirely responsible for payment. That holds true even if, for example, the victim worked in government shipyards where asbestos use was required by law.

Colorado Sen. Gary Hart has sponsored a bill that would change that. It would set a one-time commission to establish how to assess responsibility. In some cases, but by no means most, that would result in government sharing the cost. State worker compensation boards would then decide who is eligible for payment, and the size of the award.

So far, the most vocal critics of the proposal are lawyers specializing in asbestos-related personal injury suits. In fact, more than 50 of them met last weekend at a South Carolina resort to counter what they call Manville's "fraudulent" bankruptcy claim, which they see as a ploy to force a solution along the lines Senator Hart suggests.

A Boulder lawyer specializing in such cases, J. Conard Metcalf, argues that there is no difference between Hart's proposal and government-backed loans to Chrysler in 1979. One difference, however, is that the government wasn't responsible for Chrysler's going broke, while it apparently bears some responsibility for allowing so many workers to suffer so much exposure to asbestos.

The Hart bill is not a bailout. It doesn't absolve Manville of financial liability, or limit the number of workers who can collect — so long as they are disabled. The bill merely tries to put the blame where it belongs, and, by establishing more uniform settlements, give asbestos manufacturers a sounder basis on which to plan. That doesn't seem too much to ask.

St. Petersburg Times
St. Petersburg, Fla., September 4, 1982

The exposure of workers years ago to a silent, unseen killer is at issue in one of the most controversial bankruptcy cases in U.S. history. The insidious dealer of death and disease is asbestos. The company that filed for reorganization under the federal bankruptcy laws is the Manville Corp., the nation's largest manufacturer of the mineral fiber used in thousands of insulation and industrial products.

No American business has ever been confronted with as many personal injury lawsuits as Manville. There are 16,000 cases pending against the company in which the claimants contend that Manville provided inadequate information about the danger of working with asbestos and insufficent protection against the hazards.

When its tiny fibers are released into the air, asbestos is a potent carcinogen. Decades after being exposed to it, a person who inhales asbestos fibers can develop several forms of cancer and asbestosis, a chronic lung disease that can be disabling or fatal.

THE FILING under chapter 11 of the bankruptcy act halts the lawsuits against Manville and prevents new ones from being brought while the company figures out how to handle its liabilites.

The Manville case raises some troubling questions.

Is the company abusing the bankruptcy laws? It seems ludicrous that a company with more than $2-billion in assets and hundreds of millions of dollars worth of insurance to pay the claims is able to use the bankruptcy laws as a shield against litigation.

One of the most deplorable aspects of the Manville case is that the automatic freeze on all pending lawsuits prevents plaintiffs from taking crucial testimony from witnesses. One witness with testimony that could be highly damaging to Manville is 83 years old and suffers from heart disease. According to a lawyer representing about 1,000 of the plaintiffs, the witness was going to testify that in the 1940s he had met with the president and counsel of Johns-Manville (as the company was known until recently), and that they had been aware that their product was injuring and killing employees.

Several lawyers representing asbestos victims are planning to challenge Manville's eligibility under chapter 11. But Manville is only one of hundreds of companies being sued by people suffering occupational and environmental diseases. Congress should examine the bankruptcy laws and make changes, if necessary, to insure that companies can't use the laws to shirk their responsiblities to people harmed by their products.

Are the financial problems created by environmental and occupational diseases too big for companies to handle?

Lawsuits filed by asbestos victims illustrate the enormity of the problem. From 1940 to 1980, about 27-million workers were exposed to asbestos. Dr. Irving J. Selikoff, chief of environmental health at Mount Sinai Hospital, has estimated that claims against asbestos companies could reach $80-billion.

Asbestos is just one of many hazardous products that could kill or injure millions of Americans exposed to them, including benzene, DES, Agent Orange and radioactive materials. That means today's workers are likely to be the victims and litigants 20 or 30 years from now. And it seems highly unlikely the companies alone will be able to compensate all those injured.

Congress is planning to hold hearings this month on Manville's decision to file for reorganization and the more general asbestos liability problem. The big question is who would pay compensation. Rep. George Miller, D-Calif., is sponsoring a bill that would require asbestos manufacturers and suppliers to set up a compensation fund. The industry is lobbying for a bill that would require asbestos companies to make only half the contributions, with the federal government supplying the other half.

COMPANIES THAT make hazardous products should bear the financial responsibility for compensating people injured and killed by them. The federal government already pays victims of occupational diseases at least $3-billion a year in disability and other benefits. That's a big burden for taxpayers. But the federal government may have to set up more compensation funds because some negligent companies simply will not have adequate assets to pay all the victims.

Are the health problems associated with hazardous products being addressed adequately?

No. What is needed, according to officials interviewed by *New York Times* reporter Philip Shabecoff, is a comprehensive policy that deals with the compensation of victims, the monitoring of people exposed to hazardous substances, a program of treatment for those who fall ill, greatly expanded research into the causes of environmentally related diseases and programs to prevent such diseases from being contracted.

The Manville case has focused attention on a national problem of great magnitude that is only beginning to be understood. For the millions of present and future victims of hazardous substances, it's crucial to address the problem now.

THE WALL STREET JOURNAL
New York, N.Y., August 30, 1982

Product liability lawyers have a name for the kind of case that forced Manville Corp. to seek refuge in bankruptcy court last week. It's called a "toxic tort," or a damage claim resulting from exposure to some toxic substance. It's a burgeoning category in a burgeoning field of law. But there must be a better way for victims to be compensated for their losses and manufacturers held accountable for the safety of their products and workplaces.

The Manville case may well prove to be the dramatic event needed to call attention to the waste and injustice that have resulted from the failure of the legal system to deal with product liability in an orderly manner. A very large and financially strong company chose Chapter 11 bankruptcy because it was facing over 52,000 lawsuits and potential damage awards of $2 billion arising from health injuries associated with years of making asbestos products.

This is a case of a company attempting to use law in an innovative way to counter the innovative ways law has been used against it. In the hands of modern lawyers, judges and juries, the old English Common Law concept of torts, wrongful acts that entitle the victim to damages, has been stood on its head. In some judicially liberal states like California, a "wrongful act" may consist of nothing more than having fulfilled a government contract years ago to protect seamen from shipboard fires, using a material that was later found to pose health risks to the people who installed it.

Some of the Manville cases are of that type, dating back to worker exposures in World War II shipyards, when asbestosis seemed a minor hazard compared to those that would be risked by sailors on fighting ships. That is not to say that corporations have always been as careful as they should be about work hazards. Damage suits, along with union pressures, safety legislation, workmen's compensation claims and increased medical understanding of occupational hazards all have combined to raise the corporate consciousness. We certainly are not inclined to argue for the abolition of tort law.

On the other hand, the Manville case illuminates serious problems. Lawyers have made product liability a happy hunting ground for fees and, as Manville argued in full-page newspaper ads last week, the lawyers in these cases often fare far better financially than the victims they represent. Fee limitation might reduce the number of speculative filings.

Findings and awards vary widely from case to case and state to state. The Nebraska Supreme Court last year upheld a state workmen's compensation board award to the widow of a man who had installed asbestos insulation for some 40 firms over 35 years. Lacking a more equitable formula, the board had simply assessed the last company the man had worked for. Statutes of limitations also vary widely from state to state, making it possible for Californians to collect large damages, for example, on claims that could not even be filed in New York. A uniform code may be the answer.

In cases where a widespread hazard has been belatedly discovered, separate court suits are highly inefficient in economic terms. Aside from plaintiff lawyer fees and court costs, product liability cases have piled large new legal costs on business corporations. Newsweek not long ago came up with the estimate that the Fortune 500 spend some $2.5 billion a year on lawyers. Consumers pay for corporate law just as they pay for the steel, machinery or engineering talent corporations use. Some chemical companies, wary of product liability suits, even have set up "product stewardship" departments, at some cost, to monitor the safety of products their customers make with their chemicals.

In the asbestos field, Congress has been groping for a more efficient way to deal with claims, mainly through an industry-financed fund. This effort, however, also has become adversarial, pitting Democrat George Miller of California against Manville et al. Congressman Miller wants to exclude existing suits. Manville wants them covered and also wants the government to pay part of the cost. Clearly, such funds can become political footballs, as black lung compensation for coal miners has demonstrated. But it should not be beyond the wit of Congress and industry to come up with an efficient way of compensating asbestos victims fairly.

Finally, a word about asbestos itself. It has about 3,000 commercial and industrial uses, some of them vital to human safety. We can't see why the producer of such a valuable product should be forced into bankruptcy for self-protection. Tort law has become a vast tangle, and whatever else Manville has achieved, it has called attention to the need for untangling it.

THE CHRISTIAN SCIENCE MONITOR
Boston, Mass., September 13, 1982

A price cannot be placed on justice. But the cost of justice need not be as high as the soaring figures for litigation in many cases of harm to individuals from manufactured products.

Asbestos is now the center of attention as its largest American manufacturer, the Manville Corporation, resorts to bankruptcy for relief from the financial drain of thousands of lawsuits. But all products would be covered by pending legislation to bring order out of the legal hodge-podge that encourages waste and exploitation. The effort to perfect and pass such legislation is the more urgent as Americans become more alert to the commercial use of hazardous substances and to their rights when victimized by them.

The problem has risen to the point of persuading conservative Republican Senator Kasten of Wisconsin to seek a federal solution when he might ordinarily be expected to lean toward states' rights. He has found a range of support from Democrats Glenn and Inouye to his own party's Stafford and Hatch. His product liability bill would preempt the present variety of state laws and require all states to have laws meeting the same standards.

These standards, to be sure, are subject to debate. But the goal of consistency is important. Its achievement could reduce what has been called the "frictional" costs — the whole array of legal and court costs — in the American system for settling injury claims. For example, one likely result would be fewer frivolous, nuisance, or exploitative suits.

Such suits are already discouraged in Britain, where the losing side pays the court costs of both sides. A disadvantage is that serious and legitimate claims may not be pursued because of the high cost of failure.

In the United States a defendant company such as Manville customarily pays its own legal fees win or lose. The plaintiff's lawyers operate under a contingent fee system. They receive nothing if they lose but perhaps a third of the award if they win.

Thus there is an incentive to search out cases and go for high awards. At the same time there is an incentive to inform victims of their rights — and victims who could not otherwise afford legal counsel can obtain it on the contingent fee basis. Wastefulness can arise when many claimants with essentially the same case go to court individually instead of in a group.

To establish some uniformity in the situation, the government in 1979 published a model product liability act for voluntary adoption by the states. There was little response.

The present proposed legislation would in effect impose a model act. It addresses a legal development that has heightened the challenge to find consistent standards. This involves the law of torts, which applies to damage or injury suits. The upshot has been to ease the plaintiff's burden in many cases. In the past the plaintiff had to show the manufacturer was negligent; now it is sufficient to show that the product itself is faulty.

Some states have allowed this strict-liability test across the board instead of in certain categories of cases. The Kasten bill would have all states accept the same categories, with negligence still having to be proved in certain types of cases. Among the elements to be considered are defects in design, defects of construction, and manufacturers' warnings and warranties.

All this may sound complicated enough, but it is simple in comparison with the whole legal situation to be addressed. If the system cannot be made to work with fairness and efficiency, pressure could grow for federal alternatives.

Already, in the case of asbestos, the government is being sued along with manufacturers because many of the claimants are shipyard workers exposed to asbestos building vessels for World War II. And there are proposals for a federal compensation fund — paid for by industry, taxpayers, or a combination of the two. If such came to pass, it is not hard to imagine attempts to provide the same for victims of other product failings who, if fewer in number, are no less deserving.

Do Americans want to go down the road to universal indemnification by government? It would hardly seem so when large cutbacks in social programs are being accepted.

Thus it becomes all the more important for the existing system to operate as responsibly as possible. It is a task not only for legislators but for the courts and the legal profession that serves them.

the Charleston Gazette
Charleston, W. Va., September 11, 1982

WHATEVER one may think of the Manville Corporation's strategy for avoiding lawsuits, one is forced to agree with its proposal that government and the asbestos industry combine resources to create a program to provide compensation to victims of asbestos-related diseases.

The plight of asbestos workers was virtually ignored while efforts were being pressed to compensate miners whose lungs had been ruined by inhalation of coal dust. Yet today it is generally agreed in the medical community that continued exposure to asbestos can leave a worker with conditions as bad, or worse, than those created by black lung.

In approving the Manville proposal, one need not assume either that the firm is trying to lay off some of its own obligation onto the government or that its concern is wholly charitable. The man or woman suffering from asbestos-induced cancer should be the primary object of national concern. We who have seen the ghastly effect of black lung should be especially understanding.

Manville, a major asbestos manufacturer, has filed for reorganizaton under the bankruptcy law because of more than 16,000 lawsuits filed by persons suffering from asbestos-related diseases. Another asbestos manufacturer, UNR Industries, has taken similar action for the same reasons. The proceeding is unprecedented in that both firms are financially healthy, and some government officials believe the bankruptcy law is being improperly used.

At an average disposition cost of $40,000, and with more than 30,000 more suits in the offing, Manville indeed faces great peril. More important than the manufacturer's position, however, is the position of asbestos victims who, if Manville should be strapped by judgments, would never receive compensation.

There is some justification, other than simple humanitarianism, for government cooperation. Many of the suits against Manville were filed by World War II shipyard workers who were fulfilling defense contracts which stipulated levels of asbestos to be used. In any event, if a black lung fund is fair, an asbestos fund is fair.

The Washington Post
Washington, D. C., August 29, 1982

IN THE ASBESTOS cases, the conventional liability suit works dramatically badly. The system is breaking down. The Manville Corporation's decision to file for bankruptcy is a warning that the rapidly growing burden of litigation cannot be handled through the normal processes of the courtroom. One reason is the sheer volume of it. Some 16,500 suits have already been brought against Manville alone, and, by the company's estimate, it can expect another 32,000. But the greater reason is the nature of cancer, which may appear decades after exposure to asbestos and leaves no clue as to which exposure, to which product, may have caused the disease.

Plaintiffs have trouble collecting from companies that may be liable, and the companies have trouble collecting from their insurance carriers. The only people who can be sure of extracting much benefit from this procedure are the lawyers. Manville says that in the first half of this year it spent $8.6 million on the asbestos litigation, of which only one-fifth actually went to the people who had been injured. Where did the rest go? Into legal fees, of course—Manville's and the plaintiffs'. These cases reveal the American legal system at its worst. It is intolerable that, of the millions spent and perhaps billions to come, such a wretched trickle of compensation should reach the unfortunate people who have actually suffered from the terrible diseases that asbestos caused.

There's a better way to handle it. The model ought to be the workmen's compensation system, with a couple of important differences. Congress is already considering several bills, and Manville's flight to the bankruptcy court is a signal that there isn't all the time in the world to make decisions.

The right solution is a special claims board that can make prompt awards, under a uniform nationwide standard, out of one consolidated fund.

Who contributes to the fund? To begin with, the companies that have produced asbestos, and asbestos products, over the past generation. Next, their insurance companies, to whom it would be worth quite a lot to buy their way out of 30 years' litigation ahead over which of them is liable for what exposure. Finally, the federal government. In some instances, federal regulations actually required the use of asbestos products. If the industry can be blamed for reacting slowly and reluctantly to the mounting evidence of danger to health, the same thing can be said of the federal government.

The present reliance on individual suits means that one person may receive a large judgment if he was exposed while working for a profitable company —i.e, Manville—while another person, with similar illness, gets nothing because he worked for a company that has since gone out of business. One person may be well compensated, while another gets nothing because he can't prove the circumstances of his exposure in, say, the 1950s.

A national claims board can extend equal compensation to people suffering equal injuries without plunging into the questions—frequently unanswerable—about which exposure, or which product, may have been responsible. That is justice of a kind that the courts, in these cases, are incapable of rendering. The victims of asbestos are people who have suffered grievous harm. They are entitled to compensation without long delays, without waiting for successive court appeals and without having to turn over two-thirds of their settlements as lawyers' fees.

THE LOUISVILLE TIMES
Louisville, Ky., September 2, 1982

No tears need be shed for the giant, and financially healthy, Manville Corp., which last week dumped its heavy burden of product liability lawsuits in the lap of a bankruptcy judge.

However, the case does suggest that a better system needs to be devised for compensating the victims of occupational disease.

Manville, the largest producer of asbestos in the western world, is not broke or even in imminent danger of becoming so. But it has been hit by thousands of lawsuits filed by workers who suffer from cancer or other diseases caused by tiny asbestos fibers that have lodged in their lungs.

Tens of thousands of additional suits are likely, since asbestos-related diseases often don't appear for years after exposure. Company executives claim the anticipated cost of judgments and legal expenses will nearly equal Manville's $2 billion in assets.

The bankruptcy petition has the effect of freezing lawsuits now in the courts and stopping new ones from being filed. A bankruptcy judge may eventually have to work out settlements with thousands of plaintiffs, although a recent Supreme Court decision makes even that uncertain.

The company would deserve some sympathy if no one had had any inkling what asbestos does to the human body. But the industry, according to documents uncovered recently, had evidence in the 1930s that the material could severely harm the workers who handled it. Because little was done by industry and government to protect millions of workers, or even warn them of possible hazards, some nasty chickens are coming home to roost.

Manville's approach to limiting its legal obligations, and protecting its stockholders, therefore, is at best a serious abuse of laws intended to help companies in true financial distress.

The possibility that chemical and drug companies facing large numbers of injury claims might try to follow the Manville example makes the case even more alarming. The bankruptcy system should not exist to shelter profitable businesses from the tragic results of their carelessness or venality.

U.S. Sen. Robert Dole, whose subcommittee has jurisdiction over bankruptcy law, charitably called Manville's petition "unusual and dubious." While he did not suggest any specific changes in the law, Congress clearly must act to stop others from following the same route.

That having been said, it's also important to recognize that conventional lawsuits are a costly and far from satisfactory way to deal with thousands of claims resulting from exposure to hazardous products. Some disabled persons inevitably end up with little or nothing while others win more than the seriousness of their injury justifies. Because the diseases take so long to develop, establishing the time and place of exposure is often difficult.

Worst of all, the proceedings often funnel large sums of money into lawyers' pockets, but provide relatively little to the victims.

Indeed, one of Manville's goals is to pressure the federal government to set up, and contribute to, a compensation fund for the victims of asbestos. The industry's argument is that government is obliged to help because as many as a third of those stricken with asbestos-related diseases were exposed to the material in shipyards during World War II.

The trouble is that the taxpayers' obligation would be virtually open-ended at a time when federal spending on health care is being reduced. Moreover, Manville and companies with similar problems would have little incentive to check the safety of their products before marketing them.

A more attractive plan proposed by Rep. George Miller of California would set up a compensation program funded entirely by the asbestos industry and the insurance companies involved. Mr. Miller plans hearings this fall.

In the meantime, as Sen. Dole seems to recognize, Congress must take steps to assure that the Manville "bankruptcy" is one of a kind.

The News and Courier
Charleston, S.C., September 14, 1982

There is an old-fashioned cautionary tale and a powerful moral to be drawn from the current plight of the asbestos industry. Executives of asbestos companies which are faced with bankruptcy or ruin by a flood of lawsuits from victims of asbestosis, an incurable, killing disease, surely had mothers who must have told them when they were children that it is always better to own up than hide the truth.

Asbestos, it is now clear from evidence presented in thousands of lawsuits that have already been decided, has been a known threat to health for almost half a century. If the asbestos industry had heeded the small, still voice of conscience and made a moral commitment to find out what asbestos dust did to human lungs — and had acted on the knowledge — then the worst industrial health disaster in history could have been avoided. The evidence available today suggests that the attitude of executives in the asbestos industry as far back as the 1930s was to try and cover up the facts, hiding the results of research. They deluded themselves into believing that they were protecting their businesses but all the while they were storing up trouble for the asbestos industry. It was a tragically mistaken attitude. The asbestos ostriches of the 1930s and 1940s exposed an estimated 27 million people to the disease. Their lack of moral sense, or, perhaps, over-developed business sense, has caused something very like a curse to be visited on the industry. In medieval morality plays, people paid for their sins. The asbestos industry faces liabilities for the sins of omission of its executives which have been estimated at between $40 and $80 billion.

Recoiling in horror from the prospect, the Manville Corporation, which is the biggest company under the shadow of the curse, filed for bankrputcy under Chapter 11 last month. Although it is one of the soundest companies in the United States, it knows it will not be able to withstand the expected siege of litigation. The company described itself as already "overwhelmed by lawsuits" and quoted an epidemiologist's estimate that it could be sued by some 52,000 victims of asbestosis. As each case costs an average of $40,000, the Manville Corporation could face a bill for legal costs and compensation of over $2 billion.

The Manville Corporation's voluntary bankruputcy, which followed another Chapter 11 filing by UNR Industries in July, was not solely intended to stave off financial disaster. The company, along with the entire asbestos industry, is also trying to make a case for a government bail-out. Congress has been under intensive lobbying to pass legislation to set up a compensation fund for victims of asbestosis. In the meantime, bankruptcy offers the asbestos companies a shelter from the lawsuits. They do not dispute the need to compensate victims of asbestosis. They argue that they should not bear the brunt of cost of compensation.

There is something in what the asbestos manufacturers claim. The government should bear some responsibility for the millions of workers who contracted the disease while working in the shipyards during World War II. Workers' compensation has proved to be woefully inadequate. But the asbestos industry and their insurance companies, which are locked in litigation over their respective liabilities, should be disabused of the idea that the taxpayer should foot the bill. The compensation fund is needed all right, but it should be mutually funded by the asbestos industry and the insurance companies involved. The government should provide a safety net for those not covered by the general fund and play a supervisory role so that victims of asbestosis do not have to sue.

The costs of litigation have been described as "truly outrageous" by the Manville Corporation. Victims receive less than a third of the money paid out in suits. There seems to be general agreement that the legal costs have been abusive. It has been established beyond doubt that asbestos dust can kill. There is no justification for lawyers making another killing. They do not seem to have grasped the moral in all this, either.

The Pittsburgh PRESS
Pittsburgh, Pa., October 9, 1983

As the health hazards of asbestos have become evident, U.S. consumption of this otherwise useful fireproofing mineral has fallen to less than a sixth of what it was 10 years ago.

But although industry has developed a variety of substitutes for asbestos, more than 140,000 tons of it are still mined annually.

Now the Environmental Protection Agency intends to crack down on the products that still use asbestos.

Calling asbestos a proven cause of cancer, the EPA proposes to: (1) ban it entirely in such construction materials as cement pipes, roofing and floor coverings, (2) place a limit on the amount of asbestos used for all other products and, (3) set up a timetable to phase out all but a handful of minor uses over the next few years.

This is one environmental issue there is little disagreement about, and we welcome the EPA's action.

Asbestos is an especially insidious, slow-acting poison. Its microscopic fibers, lodged in the human body, have been implicated in a variety of diseases, particularly cancer of the lungs and the lining of the lungs.

Questions surrounding asbestos already in the environment are, however, a matter of controversy — mostly financial.

The U.S. Department of Education, for example, estimates it will cost about $1.4 billion just to remove asbestos from 14,000 schools built or renovated after 1940.

But the EPA's move to ban or phase out asbestos will help assure a safer environmental future for Americans.

EVENING EXPRESS
Portland, Maine, September 27, 1983

If there's one thing we've learned from the long, sad history of illness related to asbestos, it's that knowledge has power only when the knowledge is shared.

Incredible as it may seem, the asbestos industry had been aware of dangers associated with inhaling the thread-like mineral since 1898. But it was not until recent decades, when knowledge was shared, that the public demanded limits on the use of asbestos, including removal from public buildings (some in Portland) where it might pose a threat.

Among public buildings where asbestos was once used for fireproof insulation are many public and private schools. According to the Environmental Protection Agency, about 66 percent of the nation's schools have failed to completely identify and report its presence.

Yet, despite the lack of complete reporting, the Service Employees Union earlier this year estimated that 3.6 million American children were exposed to asbestos. Because asbestos can pose a serious threat of cancer and other diseases, varying widely from person to person, it's imperative that schools, as a first step, comply with all requirements of the EPA report.

Leroy Nisbett, director of the Maine Department of Education's School Facilities Division, says the department has "blanketed" Maine superintendents with information about reporting on asbestos. Nisbett estimates the reporting in this state is "close to 100 percent compliance."

Fine. But the object is not just to gather but to share the knowledge that emerges from the reports. EPA Administrator William D. Ruckelshaus agrees. He has directed agency officials to send out press notices where local schools are found to be in violation of the rules that govern safe use of asbestos. Public awareness, he reasons, should help get potentially unsafe conditions corrected.

Certainly the chance is better than in the long, dark decades when the dangers of asbestos were ignored.

ST. LOUIS POST-DISPATCH
St. Louis, Mo., November 4, 1983

The Occupational Safety and Health Administration has issued an emergency rule to immediately reduce by 75 percent the permissible exposure limits for asbestos fibers in the work place. The first such action taken by the Reagan administration, it was warranted by the hazard faced by workers in manufacturing, construction and shipbuilding.

OSHA has found that 50,000 employees are exposed, over an average working day, to levels of asbestos fibers ranging from 500,000 to 2 million per cubic meter of air. Labor Secretary Donovan, in announcing the OSHA action, declared that the agency predicts "three excess cancer deaths per 1,000 workers exposed for one year at the current permissible exposure level." The new standards, OSHA believes, will save 200 lives a year among the exposed workers.

Asbestos, which is widely used as a fire retardant and as a reinforcing agent in construction work, is known to cause, in addition to cancer, a lung disease known as asbestosis. Three manufacturers of the substance already had filed for bankruptcy to protect themselves against a deluge of damage suits from victims of asbestos-related diseases. Asbestos, in short, is one of the deadliest threats to the safety of the industrial environment. OSHA has acted responsibly to minimize the danger.

THE LINCOLN STAR
Lincoln, Neb., March 1, 1983

When sprayed-on building materials containing asbestos became available following World War II, they were "an architect's dream," a federal Environmental Protection Agency official said last Wednesday.

They were easily applied, and provided excellent soundproofing and insulating qualities for ceilings and walls. They would not burn, and could be applied over almost any type of surface. Best of all, they were cheap.

Asbestos-laden materials became so popular that they were the rule rather than the exception in large building construction during the 1950s, and were used even in many residential dwellings, as well.

By the late 1960s, however, the dream was becoming a nightmare. Asbestos had been linked with several forms of cancer and other diseases, and in 1972 the EPA banned the use of "friable" asbestos materials — materials that can fragment and release the fibers into the air.

Largely because of the post-war baby boom, there were many school buildings built during the 1950s, and asbestos-containing ceiling treatments were common. Lincoln Public Schools were no exception.

During the past few years, the EPA has been studying the problem of asbestos in schools, and last year published new regulations giving schools a year to inspect and analyze friable material for asbestos content and make their findings public. The regulations do not, however, require that the materials be removed. .

As one school staff member cynically put it, "The law just says you have the right to know what you're dying of."

In compliance with these regulations, the Lincoln school system hired a consultant who inspected every one of its 50 buildings and found sprayed-on asbestos materials in 29 of them. Its removal could cost between $1.5 and $4 million.

From what Robert Den Hartog, associate superintendent for business affairs, told the school board Wednesday, however, it appears likely that the school system will begin the job of removing the materials as soon as possible.

It won't be an easy task, and the board will have a hard time coming up with $4 million, if that's what it costs.

We urge the school board, however, to place the removal of asbestos at the top of its list of capital improvements for the coming year. It would be preferable to postpone any major building renovations or additions until after the asbestos problem has been solved.

As Den Hartog suggested, the best approach would be to start with areas where students and staff are the most likely to touch or disturb the sprayed-on material, such as in low-ceiling hallways or multi-purpose rooms where balls bounce off the ceilings.

The eventual removal or sealing-up of all the potentially hazardous areas is strongly urged, however, for the safety and peace of mind of the students who are using the schools as well as staff members and the public.

We also urge the state Legislature to remove the costs of asbestos eradication from the 7 percent lid on local budget increases, because this is indeed an "extraordinary problem," as State Sen. Don Wesely of Lincoln has said.

While the schools are cleaning up the mistakes made by builders in the past, it would behoove local businesses, apartment owners and others to inspect their buildings, as well.

The EPA zeroed in on schools because, as one official put it, children "have such a long lifespan ahead of them." But if we eliminate asbestos from school buildings, and still have the potentially deadly fibers entering the air from other sources throughout the community, the problem will be far from solved.

The Providence Journal

Providence, R.I., November 8, 1983

The Reagan administration's decision to toughen its stand against exposure to asbestos in the workplace is a welcome concession that further neglect of this problem could cost many lives.

Labor Secretary Raymond Donovan announced last week that an emergency standard had been issued, effective immediately, reducing the permissible exposure level over an eight-hour period from 2 fibers per cubic centimeter to ½ fiber. He said risk studies showed that under the current standard "three excess cancer deaths per 1,000 workers exposed for one year" would occur. Further, even those subject to short-term exposure would face "grave risk of danger."

Asbestos, a mineral of extraordinary heat-resistant properties, has been used in thousands of construction and consumer products. From 1950 until it was banned in 1973 sprayed-on asbestos, the most dangerous type, was used commonly to insulate pipes and fireproof homes, offices, factories and schools. Over 40 years, an estimated 11 million to 12 million workers were exposed.

OSHA (Occupational Safety and Health Administration) has shown increasing concern about the safety of construction workers at sites where older buildings were being demolished. Asbestos that flakes or powders is easily inhaled or ingested, causing the chronic lung ailment asbestosis and sometimes cancer.

Soon after it was created in 1972, OSHA adopted an asbestos standard. Although much has been learned meanwhile about the material and its danger to humans, the standard has not been changed since 1976. One startling finding is that the mortality rate is 53 times higher for smokers who are exposed to asbestos. The emergency standard issued last week comes none too soon, therefore.

The question now is whether the new standard will be strictly enforced. OSHA has included considerable flexibility in the new regulation relative to the compliance methods employed by the manufacturing, construction and maritime industries, which are covered by the rules. These might include engineering controls, altered work practices, providing workers with masks and respirators and removing fibers from the work area with suction equipment.

A strong standard with weak enforcement would do little to protect the 375,000 workers the Labor Department says face "grave danger." It remains for OSHA to set up an inspection program that will see to it the new rules are observed in fact and not in the breach. The first step has been taken and that is good, but the second is even more crucial to the workers involved.

THE ARIZONA REPUBLIC

Phoenix, Ariz., November 4, 1983

THE Reagan administration continues to change direction in its policies, and the latest shift shows a determination to be tougher with industrial health standards.

The Occupational Safety and Health Administration has issued an "emergency" order on asbestos exposure, toughening the allowable worker exposure to asbestos fibers by reducing it by 75 percent. The legal exposure level during an eight-hour work day was slashed from two fibers per cubic centimeter of air to one-half fiber.

OSHA said 375,000 workers faced "grave danger."

OSHA officials told *The Republic* a year ago that new exposure standards would not be needed for several years. However, manufacturing representatives said the work standard could be reduced to one cubic centimeter of air. OSHA then undertook a study of its own regulations.

The turnaround began at the White House because of political fallout from the controversies at the Environmental Protection Agency and the Department of Interior.

EPA became embroiled in a political furor when it was revealed that funds were not being spent to clean up various environmental problems. Attention also focused on relationships between industry executives and EPA officials. White House aides also were alarmed by the political fallout caused by Interior Secretary James Watt.

EPA chief William Ruckelshaus has already begun reversing policies, and William Clark, who will shortly take over at Interior, also is expected to make some changes.

The White House has finally recognized a political truth — the public does not want a relaxation in regulations that are designed to protect health.

It also has at last understood that environmental policies cross party lines — Republicans are as concerned as Democrats.

Some critics perceive the administration to be anti-environment. That's unfair. Rather, it has tried to redress some of the excesses of the Carter years and bring greater balance to such decisions.

Detroit Free Press

Detroit, Mich., November 4, 1983

THE LABOR Department's emergency rule tightening restrictions on the amount of asbestos fibers workers can breathe ought to become permanent — if only to highlight the grave peril in *any* exposure to asbestos fibers. The hazards of breathing asbestos have been well documented, and recent studies suggest the dangers may be even greater than previously thought.

In announcing the emergency rule, Labor Secretary Raymond J. Donovan said the Occupational Safety and Health Administration (OSHA) predicted "three excess cancer deaths per 1,000 workers exposed for one year at the current permissible exposure level." Studies also show "risk of grave danger" to people with even short-term asbestos exposure, the agency noted, adding that the mortality rate was 53 times higher for smokers exposed to asbestos.

The heightened concern over the daily asbestos exposure standard — now reduced from two fibers to half a fiber per cubic centimeter of air — reflects new research findings on the dangers of breathing particles of the woven mineral fiber. More than 200,000 of the nine million surviving workers exposed to asbestos during the past 40 years will die of asbestos-associated cancer by the end of the century, an internal Department of Labor report concluded two years ago.

The debate over what to do about asbestos has been complicated by its widespread industrial use, the difficulty of finding low-cost replacements for some of its applications and the fear of increased litigation by exposed workers. Eventually, these concerns may cause changes in workers' compensation programs and product liability laws.

The primary problem, though, remains safeguarding workers exposed to asbestos dust. Safety standards have improved markedly since the 1960s, and the most dangerous applications of asbestos, particularly spraying, have been banned almost everywhere. But in face of evidence that even low-level exposure to asbestos can trigger the development of disease, OSHA acted wisely in tightening restrictions. During the 60-day period for public comment before the rule becomes permanent, citizens ought to let the agency know they appreciate such protection.

The Oregonian

Portland, Ore., February 19, 1984

The U.S. Environmental Protection Agency appropriately has sounded an alarm in its claims that Pacific Northwest schools are failing to comply with health and safety regulations designed to protect children from asbestos exposure in school buildings.

That does not mean, however, that the federal EPA is right in its implied contention that Northwest school buildings are hazardous to student and teacher health. The truth no doubt lies somewhere between the EPA's forceful rhetoric and school districts' overly defensive responses that satisfactory checks for asbestos problems have been made and that corrective action, where applicable, is under way.

For Oregon, this is clearly a case in which an independent audit of public schools by either the state Health Division or the state Department of Education would be helpful. Parent groups and teacher unions, too, have a stake in this matter and should press school officials for a public accounting.

It is not reassuring to parents of school children or school employees who work in these buildings to learn that one undisclosed Northwest school district certified that no asbestos-related problems existed, while a surprise spot-check by the EPA found evidence of asbestos exposure widespread in two buildings in that district.

Nor is it especially reassuring to hear an Oregon Department of Education official say that every school district in the state had assured her office that checks for asbestos problems were made when the Portland School District admitted that at least two principals made "honest mistakes in failing to post notices" to advise students of the presence of asbestos in those schools.

To the Portland School District's credit, it has embarked on a $1 million asbestos cleanup and treatment program in which masonry specialists are attempting to correct, albeit with patchwork approaches, glaring asbestos problems.

Whether the EPA, in this get-tougher-than-usual election year, is overreacting about the Northwest's asbestos problems, or whether schools are underreacting to a well-documented threat to public health, citizens should be reacting with anxiety to uncover the truth.

Asbestos exposure is a menace in this nation. Asbestos has been found to pose a serious threat of cancer and asbestosis, a progressive lung disease. Congress, in 1980, required all school districts to survey their buildings for the presence of asbestos and to notify principals, employees and parent-teacher associations if any was found. Schools also were required to post the results of the inspection in the school.

That some schools in Portland and perhaps elsewhere in the region have disregarded any parts of these requirements builds little confidence among citizens that the asbestos threat in the Northwest, whether slight or EPA-magnified, is being taken seriously enough.

THE ATLANTA CONSTITUTION

Atlanta, Ga., March 13, 1984

The Reagan administration's attempt to reduce the exposure of industrial workers to asbestos was halfhearted at best: too little, too late, and now, according to a federal appeals court, too poorly documented.

The 5th Circuit Court didn't question the need for a tough, new workplace standard. On the contrary, asbestos "doubtless may present a grave danger to workers," it said.

But it took issue with the Occupational Safety and Health Administration's issuance of an "emergency rule" — a backdoor approach that bypassed normal procedures — citing the agency's failure to demonstrate that an emergency existed.

The administration could hardly have expected otherwise. OSHA chief Thorne G. Auchter had for three years behaved as if there were no emergency, putting off the "urgent" recommendations of government scientists and his own staff — and telling a House subcommittee less than a year ago that there was no basis for emergency action in the interim.

In contrast with his predecessors, who labored to lower the permissible asbestos content of workplace air, Auchter took issue with growing evidence that exposure to even minute concentrations of asbestos fibers led to debilitating lung diseases, including cancer.

But instead of proposing new standards under the agency's normal notice-and-comment rule-making procedures, as he had been urged to do — a route that, the appeals court volunteered, might have withstood a legal challenge — Auchter bypassed the hearing process altogether, precipitously publishing an "emergency rule" last November that drew an industry challenge.

Auchter, who plans to resign later this month, said that the Labor Department would try again "within a month" to come up with an acceptable asbestos rule, but expressed little hope that it would succeed. Indeed, how can it, in the absence of a strong commitment on Auchter's part to protect the more than 375,000 industrial workers at risk?

By OSHA's own estimates, the proposed rule might have saved 80 lives in six months. Yet the agency failed to make its case. The public is left to wonder whether the agency, even claiming an emergency, ever meant to make its case.

Post-Tribune

Gary, Ind., January 10, 1984

The asbestos scare at Peifer School in Schererville is not the kind of issue school boards and administrators are trained to handle. Closing the school for the rest of this year was a drastic action. Based on what the officials knew about the problem, it was the right decision, we believe.

There may indeed be no health hazard from the flakes coming from the ceiling. Health officials earlier had said there was no danger, but that was before some small leaks were discovered. There will always be differing opinions even among experts on such matters. What is a school board supposed to do when the evidence is not absolute either way? If there is to be an error, let it be on the side of caution, even if, as in this case, considerable inconvenience will result. Faced with that kind of a choice, safety comes first.

Peifer students will be bused to Orchard Park School in Highland, a school that has been closed since 1981. Some parents don't like that prospect, and we don't blame them. But the decision has been made — a school reasonably close has been found, and now the education of those 400 or so students should be resumed as promptly as possible.

What is learned from the Peifer experience should help health and school officials handle any similar problem. More important, perhaps, parents and the school board will understand each other's mutual concerns better. Everyone wants what is best for the children.

The Kansas City Times

Kansas City, Mo., February 13, 1984

Public employees at the State Office Building in Downtown Kansas City must think they have been allowed to slip through government's cracks. Since mid-1981 it has been known that dangerous asbestos fibers composed the insulation there, but nothing has been done. The estimated cost of $8 million to clean up the asbestos is the reason.

The governor's proposed $225 million capital improvements bond issue last year contained most of the needed money, but it was rejected by the General Assembly. The much smaller $50 million issue finally approved did not include funding for this work.

A lot of other important projects were lost when the issuance of the bonds was disapproved, but the State Office Building problem is a special situation. Although some experts do not believe there is an immediate danger to workers, others disagree on what level of asbestos actually is safe. Generally asbestos-related diseases do not show up for years after exposure. Federal standards on asbestos levels are considered too weak, and attempts to strengthen them have been thwarted by an industry lawsuit.

Even if the danger is not great, public workers should not be required to work under even minimal risk conditions. As with the dioxin scare in eastern Missouri, too little is known about the long-term effects of exposure. Public employees deserve as much government assistance as the residents of Times Beach, Mo., received when confronted with dioxin contamination. Even if the bonds are not approved, state government needs to come through with the money for this cleanup.

RAPID CITY JOURNAL—
Rapid City, S.D., May 9, 1984

Tuesday's Associated Press report that the Environmental Protection Agency is looking for ways to remove asbestos from more than 14,000 schools and public buildings was of special interest to Rapid City readers.

Asbestos has been in the news here since last week, when concern was raised that a water-soaked asbestos ceiling at West Junior High might be a health hazard. Corrective measures have been taken and tests showed no asbestos fibers in the air, although school officials still had to deal with an offensive odor not related to the asbestos scare.

But the incident demonstrated that the Rapid City Area School District has been in the forefront in eliminating the dangers of the cancer-causing material from public schools, either by removing materials containing asbestos or sealing asbestos fibers so they cannot escape into the air and be inhaled.

Such is apparently not the case in many places around the nation, acknowledged the EPA, which admitted its voluntary program to remove the health risk from public buildings has failed. "We are looking for ways now to make it work," said an EPA spokesman.

Asbestos was used in insulation and building materials in schools and other buildings until the EPA, recognizing the danger, banned it in 1978. In May 1982 the agency required education officials to inspect their schools for "friable" (dry, easily crumbled) asbestos. notify parents and teachers and post health warnings. The idea was to put pressure on schools to deal with the hazard.

The problem was that many public agencies didn't have the money to eliminate or cover the asbestos.

To their credit, Rapid City school officials recognized the problem in 1979 and have been dealing with it ever since. So far some $120,000 has been spent to remove or cover asbestos in the 13 schools in the district in which it was found. West and South Junior Highs and General Beadle Elementary School are the only schools in which work remains to be done, and that's scheduled this summer.

The only fault the EPA found in an inspection last September was inadequate warning signs and record keeping, which school officials say has since been remedied.

A number of pulmonary problems, including a fatal form of cancer, have been linked with the inhaling of asbestos fibers, although several reports have said only people exposed to large amounts of the mineral risk getting cancer.

Still, the asbestos problem is a graphic example of a material once thought to be safe that has proven otherwise. And the Rapid City Area School District has set a good example by taking steps to eliminate the health hazard.

Wisconsin State Journal
Madison, Wisc., April 17, 1984

In 1980, when Madison schools were surveyed for asbestos, it was found in 13 buildings. The school district ordered $200,000 in remodeling work to control the substance.

In air tests conducted at four schools during remodeling in 1981, asbestos levels were found to be far below the legal limit designated by the federal government.

Now, new concerns have risen. Wear and tear in buildings have exposed asbestos again in some places.

The presence of small amounts of asbestos fibers in the air breathed by teachers, students and administrators is no reason for panic, but it is cause for prudent action by the administration and School Board.

That action already has begun. The administration asked the state to conduct new tests of asbestos levels in the 13 buildings — nine of them used for schooling — where the substance is known to be present.

The state has agreed. Further, it will recheck the district's 33 other buildings to make sure they do not contain asbestos.

We suggest these additional steps:

✔ The district should study the logistics of moving classes located in rooms with asbestos exposure to problem-free rooms in the same buildings. If relocation can be done easily, it should be done now to reduce worries. Difficult relocations should be undertaken if test results justify concern.

✔ The district should begin preparing now to completely remove asbestos from its buildings.

It's unlikely the new state tests will show asbestos levels anywhere near federal standards. But concern arises because there has been no scientific consensus on what levels of asbestos exposure are acceptable from a medical, as opposed to legal, standpoint.

Painting or tiling over asbestos surfaces is a short-term solution, but wear and tear eventually uncovers asbestos again. Until asbestos is removed, health worries — justified or not — will be a recurring problem.

The district should consider a gradual, phased program of removal, giving highest priority to buildings with the most asbestos and most children. Temporary repairs can be made in lower-priority buildings and areas until their turn for removal comes up.

Howard Sampson, the district's assistant superintendent for business services, estimates that removing asbestos from the 13 buildings where it is known to exist would cost about $1 million.

With some clever budgeting by the school district, and a phased program, that cost can be a bearable price to pay for peace of mind among teachers, parents, students and school-district officials.

AKRON BEACON JOURNAL
Akron, Ohio, February 7, 1984

THE HEALTH hazards posed by asbestos are well known. Once widely used in the construction industry for fireproofing and insulation, asbestos has been linked to serious lung disease and a deadly cancer known as mesothelioma.

With medical evidence mounting, the federal government recently moved to place even tighter restrictions on asbestos exposure for workers in certain industries.

But before that ever happened, the Environmental Protection Agency ordered every school district in the country to inspect and report on any asbestos in their buildings.

Results of those inspections were due last summer. If an internal EPA report is accurate, the schools have earned an "F" for effort.

According to the EPA document, nearly half of the nation's schools have failed to notify parents and staff of inspection results.

And of 2,632 schools in which asbestos has been discovered, less than 20 percent have developed plans to deal with the potential health threat.

In describing the situation, one EPA administration said, "We've got problems." An understatement indeed.

In fact, the federal agency is a big part of the problem. Its inaction and lukewarm bird-dogging of the program have contributed to a flagrant disregard of the law.

Locally, the track record appears better. The University of Akron and school districts in Akron, Cuyahoga Falls, Tallmadge and Stow have gone to great lengths to explain the results of their asbestos inspections, and to correct immediate health hazards.

But this type of response may be the exception. Whether out of fear that parents will panic or, more likely, fear of huge repair bills, many schools have moved slowly or ignored the directive.

A giant loophole in the EPA regulation is perhaps the leading cause of this sorry state of affairs. There is no requirement that asbestos be cleaned up. All schools need do is make a single inspection and report the findings to parents and teachers.

At last, the obvious is under discussion. EPA officials may require annual school checks for asbestos contamination and specific steps to remove or seal the toxic material.

Without some financial help — at least when an immediate hazard exists — schools facing major repairs may not take quick action. Yet there is nothing in the President's 1985 budget to acknowledge the problem.

In light of the glaring flaws in the school asbestos program, Congress should demand tighter enforcement and provide money to help schools correct the most pressing cases.

After three years of tiptoeing around the law's intent, the EPA and the schools must learn that this nation takes the health of children and school staff most seriously.

THE SUN
Baltimore, Md., March 13, 1984

Asbestos in the schools, posing at least a potential health hazard to children and staff, is a nationwide problem. If hazardous conditions exist in a Baltimore city school within two years of the system certifying itself free of problems, this does more than raise doubts about the assurances relating to other city schools. It raises doubts about such assurances everywhere.

To be free of problems is not the same as being without asbestos. Asbestos can flake or be scattered later on, creating problems that did not exist earlier.

The conditions found by the White Lung Association in five city schools, among 13 inspected, are grounds for concern and correction. They show that the city and the school system cannot afford to remain complacent, believing that all potentially dangerous asbestos conditions have been corrected. Clearly, this was not so.

Since the results of this investigation appeared in The Evening Sun the city school system has taken a proper attitude. It is not trying to cover up the problem.

Removing and replacing asbestos requires expertise. People who handle asbestos should know what they are doing. So should organizations.

A problem that is statewide and nationwide becomes to some extent less a local than a state and national problem. The city school system should err if at all on the side of health and safety in dealing with the asbestos that remains. And it should have the help of the Maryland Department of Education and federal Environmental Protection Agency in doing so.

The Hartford Courant

Hartford, Conn., May 19, 1984

It's been known for some time that asbestos in schools endangers the lives and health of hundreds of thousands of students, teachers, administrators and other staff members. What hasn't been known is who will do the pushing needed to get the cancer-causing stuff removed or sealed. Let's hope the U.S. Environmental Protection Agency's current hearings convince the agency that it has that responsibility.

So far the EPA's record on the asbestos hazard has been dismal. Its only significant action has been to require inspections of schools and notification of parents and employees. The agency assumed that awareness of the risk would generate sufficient pressure for corrective action.

The assumption was wrong. Little has been done, largely because cleaning up asbestos can be expensive and the EPA hasn't demanded it. Parental apathy might also be a factor. Now the agency is conceding that voluntarism isn't the answer. Its hearings, one of which will be in Boston, seem to be an effort to consider alternatives.

Money is a stumbling block, but it's not insurmountable. Cleanups might be financed by local or state bonds, low-interest federal loans or grants from a new source like the toxic waste superfund.

But what's needed most is a firm declaration by the federal government, the nation's chief protector of public health, that the hazard posed by asbestos in schools must be eliminated. There's no reason to ignore the problem in other public buildings, for that matter.

The EPA hearings will serve a useful purpose if they inspire plain talk like that of Irving Selikoff, an expert on health effects of asbestos:

"We're gazing at children being exposed to asbestos. We're seeing teachers, personnel being exposed. We don't really need any additional research about what can happen. What we need now is controls."

THE INDIANAPOLIS NEWS

Indianapolis, Ind., April 23, 1984

One disabled Indiana worker said of the symptoms of his disease: "Well, the best way I can describe it is, it's a sensation very much like drowning. Sometimes, you just feel like you are suffocating."

The symptoms decribed are the symptoms of asbestosis. It is an affliction common in varying degrees to workers who inhaled asbestos dust in auto parts factories, in the hulls of ships, in refrigeration equipment, on construction sites and elsewhere.

According to a U.S. Department of Labor study, an estimated 676,000 Americans whose lungs have been exposed to asbestos could die during the next two decades.

Asbestosis has spawned massive litigation in behalf of disabled workers who claim that firms which mined, manufactured and utilized asbestos failed to issue warnings or employ safety measures for workers handling asbestos even after the dangers of asbestos inhalation were known. Nationwide, at least 24,000 lawsuits have been filed and asbestos cases have clogged the federal court dockets in both of Indiana's district courts.

A recent Rand Corp. report pegs the potential liability to the firms being sued and their insurers at anywhere from $4 billion to $38 billion. One of the nation's largest asbestos miners and manufacturers, the Manville Corp., has instituted Chapter 11 bankruptcy proceedings because of the liability facing it. Some insurers contend their limits of liability have been reached.

If asbestos manufacturing firms are facing bankruptcy, victims of asbestosis are facing worse.

They are facing statute of limitations rulings on products liability that vary from state to state and are often restrictive. Many plaintiffs have been denied recourse in the courts because of the long interval between their last exposure to asbestos and the time when the disease manifests itself.

Fewer than one in four court claims have been settled to date. Although the average settlement is about $100,000, the victims net only $30,000 to $50,000. The lawyers are getting the rest.

According to the Rand Corp., only one third of the money spent by asbestos companies and their insurers to defend against claims in court has gone to the victims. The remainder, nearly $400 million, has gone to pay attorney fees.

Congress is attempting to draft a new compensation system for workers who suffer from asbestosis. The House Subcommittee on Labor Standards has spent six years working on a compensation program which would be administered by the government and financed by the asbestos industry.

Meanwhile, the Asbestos Claims Council, a coalition of asbestos manufacturing firms and insurers, has been working on mechanisms for creating a fund to enable asbestos victims to collect on claims without resorting to the courts.

Both efforts are mired in controversy. And while the wrangling goes on, workers and their families suffer, legal payments mount, the government pays about $3 billion a year to disabled workers through Social Security disability, veterans' benefits, Medicare, Medicaid and welfare benefits and more companies face bankruptcy.

It is time for Congress to halt the debate over the particulars of a compensation program and come forward with an overall strategy for dealing with occupational diseases.

The issue yesterday was black lung disease. Today it is asbestosis. Tomorrow it could be silicosis or any number of other afflictions lurking in this post-industrial age of chemicals and radiation.

THE TENNESSEAN

Nashville, Tenn., June 1, 1984

SEN. Jim Sasser has succeeded in amending the Senate education bill to provide loans and grants to help the nation's school systems identify and remove asbestos insulation in walls and ceilings.

The amendment would provide $50 million the next two years and $100 the following four years. The measure also transfers the responsibility for locating and removing asbestos from the Department of Education to the Environmental Protection Agency.

The legislation is a step in the right direction. But it still has to be approved by the full Senate and the House, and the job is likely to take longer than six years and more money than has been approved for it.

Senator Sasser said indications are that $1.4 billion may be needed to remove asbestos from all schools which have it. Asbestos, which is believed to cause lung disease, was widely used as insulation in the 1940s and 1950s and is still firmly embedded in school buildings all over the country. In Tennessee alone, Senator Sasser said, $6 million has been spent and an estimated $46 million more is needed to remove asbestos from school buildings.

Many school systems in Tennessee and other states can't afford the cost of removing the asbestos from their buildings. Others can and some may elect to go ahead and remove their asbestos without relying on federal assistance. The systems that don't have the money to remove the asbestos on their own should receive federal assistance as soon as possible. Congress should act on the matter without further delay. The health of children is at stake.

Lead:
EPA to Cut Gasoline Levels

Lead, one of the most important raw materials in modern society, is also one of the most pervasive toxins in the environment. The symptoms of acute lead poisoning were recognized as long ago as the first century A.D.—convulsions, coma and death—but cases of acute poisonings in the United States are now rare. The effects of chronic exposure to lead, however, are also frightening. The metal accumulates in bone and soft body tissues, and works most strongly upon the blood-forming, nervous and kidney systems. Exposure even at low levels can cause fatigue, headaches, poor appetite, clumsiness and diminished mental capacity. Young children are particularly at risk, since their nervous systems are not yet completely developed; studies have shown a correlation between high blood levels of lead and learning disabilities, mental illness and lowered intelligence in children. There is also some evidence that lead may affect the reproductive systems, disrupting the ovarian cycle in women and producing impotence and lowered sperm counts in men.

The most widespread airborne sources of exposure to lead are auto exhaust fumes and industrial facility emissions. Other sources include lead-based paints, thinners and solvents, and cigarette smoke. (In the 1960's, the problem of lead poisoning in children who were eating chipping paint from the walls of older, deteriorating housing came to light and received much publicity.) A growing worry is that the level of lead in foodstuffs is increasing to dangerous levels. Lead is absorbed by plants through their roots from the soil and through their leaves from the air. It may also be transferred from machinery to processed foods.

The Environmental Protection Agency began restricting the lead content of gasoline in 1973, and has recently announced its intention to further reduce the maximum allowable amount. (Lead raises the octane level of gasoline and reduces the problem of engine "knock." Without it, larger amounts of crude oil or more expensive additives are needed to produce fuel with a comparable octane level.) In its announcement, the EPA acknowledged that studies had shown a nearly exact correlation between lead levels in gasoline and the average lead level in the blood of the U.S. population. It also reported that approximately 97,000 children currently require medical attention because of lead in their bloodstreams. Although the EPA had halved the maximum lead content of gasoline in 1982, to 1.1 grams per gallon, the level of lead exposure did not subsequently decrease as far as expected. Part of the problem can be traced to the widespread practice, among owners of newer cars equipped with antipollution devices, of "fuel-switching." The EPA estimates that 13% of vehicles on the road have been illegally altered to accept the cheaper leaded fuel.

BUFFALO EVENING NEWS
Buffalo, N.Y., April 4, 1982

The Environmental Protection Agency is to be congratulated for once on its decision to maintain its strict regulations on the amount of lead in gasoline. The decision is in welcome contrast to some of the anti-environmental moves made by the EPA under the Reagan administration.

The proposal to relax the rules on lead had come as part of an overall administration plan to re-examine many federal regulations and repeal those that were found unnecessary. Some of the restrictions on lead were to have been dropped on the grounds that they were no longer necessary, since the use of leaded gasoline in the nation was declining anyway. All new autos are designed to use only unleaded gasoline.

Everyone agrees that lead in the atmosphere poses a serious threat to health — especially to children. A new federal report concluded recently that "the problem of pediatric lead poisoning in the United States has not been solved." It said that 26,500 of 502,900 children screened in 1980 had toxic levels of lead in their blood and that continued efforts to reduce lead contamination would increase children's "margin of safety."

The use of unleaded, rather than regular, gasoline has not been increasing as much as EPA officials had expected. One reason for this is that new car sales have plummeted, and many older cars, which use leaded gasoline, remain on the road. Another factor — one hard to evaluate — is the illegal use of leaded gasoline in newer cars.

About a fifth of the lead released into the environment comes from gasoline, and EPA Administrator Anne M. Gorsuch is wise to recognize that the fight to free the air of lead is not yet won. Lead poisoning is known to cause birth defects and when absorbed into the body through breathing or eating can result in slow and irretrievable brain damage. The curbs on lead in gasoline should be maintained until more progress is made in removing this poison from our environment.

The Washington Post
Washington, D.C., March 22, 1982

NINE YEARS AGO, the federal government began a program to reduce the lead in gasoline by stages. Starting with the product of the largest refiners, the amounts of lead allowed in gasoline have gradually been lowered. The last stage, application of the tightest standard—that is, the one permitting the least lead— to the product of small, independently owned refineries was to have taken place later this year. But now the Environmental Protection Agency has proposed to extend the October deadline and is considering whether to raise the allowed levels or rescind the lead standard altogether.

The impetus for this move comes from the independent oil refiners, who have opposed the lead standard from the beginning. For the first time in four administrations, they have found a receptive ear at the regulatory relief task force headed by Vice President Bush. In the face of overwhelming evidence to the contrary, the refiners no longer argue that lead does not constitute a major health hazard. Their argument is now based on the inequities created by loopholes in existing law.

The exemption for small refineries also covers "blenders," businesses that are not bona fide refineries but that can make a profit by buying gasoline and mixing it with high concentrations of lead. The refiners' objection to this perversion of the law's intent is valid. But their proposed solution is itself perverse. The blenders' loophole will vanish automatically if the October deadline is observed. The refiners also object to the exemption allowed for imported gasoline, but that also could be easily corrected.

Finally, the refiners argue that leaded gasoline use will gradually decrease anyway, since most new cars require unleaded gasoline. But depending on what EPA decides to do—drop the standard or relax it for some or all refiners—the amount of lead in the air will certainly stop declining and, at least for some years, increase. There is little reason to think that resulting savings to oil refiners would be worth the added health risks.

Lead is one of the most potent poisons in the environment. The blood lead level considered by doctors to be safe has dropped by half in the last 20 years, and is still being lowered. Quite apart from the disastrous effects of acute lead poisoning, effects on adult reproductive systems are being found at levels not too much higher than those found to occur in average urban dwellers. The effects on children's mental development may include lifelong, irreparable damage. Recent studies, though not definitive, have turned up evidence that lead levels found in "average" urban children result in lower IQ and various kinds of behavior associated with learning defects.

On the basis of evidence available two years ago, the National Academy of Sciences concluded that "a serious effort should be made to reduce the baseline level of exposure to lead for the general population." Reducing lead in gasoline is the easiest and cheapest way to do that. EPA should look long and hard at this "relief" plan: relaxing the lead standard is not a safe or sensible thing to do.

DESERET NEWS

Salt Lake City, Utah,
August 3-4, 1982

For the second time in less than a decade, Washington evidently is going to tell gasoline producers to get the lead out.

This move reflects the extent to which public health is still being impaired by lead spewed into the air after the gasoline containing it is burned by car engines.

And it reflects some degree of the flexibility of the Reagan administration, which previously had indicated present lead controls should be relaxed because they were bankrupting small refiners.

Even so, the administration's about-face still looks like a case of using the wrong methods to achieve an otherwise worthwhile objective.

Several years ago, the U.S. Environmental Protection Agency ordered the refiners to reduce the amount of lead in their gasoline by 60% from 1975 to 1979.

This week, the EPA indicated it will propose new regulations mandating the removal of an additional 31% of the lead in gasoline over the next eight years. This would be accomplished by applying only to regular gasoline a limit that now is applied by averaging a refinery's production of both leaded and unleaded gasoline.

The public has a right to be concerned about the amount of lead in the air it breathes. About 90% of the lead in the atmosphere comes from leaded gasoline. Lead poisoning can cause kidney and brain damage as well as anemia and mental retardation. In extreme cases, it can cause death.

The most urgent part of this problem is the exposure to children. They are less tolerant of lead than are adults, and they are more likely to come in contact with contaminated roadside dust while playing.

Repeated surveys have indicated that a significant number of American children are absorbing excessive quantities of lead. One recent government study said 675,000 pre-school children alone currently have too much lead in their blood. If the government had gone ahead with its previous plans to relax controls on leaded gasoline instead of tightening them, another study predicted an additional 500,000 cases of lead poisoning among pre-school children in 1983.

But there are other ways of dealing with this problem. The use of unleaded gasoline is expected to keep increasing because more and more cars cannot use regular fuel, which damages their pollution control devices. Some motorists, however, are disconnecting those devices so they can continue using leaded gas. Ethyl and duPont are sid to have designed tailpipe devices to filter out lead emissions. Detroit also is working on a car engine able to use leaded gas and still not pollute.

Shouldn't the government base its lead limits on what comes out the tailpipe rather than on what goes into the gasoline?

THE SACRAMENTO BEE

Sacramento, Calif., March 26, 1982

Three years ago, a medical researcher at the University of Pittsburgh named Herbert Needleman reported the results of an imaginative study whose conclusions had long been suspected: that there is a direct correlation between the amount of lead a child ingests and his or her problems in thinking, learning and behavior. Those findings have since been replicated by other research in Germany and England.

Roughly a year later, Vernon Houk, acting director of the federal Center for Environmental Health, discovered that between 1977 and 1980, the period when, because of federal regulations, the lead content of gasoline dropped sharply, the mean level of lead in blood samples taken from children in 60 U.S. cities declined by more than 25 percent. "We know," said Houk, "that this was a real reduction (and) the only thing we know that changed in the environment during that time was that there was a 30 percent reduction in the amount of leaded gasoline used in this country."

Leaded gasoline is not the largest source of lead in the environment; the chief sources are paint, soil and dust. But according to the journal Science, airborne lead, which comes primarily from auto exhausts, may account for 20 percent of the lead found in blood samples; it is also the most widespread source of lead, and is the most easily controlled. Nonetheless, EPA, the federal Environmental Protection Agency, following the anti-regulatory course of the Reagan administration, and under strong pressure from some refiners, is now considering dropping federal controls on lead in gasoline.

The controls do not prohibit the use of lead, but they restrict it in such a way that large refiners may not produce gasoline that contains, on average, more than .5 gram of lead per gallon. Since half the gas they make is unleaded, leaded gasoline contains about one gram of lead per gallon.

So far, small refiners have been exempt from those requirements, but if they're not changed, they will have to begin complying in October, something which they claim would be prohibitively expensive. At the same time, an economic study done for EPA indicates that if all federal limits on lead content were lifted, the major refiners would gain a windfall of $133 million in the next two years while small refiners would gain about $62 million in 1983.

When the refiners first went to the administration with their complaints about the lead standard, they were asking for what Science calls "a modest revision of definitions and deadlines ... (But) the warm reception they received encouraged them to ask for more. They said they wanted complete abolition of the lead standard."

In light of what's suspected about the connection between lead and a variety of serious medical problems — and in light of what's already known — it would be unconscionable to abolish the lead standard. Houk says that no childhood disease begins to approach lead poisoning in the breadth of its impact, an impact that appears to become greater with poor nutrition and which may therefore get worse as nutrition programs are reduced. "We have demonstrated that we can control lead in gasoline, and it just seems prudent that if you can do it, you should."

EPA holds hearing on the matter April 1 and 2. There is no question about what the decision should be.

The Morning News

Wilmington, Del., April 28, 1982

UNDER THE Reagan administration, there has been a general softening of requirements for a clean and safe environment.

The lowering (by legislation or regulation) of many standards has been justified on grounds of "keeping government off our backs" and "helping the economy along." In some instances, there is merit to these justifications; in others, there is none.

Into the "no-justification" category falls a proposal by the Environmental Protection Agency to drop its regulations regarding limits on lead in gasoline. Such a change would "risk intellectual impairment of large numbers of children," according to the American Academy of Pediatrics.

Dr. Glenn Austin, president of the AAP, points out that atmospheric lead, mostly from leaded gasoline, has overtaken flaking paint as a significant source of lead poisoning. Children — and adults too — inhale atmospheric lead from road dust, and lead absorption can impair intellectual development in children as well as cause serious illness.

Studies have shown that the average blood-lead levels in children decreased proportionately with the phasing in of unleaded gasoline. That process could be reversed if the EPA changes the requirements for lead in gasoline.

For the national government knowingly to sanction any step that could impair the healthy development of our youth would be foolish indeed. The pediatricians' advice should be carefully evaluated before the EPA makes any changes.

• • •

POSTSCRIPT: The Delaware Adolescent Program, Inc. (DAPI) has never had a really easy time of it. Money has always been a problem. Recently just finding a place to run the program has been tough. Now, this exemplary program for adolescent pregnant girls and their babies has been visited by one of the worst fates that can befall a Delaware organization: Rep. Al O. Plant has decided that DAPI needs his inimitable "help," including a board of directors more to his liking.

FORT WORTH STAR-TELEGRAM
Fort Worth, Texas, August 5, 1982

Lead in gasoline improves the performance of automobile engines. Lead in the bloodstream can reduce the performance of a child's brain. Enough of it in the blood can kill the whole human engine.

No reasonable person would admit to placing a higher priority on automobile engine performance than protecting the mental capacity and physical health of children. That is why so many Americans, not just environmentalists and public health professionals, were so concerned about the Environmental Protection Agency's announced proposal early this year to relax its rules aimed at reducing the lead content in gasoline.

The proposed relaxation of the lead reduction rules were part of the Reagan administration's thrust toward reducing the trouble and expense to industry of excessive regulation. After examining the evidence on relationship between high levels of lead in the air and the public health, the EPA fortunately has reversed itself and is proposing regulations to speed up the process of getting the lead out of gasoline.

Leaded gasoline is not the only source of such contamination. The Texas Air Control Board has released findings of a study it recently completed in Dallas that shows that a lead smelting plant in that city is generating toxic lead emmissions that are more than twice the safe level permitted by the EPA. About 300 pre-school children living within a half mile of that and another lead smelting plant were found to have high levels of lead in their blood in a study conducted by the city of Dallas in 1981.

Some environmental officials believe that thousands of West Dallas residents could be asbsorbing dangerous levels of lead from the smelting operations. It is crucial that the EPA and the TACB monitor such situations and take appropriate action to correct them.

The EPA's new proposals to speed up getting the lead out of gasoline is an indication that the agency now is placing due emphasis on the overall lead pollution problem, which, according to a recent survey conducted by the National Center for Health Statistics, affects about 675,000 children. Some environmentalists, however, have expressed fears that there might be too many exceptions in the new rules.

Every one who cares about the physical and mental health of children should take exception to any exceptions that would impede the process of getting the lead out of the air.

The Houston Post
Houston, Texas, June 4, 1982

Lead poisoning is draining the intelligence of small children in cities throughout the country. One in 25 preschoolers, from all levels of family income, has enough lead in the blood to do irreversible brain damage. Lead is all around us. One tankful of leaded gasoline throws as much as 2 ounces of lead into the air through the exhaust pipe. The rooms and halls of 28 million buildings are painted with lead-based paint. Small children metabolize lead more easily and store it longer than adults.

It is the nation's children who are being harmed. It is the nation's adults who are paying the present costs. But in the long run, if the poisoning continues at its present rate, the nation itself will be weakened through generations of less intelligent adults.

For years lead was the chief anti-knock additive in gasoline. In 1979 the Environmental Protection Agency required that the lead content be phased down to the present 0.5 grams per gallon. With the Reagan administration call to get the government off our backs, the Presidential Task Force on Regulatory Relief zeroed in on the low-lead regulation. Because several smaller refiners complained that they could not meet the 1982 standards and stay in business, EPA Director Anne Gorsuch considered exempting them. When large refiners protested that they would be at a competitive disadvantage if small refiners were exempt, it was proposed that the EPA simply scrap the standards on the lead content of gasoline.

But scientists fought back in vigorous protests to Congress and to the EPA. And to the surprise of Washington, many of the major refiners did not show the expected enthusiasm for a lifted lid on lead. They had accepted the original regulations. It was the unfair exemption to competitors they had objected to. Small refiners who have already met the standards are siding with the big refiners on this point.

Meanwhile, studies done for the EPA have shown a strong correlation between lead levels in gasoline and lead poisoning in children. And the cost-benefit analysis required by the administration for all major regulations has shown that to eliminate the standards would save industry about $100 million a year, but would cost the nation from $140 million to $1.4 billion to treat the additional 200,000 to 500,000 children likely to get lead poisoning. A University of Illinois study estimates that lead poisoning now costs the country up to $1 billion a year. Eighty percent of this goes to special education for children whose brains have been impaired, 20 percent in lost productivity from workers who were exposed to lead in childhood. Surely this is one federal regulation that should not be dismissed.

DAYTON DAILY NEWS
Dayton, Ohio, June 1, 1982

By the tens of thousands the children are poisoned. Nowadays the poison doesn't usually kill them as it used to. But it damages their brains, their ability to learn, as well as their health.

The poison is lead.

The federal government's progress against lead poisoning over the past decade slowed down last year. Then President Reagan's Environmental Protection Agency told gasoline manufacturers they could use a formula that would permit them to have 10 percent more lead in their fuel. That would spew about 10,000 tons of lead into the air, according to one consumer group.

Now the EPA says it is reviewing that rule interpretation. It ought to. For this administration also is cutting back funds for programs that screened children to test the amount of lead in their blood.

Though the level of lead in youngsters' blood has been reduced 26 percent since federal controls, the damage is still extensive. Some 43 percent of black children in poor city neighborhoods and 23 percent of white suburban children were found to have too much lead in the blood. Studies also are showing it takes less lead than previously thought to hurt learning ability. And there isn't much success in reversing the damage if the threat isn't found early. Prevention has worked; cures haven't.

The Wall Street Journal recently cited a study estimating that lead poisoning costs the nation $1 billion a year, most of that in special education for children with learning disabilities and most of the rest in a loss of productivity among workers exposed to too much lead when they were young.

That's far too great a price to pay for the convenience of industry and motorists, who have safer alternatives to heavily leaded fuels.

Dayton, Ohio, August 10, 1982

The clean air buffs were caught off guard. In an unexpected switch, the Environmental Protection Agency was reported ready to propose new rules to get more lead out of gasoline faster.

For that, a lot of praise is due, though praisers these days have turned cautious to make sure they are not being taken. Interior Secretary James Watt got a hearty round of applause from conservationists, too, when he first announced a plan to protect wilderness lands; when the plan turned out to put those wildernesses in jeopardy later, the applause became an embarrassment.

If the rules are as improved as reports say they are, though, EPA can take credit for having enough sense to abandon its previous plans to weaken lead regulations. Excessive lead whacks the brain. Lead-laden air especially hurts children who have to breathe it, especially in inner cities where car and truck smoke hang around. Studies have shown how severely lead can hurt mental performance and even cause retardation.

Other studies also have shown that lead in the blood has dropped at the same rate that the lead in gasoline was reduced. That is a direct benefit of government action.

Some deregulators in the Reagan Administration had supposed that "the marketplace" would take care of the lead problem, that since more cars were using unleaded fuel, regulations could be eased. That assumption can't be trusted. Manufacturers can still turn out cars that use leaded gasoline if those cars can meet certain air quality standards.

The best way to protect people is to keep reducing the amount of lead in leaded gasoline, as the EPA is now proposing to do.

Des Moines Tribune

Des Moines. Iowa, June 14, 1982

Lead poisoning in children, which can lead to irreversible brain damage, is not the problem it was in Des Moines a decade ago. But nationwide, it's a greater threat than it was earlier thought to be, the National Center for Health Statistics reports.

Four percent of preschoolers, from the middle-class suburbs as well as inner-city neighborhoods, have dangerous levels of lead in their bloodstreams. That's 50 percent more than health officials had expected. But in Des Moines, public-health nurses are finding high levels of lead in the blood of only one-half to one percent of children checked at well-baby clinics.

The poisoning is usually traced to paint, although auto exhausts contribute to the problem. Children ages 1 to 3, who commonly put in their mouths anything they find, sometimes eat paint that flakes off walls, moldings, stairs and railings. Before World War II, house paint contained high concentrations of lead; hence the problem is greatest in older homes. Federal law now prohibits sale of paint with high lead content.

In the early 1970s in Des Moines, public-health officials conducted blood-lead checks on as many as 3,000 to 4,000 children per year in the "lead belt" of older homes in the inner city. Dangerous levels were found in as many as 12 percent of those checked, said Dr. Julius Conner, Polk County Public Health director. Owners of homes where high-lead paint was found were told to remove the paint or to panel over areas where such paint was less than four feet from the floor or in other areas where small children could reach it. Housing inspectors still routinely check for lead.

Meanwhile, the U.S. Environmental Protection Agency forced oil refiners to reduce the amound of lead in gasoline. An NCHS study showed that, between 1976 and 1980, there was a 36-percent drop in the level of lead in children's blood, directly related to the cleanup of auto emissions. That could change. A Reagan task force on regulatory relief wants the EPA to allow more lead.

The measurable cost of lead poisoning in the United States has been estimated by the University of Illinois to be $1 billion per year, most of that going for special-education classes for children mentally impaired by lead poisoning. Not measurable is the pain and anguish it causes. Devra Lee Davis, director of the Environmental Law Institute, said: "Lead won't kill these kids. It will make them a little dumber, make life a little harder. . . . Lead just drains the quality of their lives."

Contributing to that drain in quality is an unacceptable price to pay to sweeten the profits of the oil refiners.

The Register

Santa Ana, Calif., September 28, 1982

Recently the Environmental Protection Agency did an about-face and tightened lead restrictions on gasoline. A regulatory review commission headed by Vice President Bush had recommended *relaxing* the rules to enhance the mileage and performance of American automobiles.

This EPA action bought momentary relief for Anne Gorsuch & Co. from their persistent critics in the environmental lobby. The lead-in-gasoline issue had an emotional undercurrent stemming from studies showing that slum children were being condemned to lifetimes of mental torpor by the concentrations of the element entering their bloodstreams and thus their brains. Its presence in high-octane gasoline was *ipso facto* to be dreaded.

Just when we thought the *Wall Street Journal* was alone in refraining from hosannas over the EPA shift, along comes the latest *Access to Energy*, a monthly free-market polemic published by Petr Beckmann, a retired University of Colorado professor.

The essential scientist, Beckmann argues that the EPA move was dictated by politics but not by science. He is as concerned as anybody about the health of poor children surrounded by auto exhausts in urban areas and thought to be particularly prone to harm by gasoline lead. But, he points out, a thorough analysis discounts the role of automobile sources in the lead poisoning of children.

In a brief but well-footnoted account,

Beckmann shows that environmental activists were looking at only half the picture when they argued that lead levels in urban youngsters had fallen since the phase-out of lead from gasoline began. In fact, those levels were falling even as gasoline lead was approaching its peak usage in 1970.

The real villain in the medical story appears to be the lead-based paint still peeling off the walls of old tenements, where it is ingested by young residents.

The political story is a different one. Big Media bought the environmentalist pseudo-science, as usual, and created a climate ripe for domestic consumption of the EPA's ultimate decision. Indeed, as a deputy to Gorsuch framed the issue in a memo to her, "The reaction of the general public will be positive....No special health effect report was prepared because the main part of the package constituting final rule making was *not promoted primarily by significant scientific considerations*" (emphasis added).

Thus, Beckmann would argue, we will burn more gasoline driving cars that emit higher levels of other pollutants, all in slavish devotion to a politically potent but unscientific notion. The economy suffers from such perverse behavior, and when that happens, the opportunities for children (and others) to escape slum conditions — and the old paint that deadens their intellects — are diminished. *That* is an environmental tragedy.

Detroit Free Press

Detroit, Mich., August 27, 1982

A PERSON should listen to his enemies, says the proverb, because they will tell him things his friends won't. That certainly holds true for the Environmental Protection Agency (EPA) whose foes in the environmental movement pushed it to change its mind about a perilous proposal allowing more lead in gasoline.

Had the EPA gone ahead with its original plan to weaken the rules on lead levels in gasoline, it could have had disastrous effects. Lead poisoning is not one of those obscure maladies encountered only in the pages of medical journals. Some half-million U.S. children under the age of five suffer from undetected lead poisoning, which can damage the brain and kidneys.

The tradeoff the EPA first wanted to make was both cruel and unconscionable: More profits for the small refineries that claimed regulations were driving them out of business in exchange for less protection for children from the possibility of absorbing air-borne lead.

But the EPA has made a lot of enemies among environmentalists who closely scrutinize and respond to its proposals. When

they flooded Washington with objections to dropping the lead limits, that proposal was replaced by one that imposes stricter rules on lead levels. The new proposals, in fact, would cut airborne levels by 31 percent in the next eight years.

The need for lead limits has not vanished during nine years of restrictions. Lead poisoning of children, even at the relatively low levels found nowadays, remains a potentially serious hazard. It is a particular menace to inner-city youngsters exposed to many cars and old, substandard buildings painted with lead-based paint. Even slight increases in any one source of lead can push such youngsters over their lead-tolerance threshold.

We can thank the environmental groups that battered the EPA with objections and statistics for keeping the EPA from abandoning a needed health and safety measure. And perhaps we should thank the EPA for introducing outrageous proposals that help keep the environmental movement alert and strong. In this case, at least, the EPA's enemies were really its best friends — and the country's children as well.

THE ATLANTA CONSTITUTION
Atlanta, Ga., August 25, 1982

The Environmental Protection Agency acted correctly Monday in proposing regulations that would require continued reduction in the amount of lead that refiners may include in gasoline. But why did EPA have to be forced to do what is so obviously right?

Scientific study after study has shown that airborne lead is a killer and crippler, especially of children, in urban areas where the air is already polluted. Children and adults who are exposed continually to lead pollutants show various long-term health problems, including decreased mental capabilities, even retardation.

During the 1970s, the government began a program to remove most lead from gasoline. An initial step was to require new car engines to be designed to burn unleaded gasoline. Older cars' engines need leaded gasoline in order to perform satisfactorily. But these cars, by normal attrition, eventually will be replaced by newer autos, reducing further the amount of lead pollutants from gasoline.

Despite the known health dangers of lead, the Reagan administration announced last August that it would seek to ease the regulations that set lead levels in gasoline — in order, it said, to help small and independent refiners who couldn't afford the investments needed to meet the requirements.

Large refiners had to meet by October 1980 an EPA rule that each refined gallon of gasoline contain no more than 0.5 grams of lead. Small refiners were scheduled to meet the requirement by October 1982 before the Reagan administration proposed they be excluded from the rule.

The proposed EPA regulation would still permit some small, independent refiners to include as much as 2.5 grams of lead in each gallon of gasoline, but the rule also decreases the number of refiners in that category from 159 to 74. Other refiners still have to meet the 0.5-gram level.

The new rules, though a compromise, are far better than what the president first proposed. EPA officials emphasize that they adopted the changes after receiving hundreds of comments from the public opposing the administration's plan to degrade the lead rule.

Good. That shows flexibility. But why is it so often necessary for the public to demand the basic protections which the administration is supposed to provide in the regular course of its sworn duties?

The Dispatch
Columbus, Ohio, August 26, 1982

THE U.S. Environmental Protection Agency (EPA), in a responsible reversal of announced intention, is imposing stricter limits on the amount of lead that can be mixed into gasoline. Although the change is provoking howls of protests from oil industry officials, the din is being drowned out by the sighs of relief from those trying to combat lead poisoning in children.

The administration had targeted the lead-level rules last year as one area where the regulatory burden on business could be eased. But after examining the issue more closely, the EPA decided against a relaxation. Government studies found that 4 percent of American pre-school children presently have excessive levels of lead in their blood. Lead poisoning can cause kidney and brain damage, as well as anemia and mental retardation. One of EPA's own consultants reported that if the lead regulations were eased, another 500,000 cases of lead poisoning could result in 1983.

The lead is mixed into gasoline to improve its octane level and, therefore, its performance. Present regulations impose a level of 0.5 grams of lead per gallon over a refinery's entire production of both leaded and unleaded gasoline. Since the production of unleaded gasoline is expected to increase, the regulations would permit more lead to be added to leaded gasoline.

The new regulations will apply only to leaded gasoline, and the EPA estimates that the new restrictions will cut by 31 percent the 326,000 tons of lead that current regulations would have allowed refiners to use over the next eight years. Small refineries, the ones most troubled by the regulations, are being granted an extension of their exemption on the lead-level regulations. The new regulations are scheduled to take effect on Nov. 1.

The EPA's change of direction was not an easy one to make. The agency, along with the entire administration, is committed to eliminating rules and regulations which are unnecessary and which place a heavy burden on businesses and individuals. Hundreds of rules have been identified as candidates for removal and a special task force is looking to see whether they can, in fact, be cut.

The lead-level switch demonstrates that the administration is moving carefully in its drive to eliminate unnecessary regulations. This will enhance its credibility when it moves to rescind other regulations deemed to serve no useful purpose. The EPA deserves commendations for its action.

The Chattanooga Times
Chattanooga, Tenn., August 4, 1982

News being defined as the unexpected, the U.S. Environmental Protection Agency made news this week. Top officials in the federal agency propose to tighten standards on lead in gasoline, speeding the phased reduction of lead content significantly. Even more surprising, EPA administrator Anne Gorsuch is reportedly expected to approve the stricter rules.

The Reagan EPA has not been known for initiatives into stricter environmental control, but for movement in the opposite direction, toward the goal of lessening federal regulation of industry. Environmental groups were understandably cautious in their praise of the proposed regulation changes. Spokesmen said the worth of the stricter standards could be compromised by exceptions not yet outlined.

While Chattanooga does not have a demonstrated problem with lead in the air, the EPA initiative — if it is what it appears to be — is welcome news. Concentrations of lead pollution, highest in urban areas where traffic congestion fouls the air, take their greatest toll on children. A recent survey by the National Center for Health Statistics found that about 675,000 children in the U.S. had elevated levels of lead in their blood, a condition which has been found to cause mental retardation in the young.

The EPA had previously signaled its intention to relax lead regulations but is said to have reversed itself in light of medical evidence of the dangers presented by lead pollution. That is as it should be. The purpose of the agency is to protect the public health, regardless of its commitment under this administration to a goal of decreasing environmental regulation.

News reports several weeks ago revealed that a public relations expert — a friend of Administrator Gorsuch — has been drawing $221 a day to help the EPA counteract its image of being soft on polluters. Responsible regulatory reactions, such as the one on lead in gasoline, will do more to help that image than the continued waste of tax dollars on PR advice.

The Seattle Times
Seattle, Wash., August 12, 1982

GET the lead out used to mean hurry up, but for awhile it looked as though the Environmental Protection Agency was doing exactly the opposite in the effort to remove lead from gasoline. Commendably, EPA now appears likely to stand by, or even speed up, the timetable for getting the lead out.

Earlier this year, the agency proposed weakening and possibly abolishing the program to phase down the lead content of gasoline gradually. Lead is added to increase octane ratings, but it does not burn in internal-combustion engines and ends up polluting the nation's air, land and water.

Children who live near busy streets or highways — especially in inner cities — breathe lead in the air, or they ingest it from playground dirt. Lead accumulates in the body and has been linked with mental retardation and other health problems.

A study by the federal Center for Disease Control found a clear correlation between blood lead levels and lead content in gasoline. Another survey by the National Center for Health Statistics found about 675,000 American children with high levels of lead in their blood.

At congressional hearings last April on lead standards, a panel of doctors and scientists urged the administration not to allow greater amounts of lead in gasoline, citing the potential danger to public health and the environment. "To reverse direction is just appalling; it is insane," said Dr. Herbert Needleman of the University of Pittsburgh School of Medicine.

Last week, an EPA official disclosed that the agency would scrap any plans to weaken or drop the lead-phase-out rules, and might even recommend a 31 percent greater reduction over the next eight years than current rules require.

If the agency sticks to this pledge, it would merit praise for giving priority to medical and health considerations — which would be one small step toward polishing EPA's tarnished image.

THE KANSAS CITY STAR
Kansas City, Mo., October 27, 1982

The longer the Environmental Protection Agency looks at the potential health problems involved in burning leaded gasoline, the more it is inclined to limit the usage of this engine-performance additive. In a turnaround from the administration's usual policy on regulatory relief, the EPA now has decided to tighten the rules on leaded gasoline even more than it did in August. Quite a change from a year ago when the issue of leaded gasoline was on the Reagan "hit list" of restrictions considered economically damaging to the refining industry.

But at hearings on the question the EPA heard from the National Center for Health Statistics that 675,000 children in this country have elevated lead levels in their blood, with 18.6 percent of the black children in some inner-city areas so affected by excessive exhaust fumes. So in August the EPA opted to reduce allowable lead content by 31 percent over the next eight years.

The old regulations had set lead limits in terms of all gasoline produced, but that enabled refiners to increase the content as they turned out more unleaded fuel for newer model vehicles. The new rules allowed large refiners 1.1 grams of lead per gallon and smaller ones (less than 10,000 barrels a day) 2.5 grams. Now the EPA has further cut small refiners to 1.9 grams and even that differential will be phased out in eight months.

Last summer the big refiners complained that the differential gave the smaller operators a competitive advantage. Now the small refiners declare the latest changes will drive some of them out of business, cost 3,000 to 5,000 jobs and boost gasoline prices. Such arguments once would have been persuasive to Reagan regulatory agencies. But when the EPA heard testimony about mental retardation effects on children with lead poisoning, it chose instead to get tough on lead content in gasoline. Now it is essential to follow through with checks and inspections for compliance.

THE COMMERCIAL APPEAL
Memphis, Tenn., October 30, 1982

EVEN THE Reagan administration — not characterized by a finely honed environmental conscience — was concerned over a survey showing 675,000 preschool children with excessive levels of lead in their blood.

That survey is credited with changing the course of Environmental Protection Agency rules that limit the amount of lead in gasoline. Last year similar rules won a prominent spot on Reagan's regulatory hit list. But now, the EPA will begin enforcing new, stricter standards effective Monday.

The tighter controls will reduce the lead level by 34 per cent by 1990, compared to the projected level under present standards.

Almost any move to improve our environment costs dearly, with industry — and ultimately the consumer — making the payments. Siphoning the lead from our systems won't be any different.

Yet scientists who study toxins have given clear warning. They insist increased lead levels can suppress fertility in adults and retard brain development, intelligence and learning ability in children.

Those findings put the cost of limiting lead in gasoline into perspective. Even the current, deregulation-prone administration gets the message. Slowly poisoning our children seems to be the only alternative to paying higher prices for gasoline with less lead.

ENVIRONMENTALISTS, in fact, argue that the new rules don't go nearly far enough. They want to see the elimination of all lead in gasoline. EPA's proposal only "preserves" the current situation of lead usage, they say.

On the other side of the fence, the owners of small refineries say the new rules could put them out of business, boosting the fortunes of a few mammoth refineries and unemployment. Their predictions for gasoline price increases are higher, too; they warn that the consumer may pay as much as 3 to 5 cents a gallon more.

Lead boosts the octane level of gasoline inexpensively, bypassing the use of a catalytic reformer that does the same thing but loses a percentage of the gasoline while performing. The small refineries say that's a loss they can ill afford.

The new rules do make concessions to small refineries, which find it harder to absorb any extra costs. For the next eight months the limits applied to them will be more lenient, allowing time to phase in processing changes.

As with most significant environmental questions, there are no easy answers. Small business and the consumer apparently will suffer the most from the proposed changes. Yet ignoring the problem poses a greater, more costly threat.

The Courier-Journal
Louisville, Ky., November 5, 1982

CITIZENS concerned about public health and the environment have had few opportunities to cheer the Environmental Protection Agency since the Reagan administration took charge 22 months ago. All the more reason to applaud the EPA for scrapping plans announced earlier this year to let refiners add more lead to gasoline.

Admittedly, the reversal came awfully close to election eve, and only after a storm of public protest. Even some in the refining industry objected on public-interest grounds — as opposed to the "unfair competition" argument that drew more complaints — to EPA's proposal that small refiners be given a special break in adding this highly toxic metal to their gasoline.

But the most important argument was a scientific finding, buttressed by other studies, that 675,000 American children had elevated levels of lead in their blood — much of it attributable to airborne lead from gasoline. The Center for Disease Control also found that improvements in blood levels corresponded closely to a decline in the lead content of gasoline produced under requirements of the federal Clean Air Act. (Excessive lead in the blood can cause mental retardation and other serious afflictions, as physicians and others were quick to tell the EPA.)

It's cheaper in the short run to boost gasoline octane levels with lead than by other means. But questions of morality aside, nothing can be more expensive than ignoring the physical and mental health of children. So the end result, a tough rule that treats all refiners equally and will reduce the level of airborne lead considerably, is something worth celebrating.

Minneapolis Star and Tribune
Minneapolis, Minn., November 5, 1982

So casual is the Reagan administration's approach to environmental responsibilities that we worried last summer when the Environmental Protection Agency decided to rewrite rules about lead levels in gasoline. Images of lead-laden children came to mind. What poisons would the EPA now permit industry to spew forth? The outcome is surprising.

The new rules governing leaded gasoline, which took effect this week, are a model of responsible environmental regulation. The standards are designed to ensure reduced levels of airborne lead. By resisting the anti-regulatory tide and attacking the lead problem with the vigor it deserves, the EPA has significantly reduced the risk of lead poisoning in children.

Earlier this year, the EPA considered relaxing present limits on lead in gasoline. But new medical evidence showed that 675,000 preschool children have excessive levels of lead in their blood, mostly attributable to automobile exhaust. And the agency's surveys showed that current lead rules are so badly written that lead levels in the air are rising.

The EPA responded in August by proposing new rules that would reduce airborne lead levels 31 percent within eight years. The plan set gasoline lead limits for large refiners at 1.1 grams per gallon and for small refiners (those producing less than 10,000 barrels a day) at 2.5 grams.

The final rules implemented Monday go even further. When large oil companies complained that the break for small companies would result in a price advantage, the administration tightened the small-refiner standard to 1.9 grams per gallon. The small-refiner break will be phased out altogether in eight months, when every refiner will have to meet the new 1.1-gram rule.

The new rules replace an ineffective, loophole-ridden system. They will eventually improve competition by eliminating the advantage given small companies. And by forcing lead reductions, the rules are sure to help protect America's children from lead poisoning. At a time when many crucial environmental regulations are threatened with repeal, the EPA's hard-line approach to the lead problem is encouraging.

THE SUN

Baltimore, Md., June 1, 1983

The need for preventing lead poisoning in children only occasionally enters the public consciousness as a major urban problem. Then awareness recedes in favor of business as usual: doing little by way of prevention, while treating the most seriously injured children after the damage is done. A bill before the City Council would mandate a costly lead-paint removal program. While the bill may be too ambitious in scope, it does promise an ongoing, public debate on this vital topic.

A series of articles in The Sun by David Brown reveals that as many as one-fifth of black inner-city Baltimore youngsters risk brain damage from lead poisoning, mostly from paint. Symptoms include mental retardation and behavioral problems. About 438 children with clinical lead poisoning were found in the city in 1982. Some 100 of them required hospital treatment.

But that probably is just the tip of the iceberg. Lower levels of lead poisoning cause subtle problems, such as small losses in intelligence. Traditional abatement — removing part of the old lead-based paint in houses and instructing residents in the need for frequent, rigorous clean-ups — isn't always as successful in reducing lead levels in children as once thought. More radical steps — replacing entire window casements and sills, for instance — may be necessary. The cost, though, to landlords and property owners could be prohibitive.

Councilwoman Mary Pat Clarke, a co-sponsor of the bill to require more thorough lead removal, believes the city should make this task one of its highest priorities, a cause worthy of floating bonds and creating loan funds, if necessary. Indeed, as Dr. Evan Charney, a Sinai Hospital lead poisoning expert points out, lead-poisoning on the scale that afflicts the inner city would not be tolerated for a minute in affluent suburbs.

But how to go about removing the danger is a most difficult problem. Some experts fear, for instance, that a massive lead-paint removal project could create dangerous clouds of lead-bearing dust. And at the moment, no one has developed accurate figures on the cost of a full-scale removal project, or how to finance such an undertaking. Perhaps the bill soon to be debated in the City Council will serve as a starting point for deciding how we can best accomplish the removal of lead-based paints as a hazard to Baltimore children.

The Dallas Morning News

Dallas, Texas, March 18, 1983

CONGRESSMAN Elliot Levitas has a proved knack for making headlines with charges of sloth and slovenliness at the Environmental Protection Agency. But are all Levitas' interpretations of EPA behavior as strained as in the Dallas lead-smelter case?

An Associated Press photograph shows Levitas, grim, half-standing, a presumably damning document in his hands, as he accuses EPA acting administrator John Hernandez of indifference to the plight of black children exposed to lead contamination in West Dallas. Hernandez spreads his hands in expostulation.

Unfortunately, the facts of the case don't quite live up to the drama. That lead contamination exists in West Dallas, everyone has acknowledged in the two years since The Dallas Morning News brought the matter to public notice.

The unresolved question is what to do about it. Levitas angrily accuses Hernandez of squelching a deal whereby the companies responsible were to clean up lead-contaminated soil. Indeed, the deal didn't go through. But, as Al Smith used to say, let's look at the record.

To begin with, the deal was far less firm than Levitas implied. In fact, it was no more than tentative.

Why didn't it go through? The main reason is that the facts were not all in. Hernandez' imputed intervention occurred 18 months ago. Eighteen months ago, the City of Dallas had barely begun blood tests to determine whether neighborhood children were adversely affected by the lead. The test results, announced in April 1982, showed that a health problem existed. But the tests couldn't and didn't pinpoint the source of contamination.

The City Council, though obviously concerned about the problem, refrained from urging immediate action against the smelters. For one thing, it seemed unwise to precipitate public panic where no certain cause for panic was known to exist.

A subsequent study by EPA and the national Center for Disease Control was released last Feb. 1. While it showed lead toxicity in 5 percent of black preschool children living near the RSR Corp. lead smelter, it showed, almost paradoxically, that preschool black children in Dallas had lower blood-lead levels than the national average. As for the Dixie Metals smelter, it was found that air-lead levels around it were safe.

The disease-control center recommended that in the RSR smelter area, the soil be quickly covered with sod.

Levitas seems to think that the EPA should have gone after the smelter owners first and investigated afterward. That may qualify in some quarters as good politics, but it isn't very good sense, unless charges are to be equated with proof.

As a matter of fact Levitas and other critics of the EPA keep making just that equation as they investigate the EPA. It was charges — never proof — of sloppiness that drove Mrs. Burford from the EPA. Now, as we see, her successor of two weeks is already on the hot seat.

This sort of inquisition is more than wearisome — it's harmful to rational discourse. It is apt to hinder rather than advance the search for a real solution to a problem that everyone wants solved.

Times-Colonist

Victoria, B.C., June 14, 1983

Lead, a soft, bluish-white metal that has been used by man since the beginning of civilization, has caused untold human suffering through the ages. Lead cooking vessels were used in ancient Rome as were lead pipes to carry drinking water, practices which undoubtedly killed many Romans.

Lead-related ailments increased in recent times as more industrial uses for lead were discovered, especially as a gasoline additive (tetraethyl lead helps engines to run more smoothly) and to color paint.

The trend to unleaded fuels has had a dramatic effect on lead levels in humans' bloodstreams.

The New England Journal of Medicine reports that lead levels in humans decreased by 37 per cent between 1976 and 1980, a period in which unleaded gasoline became widely available.

While encouraging, research shows there is still a long way to go. For one thing, the body simply has no tolerance for lead; there is no "safe level."

In submissions on phasing out lead use to Environment Canada, some oil companies cited the lack of conclusive (Canadian) medical evidence about the effects of lead on health.

But the environmental group, National Survival Institute, doesn't think the government should wait for Canadian medical evidence.

"Why re-invent the wheel?" asks NSI spokesman Jeannette Ackroyd. "Studies have been done. Are our children different from their U.S. and European counterparts? And what do the oil companies consider to be conclusive medical evidence?"

NSI's ideas of offering unleaded gasoline at a cheaper price than leaded fuel and phasing out lead use within a limited time frame are worthwhile. Environment Minister John Roberts should do his best to sell them to cabinet.

The Miami Herald
Miami. Fla., March 25, 1984

EVERYONE wants clean air, but the Senate Environment Committee got a bit carried away on the subject the other day. If the panel gets its way, a motorist who puts leaded gasoline into a vehicle designed to run on unleaded fuel will be guilty of a Federal crime. So will anyone who disconnects the pollution-control device on his auto.

If the criminal is but a hapless individual car owner, his Federal crime will cost him no more than a $2,500 fine. If he happens to be an auto dealer, mechanic, fleet manager, or service-station attendant, however, woe unto him. Then the penalty could run up to $10,000 per violation. Agents from the Environmental Protection Agency (EPA) would be empowered by this panel to slap such fines out on the spot, subject of course to appeals to courts.

No one would deny that air pollution — specifically from leaded gasoline — is a serious problem deserving Government regulation. But this approach is a tad tyrannical.

Most autos produced over the past 10 years contain catalytic converters. These anti-pollution devices filter the vehicle's exhaust. Leaded gasoline ruins the converters, however, so cars equipped with them must use unleaded fuel.

Unfortunately, unleaded fuel is a bit more expensive than leaded, so many people trying to save a few cents either have disconnected their converters or simply ruined them by using leaded gasoline. The result — lead-laden auto exhaust polluting the air. Lead poisoning causes brain damage and other health problems, especially in children.

It already is illegal for auto manufacturers or dealers to disconnect any pollution-control devices, and properly so. The Senate panel would extend the prohibition to individual car-owners.

Desirable as the goal is, this is an unwise way to achieve it. A far-simpler, more-effective, and much more desirable means to such worthy ends would be simply to outlaw leaded gasoline. Current law and EPA regulations aim at phasing leaded gasoline off the market by 1992. EPA Administrator William Ruckleshaus believes that timetable could be sped up.

That's what Congress should mandate. That would eliminate the problem at the source. It would avoid the foregone futility of trying to have Federal agents monitoring gasoline pumps around the nation lest some devious motorist grab the wrong hose.

A slap on the wrist to those well-intentioned but overzealous Federal regulators on the Senate Environment Committee who got carried away 16-2 with this foolishness.

The News American
Baltimore, Md., March 1, 1983

At about the time emissions testing began in Maryland, the Environmental Protection Agency predicted that a lot of cars, initially earmarked for unleaded gasoline, would be found to be using the leaded stuff. Many drivers, the EPA said, foolishly have been using leaded because it is cheaper and because they wrongly believe they will get better engine performance. What they will get, the EPA argued, will be bills for tune-ups and oil changes that will far exceed the savings in gasoline costs.

But now the EPA has come up with new ammunition that it intends to use for an apparent campaign to ban all leaded gas. EPA Administrator William Ruckelshaus says his agency has built a strong case that exhaust from cars using leaded creates a health problem for children living in congested areas — that it actually lowers their IQs.

Environmental and health groups have long pressured the EPA to do something about leaded gas, and the efforts have been bolstered in recent weeks by the publicity over ethylene dibromide (EDB). EDB, widely used as a pesticide, has been banned for most agricultural uses. More than 90 million pounds of it a year, however, are used as an additive to produce leaded gas, to help prevent the lead from accumulating in engine parts. The environmentalists are arguing that the elimination of leaded would dry up the market for EDB.

The arguments are persuasive. Certainly if leaded gas is a health threat it should be banned. And if, in the long run, it will cost the motorists more money, there should be no valid objection from that quarter.

Initially EPA was considering a rule to ban all leaded gas by 1990. Now it wants to speed up the process, and it might as well get on with it.

Richmond Times-Dispatch
Richmond, Va., April 4, 1984

It has been scientifically established that inhaling or ingesting lead can be harmful to one's health. Should we therefore bulldoze into oblivion all old buildings whose walls are covered with layers of lead-based paint? They are, after all, a major hazard. Wrote U.S. News & World Report in its March 19 issue:

"Common household dust often is the source of lead poisoning in children, according to pediatricians at Sinai Hospital in Baltimore. While most cases of severe lead poisoning are caused by ingesting chips of lead-based paint, dust containing tiny paint particles is the culprit among the vast majority of children with mild or moderate cases of poisoning."

Razing all structures contaminated by leaded paint would be an extremely expensive and disruptive way to solve the problem. Who would reimburse the property owners, and who would find homes for all the displaced people? Surely the Environmental Protection Agency would never advocate such a remedy.

Or would it? If the agency's staff can recommend burdening American motorists with millions of dollars in extra fuel costs to get rid of leaded gasoline, the health hazards of which are debatable, why should it be considered incapable of recommending the destruction of millions of dollars worth of lead-tainted buildings, the hazards of which are certain?

The EPA staff seemed to favor a total and immediate ban on leaded gasoline in a report publicized last week. As the basis for that view, the staff offered what appeared to be a wildly exaggerated claim that eliminating lead from gasoline would prevent lead-induced brain wave changes in 1.5 million children and reduce the number of children requiring medical care from lead-related illnesses by 45,000. In an attempt to reassure motorists, the report said a ban on leaded gasoline would reduce automobile maintenance costs and add no more than 2 cents per gallon to the cost of fuel.

Those figures were promptly challenged by the nation's top expert in leaded gasoline — the Ethyl Corporation of Richmond, which invented tetraethyl lead 60 years ago. In an appearance before the House subcommittee on public health and environment, Lawrence E. Blanchard Jr., vice chairman of Ethyl, declared that years of research have failed to discover "any harmful effects from lead from automobiles in the general atmosphere." And the cost to the motorist of eliminating lead from gasoline could be five times higher than the EPA staff's estimate.

Adding lead to gasoline has saved motorists millions of dollars over the years. Less crude oil is required to produce a gallon of leaded gasoline than to produce a gallon of unleaded fuel, and lead additives increase the octane of gasoline. Mr. Blanchard estimated the elimination of lead would add 5 to 10 cents per gallon to the cost of fuel. Moreover, there are about 50 million cars and trucks with engines designed for unleaded gasoline, and at least 30 million of these vehicles will remain by 1987. They can use unleaded gasoline, but it could be severely damaging if used for prolonged periods. Indeed, Mr. Blanchard reminded the subcommittee, General Motors has warned in congressional testimony that "'many vehicles on the road today require lead to avoid catastrophic engine failure.'"

The use of gasoline lead has declined sharply since 1970, when the EPA embarked upon a program to phase out the additive over a period of years. The program was boosted by the development of cars with catalytic converters, which require unleaded gasoline. As older vehicles disappear from streets and highways, the use of gasoline lead will continue to drop.

Congress and the EPA should allow this gradual process to continue. A total and abrupt ban on leaded gasoline would be unfair, uneconomical and unnecessary.

EVENING EXPRESS
Portland, Maine, May 9, 1984

This is National Clean Air Week, aimed this year, according to its sponsors, at drawing public attention to an aspect of air quality "that the individual can do something about."

Efforts are being made this week to convince automobile owners not to tamper with emission control devices or put leaded gasoline in vehicles designed to run on unleaded fuel.

Fine. Education is one way to improve the environment. But government regulation could be even more effective in this case.

We know, for example, that motor vehicles account for more than half of all air pollution in the nation. We know that automobile emissions account for at least 90 percent of all carbon monoxide and airborne lead emissions in U.S. cities.

The question is, what are we doing about it? Until now, not enough.

That may be about to change. Officials of the Environmental Protection Agency have announced they will propose "something like a ban" on leaded gasoline sometime this summer.

That's good, provided that "something like a ban" means, at the very least, a sharp curtailment of the use of gasoline containing lead in this country.

Knowing what we know about the poisonous quality of lead in the air, an outright ban would probably be better. Nevertheless, action in this area—is long overdue.

It seems clear that until leaded fuel is outlawed, many motorists will continue to maneuver around devices aimed at encouraging the use of the more expensive unleaded gasoline.

Education can help—especially in demonstrating to motorists the false economy of removing emission control devices from their cars—but in this case simple education clearly is no substitute for a strong public policy in the continuing battle for clean air.

The Birmingham News
Birmingham, Ala., May 9, 1984

The Environmental Protection Agency has announced that it will soon propose a ban on sale of regular, leaded gasoline. A long trail of public hearing and the like would have to be followed before such a ban would actually take place, but the EPA makes a strong case that it is needed.

For one thing, according to EPA Administrator William Ruckelshaus, there is widespread evidence that many motorists have tampered with their cars designed to burn only non-leaded fuel, with a consequent damage to air pollution control efforts.

Cars and trucks manufactured since 1975 have been required to have catalytic converters, which are destroyed by leaded fuel. However, an EPA survey found that 13 percent of the cars had been altered to use cheaper leaded fuel. This both destroys the emission control system and fouls the air — and the EPA says that vehicle exhaust still accounts for more than 90 percent of all the carbon monoxide and airborne lead in U.S. cities.

For another thing, recent studies have emphasized the danger that lead poisoning poses. The Centers for Disease Control in Atlanta says damage to health — and especially brain damage to youngsters — begins to occur when as little as 10 millionths of an ounce of lead is found in a quart of blood.

Finally, Ruckelshaus says that those who illegally switch to leaded fuel actually practice "false economy," since using the cheaper fuel damages car engines designed to burn unleaded. The Motor Vehicle Manufacturers Association has estimated that such damage and repair costs average about 19 cents per gallon of gas, as opposed to the about 7 cents per gallon saved at the pump.

All of which seems to make a strong case for banning lead entirely. It will be instructive, however, to hear opposing viewpoints in the controversy that is certain to erupt when EPA makes its proposal official.

The Pittsburgh
PRESS
Pittsburgh, Pa., March 17, 1984

Leaded gasoline is already slated to be almost completely phased out of the nation's gas tanks by 1990. But the Environmental Protection Agency is now considering speeding up that schedule.

One reason it cites is the mounting evidence of the health danger of lead in the environment, particularly to small children.

Although the amount of lead discharged into the air, mostly from vehicle exhausts, has been reduced to about a quarter of what it was in 1971, the agency knows more about lead than it did before. It now has "really solid numbers" about lead's threat to health, says EPA administrator William D. Ruckelshaus.

Another reason is the EPA's estimate that some 17 percent of U.S. motorists, whose cars are supposed to be fueled with unleaded gas only, are improperly using leaded. They do this both to take advantage of lower prices for leaded and in the mistaken belief that they will get better performance and economy.

There is no question that the environment will be better off when leaded gasoline is a thing of the past. However, there are still millions of cars on the road that were built to use it and many of them will be around for a number of years. The existing phaseout schedule is, in fact, set up to take that into consideration.

Furthermore, no matter how much or how quickly lead content is reduced, as long as there is a significant price differential between leaded and unleaded gas, those car owners who shouldn't be using leaded now will continue to do so, even if it doesn't do their cars any good.

Why not a tax on leaded gas to wipe out that price differential? If the EPA's own estimate is correct, this could result in an immediate reduction in the use of leaded fuel by as much as 17 percent, while still leaving it available for those whose cars need it.

THE BLADE
Toledo, Ohio, January 23, 1984

EVIDENCE that growing numbers of Americans are putting leaded gas in cars with catalytic converters is beginning to worry the federal environmental protection folks. Leaded gas destroys the converters and leads to a significant increase in emission of pollutants from vehicle exhausts.

Some motorists have the notion that leaded gas will give their buggies the zip they used to enjoy in the days when high-compression engines gulped high-test leaded gasoline. They should think again. An EPA spokesman said that motorists have the wrong idea if they think their catalytic converter-equipped cars will run more efficiently on leaded gas. The agency cites oil-industry tests that show unleaded gas is easier on cars, resulting in lower bills for such items as spark plugs and carburetor and exhaust-system repairs.

The EPA has no authority to crack down on individual motorists who switch to leaded gas in cars equipped with the catalytic converters, although it is illegal to do so in 40 states. The federal agency did fine the city of Philadelphia $327,000 for disconnecting pollution-control devices on — of all things — 131 police cars.

Greenville County, S. C., was fined $630,000 for tampering with some of its vehicles and filling them with leaded gasoline. And in Detroit the EPA nicked 17 gasoline blenders a total of $550,000 for putting too much alcohol in unleaded fuel as an octane booster. An improper alcohol-gasoline blend can increase emissions of pollutants.

The real problem, if EPA estimates are correct, is that 13 percent of American motorists are pumping lead into unleaded-only cars. The reason is simple enough. There is a five or six-cent price spread between leaded and the more expensive unleaded fuels. Oil company spokesmen maintain that it costs more to refine unleaded fuels, but to many motorists the explanations are unconvincing.

There are enough clunkers on the road to make it necessary to continue selling leaded gas, but one way to deal with the problem is to impose a federal or state tax on leaded fuel that will raise its price above that of unleaded gasoline. That would avoid the more drastic solution of banning the sale of leaded gasoline.

The only way to make certain that a catalytic converter is working the way it is supposed to would be through periodic inspections, whether once a year, twice a year or whatever experience would dictate. The efficiency of a converter can be determined quickly and simply; if it is not functioning, it makes no sense to keep it attached to a car that is violating air-quality standards.

In any case it is impractical to try to police motorists who find ways to bypass the unleaded-fuel collars on gas tanks or who induce mechanics to disconnect or disable their catalytic converters.

To burn leaded gas in cars designed for unleaded fuel unquestionably is a foolish way to try to economize. But the way to make that plain to everyone is simply to raise the price of leaded fuel and thus eliminate the incentive to do so.

Auto Emissions: Hydrocarbons, Carbon Monoxide, Nitrogen Oxide

One of the most controversial sections of the Clean Air Act concerns the standards and deadlines for pollutants from automobile exhausts. Under the legislation passed in 1970, the levels of three pollutants—hydrocarbons (HC), carbon monoxide (CO) and nitrogen oxide (NOx)—were required to become less each year, through the use of catalytic converters and other pollution control devices. The yearly standards were designed to clean up 90% of the hydrocarbons and carbon monoxide, and 75% of the nitrogen oxides emitted from vehicles, in order that cities could meet the national ambient air standards by 1975. In 1977, the Clean Air Act was amended to delay the deadline until 1982, or, for areas with severe auto-related pollution, until 1987. (See pp. 122–133.)

Those areas of the country that still could not meet the clean air standards by 1982 were designated "nonattainment areas," and as such were required to establish vehicle-inspection programs. These unpopular programs are intended to identify, through on-the-road inspections, those individual vehicles which still violate current emission standards, and to catch drivers who have damaged pollution control devices on their cars through the illegal use of leaded gasoline. (See pp. 66–77.) The EPA is empowered, if the state refuses to establish vehicle inspection programs, to cut off highway funds and other kinds of federal aid.

The health risks of breathing heavily polluted air include an increased risk of lung cancer and other respiratory illnesses such as emphysema, asthma and bronchitis. Although automobile emissions account for approximately half of all air pollution, they generate a much lesser percentage of the health hazards associated with such pollution. Nitrogen oxides, which are emitted by coal and oil-burning plants as well as by vehicles, increase the risks of acute respiratory disease; in combination with hydrocarbons, they also form urban smog, which exacerbates chronic respiratory illnesses. Carbon monoxide, when inhaled from heavy traffic, reduces the body's supply of oxygen by replacing the oxygen in the red blood cells. This can cause dizziness, unconsciousness or death by coronary disease. The particulates released in auto exhaust—solid particles of dust, soot or smoke—can lodge in the lungs or pass into the bloodstream; these often carry traces of toxic metals and carcinogens.

SAN JOSE NEWS
San Jose, Calif., April 10, 1980

SINCE last July, California has been under federal orders to adopt a system of annual automobile emission inspections. So far, the Legislature has rushed to obey that mandate with all the eager alacrity of a nine-year-old heading for an appointment with the dentist.

However, there are a few faint but hopeful signs that the state's lawmakers are finally perceiving the necessity of obeying the law and the wisdom of protecting the air.

A bill by Assemblyman Victor Calvo, D-Mountain View, which would provide for the establishment of annual inspection programs in the state's high-pollution areas, was approved by the Assembly Transportation Committee on Tuesday and now heads for the Ways and Means Committee, where a vote is expected late this month or in early May.

Calvo's bill, AB 2145, would permit officials of Air Quality Management Districts in locales which exceed federal air pollution standards — including the Bay Area — to request the creation of annual inspection programs.

Another pending bill by Sen. John Foran, D-San Francisco, would create an annual inspection program — but it also would lower California vehicle emission standards to the federally dictated minimum. That's a very dubious trade-off; the state Air Resources Board has calculated that the Foran measure actually would produce a substantial net increase in air pollution.

Under Calvo's bill, annual inspections in the Los Angeles area, where the smog is especially bad, would be done at special stations operated by a private firm under contract with the state. Other areas could choose between that "centralized" type of system or a "decentralized" one in which state-certified private mechanics would both perform the inspections and make any needed adjustments or repairs.

Under the centralized system, car owners would pay a modest fee, probably $11, for the initial inspection. If their car flunked they would have it fixed by a mechanic of their choice and then re-inspected at the central station for a lesser fee.

In theory, the centralized approach offers fewer opportunities for incompetence, carelessness and simple chicanery, and experience seems to bear the theory out. According to the Air Resources Board, a year-old centralized program in Los Angeles has been 10 times more effective in reducing pollution than an earlier program that relied on state-licensed private garages. A similar program in Phoenix, Arizona reportedly achieved a 25 percent reduction in carbon monoxide levels in three years.

However, in some situations a decentralized system might be more practical — for example, in relatively rural areas — and Calvo's bill is wise in letting each area make its own choice.

The main thing about the Calvo bill is that it would create an annual inspection system without trading away the state's admirably tough emission standards. Federal law requires an inspection program, common sense supports it, and sound environmental policy demands it. It's time for the Legislature to stop dawdling and pass it.

Rocky Mountain News
Denver, Colo., March 9, 1980

TO hear some members of the Legislature tell it, the U.S. Environmental Protection Agency is about the biggest villain to come down the pike since Alferd Packer ate some compatriots in the Colorado Rockies.

They contend — speciously — that EPA is devouring state self-rule with its insistence that something be done about auto air pollution, which is weakening the hearts, damaging the lungs and doing God knows what other damage to Front Range residents.

EPA, of course, is just doing what it's paid to do: Enforce environmental laws passed by Congress. And Congress has passed a law requiring states to reduce air pollution. The original deadline for new Colorado standards was 1975. Three times since then, that deadline has been extended, with the state itself setting the most recent deadline. Finally, EPA said enough is enough; meet the new deadline by taking some action — or federal funds for highway and sewer projects in the state will be withheld.

Congress was right and EPA was right, but they are not the real issues. The real issue is public health. The Legislature — whose expressed shock at all this is a farce, considering that it's known what might happen for years — should have enacted anti-pollution legislation long, long ago, even without EPA giving it a hard time. It's not unusual, by the way, for Congress to establish standards in one area or another (education, for instance), or for federal agencies to threaten to hold back goodies. That's the only way they have to force state compliance with federal laws. Usually, the agencies hold back far less than they threaten, if anything at all.

Maybe it sticks in the craw a bit that EPA has focused on a specific law it says it wants the Legislature to pass. But reading between the lines makes it clear that EPA is flexible — that what it's really seeking is meaningful action, and that it will happily let Colorado off the hook if the Legislature comes up with any serious bill that demonstrates a good-faith effort.

For a decade now, it's been clear that something had to be done about Front Range pollution. In fact, air pollution probably ranks as one of two or three of the most important issues Colorado faces. And what anti-pollution laws are on the books? None.

Some legislators — probably those who have been breathing Denver air too deeply — say, well, nothing needs to be done because Denver air is going to be getting better in another half decade or so; Detroit, they point out, is going to be required to make cars that pollute less. That argument is the equivalent of saying swimming pools shouldn't hire lifeguards this year because statistics show people will swim better next year. Those statistics could be wrong, and anyway, what about the interim? And even the most optimistic projections of reductions in Denver pollution don't forsee a time when the problem will have gone away. With the exception of a handful, our legislators have an insufficient sense of urgency about air pollution. Maybe the EPA's action will help.

The Courier-Journal
Louisville, Ky., April 11, 1980

GOVERNOR BROWN wasn't exaggerating when he called the General Assembly's neglect of an automobile emissions testing program "the most irresponsible" failure of the session. The legislature recessed last week and will return for final meetings Monday and Tuesday.

If lawmakers let the final two days go by without action, Kentucky may stand alone among the states in defying the federal Clean Air Act, and in refusing to protect the health of its citizens. Under the law, 29 states where urban pollution levels are high are being required to set up programs to inspect auto emissions.

Prior to 1980, 24 states had enacted laws to do so. Virginia, which set up a program to cover Richmond and the urban counties adjacent to Washington, is among the more recent to fall in line.

In three other states, legislation is pending. That leaves Indiana, which is working to set up a joint state-county plan, and Kentucky, where, if last-ditch efforts fail in the legislature, nothing at all is in the works.

Perhaps, if no bill is passed, effective local action might be taken to set up an inspection program. But that seems sure to be more expensive and tougher on local communities than a state program bolstered by Governor Brown's offer to help with the expense. And local inspection enforcement, as in Cincinnati, has all too often been a flop.

The need, in terms of the quality of life and health in Jefferson County and Northern Kentucky, is obvious. Properly run inspection programs are the best means of making a serious improvement in air pollution. The prediction that airborne hydrocarbons, which cause summer's choking ozone pollution, could be cut by one-third has been borne out by other cities. Carbon monoxide could be reduced even more drastically.

Legislators who can't understand that surely can understand the federal law that requires cutting off all federal support to community development. Air pollution permits required to start new industries also would be denied.

A loss of $140 million

The fund cutoff, for a beginning, could cost Jefferson County $87.2 million in highway funds and $2 million for sewer construction. The total in the Kentucky counties involved is estimated at more than $140 million. Louisville officials, who are pushing Governor Brown and the legislature for action, realize the community can't stand such blows.

The legislative stalemate has come about because the delegations from Jefferson and the three urban counties opposite Cincinnati (Boone, Campbell and Kenton) think it's bad politics to clean up the air. There are some honorable exceptions, such as Jefferson County's Senators Eugene Stuart and David Karem, and Northern Kentucky's Representatives Terry Mann and Art Schmidt.

But most have reacted like the chairman of the Jefferson County delegation, Representative Carl Nett, who still says "I don't think the legislature is going to let the EPA tell them what to do." Representative Nett is intelligent enough to know that that's exactly like a speeder saying no traffic cop is going to tell him how fast he should drive.

The U.S. Environmental Protection Agency deserves some abuse, not for enforcing the law, but for waffling and shilly-shallying about the penalties. The most recent bungle was a sudden announcement that four predominantly rural counties in Northern Kentucky — Gallatin, Grant, Owen and Carroll — as well as the three urban counties, are among the counties that must have inspection programs.

Inclusion of the rural counties, apparently because they're included with the urban ones on a bureaucratic air quality map, certainly is no help to those who would like to get a law passed for the counties that need it. Those counties aren't even classified as non-attainment areas.

Now the EPA, somewhat lamely, says the status of the rural counties is "negotiable." But there is no legitimate complaint against EPA for enforcing the law in areas, like Jefferson, Boone, Campbell and Kenton counties, where the law clearly requires it.

If we don't want laws against speeding or air pollution, the proper place to complain is to the legislative body that enacted them. In the case of air pollution, that source is Congress, which directed that areas seeking federal development funds meet clean-air standards.

There seems only a slim chance the General Assembly will recover in the two remaining days of the session and create an emissions program the easy way. Trying to do it locally will be much more difficult. But it will have to be done unless Jefferson County and Northern Kentucky want their economies to become as stagnant as the summer air.

DAYTON DAILY NEWS
Dayton, Ohio, January 2, 1980

It is the modern dilemma: Every solution is a new problem. Witness the struggle of the federal Environmental Protection Agency to set emission standards for diesel cars and light trucks.

Diesels are catching on with automakers and buyers alike for good reason. They get excellent mileage (thus helping solve the oil-import problem), rarely catch fire in collisions (and so improve safety) and emit little hydrocarbon or carbon monoxide (a boon to the environment, therefore).

The catch — and doesn't there always seem be one these days? — is that diesels spew out much heavier particulate emissions, soot, than do the familiar gas buggies. That's where EPA comes in.

No one is quite sure whether particulates from diesels pose a serious problem. Some environmentalists fear the health consequences, others merely denounce the dirt. Neither effect is pleasant to contemplate, though if the problem is only dirt, there could be some latitude in deciding what to do about it. It might be cheaper and more efficient just to wash it off the streets, for example, than to require expensive gimcrackery of questionable reliability on each vehicle.

For 1981 models, EPA has proposed standards that only the VW Rabbit diesel currently meets but that probably can be met by other manufacturers. The necessary turbochargers would run up the prices of vehicles $200 to $300 and would somewhat cut the diesel's mileage advantages.

The trade-offs may be arguable but are not unreasonable. For 1985, however, EPA is considering standards that industry observers agree no manufacturer currently knows how to meet and that could wipe out the oil savings altogether. That is clearly over-zealous.

EPA is caught, in this, between conflicting social imperatives. If one must give, the give this once should be in air standards. The diesel is a promising technology, currently offering more solutions than problems.

The trick for EPA, while keeping diesels from fouling the air wantonly, is to allow the technology to develop. The fuller opportunity it has to evolve as a gas-saving solution, the greater the chances will be a bit later for the technology to refine its air-saving possiblities as well.

The Wichita
Eagle-Beacon
Wichita, Kans., June 6, 1981

Congressman Dan Glickman, D-Kan., is right to call for a review of the impact of automotive catalytic converters on the environment as the Clean Air Act comes up for reauthorization this year.

Allegations that the pollution control devices hurt fuel economy have been common ever since they became standard equipment on many cars in 1975. There is considerable evidence to back up such claims, although, in the last year or so, advances in computer-controlled emissions feedback systems, coupled with more sophisticated converters, reportedly have made better gas mileage possible on some cars.

A more serious claim is that the devices actually have contributed more to pollution than they have reduced it. There have been allegations that catalytic converters are responsible for higher sulfur emissions than otherwise would be the case and, thus, have been a major contributor to the acid rain problem that plagues the industrialized eastern regions of the United States and much of Canada to the north.

Catalytic converters never have proven popular with American motorists, but generally were considered something that had to be tolerated in the effort to clean up the worsening air problem. If they now are implicated in making matters worse, that is all the more reason for a careful analysis of their role in reining in pollution. The Environmental Protection Agency and the House Energy and Commerce Subcommittee on Health and Environment should give the matter careful attention, as requested by Mr. Glickman.

The Washington Post

Washington, D.C., November 17, 1981

CONGRESS IS considering an interesting "first" in the 20-year history of environmental cleanup. Last year, the emissions standards for all three of the chief automobile air pollutants—hydrocarbons, carbon monoxide and nitrogen oxides (NOx)—were finally met. One year later, the question now being debated is an auto industry proposal that would undo this considerable achievement by rolling back the carbon monoxide and NOx standards to twice their current levels.

Congress' original 1970 goals for cleaning up auto air pollution eventually proved to be—as Detroit had predicted—overly ambitious. There were numerous delays and relaxations. But though light- and heavy-duty trucks are still unregulated, the auto emissions standards have already resulted in measurable—and in some cases major—improvements in the nation's air quality, especially in the big cities. Though performance on the road falls below the official standard, new cars must emit 90 percent less carbon monoxide and hydrocarbons and 75 percent less NOx than did pre-regulation cars. Contrary to expectations, this remarkable improvement has been made without cost in fuel efficiency. In fact, the sophisticated controls that were required actually increased mileage.

Why turn back the clock? Cleaner exhaust has not come free. Current industry estimates peg the cost of emissions control hardware at $250-$425 per car. If the standards are relaxed, some of the hardware can be removed. But the sticker price reduction will not be nearly as big as the cost, since the capital investments that have already been made must still be amortized. The saving for the consumer is likely to be in the neighborhood of $100. Another major rationale for the rollback is that the higher NOx standard is just the level current diesel engines can achieve without additional controls. General Motors, in particular, is counting on diesels to meet fuel economy standards. In fact, a number of studies—most recently one made by the Congressional Research Service—have concluded that only GM and Japanese and European manufacturers are likely to benefit at all from the rollback, while the ailing American companies would be further weakened.

Any proposal to relax health-related environmental standards should have to demonstrate pretty convincingly that the standard is unnecessary, technically wrong or impossible to meet. The criteria should be correspondingly higher in a case—such as this one—where a standard has been met and been shown to be effective, and where a sizable capital investment has already been made. The proposal to roll back auto emissions standards does not fit the bill.

THE BLADE

Toledo, Ohio, September 28, 1981

AMERICAN auto makers have their own economic survival in mind in their efforts to encourage Congress to ease the present auto-emission pollution laws. But aside from the economics involved, there are no compelling reasons for the laws as they apply to auto-emission standards to remain as is on the federal books.

Last week the industry made its case before a congressional subcommittee that the 1981 standards governing the discharge of carbon monoxide and nitrogen oxides — standards that are scheduled to remain in effect indefinitely — are unnecessarily stringent. The auto makers contend that anti-pollution equipment costing between $80 and $360 per car can be removed if Congress acts quickly to delay implementation of the 1981 standards set by the Clean Air Act.

Making American cars cheaper and thus more competitive against the flood of imports is an especially desirable goal, not least of all because of the need to preserve jobs in auto-related industries, including several thousand in Toledo.

Another case for more realistic rules was made recently in a study which pointed out that completing the task of removing the last traces of a pollutant can be more expensive than removing a majority of it (say, between 80 and 90 per cent). Auto makers argue that producing virtually pollution-free cars imposes an unrealistic cost burden on car buyers.

On top of the fact that Americans are fed up with costly, inefficient pollution devices, the environmental hazards that modified standards would produce are far less severe than they might seem at first blush. Since the Clean Air Act was approved in 1970, each year the number of older, polluting cars on the road has declined. So, even though auto makers now seek to loosen some of the rules, air quality should continue to improve as the nature of the nation's auto fleet changes.

At a time when most American car makers are fighting for survival, attempts to inject more reason into the Clear Air Act's standards are especially welcome. The proposed modification is just one more step that should be taken to help Detroit regain its pre-eminent position against overseas challengers.

Oregon Journal

Portland, Ore., October 16, 1981

America's air is becoming cleaner, thanks to laws requiring automobile makers to meet air pollution standards. But the country is at a crossroads, and Congress must decide if pollution standards will be rolled back.

The American auto industry is leading an assault on auto emissions provisions in the Clean Air Act. The act expires this year, but the standards for auto manufacturers will continue to be raised. For example, in 1984 all cars will have to meet a stringent high-altitude standard designed to reduce pollution in mountainous areas.

The industry's main selling point: Progress has been made in ridding the air of pollutants, but the gains scheduled to be made in the next few years will be expensive without making that great a contribution to improved air quality.

The Reagan administration may be sympathetic to the auto industry, which has been hit hard by high interest rates and overseas competition.

But the argument really is the same one made by the auto industry when it first faced tightened pollution standards — the price of motor vehicles will rise and the improvement in the air won't be that great.

The worst Congress could do is to go along with a General Motors plan which would average emissions under a so-called "bubble" system. Averaging would permit some autos to exceed the standard while the lost ground was made up on other models.

One possible result is that one auto maker, let's say GM, would go after the small-car market by dropping prices and pollution control devices on small models and charge more for larger models. The lack of air pollution control equipment on a vehicle would become a selling tool that could attract more buyers to small, polluting vehicles. The average wouldn't occur, and air quality would deteriorate.

Air quality isn't an area where we should keep compromising. Auto emissions still are the No. 1 single polluter. Gains are being made in cleaning the air, and they should continue as long as the standards don't become an unbearable burden. And the present standards are far from being unbearable.

TULSA WORLD

Tulsa, Okla., October 12, 1981

SOON Congress will resume the battle over clean air, with auto emission standards the key issue.

On one side will be the odd alliance between the environmental extremists and the makers of auto pollution equipment. Against them, predictably, is the U. S. auto industry.

With 1970 as the base year, U. S. automakers have succeeded in reducing exhaust pipe pollutants by something like 90 per cent. Under current law, emission standards will continue to be tightened to reduce emission pollution a few more percentage points.

The automakers say they have the technology to comply but are asking a question that should seriously be considered by both Congress and the American people; is the extra expense worth the few added percentage points?

For example, Ford Motor Co. estimates the 90 percent reduction in emissions has been achieved at a current cost to consumers of $500 per auto. Gaining the few extra percentage points would cost another $300 per vehicle.

Ford and the other automakers contend the emissions battle and air quality have been substantially won, and that as autos older than 1974 begin passing from the highways the overall pollution will continue to fall.

Therefore, they see no reason to force the rigid controls set out in the current act. Why, for example, should all U. S. vehicles beginning in 1984 be designed to meet strict emission standards for mile-high Denver?

By holding the standards at the 1980 level, automakers contend they can eliminate a lot of the power-robbing, fuel-eating equipment under the hood of present vehicles because of advances in electronic control of fuel mixtures and achieve not only a reduction in initial cost but greater fuel efficiency and better performance.

Yet environmentalists will portray this as a "roll back" in emission standards, as an abandonment of the commitment to clean air.

It is nothing of the kind. It is an attempt to get Congress to re-evaluate the clean air law in the light of common sense.

The Cincinnati Post
Cincinnati, Ohio, September 30, 1981

Congress ought to be wary of the auto industry's request to relax anti-pollution requirements for cars and trucks.

The changes would approximately double the amount of carbon monoxide and nitrogen oxides that could be emitted from exhaust systems.

It would be one thing if the companies were asking for relaxation of unmet requirements. But what they want is a rollback of emission standards that already are being met.

Supporters of the rollback argued before a congressional committee the other day that there would be no decline in air quality. How can discharging more pollutants into the air fail to make the air more polluted?

The auto companies also claim that savings resulting from leaving some of the anti-pollution devices off cars would be passed on to customers. That's dubious, too.

We don't believe in loading unnecessary requirements, such as airbags, on auto manufacturers. But it would be a step backward to relax anti-pollution standards that are being complied with and whose costs already have been built into auto prices.

The auto industry's big problem is not clean-air standards but rather its burgeoning labor costs. The automakers ought to spend more time trying to figure out how to get their labor costs down and less on badgering the government to reduce clean-air standards and erect trade barriers against foreign imports.

The automakers certainly haven't made a case so far for changing the carbon monoxide and nitrogen oxide standards. We doubt that they will.

St. Louis Globe-Democrat
St. Louis, Mo., August 29, 1981

Despite what might be heard from environmentalists, the law of diminishing returns has not been repealed.

It is working with a vengeance in the government program that continually requires automobile makers to upgrade emission-control equipment to reduce auto pollutants.

For example, in 1968 the auto emission controls required to reduce the hydrocarbon and carbon monoxide emissions 50 and 62 percent respectively cost only an average of $16 per car. By comparison, the cost in 1980 of reducing the oxides of nitrogen, hydrocarbon and carbon monoxide emissions averaged $118 per car.

Now the new, much more stringent set of emission control standards mandated under the Clean Air Act that would require a 76 to 96 percent in these emissions are adding an average of another $400 per car.

The question being raised by the National Academy of Sciences as well as by the auto industry is whether this costly effort to control the last increment of pollutants is worth the money.

A recent report of the National Academy of Sciences suggested that the carbon monoxide standard on new cars need not be more stringent than the 1980 requirement of 7 grams per mile (as opposed to the 3.4 grams per mile called for in the 1981 standard) in order to achieve the results called for in the Clean Air Act.

This finding apparently is based on the fact that carbon monixde levels have been dropping at about 7 percent per year nationwide. The CO levels have been reduced 96 percent since 1968.

The same appears to be true on the nitrogen oxide levels. They have fallen 76 percent since 1968 and projections indicate they would continue to drop under the 1980 emission control standard of 2 grams per mile, and thus make the 1981 standard of 1 gram per mile unnecessary.

If Congress agrees with the automakers that the 1980 standards on carbon monoxide and nitrogren oxide are sufficient to meet the goals of the Clean Air Act, then it should promptly consider rescinding the much more stringent 1981 standards (some of which haven't gone into effect as yet due to waivers granted by Congress). This would save buyers an average of $300 on a new car.

Time is an important factor for the carmakers because they already are in the design phase of the 1983 models. Relatively fast action is required — no later than the end of this year — to enable the auto companies to avoid loading the 1983 models with the costly equipment mandated under the more stringent 1981 controls.

The San Diego Union
San Diego, Calif., December 16, 1981

The Environmental Protection Agency is backing away from its demand that California require the annual inspection of smog devices on automobiles or face the loss of federal funds under various grant programs. This is no reason for the state Assembly to back away from passage of the smog-device inspection bill that was passed by the Senate last August.

The requirement for annual smog inspections was laid down by the EPA during the Carter administration. The new administration is pointing out that the Clean Air Act empowers the EPA to require only "periodic" inspections of air pollution controls. The agency therefore may settle for something less than checkups on an annual basis, such as inspections every two years.

With $850 million in federal grant funds hanging in the balance, the Senate last summer passed a bill calling for inspection and possible repair of smog equipment as a condition for renewing the annual registration of motor vehicles. We think the principle of annual inspections is a sound one, and that provision should not be changed when the Assembly considers the bill early next year. Why dilute the effectiveness of the program?

The case for annual inspection of motor vehicle smog devices does not rest on a directive from Washington but on the nature of the air pollution problem around major metropolitan areas in California. A car with a poorly maintained anti-smog system can produce two to three times more pollution than it should. Faulty equipment also reduces gas mileage and can even affect the health of the occupants of a car.

California was battling its air pollution problem long before the federal Clean Air Act was passed a decade ago and before the EPA came into existence. The state has required sophisticated pollution control equipmemt on cars sold here since 1964. Annual inspection of that equipment should have been required long ago, and a change in the fine print of a federal regulation is no excuse for delaying the enactment of the inspection program any longer.

THE ARIZONA REPUBLIC
Phoenix, Ariz., October 16, 1981

THE American auto industry has done as much or more in controlling emissions than any other industry.

Now automakers want Congress to relax 1984 standards on reducing carbon monoxide and nitrogen oxide emission levels.

The Environmental Protection Agency supports Detroit automakers.

The new legislation would allow new vehicles to emit about twice as much pollution as the original emission control plan for 1984 permitted.

However, the industry emphasizes that, as new cars replace older vehicles, there will be a continuing decrease in auto emissions.

In the Phoenix metro area, for example, carbon monoxide and other emissions clearly show a downward trend, and the Valley will meet the national ambient air quality standard set for 1987.

The automotive industry, particularly General Motors, argues that even with less strict standards the original act will still fulfill the nation's air quality needs.

GM maintains the amount of emission-control hardware that has been legislated simply is unnecessary. Consumers could save as much as $300 per auto if some unneeded equipment were removed from certain models.

GM maintains that relaxation of emission standards could save its customers $1 billion over the new model year.

It wants Congress to agree this year to relaxation of standards since delay until next spring means the company may not be able to make the equipment changes for 1983 and some 1984 models.

It is doubtful such amendments will be passed in time, particularly since most of the debate will center on highly technical chemical and medical data.

Most Americans agree the original 1970 act was a good one.

But the question is just how far industry and consumers should be expected to go in quest for perfection in air quality.

Pittsburgh Post-Gazette
Pittsburgh, Pa., March 18, 1982

Pennsylvania legislators like to believe that, in their campaign to stop federally-mandated auto-emission inspections, they stand right alongside President Reagan in stripping government to the bare frame. But they should not be surprised when Pennsylvania is pushed aside once the Reagan budget starts to roll.

For a second time, the state's objections to a program of emission inspections have received little sympathy from the federal government. This week the Environmental Protection Agency indicated that it will withhold more than $500,000 designated for other air-pollution control efforts in Pennsylvania as a reprisal for the state's obstructionism on emission inspections.

Those lost funds pale in comparison to the $302 million in federal highway revenue also blocked by the legislators' obstinate refusal to pay for the state's very small responsibility for training the mechanics who will administer the tests. And as a further indication of its belief that Pennsylvania hasn't got a very good reason for refusing emission inspections, the EPA also will refuse permission for the construction of any new industries that are major sources of pollution.

The Legislature's obstruction of the program is intended to save motorists an additional fee for annual measurement of the gases emitted from their auto's exhausts and the cost of mandatory tune-ups for cars that fail. The examinations will help to keep automobiles in metropolitan areas from producing volumes of exhaust that are dangerous to public health and will serve to maintain the complicated emission-control systems in new automobiles.

The new requirement could be met during the present mandatory mechanical-maintenance inspections so that motorists would experience no real inconvenience. And the Legislature could also reduce the scheduled inspections from two to one. As for costs, the Group Against Smog and Pollution says that such programs in other states typically cost $5 for inspection and an average of $29 for tune-ups to cars that don't meet the standard. The National Commission on Air Quality notes that emission inspections increase fuel efficiency and so save auto owners money on gas.

The Legislature has earned a contempt citation from the U.S. District Court in Philadelphia as a reward for its pointless grandstanding. Unwittingly, it has also played into the hands of the Reagan administration. In the midst of a campaign to strip the EPA budget to the bare bone, the administration will certainly be more than willing to oblige any grant recipient offering an excuse for the withholding of federal funds. In turn, that loss of funds will have serious impact on the state's overall effort to control air pollution.

That is a far different outcome than lawmakers expected and a further indication that the self-interest of Pennsylvania ought to compel the Legislature to comply with the ruling of the federal courts.

The Register
Santa Ana., Calif., February 11, 1982

The federal and state governments are dickering about whether tyranny every other year would satisfy the Environmental Protection Agency. The EPA is still holding some building permits hostage as a none-too-subtle inducement to the state legislature to impose some sort of increased inspection program on California motorists. The state Senate passed a bill requiring annual smog inspections, but the Assembly has wisely refrained from approving it. Now the EPA has said that *maybe* — not definitely, but maybe — inspection every other year would satisfy its hunger for useless regulations. The governmental game-playing continues.

It's important to understand that the debate is really about bureaucratic games and not cleaner air.

For starters, Southern California would exceed EPA's standards for ozone particulates if the area contained no cars and no people. The EPA claims the power to demand vehicle inspection for areas with more than .12 parts-per-million in ozone pollutants. Ozone pollutants from natural sources give us a constant level of .03 to .05 ppm, and on sunny days with pollutants trapped by our mountains and an inversion layer, the area would exceed the EPA standard strictly from pristine nature. In fact, this is true of 88 percent of the land area of the United States. EPA's standards are literally unattainable.

By setting unattainable standards and arrogating to itself the right to demand laws of areas that don't attain them, the EPA has made itself, in effect, a dictator over state governments.

Second, even an annual inspection program will make *no noticeable difference* in the quality of the air. According to the California Air Resources Board, the best we can expect from a vehicle inspection program is a reduction of 9 percent of the hydrocarbons and 6 percent of the nitrous oxides. That's an optimistic estimate. A more realistic estimate is that the program would reduce these pollutants by less than 5 percent.

Third, by the time any program is implemented it would be obsolete. If an inspection program were mandated now, it wouldn't go into effect until 1984 in the South Coast Basin. By that time most of the older cars, which constitute the biggest problem, will be in the junkyard. But California motorists will be saddled with the cost (estimated at $15 to $45 per car per inspection) and the inconvenience of an inspection program from then until. . . when?

Finally, the Clean Air Act of 1977, which (by a rather farfetched interpretation) gives the EPA the authority to demand inspections, must come up for revision this year or cease to exist. The EPA itself has offered changes in the Clean Air Act that would eliminate these mandatory sanctions from the law. It is a better than even bet that the next version of the law will not contain the mandatory sanction provision. If California is foolish enough to pass an inspection law now, motorists will pay needlessly for years to come.

EPA's coy little suggestion, that maybe an every-other-year inspection program would satisfy it, is simply bureaucratic game-playing, an effort to get some sort of law on the books while EPA still has a shred of authority. The issue is power, not clean air. The state Assembly should continue to resist this blatant blackmail.

THE SUN
Baltimore, Md., November 30, 1982

Now that state legislators have wasted $6 million of taxpayers' money to postpone a politically unpopular step until after the elections, it is time to face facts. Any further tinkering with the scheduled auto-emissions inspection program would only lead to further, needless expense.

Early this year, to recapitulate, there was a groundswell of opposition in the legislature to the emissions-testing program, primarily because it would have inconvenienced drivers in the Baltimore-Washington area. To defuse the issue, Governor Hughes agreed to a one-year delay. But that meant an existing deal with a firm to build and run 10 testing stations had to be renegotiated—at a cost of $6 million in extra fees.

Yet now that the state is ready to proceed with this new arrangement, some lawmakers are pushing for revisions of the law so cars are inspected every two years rather than annually. If such a change is made, it will destroy months of tortuous negotiations. It certainly will not lessen the price tag of the $67 million program.

But there is a more ominous problem. The reason for the state auto-emissions law in the first place is the federal Clean Air Act, designed to purify polluted air in industrial cities like Baltimore. By 1987, Maryland must comply with federally required clean air standards, and auto-emissions testing was supposed to help the state move in that direction. If lawmakers succeed in weakening the testing program, another method must be found to clean up the air: A clampdown on industrial pollution and possibly a ban on new industries. The loss of jobs and the damage to the state's economy would be staggering.

Those who pushed so hard for the one-year delay in the auto-testing program claimed Congress and President Reagan were on the verge of repealing key parts of the Clean Air Act. Why impose an unpopular program when it may not be necessary? That line of thinking turned out to be specious. As a special General Assembly panel discovered, emissions-testing is very much a part of federal law and is unlikely to be changed. By failing to meet the federal standards, the state risks a federal cutoff of $200 million in highway funds. Even a revision in the frequency of auto inspections might jeopardize federal aid.

With the election safely behind them, legislators can afford to face this unpopular issue. Fighting the inevitable has been costly and will be more so in the future. It is time to get on with auto-emissions testing. It could make the Maryland environment a little bit cleaner for everyone.

The Houston Post

Houston, Texas, November 22, 1982

For years Houston has managed to fight off some of the stiffer pollution control demands of the Environmental Protection Agency by proving that this city's problems are quite different from those of Los Angeles with its much greater number of cars and trucks. But as our vehicle registration grows, as our cars and trucks add more and more pollution to the air, we are asking for trouble unless we tighten our own community discipline. The State of Texas is stepping in to lend a conducive hand.

Harris County is the only one in Texas that does not meet federal clean air standards. This will surprise no Houstonian who on a clear, sunny day can see blue sky above but only gray haze at street level. We owe our unsavory record to the fact that Harris County also has more automobiles than any other Texas county. Now the Texas Air Control Board, in a plan worked out with the EPA, has given fair warning. Starting in January, cars and light trucks will be subject to pollution inspections to see if the pollution control devices have been disconnected or tampered with. No car will receive an inspection sticker until its pollution control devices are repaired and working.

One owner out of five in Harris County has disconnected or tampered with the emission control devices, twice the national average. To bring that one in five into compliance should make a marked improvement. The board believes that carrying out the inspections and enforcing the laws will cut ozone pollution by 45 percent in the next five years, by 71 percent after 1987. We must hope so. The county's levels are now 40 percent above federal standards.

But the air board and the Department of Public Safety will also ask the Legislature to pass strict laws to prohibit removing or altering the manufacturer's emission controls and to prohibit people from using self-service stations to pump leaded gasoline into vehicles designed to run on unleaded. Many car owners and even many mechanics still share the obsolete notion that pollution control devices raise operating costs and lower performance. But the automobile industry says that as technology has improved, the new cars have high performance and fuel economy and low pollution so long as they are properly maintained. Because Houston has an above-average percentage of new cars, it should be in better position to meet federal standards than many other cities. It is time that we did so. Not to do so is to invite the EPA to cut off federal funds for our highways and ban construction of new ozone-related industries. Besides, who wants to breathe dirty air in a city that has so many clear, sunshiny days?

THE WALL STREET JOURNAL

New York, N.Y., April 5, 1982

The papers tell us that the President isn't likely to succeed in modifying the Clean Air Act this year. It seems that environmental groups have intimidated Congress, which would prefer to tiptoe around this delicate issue as long as possible. Sen. Robert Stafford, chairman of the Environment Committee, went so far as to warn some businessmen recently that the Senate won't act unless they "turn down the heat" on Congress.

With all due respect to the sensitive Mr. Stafford, we recommend that the heat be turned up to a boil. The auto industry — and the country — greatly need a revision in current Clean Air standards, especially for car and truck emissions. The Environmental Protection Act helped improve the quality of air with its early emission standards. But those initial goals have long been met. What we are often faced with now are unrealistic standards that gain only a marginally cleaner atmosphere at great cost to industry, workers and car and truck buyers.

A case in point is the EPA's diesel-fuel emission standards. Not so long ago the EPA didn't have any such standards. But rising gasoline prices spurred Americans to buy more cars that used the cheaper diesel fuel. So the EPA watchdogs fretted because diesel-powered vehicles spit out some 30% to 70% more "particulates"—or soot—than gas-powered cars.

The EPA's initial soot standards went into effect last fall. But those rules—allowing 0.6 gram of particulate emission per mile for cars and light trucks—aren't a problem. Most vehicles can now meet them. The problem is that the EPA intends to tighten those rules to 0.2 gram per mile by the 1985 model year. And it intends to impose initial standards in 1986 for heavy-duty trucks.

Such goals don't take account of the real world. For one thing, the technology to meet those tighter rules doesn't yet exist. General Motors and other diesel makers are working on a "trap oxidizer" that catches and burns off soot before it escapes from the tailpipe. But so far they haven't been able to make a trap that can last the 50,000 miles the EPA says it must by 1985. Worse, GM says the trap doesn't seem to work at all for heavy-duty trucks. And even if the traps were perfected, GM estimates they would add as much as $1,000 to the price of a new diesel car and some $2,500 to the price of a heavy-duty truck. Trucks would also suffer a power loss, which has important economic consequences in light of the fact that more truck hours would be needed to do the same amount of work.

Still, those costs might be bearable if the health hazard were serious enough. Incredibly, though, even the EPA admits that tests on animals have yet to prove any serious problem traceable to diesel exhaust. The danger, it turns out, is mainly cosmetic—some dark air.

The same can't be said for the side effects of the methods to control soot. An EPA spokeswoman admits that current methods increase an engine's output of nitrogen oxide, which environmentalists know to be potentially far more dangerous than soot. Since current EPA rules envision further tightening of nox emissions by 1984, the technical problem will thus grow more complex. Moreover, a recent study at Johns Hopkins School of Medicine shows that exhaust recycling, used now to control nox emissions and one of the few technologies that seems to offer possibilities for tighter controls, has problems in diesel applications. Specifically, pyrene, an inherent product of diesel combustion, and nitrogen dioxide, also formed during combustion, combine to form a series of compounds. At least one, called 1-nitropyrene, has produced mutants in bacterial tests and is thus suspected of being a carcinogen.

With so much uncertainty, one might expect the EPA to sympathize with auto makers' requests to delay the diesel standards. Indeed, the EPA under its new management has shown a greater consideration for cost-benefit equations. But it has not yet relented on this issue.

Part of the problem obviously is Congress, which still is heavily influenced by an environmental lobby that is seldom willing to make concessions. That is why Sen. Stafford, a Republican from a state (Vermont) where environmentalism is a powerful political force, is dragging his heels. The politics of air quality has never been easy, but if we end up with engine gadgets that cause more health problems than they solve, we'll know who to blame.

Chicago Tribune

Chicago, Ill., August 2, 1983

With William Ruckelshaus making good on his vow not to let Illinois off the hook on establishing an auto emission testing program, the focus now must be on the best way to carry out the requirement. Further delay only would be fighting the inevitable.

The U.S. Environmental Protection Agency administrator put the squeeze on a balking state EPA by threatening to withhold federal funds unless Illinois got in line with requirements of the federal Clean Air Act. The state EPA, under pressure from Gov. James Thompson, capitulated after years of steadfastly maintaining that such a costly program is unnecessary because the air is getting cleaner.

It's not clear who is right in the dispute, although federal officials had on their side statistics showing that already this year there have been 19 carbon monoxide air pollution violations, making this potentially the worst year for air in Illinois since 1979. That supported some environmentalists' claims that recent years of improving air quality had as much to do with favorable weather conditions as pollution control.

Regardless, the dispute now seems academic. Under the agreement reached with the federal EPA, the state must have in place by 1987 a testing program that will affect motorists in Cook, Du Page, Lake and Kane counties, as well as the East St. Louis area.

The goal is to identify vehicles that are polluting the air, require the necessary tuneups or repairs and ultimately reduce ozone pollution by 25 percent and carbon monoxide pollution by 35 percent. Enforcement would be tied in with the annual state license plate renewal.

Testing programs can be effective. Studies have shown that some simple engine adjustments can reduce carbon monoxide emissions by 60 per cent and hydrocarbon emissions by 45 percent. Improved fuel economy is a side benefit.

In setting up a program, the state EPA and legislature must be aware of the pitfalls, especially the lessons learned in Chicago's disastrous emission testing experiment. That program—condemned as wasteful and useless—collapsed in 1979 because it involved only the city, was voluntary, did not require repairs to autos that failed and became a patronage wasteland.

For any system to work, it must be mandatory, encompass the metropolitan area and provide enough testing stations to be convenient and efficient for motorists. Critically important is the need for strict licensing and monitoring to prevent testing stations from becoming corrupt.

For guidance, Illinois can look to New Jersey, the nation's acknowledged leader in auto emission testing and the first to make it mandatory in 1974. New Jersey runs its own testing stations, but licenses 4,800 private garages for repairs and retesting—and checks the integrity of the garages by sending inspectors through with deliberately altered cars.

With 3 million cars in the state, 20 percent fail annually, but repairs average less than $40 per car, and most require only a minimal fee for simple carburetor adjustments. The result: a 40 percent decrease in carbon monoxide pollution at 18 measuring sites.

The New Jersey experience is evidence that the system can work. Now that Illinois has no choice about whether to test, it is time to follow the best examples and do the thing right.

ST. LOUIS POST-DISPATCH

St. Louis, Mo., July 26, 1983

Illinois officials have bowed to the dictates of the federal Environmental Protection Agency and will begin emission testing on automobiles in heavy pollution areas by 1987. As in Missouri, state officials had many valid questions about the need for a testing program that is likely to be expensive; but, faced with the cutoff of federal aid and restrictions on industrial growth, the state had little choice.

The Illinois EPA has said that it does not know how the testing will be carried out — whether by state or local operations or through a program using local service stations. The state should consider the latter and should mandate a safety inspection program statewide. Proposals for car safety checks have repeatedly died because of the opposition of Chicago legislators who claim that such inspections would penalize the urban poor. (The burden on the rural poor is apparently not important as pickup trucks have had to pass safety inspections for some years.) Now that cars in the Chicago area and in Madison and St. Clair counties are to be tested for emission standards, they as well as all others in the state should also be checked for safety problems.

Passing the safety inspection can be inconvenient and, sometimes, expensive, as Missourians will attest. But these checks, by ensuring that brakes, tires, exhaust systems, etc., are kept at a minimum maintenance level, also save lives and even larger amounts of money that might otherwise be lost through a needless accident. The Illinois Legislature should order that they begin statewide along with the emissions tests.

Herald News

Fall River, Mass., March 2, 1983

The prospect of the new, more elaborate system of automobile inspections beginning in April has alarmed some gas station owners to the point where they are predicting "complete chaos" when the system goes into effect.

The auto inspection program will be run by the Department of Environmental Quality Engineering and the Registry of Motor Vehicles jointly because the new system requires testing of tailpipe emissions.

According to the new law, each automobile will be inspected and tested once a year at a cost of $10 to the operator, with the routine safety inspection to be done at the same time as the new emissions test which requires the installation of expensive equipment to analyze the gases from auto tailpipes.

Approximately 2,000 service stations throughout the state will be equipped to perform the inspections.

In order to make the inspection of all the autos in the state feasible, a system has been worked out so that those autos with license plates ending in the number four will be inspected in April, those ending in five in May, and so on.

The difficulty seems to be that only 316 stations had been approved for the inspection process as of February 24, and 709 analyzers are on order.

Given the slowness of delivery, and the fact that less than a month remains before the new inspections are scheduled to begin, it is not difficult to understand why the service stations are predicting confusion, if not actual chaos.

State officials are reported as saying the program will begin on April 1, as scheduled, in spite of protests by service station operators.

Meanwhile, the legislature is reported as unlikely to postpone the date when the new inspection system will go into effect because the Environmental Protection Agency has warned that pushing the date back will mean that the state will face suspension of several federal aid programs.

Highway aid, sewerage grants and air pollution control programs could be suspended. Furthermore, the state might be forbidden to allow some industries, including some high technology firms, to expand.

The possibility of these sanctions is certainly a weighty argument in favor of not postponing the new program.

Yet the practical realities cannot be ignored, even by the EPA.

If in fact far fewer than 50 percent of the 2,000 stations supposed to do the inspecting and testing had even ordered the necessary equipment a week ago, then it is hard to see how the program can go into effect on time.

It is all very well for the federal and state officials to agree that it should, but if the logistical problems are insurmountable, then the officials will have to adjust to reality and postpone the starting date.

Nor would it be in any sense fair for the state to be penalized by the federal government for this reason.

If the public is to be expected to comply with the rules of the new system, and is to be penalized if it does not, then the 2,000 service stations must be ready to carry out the emission tests as well as to make the safety inspections.

If those stations are not equipped, then it is not reasonable to think the program can go into effect.

Neither the EPA nor the Registry, nor for that matter the Legislature, can afford to ignore the practical complaints of the service station operators who will be expected to carry out the program.

Nor can the public be expected to comply with the new regulations unless the equipment is on hand for the annual tests and inspections to begin.

The Detroit News
Detroit, Mich., October 10, 1983

No, that's not a cheer for the governor's speech to the Legislature last week, a rather tired compendium of mostly old economic development ideas which aren't likely to produce much economic development. But we like to give credit where credit is due, and Governor Blanchard is hanging tough against the federal government on an issue that would very much affect economic development here.

The Environmental Protection Agency (EPA) is demanding that Michigan institute a costly vehicle inspection program in order to meet federal air quality standards by the deadline year of 1987. Aside from the fact that EPA Administrator William D. Ruckelshaus wants to prove he's not James Watt, there seems little to justify the federal demand. The state's air quality has been improving steadily, and the costs of such a draconian program are unjustifiable.

When they push for expensive programs aimed at the last few percentage points of pollution, environmentalists seldom give credit for progress made. By the end of this year, the federal auto exhaust standard will already have a 96 percent reduction in hydrocarbon and carbon monoxide emissions, and a 76 percent reduction in oxides of nitrogen, over a preregulated 1967 car. On the factory side, state authorities have ordered Detroit Edison to switch from coal with a 2.3 percent sulfur content to coal with 1 percent sulfur content by Jan. 1, 1985, whereas Ohio allows 2.49 percent. This will cost each Edison customer $23.50 a year.

These improvements notwithstanding, Mr. Ruckelshaus threatens to cut off all air and water quality grants and order the Federal Highway Administration to stop highway funds if Michigan does not institute an auto inspection and maintenance (I & M) program, under which all cars would be checked annually and owners would be forced to repair faulty emissions systems. No figure is available for Michigan on the potential loss of federal funds, since it is not known whether Mr. Ruckelshaus intends to cut off all state money or only that going to the bad-air region. If Southeastern Michigan is the target, the state would lose $125 million for I-696 construction alone, plus $3 million in air quality grants to the state and Wayne County.

Mr. Ruckelshaus also can deny permits for new factories. Governor Blanchard's staff says this would cost Michigan 10,000 jobs, 5,000 in four proposed new factories that would need clearances for hydrocarbon emissions, plus 5,000 highway construction jobs.

The Legislature mandated an I & M program in 1980 and directed state agencies to write the appropriate regulations. This has been done but legislators refuse to process them. Secretary of State Richard Austin calls this "de facto repeal." This legislative reversal came just in time: Mr. Austin claims an I & M program for only Southeastern Michigan would cost $449.1 million.

So far, Mr. Ruckelshaus has refused Governor Blanchard's argument that I & M is unnecessary, unworkable, and offers no measurable public benefit. EPA officials have also rejected the governor's modified plan to exempt car owners who have their engines tuned by a mechanic.

Governor Blanchard now goes to his fallback proposal, that the state prepare a standby program that would be implemented Jan. 1, 1986 only if the June 1985 air quality figures show the state has not met the standards.

The EPA should let state officials prove their argument, that Michigan can achieve compliance without wasting millions of dollars on a new program. The governor should continue hanging tough.

The Pittsburgh Press
Pittsburgh, Pa., May 25, 1983

Maybe there is somebody in Pennsylvania who can fathom the logic behind the new auto-exhaust-inspection program the state has been forced to implement under federal duress.

But at this point about the only thing that can be understood is this: Pennsylvania now has the federal funds — $419 million — which had been frozen by U.S. Judge Louis Bechtle because of the state Legislature's refusal to bow to federal pressures to apply emission controls to motorists in the Pittsburgh, Philadelphia and Allentown-Bethlehem areas.

With this money, the state can proceed to gear up the road building and upgrading that PennDot had feared it would be impossible to carry out during this construction season for lack of dollars.

Hooray! And honk-honk!

The tailpipe-inspection program, however, stands as a monument to the lack of common sense.

It will not apply to most Pennsylvanians.

It will not apply to big trucks and buses.

It will not apply to diesel-powered vehicles.

It will not, in fact, apply even to 20 percent of the vehicles in the five counties included in the Pittsburgh area (now reduced to four). Nor will it apply to those who live outside its boundaries but drive within them.

In some cases, neighbors will be treated differently simply because they have different ZIP codes. One will have to have the car's exhaust system inspected; the other won't.

Incongruities such as these should be addressed before the inspection program takes effect on June 1, 1984.

Congress should amend that part of the Clean Air Act that has led to such helter-skelter application of the law.

Either every motorist in Pennsylvania, and in the rest of the country, should be subject to such inspections — or nobody should.

THE RICHMOND NEWS LEADER
Richmond, Va., September 15, 1983

The Environmental Protection Agency, as usual, is in a snit: Americans simply will not do what is good for them, no matter how hard Big Daddy tries to lead them in the right direction.

The latest instance of this exasperating — to the EPA — perversity is the insistence of some motorists on using leaded gasoline that messes up the air pollution controls on their cars. The EPA has wrung its hands over this dreadful display of bad faith and has decided, by golly, that Something Must Be Done. The threat of hefty fines apparently is not working, so the EPA will get tough.

First, the EPA wants the states to step up enforcement against such dastardly circumvention of its regulations. States also will be encouraged to beef up inspections of cars for evidence of tampering with emission controls, either through the use of leaded gasoline or through removal of the controls. Obedient states will receive clean air credits to offset other EPA requirements.

If that doesn't work, the EPA really will flex its muscles.

On its list of possible last-ditch efforts against stubborn motorists is an outright ban on the sale of leaded gasoline. An alternative would increase the price of leaded gasoline to the higher price level of unleaded gasoline. Because this would cost all motorists from five to 15 cents per gallon more, the total cost could be close to $2 billion a year.

Of course, only 13 per cent of owners of cars requiring unleaded gasoline opt for lower-priced leaded gasoline, but it doesn't matter that the percentage of motorists who flout the EPA's wishes is small. If it comes to punishing all motorists through the imposition of higher gasoline prices instead of punishing only the small number of miscreants, then the EPA will go for it. It will operate on the same premise that Congress used in passing a withholding law on interest and dividends because a minority of taxpayers were not reporting that income — and everyone knows what happened to *that* colossal blunder.

Maybe the EPA will come to its senses as well and see the dubious principle in punishing everyone for the transgressions of a few. Such a scatter-gun approach invites contempt for the law and even more disgust for a do-good federal bureaucracy. Voters elected Mr. Reagan President to relieve them of the burden of such draconian threats, so perhaps the White House needs to remind the EPA discreetly that no one elected it King of the Road.

FORT WORTH STAR-TELEGRAM
*Fort Worth, Texas,
November 5, 1983*

Everyone for clean air shout "Aye."

Bravo. The "Ayes" have it.

Now everyone for a mandatory auto emissions inspection program shout "Aye."

We can't hear you.

Just a minute, here's an "Aye." Here we have Michael Morris, assistant director of transportation for the Regional Transportation Council. Morris maintains that Tarrant and Dallas counties will need to establish emission control and inspection programs if they are to be able to meet federal air quality standards by 1987.

In Morris' view, Tarrant and Dallas counties, which were cited last year by the Environmental Protection Agency for failing to meet air pollution standards, will have to reduce emissions by 30 to 50 percent in order to avoid punitive action by the EPA. The punishment probably would take the form of preventing the issuance of building permits for new sources of air pollutants.

Harris County has already set up an emissions inspection program. It involves visual checks by specially trained personnel and prohibitions against using leaded fuels in cars calling for unleaded gasoline, tampering with catalytic converters and selling parts that could aid in circumventing air pollution control devices.

Tarrant and Dallas counties, however, might not want to go with that kind of program. They could use the mechanized test of tailpipe emissions that is used in most states.

But, if Morris is correct, they will have to do one or the other soon in order to meet the deadline. While the EPA has backed away from such deadlines before, it might be imprudent for Tarrant and Dallas counties to depend upon its doing so again. At the least, they should make the effort to attain the air standards. The EPA probably would be lenient if it saw evidence that a reasonable attempt to achieve the kind of air quality everyone wants is being made.

Detroit Free Press
Detroit, Mich., December 4, 1983

UNLESS the federal government relents, Michigan motorists may be forced to spend several million dollars to achieve air quality goals that we'd be meeting in a short time anyway without the expense. The state just missed meeting the federal standards for ozone and carbon monoxide by the end of 1982, so Michigan must, according to the Environmental Protection Agency, establish an inspection and maintenance (I&M) program in the tri-county area or suffer sanctions.

An I&M program will cost motorists several million dollars in testing fees and repair bills for the emissions control equipment on their cars. If Michigan refuses to set up an I&M program, the EPA could impose sanctions that include a cutoff in highway and air quality funds and a ban on permits for new industry. The sanctions could keep us hip-deep in potholes for the next decade or block the construction of new auto plants here. They could also strip pollution control agencies of federal enforcement dollars — which hardly makes good sense.

We concede that there are some arguments on behalf of I&M. The auto industry was dragged kicking and screaming to develop the emissions controls that consumers are now paying for, and if the controls aren't working properly, that's a wasted investment. Regular maintenance of emissions equipment can improve your car's mileage. Still, I&M amounts to a new charge imposed on motorists at a time when the public is weary of such impositions and when more important social and environmental programs are starving for funds.

The Blanchard administration is trying to sell the EPA on some alternatives — a delay in the penalties or a compromise in which proof of a regular tune-up on your car would constitute compliance with the I&M law. But air pollution officials here believe Michigan will meet federal air quality standards by no later than 1987, even without I&M; it doesn't seem sensible to spend millions to achieve a goal we'll reach soon anyway.

There are several ways in which Michigan might be spared the problem. The EPA could be more flexible in its demands. Rep. John Dingell, D-Trenton, who heads the House Energy Committee, could probably get the sanctions section of the law amended with a bang of his gavel, but he is holding out for a comprehensive rewrite of the Clean Air Act, a task only slightly less lengthy and complicated than rewriting the Bible. The quickest and most direct solution, if EPA is adamant, is for Attorney General Frank Kelley to sue to get the sanctions overturned as unreasonable.

Without question, clean air is important, but there are more cost-effective ways to spent the money I&M would cost. Michigan and the tri-county area have a record of steady progress in reducing air pollution. In 1982, Wayne County recorded only six 8-hour periods when the federal carbon monoxide standard was exceeded and only four 1-hour periods in which the ozone standard was violated. Some county monitoring stations recorded no violations at all. Across the rest of the state, there were four violations for CO and 14 for ozone.

Yet for an occasional whiff of carbon monoxide or ozone, the whole tri-county population may have to line up to have their tailpipes checked — or see Michigan lose highway funds or new industry. The penalty seems woefully disproportionate to the crime.

The Hartford Courant
Hartford, Conn., February 4, 1983

The first month of the state's new auto emissions program has passed, and the sky has not fallen. If anything, the sky is improving as Connecticut's air gets cleaner.

In the first two, half-month testing cycles, the much-maligned program seems to be working smoothly enough.

The pace is proceeding as expected. More than 73,300 cars were tested during the first cycle, and 56,000, the second; 67,000 cars are assigned for testing during each cycle. The 16.5 percent failure rate so far is close to what was anticipated.

The long lines of the first few days have dwindled to virtually nothing, the testing procedure generally is taking place within the projected 20-minute period and the number of complaints is going down.

Considering that the program is following so close to plan, it seems premature, at best, for the General Assembly — which passed legislation implementing the program just a few years ago — to be so intent now on backtracking.

Twenty-eight separate pieces of legislation affecting the emissions program already are in line for consideration. They range from outright repeal to expansion of the program to include diesel trucks and buses, which don't emit the same types of pollutants targeted by the testing program.

Some proposals might have merit, but the Legislature should keep in mind that the testing program is an integral element in the state's plan to meet federal clean air requirements.

If the program were eliminated, or even weakened substantially, the state would run the risk of economic sanctions by the federal Environmental Protection Agency, including the loss of $250 million a year in federal funds.

More importantly, the state would reverse progress toward making Connecticut's air more healthful to breathe. Health costs resulting from air pollution amount to an estimated $120 million a year in Connecticut, to say nothing of the human misery.

The Legislature should not risk ruining a good thing before the program has a chance to prove itself.

Post-Tribune
Gary, Ind., April 11, 1984

Some drivers may turn the air blue when their cars are stopped for inspection in Lake and Porter Counties. That could pollute the atmosphere more than the emissions from their vehicles' exhaust systems. But the testers aren't bad guys — they will be carrying out the federal law.

It may sound like a nuisance, but if the inspection program due to begin May 31 reduces the numbers of vehicles that spew a lot of dirty stuff into the air, it will be a public service. Lake, Porter and Clark and Floyd counties at the southern end of Indiana are the only four in the state that have flunked federal standards on emission of sulfur dioxide from motor vehicles. That's not good for the local image, but the thing to do is to go along with the tests — to cooperate instead of trying to block the test program.

Our opinions

Most violations, the experts say, should be easily correctable, at reasonable expense. We assume there are ways to verify the ordered repairs.

There could be a bonus discovery in some cases. Drivers who have cheated by adjusting their cars for the use of leaded gasoline could be left holding the catalytic converter bag. That's their problem.

For years, the Environmental Protection Agency has been trying to phase out leaded gasoline, a health hazard in at least two ways. A serious move to cut down on dirty emissions ought to include an attempt to phase out leaded gasoline. As more cars required unleaded fuel, the use of leaded gasoline was expected to disappear by about 1992. It won't, because so many car owners are making adjustments to their vehicles and using leaded gas. Those cheaters should be stopped.

The EPA should consider putting a ban on leaded gas and make sure it's phased out in a few years. Lead from automobiles is blamed for lead poisoning, which is deadly enough. But the chemical called EDB is added to leaded gasoline to make the fuel flow better. Some of the chemical isn't burned, and that's another hazard in the atmosphere.

Switching completely to unleaded gasoline would be expensive for consumers, because the refiners would pass on the extra costs. But there would be health benefits, and the reduced cost of vehicle maintenance. There seems to be a high octane case against the use of leaded gasoline, if the government is serious about wanting cleaner air.

Houston Chronicle
Houston, Texas, February 22, 1984

The Environmental Protection Agency threw a real curve ball at Texas and Harris County last week.

After studies, hearings and long negotiations in good faith, the Texas Air Control Board and the regional EPA officials reached agreement on a plan to control auto emissions in Harris County.

Then, the agreement was rejected in Washington.

This is an issue that could affect every driver in Harris County, and it is an issue with millions in highway funds and perhaps thousands of badly needed jobs at stake.

This is an issue that looms like a dark cloud off on the horizon. Only that cloud has been there for so long many have begun to ignore its presence — which could prove to be a costly mistake.

Houston and Harris County do not meet federal air pollution requirements under the Clean Air Act. Stricter controls on emissions from industry, primarily those along the Ship Channel, are improving air quality here. There is serious question whether automobile emissions play that big a part in our pollution problems. But the federal government says the law requires a vehicle inspection program.

The question is what to require in an auto emissions program and how it will be administered. The federal government would prefer tests which call for attaching electronic devices to tailpipes to measure emissions. The state pushed, and the regional EPA officials accepted, a substitute test that would measure a vehicle's exhaust with a specially treated paper. That test, along with muffler and gas intake inspection, could be conducted at the same time as the regular safety inspections. The agreement on this plan last November seemed to resolve the problem, saving Harris County drivers the expense of the electronic tests.

Suddenly, that compromise is threatened. Texas air control officials are wondering if they have been negotiating with the wrong people all the while.

A special board meeting is scheduled here Friday. The EPA, in all fairness, should be ready at that time to state its position clearly and be ready to reach an agreement — one that it is willing to stand behind.

THE SACRAMENTO BEE
Sacramento, Calif., March 19, 1984

Mandatory smog-control inspection of vehicles, already required in many other states and countries, goes into effect in a number of California counties today, including all of Sacramento County and parts of Placer and Yolo counties. Inspections must be carried out every other year. However belated, and whatever the teething problems such a massive program is bound to encounter, it is a step toward making the air we breathe less hazardous to our health.

Nine counties and areas of 10 others which have failed to meet requirements of the federal Clean Air Act are involved. Failure to introduce smog-control inspection would have meant a cutoff of federal highway funds.

Owners of cars whose vehicle identification numbers end with an even digit will receive notifications with their 1984 motor vehicle registration renewal forms. Cars whose ID numbers end with an odd number will receive notices next year. Inspections at state-certified shops must be carried out and, if a vehicle fails to pass, repairs must be made before a certificate of compliance may be issued. Certificates will cost $6, inspections are estimated to cost an average of $20, and repairs to noncomplying cars may not cost more than $50 unless smog-control equipment has been tampered with.

Considering the potential benefits to be gained from the inspection program, the costs involved are modest. Whether the same may be said of the inevitable complications of implementing a program involving about 6 million vehicles annually remains to be seen. The Bureau of Automotive Repair, a little-known state agency responsible for setting up and monitoring the program, has more than doubled the number of its field offices and has installed a computerized inspection system wherein the machine employed, not the mechanic, will determine whether and to what extent a car fails to pass muster.

How well the system works will depend in large measure on how quickly repair shops can be certified, how quickly they can accommodate the flood of cars and the volume of repair work needed to bring vehicles that don't pass the test up to standard. It also will require a large measure of patience on the part of a public whose awareness even of the program's existence is uncertain at this point. Having adapted more readily than perhaps any other place to the motor vehicle, California now must adjust itself to one of the controls that for health's sake now necessarily goes with it.

THE ATLANTA CONSTITUTION
Atlanta, Ga., April 11, 1984

From 1979 through 1982, Atlanta's air seemed to be getting cleaner, the state's Environmental Protection Division reported. Then Mother Nature inhaled and held her breath over Atlanta. For more than a dozen days in 1983, Atlanta suffered stagnant air and the worst air-pollution conditions it had ever experienced.

As a result, the federal Environmental Protection Agency has put Georgia on notice: The Atlanta region must make significant improvements in its air quality or face sanctions that include a cutoff of all federal highway funds, a ban on permits for new industries that emit hydrocarbons and no more grants for aid in pollution control.

Pretty severe penalties, but then the problem is not to be taken lightly. Motorists in three counties — Fulton, DeKalb and Cobb —

already are required to have their autos' emissions checked annually and repaired if necessary. Look for that law to be enforced more rigidly.

Other counties, such as Gwinnett and Clayton, have been let off the hook on emission inspections. Now they may get to join their cousins in the other three counties at the inspection stations. In addition, the state is considering making the standards more stringent. It would be more difficult to pass the emissions test and get a sticker.

But these are only stopgap measures that may enable the state to avoid the harsh sanctions imposed on areas that imprudently ignored EPA warnings that their air had run afoul of the law.

The long-term solution, as Georgia EPD Director Leonard Ledbetter sees it, is expansion of public transportation throughout the metro area to get more single-occupant vehicles off the road. He also proposes more computerized traffic lights and more left-turn lanes to reduce the lines of automobiles at busy intersections. The latter items are already being carried out in most of the metro area, though often imperfectly. Resistance remains to the most obvious solution to the foul-air problem — greater and wider use of public transportation.

Clean air is a vital part of the high quality of life in Atlanta. For both health and economic reasons, it is essential that we restore and maintain the quality of the atmosphere around us. It is incumbent upon the leadership of the metro area to look past the immediate problem and plan for the long-term future to maintain a clean environment in and around Atlanta.

The Boston Herald
Boston, Mass., April 15, 1984

THERE'S an especially foolish kind of false economy being practiced, we are told, by "thousands" of American motorists who own cars that can take unleaded gasoline only.

However, leaded gas is about six cents cheaper per gallon, and these drivers use it instead — in violation of the law. In the end, they'll pay a lot more to repair the damage to their car than they saved, and if they're caught using leaded gas illegally they'll be subject to fines.

It is estimated that about one-fifth of the catalytic converters in motor vehicles have either been destroyed by leaded fuel or have been removed by owners so that leaded gas can be used. With them either out of operation or out of the car, the engines spew lead and other pollutants through the exhaust system into the atmosphere.

The hydrocarbon pollutants are bad enough, but lead is especially poisonous when pumped into the air. It affects the brain functions of children. Leaded fuels also contain the known carcinogen, EDB.

Under current plans of the Environmental Protection Agency, almost all leaded fuel will be phased out by 1992. This isn't soon enough for EPA Administration William D. Ruckelshaus. He believes banning leaded gas totally and soon might much easier than enforcing the law against those using it illegally.

Ruckelshaus is probably right, even though speeding up the phase-out by several years would cost refiners — and motorists — a heap in higher gas prices.

The alternative, though, would be worse, for as long as leaded gas is used in large volume it will continue to befoul the very air we breathe.

The Washington Times
Washington, D.C., February 1, 1984

And the EPA said, "Let there be clean air in Maryland or we'll impose a construction and sewer moratorium and withhold highway repair funds." And the targeted multitudes had a look at how the EPA planned to cleanse the skies, considered the plan to be ineffective, oppressive, and a stark case of blackmail, and passed upon it this harsh judgment: "It stinks." We agree.

We're all for clean air, but we oppose, along with almost the entire driving population of the Free State, the forced auto emission inspections which are to begin today unless Maryland wishes to suffer the costly consequences. Air standards probably won't improve. Only the octane level of motorists' blood has risen.

These programs are supposed to reduce carbon monoxide and ozone pollution by identifying vehicles that are "gross emitters" of pollutants and then forcing their owners to seek remedy. This would be nice if it worked, but the jury on emissions-testing effectiveness went out, came back, and failed to call for champagne.

In Portland, Ore., for instance, the program has been in effect since 1974, but tests show that only one of the four pollution monitors recorded any improvement, and that was but marginal. New Jersey, whose mandatory program has been in place since 1973, has seen no improvement in ozone pol-

lution. Arizona, in the tailpipe-sniffing business since 1974, reports no improvements in a recent study. Certainly, Maryland joins a grand tradition.

There are lots of reasons emissions-testing fails. Most states set dollar limits on the costs of repairs required to pass the inspections; beyond these amounts, waivers are issued. Maryland's limit is set at $50, and exempted from its program are the poor and unemployed, who have other things to spend their money on than regular trips to the car doctor. And you can't hardly get a mechanic to climb under a car for half a hundred. Good luck when the muffler rots.

Another shortcoming: These tests measure pollutants as a percentage of exhaust gas, not in total amount, which means that a small-engined town car will flunk if it's just over the testing limits, while a muscle-bound road car just under the limits will pass, although the larger car is doing a great deal more polluting.

Another problem with the Maryland plan is that the 1.8 million vehicles affected annually must be checked at one of only 10 "centralized" locations. And don't think the inspectors will hold the cars back if all four tires on the vehicle have less tread than the standard roller skate.

Maybe the new rules will improve Maryland's air. But we're not holding our breath.

The Idaho STATESMAN
Boise, Idaho, March 26, 1984

Reluctance in Kuna, Garden City and Meridian to enforce a mandatory vehicle-exhaust inspection program is a disservice to all Ada County residents — especially residents of those towns.

Not liking the federally required program won't make it go away, and by refusing to join the county and Boise in enforcing the law, those city governments would encourage non-compliance. People who live there will be subject to fines outside their city limits anyway.

Garden City Mayor Margaret Mockwitz's comments are typical of most of those resisting the program: "We're not sold on it," she said. "We're not convinced it will do what it's supposed to do. It will just cost the people a lot of money."

With inspections costing a maximum of $10 for cars built between 1970 and 1981 and a maximum of $30 for newer cars, Garden City residents aren't going to save much when faced with $25 fines for non-compliance.

The program was mandated by the federal Environmental Protection Agency because the county's carbon-monoxide pollution has not been lowered to an acceptable level. While bureaucrats are not immune from mistakes, we suspect EPA experts are in a better position to estimate the effects of the program than are Garden City's council members.

In Kuna, the city council has assumed an ostrich pose. Mayor Willard Nelson said members felt the problem would resolve itself in two or three years. Perhaps that is so, since the rest of Ada County supports the program, but Kuna has no good reason to exempt itself.

Meridian hasn't decided on the program yet, but council members reacted negatively to an Air Quality Board presentation.

We hope all three towns join the rest of the county in getting with the program. Program administrator Del Tredinnick offers the best reason for that: "We're trying to clean the air."

The News American
Baltimore, Md., January 31, 1984

Chaos is the only word we can think of to describe the current state of Maryland's unpopular auto emissions testing program. About the only glimmer of hope came from Environmental Protection Agency Administrator William P. Ruckelshaus who indicated in Annapolis Monday that, if there were sufficient reason, Maryland might be granted a delay in starting up the program. But he pointed that just because people object to it would not have any weight. He also noted that last week the EPA cut off federal highway funds to Fresno County, Calif., because of the county's delays in startng the testing.

Well, there seems to be a good many reasons why a delay would be justified. There is the Carroll County situation where a judge issued an injunction banning testing there — a decision which must be reviewed by the Court of Appeals. Howard County Monday did file for such an injuction. That's enough legal confusion alone to justify a delay.

Then there is the spate of bills that have been introduced in the legislature calling for additonal testing sites and for delays in the starting date. Whether these bills get through or not, they still reflect the legislature's concern with the program as it is now constituted.

Finally, just about everyone in Maryland is complaining about the shoddy way the state went about arranging the program and the sorry situation of only seven of the 10 testing stations ready to operate on time. And 10 testing stations weren't enough to begin with.

Secretary Ruckelshaus made it clear that if we don't have a testing program, sooner or later we will indeed lose millions of dollars for highway construction and sewage treatment. But the fact of the matter is that a day before the testing program is supposed to start, Maryland just isn't ready for it. We think it is incumbent on the governor and the legislative leaders to go to the EPA immediately and seek a delay.

THE DAILY OKLAHOMAN
Oklahoma City, Okla.,
May 13, 1984

NOBODY likes to take dictation from the federal bureaucracy, but all of us have to live with some of it. Another small dose may be coming.

Deteriorating air quality has resulted in the Tulsa metropolitan area being designated by the Environmental Protection Agency as a non-attainment area — one in which federally mandated air quality standards are not being met. Oklahoma City reportedly may soon earn the same dubious distinction from the EPA.

Contributing to the problem are those motorists who deliberately remove or tinker with the emission control systems built into their cars. Presumably they do it to use slightly less expensive regular instead of unleaded gasoline.

Tampering with or removing the emission control device is against the law, but some motorists apparently are willing to risk it. More important, though, is the fact that their cars are adding unduly to air pollution.

EPA already is empowered by law to order sanctions against non-containment areas if corrective measures are not taken. These sanctions could include withholding millions of dollars in federal funding for highways and sewage treatment plants.

This unpleasant prospect can be avoided if the state Senate will act favorably on a conference committee report already approved by the House. The measure originated as HB 1865 by Rep. Don McCorkell, D-Tulsa, and Sen. Bernest Cain, D-Oklahoma City.

Briefly, it provides for adding visual inspection of emission control equipment to the list of things already required in the state's annual vehicle inspection program, beginning July 1, 1985.

Details such as which vehicle model years would be covered are left to regulatory discretion of the commissioner of public safety. But the Legislature would have a chance to review proposed regulations before they went into effect next year.

According to the Oklahoma State Chamber of Commerce, this represents "the minimum legislation necessary to satisfy the EPA." Distasteful as federal threat may be, HB 1865 seems a small enough price to pay to avoid potentially damaging sanctions.

"HAH! NO MORE $9.25 FEES, FAIL STICKERS, $75 REPAIR BILLS, NON-COMPLIANCE HASSLES — I GUESS WE CAN ALL BREATHE A LITTLE EASIER NOW....!"

Rockford Register Star
Rockford, Ill., May 8, 1984

Federal highway money is crucial to the immediate future of us all, especially in northern Illinois where the long-awaited dream of a north-south freeway seems within reach.

We can't tolerate its jeopardy because the state is dragging its feet on meeting U.S. Environmental Protection Agency requirements.

This week, the EPA served notice it has begun proceedings to cut off millions of highway dollars because testing of auto exhaust emissions in the Chicago and East St. Louis areas has not begun. And whatever the Illinois General Assembly does at this point may be too late, an EPA official says.

The EPA, doing the job its name mandates, has found Illinois does not meet air pollution standards, particularly in the Chicago and East St. Louis areas. The state has known since 1982 that it was under a federal deadline to begin the tail-pipe testing. Now, the EPA is getting tough and it can hurt by hitting us right in the highway money allocations.

State legislation to start the tests has won committee support. But an EPA official said public hearings — the next step to fund cutoff — will be called, and the only way to forestall the cutoff order will be the start of actual testing.

If our legislators need a priority for action, we urge they place it on the testing of auto emissions.

Cigarettes: 'Passive Smoking'

The dangers of cigarette smoking have been well-documented and are familiar to most of the American public. Since the federal government first linked cigarettes to cancer in 1964, the warnings to smokers have grown more frequent and more dire. The American Medical Association has called smoking the chief preventable cause of death in the United States, pointing to its implication in a long list of serious illnesses: lung cancer, heart disease, cancers of the esophagus, pancreas, larynx, mouth and bladder, and chronic bronchitis and emphysema. The list of hazardous substances in tobacco smoke is equally long: tar, nicotine, carbon monoxide, cadmium, nitrogen dioxide, ammonia, benzene and formaldehyde. Nicotine, in addition to being poisonous, is addictive. (It is now also suspected that some of the flavoring agents added to low-tar and low-nicotine brands may produce carcinogens when burned.)

More recently, attention has been focused on a related problem; non-smokers who are regularly in close proximity to smokers may suffer the same health effects as their puffing neighbors. Research reports published in 1980 and 1981 concluded that non-smoking workers regularly exposed to smoke from the cigarettes of fellow employees suffered significant damage to their lungs, and that the non-smoking wives of men who smoked were more likely to develop lung cancer. In February, 1984, the Surgeon General issued the strongest warning to date on "passive smoking," saying that there was "very solid" evidence that nonsmokers had suffered lung disease from exposure to cigarette smoke. The report also stated that children whose parents smoked had a greater susceptibility at an early age to respiratory problems, including bronchitis and pneumonia.

The increasing awareness of the potential hazards of "passive smoking" have resulted in the passage in most states of legislation restricting or banning smoking in public areas. Hundreds of local communities have also instituted anti-smoking rules for such enclosed areas as restaurants, theaters, elevators and buses. The Civil Aeronautics Board as of 1973 required commercial airlines to provide separate seating sections for non-smoking passengers. The CAB went further in May, 1984, voting 3-2 to ban smoking on all domestic airline flights of two hours or less. The decision was reversed the next day, however, when it was decided that the ban would create major administrative problems.

Portland Press Herald
Portland, Maine, May 29, 1981

The governor has signed a bill banning smoking at indoor public meetings, including local council and school board sessions. But enforcement will be left to common courtesy and peer pressure, not police and sheriffs' deputies. And that's the way it should be.

While it is impossible to refute the argument that smoking in a crowded meeting room may annoy or even be harmful to participants who are allergic to cigarette smoke, the decision to ban smoking should be made locally.

If smokers cannot abstain during a brief meeting or slip out to a hallway for a cigarette during a lengthy session, and if others present object to smoking, there is nothing to prevent the group holding the meeting from voting to ban smoking. The state's obligation to protect its citizens does not include the right to ban smoking at gatherings of local government bodies which are perfectly capable of making that decision on their own.

Fortunately, both the governor and the Legislature recognized this in approving only a modified version of the original smoking ban measure. As proposed, violators were subject to a fine. As approved, no penalties are established. In effect, the law encourages local bodies to ban smoking at public meetings.

There's nothing wrong with that. But to mandate a ban at city, town and county meetings, regardless of the ability of local bodies to make that decision themselves, would represent undue state interference with the conduct of local government affairs.

Fort Worth Star-Telegram
Fort Worth, Texas, May 19, 1981

Smoking is a nasty, expensive, dangerous habit, and the world would be a better place if the practice had never been started.

Nevertheless, the Civil Aeronautics Board would be ill-advised to ban smoking altogether aboard commercial airlines, as some of the more strident anti-smoking forces are advocating.

There are a number of reasons such a decision would be ill-advised. In the first place, it is hard to find any real justification for the CAB to concern itself with the matter of smoking. This is the era of deregulation, of prying governmental controls loose from those areas best left to the private sector, and the issue of smoking aboard commercial aircraft is certainly one of those areas.

Also, even if the CAB or any other rule-making entity made smoking illegal aboard airplanes, there is hardly any practical manner to enforce the rule. Smoking may be a bad habit, but it is a habit shared by millions of Americans, and most of them are not going to fidget cigarette-less through a lengthy flight, no matter how well-intentioned the authority making the rules may be.

During a recent hearing on the matter before the CAB, airlines spokespersons said an outright ban on smoking would send passengers scurrying to the restrooms, which are designed for other purposes, in order to sneak a smoke, much in the fashion of adolescents trying to fool their parents.

Anti-smoking forces countered with a suggestion that airlines could circumvent such strategy by installing smoke detectors or even sprinkler systems in restrooms aboard airliners. They seem to ignore the fact that installation of such devices would boost the cost of air travel considerably. And besides, the suggestion is absurd on the face of it.

Left to their own methods, the airlines will come up with a solution to the smoking-no smoking dilemma.

The CAB is to render its decision on June 18. We hope it decides to butt out.

The Des Moines Register
Des Moines, Iowa, May 20, 1981

The Civil Aeronautics Board has an unenviable job trying to decide what to do about smoking aboard commercial airliners. Should it be banned entirely? Banned only on small planes and short flights? Should separate seating be kept? Should all rules be dropped? No solution will satisfy everyone.

Smoking would be no problem if airlines strictly segregated smokers and non-smokers, and if there were foolproof ways to contain and exhaust smoke.

Smokers may have a right to light up, but only in private places or public areas where their fumes will bother no one. In enclosed public places, the right to smoke stops where non-smokers' nostrils begin. The right to breathe uncontaminated air pre-empts the right to contaminate the air.

The argument for either a prohibition or at least more effective insulation of non-smokers from smoke is based more on health than esthetics. Research is beginning to confirm the common-sense supposition that breathing exhaled tobacco smoke is hazardous to the health of those who don't smoke.

Although an outright prohibition on smoking in public conveyances would be justified, the difficulty of enforcement argues against it. For now, the CAB should require stricter segregation and smoke-containment measures by the airlines. If that means planes have to be modified, the CAB might let airlines charge their smoking passengers a premium to pay for the accommodation of their habit.

The Boston Herald American
Boston, Mass., May 19, 1981

Anyone who has ever had a good meal marred by cigarette or cigar clouds from a nearby table will be cheered by Newton's new ordinance requiring sections for non-smokers in restaurants licensed by the city.

The ordinance is the first of its kind in Massachusetts, and among the first such in the U.S., according to Cynthia Ferguson of the American Lung Association of Massachusetts.

The city's ordinance signed by Mayor Theodore Mann, requires that all Newton eating places which seat 50 or more people must make at least 15 percent of its seats available to non-smokers.

That seems a reasonable enough split of space in the ongoing struggle between smokers and non-smokers — though we suspect many non-smokers, having secured this beachhead, will press on for more territory.

Even a cigar-smoking editor can appreciate non-smokers' desire to be free of the aroma, the smoke or both.

Newton's law provides that most American of virtues, a free choice. "Smoking or non-smoking, please?" may become as familiar a refrain in that city's eateries as at Logan Airport.

Other cities in the Commonwealth would do well to emulate Newton's commendable sensitivity to non-smokers. Newton put its ordinance into effect after several months of voluntary experimentation by many of its restaurants. That test-run demonstrated that sealing off a section for non-smokers would cause only a miniscule proportion of smoking customers to abandon their favorite restaurants.

Voluntary non-smoking sections would help in restaurants in other cities, too, even before they're made law. It would be not just good manners, but good economics.

Newton's restaurants anticipate increased patronage as a result of their increased choice for non-smokers, according to Mayor Mann.

Why shouldn't other cities claim that same benefit?

Non-smokers who want to reward both themselves and good-neighbor restaurants by patronizing them may obtain guides to Massachusetts restaurants with non-smoking sections by writing the American Lung Association of Massachusetts, 385 Eliot Street, Newton Upper Falls 02164.

And smokers, take heart. You've got a lot of elbow room left, plus a clearer conscience.

The London Free Press
London, Ont., August 10, 1983

As medical evidence mounts on the hazards of second-hand smoke, non-smokers are becoming increasingly unhappy about having to share the same air space with someone puffing on a cigarette.

Things aren't as bad as they used to be. Not too many years ago, it was considered chic to smoke; anyone with the temerity to complain about having to squint at a movie through a heavy smokescreen or inhaling the effluent from a pack-a-dayer at the office was considered a spoilsport.

But there is more public awareness of the health hazards from smoking these days. There are more public places where smoking isn't allowed, such as movie theatres, and there are others where there are no-smoking sections, such as airplanes. But there's still a lot of smoke in the air, and London Ald. Grant Hopcroft is going to try to do something about it.

Hopcroft plans to introduce a municipal bylaw to regulate smoking in public places such as restaurants, municipal buildings, hospitals, medical offices and stores.

It is difficult to understand why the medical profession has gone along with smoking in hospitals where some patients are struggling just to breathe. Some hospitals have belatedly started cracking down on smoking in patients' rooms. Individual doctors should be left to make the decision about hanging no-smoking signs in their own private offices, which aren't in the same category as "public" places.

Requiring restaurants to provide no-smoking sections is possible, but it would have to be acknowledged that some small one-room restaurants would have difficulty providing much of a haven for non-smokers, even if they had tables especially designated for them.

In principle, the idea of a municipal smoking bylaw is excellent; in practice, it could be complicated to apply and difficult to enforce. But it should be worth the time and trouble to work out the most practical options. As the old slogan says, it's a matter of life and breath.

THE DENVER POST
Denver, Colo., February 27, 1984

A PROPOSAL to prohibit smoking in food stores, snuffed out by the Denver City Council last fall, is about to get a fresh airing.

A revised version of the measure, due to be submitted to the council this week, would require supermarkets and convenience stores to post "No Smoking" signs. It also would give the Denver Board of Health authority to enforce the ban.

The Denver proposal is mild by comparison with an existing ordinance in Boulder, where the city code prohibits smoking not only in food stores, but in theaters and concert halls and at public hearings and indoor sporting events.

And compared to a bill proposed in Fort Collins — to ban smoking virtually everywhere indoors except in private homes or in designated smoking areas — Denver's looks almost timid.

Smokers in grocery stores are supreme sources of annoyance to people who like to breathe clean air — even including many people who smoke. The pollution they create may not bother shoppers who merely want to squeeze the Charmin, but it is irksome indeed to those who like to smell the cheese or sniff the celery.

Asking a smoker to lay off for the few minutes it takes to go up and down the aisles certainly can't be considered an outrageous request. The airline industry already is beginning to ban smoking on flights of an hour or less.

Denver-based Rocky Mountain Airways, a regional carrier which has had a no-smoking rule on all flights since before Christmas, reports that the positive feedback far outweighs the complaints. "We would never consider going back," a vice president of the airline said last week.

It would be futile — and silly — to declare smoking illegal merely because it's a bad habit. But a growing body of evidence indicates — and polls show the public clearly believes — that the fumes created by the one-third of Americans who smoke pose a health hazard to the two-thirds who don't.

Pollster Louis Harris reported last fall that 82 percent of the non-smokers and 55 percent of the smokers he questioned agreed that "smokers should refrain from smoking in the presence of non-smokers."

In the face of overwhelming evidence that the public would support a ban on smoking in public places, it is hard to argue that the city government should keep setting out ashtrays.

The News American
Baltimore, Md., January 19, 1984

As many as 1.5 million Americans will suffer heart attacks this year, and 550,000 are likely to die as a result. So predicts the American Heart Assocation in its annual report on cardiovascular disease, the nation's No. 1 killer. Take care of yourself, pleads the association; pay attention to what you eat, eliminate stress from your daily routine (how, pray tell?) and quit smoking.

The heart association is not trying to say that all those attacks and all those deaths are going to be caused by ciggies. But most Americans, puffers as well as non-puffers, are convinced by now (or ought to be) that the damned things can and do kill.

And more and more are becoming less willing to put up with smoke being blown in their faces and, indeed, are convinced they have the *right* not to be subjected to that smoke. And not only because they find it annoying: They see it as a danger to their own health.

These are the main reasons why 36 state legislatures have been able to pass laws that in effect segregate the puffers from the non-puffers in various places where the public gathers. Now a a coalition called the Health Majority, made up of the state chapters of the national heart, cancer, lung and thoracic societies (plus a dozen other organizations, including the city's and the state's health departments and the state medical society) will be pushing for similiar laws in the current session of the General Assembly.

The coalition reports that many delegates and senators, some of whom are non-smokers, have been reacting sympathetically when three bills being proposed are explained to them. One reason is that the

The Health Majority reports that state legislators have been reacting sympathetically when three bills being proposed are explained to them. One reason is that the bills do not attempt to tell people they can't smoke — rather they say that areas should be set aside both for people who do and who don't.

bills do not attempt to tell people they can't smoke — rather they say that areas should be set aside both for people who do and don't. House Bill 336 calls for "no-smoking sections in areas utilized by the general public" in buildings owned by the state; HB 337 would prohibit smoking on the floor of retail stores, although smokers could light up in specified non-selling areas. HB 338 would mandate designated smoking and non-smoking areas in restaurants — but not, the sponsors hasten to say, in bars.

This is reasonable legislation, and we're for it. The bills are due a hearing before the House Environmental Matters Committee in early February. Do they have any chance at all? Ten years ago we would have said "no." But now the majority of the delegates and senators at work in Annapolis are not the cigar-puffing pols of old — and we have hunch that their constituents will be more supportive than even they could imagine.

THE SAGINAW NEWS
Saginaw, Mich., February 21, 1984

The war of the noses — and throats and lungs and hearts — goes on, but the outcome no longer turns on whether smoking is bad for you. The tobacco industry may plead for an "open debate," but as health groups outraged at its new ad campaign insist, there's good reason for those warnings on every cigarette pack.

The chief remaining question is whether those who take a risk are imposing it on others. The medical evidence is suggestive, but not conclusive. Smokers should prepare to lose this fight anyway.

And a fight it is. Many nonsmokers are also anti-smoking, personally and politically. Increasingly, they are winning, as with San Francisco's new ordinance restricting smoking in the workplace.

Now, in Washington, the Civil Aeronautics Board is pondering whether to ban smoking on flights of less than an hour, or perhaps two hours at the most. In Lansing, the Legislature is studying a proposal by Rep. Justine Barns, D-Westland, to restrict smoking in virtually all public places.

There are good arguments for the CAB plan. An airplane is a very small space. Even the best ventilation systems won't fully protect passengers with allergies or other ailments seated just a foot or two from the smoking section. Except for the most hopelessly hooked smokers, an hour or so of abstinence is not too much to ask.

The state bill, though, seems excessive and unnecessary. In some workplaces, it may not be feasible. Michigan already is more attentive to the health of its citizens than other states that do not have, for instance, no-smoking sections in larger restaurants.

But why is there a need for laws to cover a situation where ordinary politeness would seem the sensible solution? "Common courtesy alone has not solved the problem," argues Peter M. Pellerito, president of the American Lung Association of Michigan.

In his statement, Pellerito offers no proof that militant smokers are engaging in large-scale defiance. No reasonable person would ignore a plea from someone who says smoke is irritating, much less from those who suffer from asthma, allergies or other ills. On the contrary, the anecdotal evidence is that anti-smokers tend to assert themselves in a sometimes abusive, even violent, manner.

The core of the attack on public smoking is that nonsmokers don't think they should have to ask for, in Pellerito's term, "the crucial right to breathe clean air." And that's a point smokers will find hard to counter. They know they're in the wrong and don't have much heart for the fight. This is one minority which may literally have to go in the closet to exercise its rights.

Rocky Mountain News
Denver, Colo., March 31, 1984

FORT Collins has lighted a fire under the anti-smoking campaign by passing the toughest smoking-control law in the state. The question is: What next?

The Fort Collins ordinance, unanimously passed by City Council without citizen objection, prohibits smoking in all enclosed public places except designated areas and requires employers to provide non-smoking places for employees who request them. The move has been applauded. And it's been suggested that other cities — Denver included — should do the same.

Not a bad idea, but cities and towns that are serious about reducing environmental health hazards might consider actually enforcing other clean-living laws languishing on their books. The two most obvious prohibit spitting in public and require picking up dog excrement.

The Denver Municipal Code is specific on those offenses. On spitting, it says: "It shall be unlawful for any person to spit upon any public sidewalk, or other public . . . place . . . ." How long has it been since you've seen a policeman tote away a spitter? And it's probably no more than a

day day or so since you stepped in spit, or on a sticky wad of bubble gum.

As for dogs, the city code says: "It shall be unlawful for any person who possesses, harbors or is in charge of any dog not to immediately remove excrement deposited by the dog upon a common thoroughfare . . . ."

When was the last time a dog owner was hauled away for letting his pet's defecation lie unscooped? A few conscientious masters clean up after their pooches, but most people are frustrated and annoyed by pet owners who stand there looking the other way.

Granted, laws on spitting and failing to clean up after dogs are difficult to enforce. But who would have thought, even as recently as 10 years ago, that anti-smokers and non-smokers were so in the ascendant that Fort Collins would expect citizens to abide by this new law with nothing but peer enforcement?

If that's possible, then concern for the environment and public health ought to curb practices as potentially offensive, ugly and hazardous as smoking cigarettes, or inhaling somebody else's.

Detroit Free Press
Detroit, Mich., May 25, 1984

THE U.S. surgeon general has issued a denunciation of smoking that stings and strikes home: Children and spouses suffer, too, he says, when smokers puff. Neither the surgeon general's remarks nor the report supporting them will settle the issue, but they will renew both the debate about smoking and efforts to limit its scope.

Surgeon General C. Everett Koop announced this week that a new 515-page report suggested strongly that non-smokers suffered lung disease from exposure to the fumes of cigaret smokers. His statement was immediately contested by the cigaret lobby and by a contributor to the report, Dr. Richard Bordow, who admitted the need for more data.

But whether or not the surgeon general's report contains hard or only preliminary data, its implication is that there may be no escape from smoke contamination for hundreds of thousands of Americans. Children of smoking parents, the report claims, have measurably more respiratory prob-

lems, including bronchitis and pneumonia, than do the children of non-smokers. A contributor to the report also stressed that an unpublished study had concluded that smokers' wives experienced a slightly higher risk of bronchitis and emphysema.

Will smokers who won't quit to reduce their chances of contracting heart disease and cancer worry about the risk to their families? Perhaps. It is one thing to risk your own premature death and reduced earnings from illness and quite another to expose infants and spouses to needless suffering.

The surgeon general's report contains only a relatively short chapter on the effects of cigaret smoking on non-smokers. This line of research ought to be pursued and amplified. Clearly, segregating non-smokers from smokers in public places isn't enough. Researchers must redouble their efforts to convince more people that smoking is one health risk they can't afford to take.

DAYTON DAILY NEWS
Dayton, Ohio, March 24, 1984

In San Francisco as of this month, every workplace must come up with a plan that specifies where and when employees may smoke. If even one employee objects to the plan, smoking must be banned entirely. The city editor of the San Francisco Chronicle has reported that in the newsroom "people are holding meetings. All kinds of animosities are festering. It's going to be awful."

Why was such a draconian measure approved by the voters? It's true, as recent travelers can attest, that Californians never talk about anything but health. If second-hand smoke was ever to become a political issue, it was likely to happen in California first. But San Franciscans, true to legend, are a live-and-let-live lot, and they might easily have decided that the anti-smoking measure

was a bit heavy-handed.

They might have, that is, until the smoking industry entered the act with an ad campaign that featured regular people talking about how we regular people can settle differences courteously and considerately, without the government butting in. The voters were not about to accept such self-serving sanctimony from an industry that tries to hook teenagers on a notorious carcinogen.

Probably the best hope the tobacco industry had was to argue that to go cold turkey on second-hand smoke would be an unhealthy shock to the system.

Why not? The tobacco folks try to tell us that the causal link between smoking and cancer has not been proved.

THE ANN ARBOR NEWS
Ann Arbor, Mich., May 10, 1984

It strikes us as odd that a state willing to go to elaborate lengths to protect the Kirtland's Warbler cannot bring itself to cleanse the air of smoke in enclosed places.

We're talking about the well-being of people and their right to work and breathe free of health hazards. When that hazard is as preventable as tobacco smoke, there is no excuse for not taking protective action.

The perils of smoking are known, documented and worrisome enough to rate health warnings on cigarette packages. A growing body of evidence is condemnatory of second-hand smoke — inhaling other people's fumes. There is the additional annoyance, some would use stronger language, of non-smokers having to wear clothes that smell of someone else's vice.

Proponents of non-smokers' rights have carried their battle to the Legislature, where a bill to ban smoking in public places except in designated smoking areas is drawing the fire of business and tobacco interests.

Public places are defined in the bill as retail stores and other commercial businesses, offices, schools, auditoriums, arenas, meeting rooms and public buses and trains. Violators would be hit with $25 fines or less.

The penalty provision isn't the real point, though. The cost of enforcement is moot because compliance is voluntary. Moral suasion and peer pressure are what will make this bill work.

Michigan is one of 36 states already restricting or prohibiting some type of public smoking. The "public places" bill, sponsored by Rep. Justine Barns D-Westland, is modeled after a Minnesota law which reportedly is working well in the nine years since it went into effect.

Opponents of the measure say it's a nuisance law and unnecessary government interference. Historically, they say, smokers and non-smokers have been able to coexist. Employer groups say their clients can protect their people if a problem arises, implying of course that the smoke-filled room isn't a problem.

That view is at variance with a variety of health organizations, including the Cancer Society and the American Lung Association of Michigan, and rising public opinion itself.

The much-publicized San Francisco ordinance restricting smoking in the workplace and the Minnesota law are examples of growing public acceptance of the idea of equal but separate. The workplace can be segregated, smokers from non-smokers, without great difficulty. Compromises can be worked out.

The mystery is why so many employers fail to see the wisdom of such separation, when the likely result is happier, healthier, more productive workers whose efficiency rises along with their morale.

Sad to say, the law is made necessary in part because appeals to common courtesy haven't worked. Polite requests to butt out go ignored. We do not tolerate a person expectorating in a public facility, yet somehow blowing smoke in a person's face is overlooked. One is clearly as offensive as the other.

Equally sad, two previous efforts to get a similar proposal through the state Senate have failed, reflecting a general reluctance on the Legislature's part to get involved in health issues. Mandatory seat belt legislation — another proven boon to health and safety — likewise gets short shrift from our legislators whose badge of courage is the non-recorded voice vote.

The Barns bill, given the chance, would raise public consciousness. Compliance would steadily improve because smokers would police themselves. They would look for a smoking section before lighting up. The net effect of having another law on the books would serve to reinforce the laws of civilized behavior.

The issue is one of health and courtesy, not someone's right to smoke. The right to breathe clean air overrides the right to smoke. The point was made that drunken driving laws don't prohibit drinking but they do prohibit drunken drinkers from hurting others. That says it all.

The Salt Lake Tribune

Salt Lake City, Utah, March 23, 1984

Only one member of The Tribune's editorial board smokes cigarettes these days though most are former smokers. The lone smoker is the tobacco equivalent of a "social drinker," lighting up for a few puffs after work and taking care not to blow the fumes in anyone else's face.

It is from this background on both sides of the smoking scene that we applaud the Civil Aeronautics Board (CAB) for refusing, by a vote of 5-0, to ban cigarette smoking on large airplane flights of less than two hours duration.

Anti-smoking forces are at the point of pushing their successes of the recent past to the edge of spiteful repression. They seem to want not only protection for themselves but, in attempting to further restrict smoking on airplanes, they seek to deny any freedom of choice to others.

The CAB decision is at least a temporary setback for this zealous campaign which claims a right to dictate the lifestyle of countless Americans insofar as smoking habits are concerned. Pipe and cigar smokers, because of their fewer numbers and their often more pungent emissions, have already been cleared from the skies by the airlines so that the identical CAB ban voted Monday only made that curb official.

Under the existing system in which seats in both first class and economy sections of aircraft are divided into smoking and non-smoking areas, there simply is no reason to further restrict the cigarette addict. True, an extra-sensitive passenger who abhors smoke may occasionally be seated so close to the "border" that some diluted fumes invade his or her pristine airspace. In our experience, however, this is a rare exception.

More common, we suggest, is the nonsmoker who will ask for space in the smoking section because that's where the "action", if any, is likely to be. As one longtime ex-smoker explained: "That's where my kind of people sit."

Regardless of why individual passengers decide to sit in smoking or non-smoking areas, the satisfying result of the unanimous CAB decision is this: The opportunity to choose is still there.

THE LOUISVILLE TIMES

Louisville, Ky., March 29, 1984

SMOKERS and non-smokers have been clamoring so loudly about their "rights" lately that it's hard for anyone who isn't strongly partisan on the issue to think straight. But the Civil Aeronautics Board somehow managed to do so when it rejected a proposed ban on cigarette smoking on all airline flights of two hours or less.

The ban, which would have affected 90 percent of all domestic flights, was sought by the same anti-smoking group that persuaded the CAB in 1973 to require segregated airline seating for smokers and non-smokers. This segregation was a reasonable compromise that allowed smokers to keep on smoking while minimizing annoyance to non-smokers. The CAB was wise to stick with that compromise and reject the proposed ban.

But segregation doesn't work on smaller planes, so the board rightly banned smoking on aircraft with 30 or fewer seats. This restriction will be resented by some smokers. But trips aboard small planes are usually short, so the deprivation should be endurable to any nicotine addict who recognizes his obligation to be fair to those who find smoke offensive in close quarters.

•

Fairness also is an issue in Brown & Williamson Tobacco Company's continuing battle to refer to federal tests in advertising its low-tar Barclay cigarette brand. When subjected to the Federal Trade Commission test that measures the nasty things in cigarette smoke, Barclays produce one milligram of tar each. But the FTC and some of B & W's competitors argue that Barclay cigarettes do better on the testing machine than when human beings actually smoke them.

That may or may not be the case. The question is one of many in a complex legal contest that involves at least two different lawsuits. But it's obviously unfair for the FTC to judge Barclay by a standard different from the one applied to other brands. If the FTC test doesn't reflect the actual amounts of tars smokers inhale when smoking various brands, then the test itself — not Barclay — should be in the dock.

Richmond Times-Dispatch

Richmond, Va., March 22, 1984

While society sometimes seems to be moving toward a decision to declare them to be such, people who smoke cigarettes are not felons. This means they have the same rights that non-smokers enjoy, including the right to travel on commercial airlines without being unduly harassed.

That right has been confirmed by the Civil Aeronautics Board's refusal to ban the smoking of cigarettes on flights of two hours or less. Such a ban, the board decided, would be "impractical and administratively burdensome to airlines" and would intrude unwisely into "the competitive market."

In other words, rather than endure such harassment many smokers would abandon the airlines for other means of transportation. The airlines would suffer financially, which is patently unfair.

This does not mean that smokers will be permitted to puff at will, to ignore the discomfort their habit might cause fellow passengers. Flying smokers will be expected to compromise. Those who travel on planes with 30 or fewer seats will not be permitted to smoke at all during their flights, since, the CAB noted, it would be impossible to segregate passengers on small aircraft into smoking and non-smoking sections. Such sections will continue to be designated on larger airliners. And people who smoke pipes and cigars will have to abstain on all airplanes for the duration of their flights, no matter how long they are.

All of this seems to result in a reasonable balance of the rights of smokers and non-smokers. And in a democratic society, that should be perfectly acceptable.

Richmond, Va., March 30, 1984

The spirit of Big Brotherly Love engulfed the legislators of the Long Island community of Suffolk County a few days ago. To protect non-smokers from smokers and smokers from themselves, the county legislature approved what may be the most restrictive anti-smoking measure in the country. And it will require business establishments to pay most of the enforcement costs.

The law would require companies with more than 75 employees to separate smokers from non-smokers. Restaurants with more than 35 seats would have to reserve one-third of them for non-smokers. Already banned in Suffolk County stores, hospitals and nursing homes, smoking also would be prohibited by the new law in all public waiting rooms. A smoker couldn't even enjoy a puff or two by retreating to the rest room, for smoking would be prohibited there, too.

Obviously, the new law will create costly nightmares for affected businesses, many of which will find it extremely difficult and outrageously expensive to rearrange their operations to keep smoking and non-smoking workers apart. One business executive denounced the proposal as "silly." Another predicted it would cause "nothing but grief." One restaurateur complained: "It's going to drive up costs. I have one big room and can't tell smoke to stay away from some tables."

But governments under the influence of Big Brotherly Love can be arrogant enough to think they have the power to order even smoke around. The Suffolk law is a manifestation of such arrogance. It represents an abusive use of governmental power. While government certainly has the right and the duty to take reasonable steps to prevent smokers from harming non-smokers, Suffolk County's law carries such action to the point of harassment. The county's objective seems to be to torment smokers into quitting.

This is patently unfair, and possibly even unconstitutional. Smoking is not a crime in this country, and governments have no right to treat smokers as criminals or as pariahs. Requiring businesses to go to extraordinary expenses to accommodate smokers and non-smokers is a confiscatory act. Such extreme measures as the Suffolk County law should be firmly resisted and vigorously challenged.

"I SAID, 'YOU HAD TO INSIST ON NON-SMOKING'!"

The Honolulu Advertiser
Honolulu, Ha., March 24, 1984

While the Civil Aeronautics Board's rejection of proposals to ban smoking on most commercial airline flights is disappointing, it points up the need for congressional action this year.

In Hawaii, the CAB's action this week confirms the status quo, where the three big interisland airlines allow only cigarette smoking in designated sections and smaller carriers already prohibit all smoking.

But the three airlines could help Hawaii where the board failed.

SINCE the CAB rejection was unanimous and the board itself is going out of business December 31, there is no chance for a change of heart there. Possibly new health findings on the damaging effects of smoking could bring fresh support for some other kind of federal action.

But the most immediate need is for Congress to act to retain existing CAB regulations on smoking. As it is they could die with the board at the end of the year, leaving health considerations up to the individual airlines.

Many travellers wouldn't tolerate a return to the old airline practice of unsegregated smoking. But, as there is confusion on fares and schedules under deregulation, so there would be variance and uncertainty on smoking rules. And that is just not desirable when you are talking about smoking, which is both a general public health danger and an acute threat for certain people.

THE CAB rejected the proposed ban on smoking on flights under two hours on grounds it was just impractical to enforce on the Mainland where some flights are within minutes of that limit one way or another, and where planes may land for short loading stops with ongoing passengers staying on board.

But in Hawaii, where all interisland flights are under two hours, a total smoking ban should provide no confusion and little inconvenience for smokers. People just wouldn't smoke on airlines, period, as they already don't within theaters and other public places for longer periods than two hours.

It is up to the three airlines, perhaps not acting in concert because of anti-trust considerations but with others following the lead taken by one, as now happens with fares.

It would strike a blow for health in Hawaii, and a saving for the airlines with no competitive disadvantage if they all acted at about the same time.

THE ATLANTA CONSTITUTION
Atlanta, Ga., March 22, 1984

In a levelheaded decision, the Civil Aeronautics Board has refused a request from anti-smoking groups to ban cigarette smoking on all domestic flights of two hours or less, a move that would have covered 90 percent of the flights in the United States.

While groups like Action on Smoking and Health, one of those asking for the ban, provide a valuable service by keeping public-health issues in the national spotlight, CAB was right in its conclusion that a smoking prohibition on major flights was unnecessary and would have been unworkable, anyway.

ASH is one of the groups responsible for the board's ruling years ago that forced airlines to provide separate seating areas for smokers and non-smokers.

Because that arrangement has worked so well, and because nearly all of the larger aircraft these days have adequate ventilation systems, further adjustments to the smoking/non-smoking regulations are simply not needed. And to use a phrase made popular by more than a few Southern politicians and good ol' boys, why fix it if it ain't broke?

CAB did, however, agree to ban smoking on planes that seat 30 or fewer, which will mainly affect the country's commuter and regional carriers. Even so, the impact on these airlines is not expected to be great, since most of them anticipated the board's ruling and have already banned smoking on their flights.

Any discomfort to smoking passengers is mitigated by the fact that most of the flights by these carriers last between 30 minutes and an hour — not exactly an eternity, even for a chain smoker.

Arsenic: Tacoma Weighs Jobs Against Health

A classic example of the kind of economic/environmental dilemma that can arise in cases of industrial pollution has been encountered first-hand by the residents of Tacoma, Wash. The Asarco copper smelter, a Tacoma landmark, refines highly impure ores in order to extract the copper contained in them. In the process, the smelter showers the area with air pollutants, including sulfur-dioxide emissions that have produced acid rain damage in nearby forests. More worrisome to Tacoma residents, however, are Asarco's arsenic emissions; this highly poisonous element is contained in large amounts in the impure ores refined by the 93-year-old smelter. Listed as a hazardous substance under the Clean Air Act, arsenic has been shown to cause cancer in humans and is suspected to be the culprit in increased rates of lung cancer in some populations living near copper, lead or zinc smelters. In 1983, the Environmental Protection Agency announced that its recommended regulations for arsenic emissions in Tacoma would risk as many as two deaths per year for each 100 people exposed to the pollutant. William D. Ruckelshaus, the agency's new Administrator, acknowledged that the emissions could be further reduced, but warned that the cost of such controls would probably be so high that the smelter would be forced to close. This would eliminate 575 jobs in the community. Ruckelshaus turned to the residents, requesting their input before a federal decision was made. Complicating the issue is the fact that the long-term effects of low-level exposure to airborne arsenic are not fully understood, although it is thought that arsenic cancers have a long latency period of up to 50 years. Abnormally high levels of arsenic have been found in the hair and urine of children attending school near the smelter, but the cancer rate for Tacoma residents has thus far been no higher than for residents throughout the state. (The operators of the smelter have said that it will shut down in 1985, citing many economic factors including the costs of instituting new pollution control measures.)

The Seattle Times
Seattle, Wash., July 20, 1983

WITH all due respect to the principles of participatory democracy, the notion of asking the general public to render an informed verdict on health hazards at the Asarco smelter in Tacoma is simply ridiculous.

Ever since its declaration three years ago that arsenic is an air-pollution hazard, the federal Environmental Protection Agency has been struggling to devise nationwide regulations and standards that would keep arsenic-related risks within acceptable levels.

The difficulty in setting standards for arsenic (and dozens of other known or suspected cancer-causing agents) occurs mostly because the technology is complex and full scientific data are lacking.

Few of us in the Puget Sound region possess advanced degrees in chemistry or other scientific credentials sufficient to render an intelligent judgment on what kinds of medical risks are involved in the discharge of more than 300 tons of arsenic a year from the Asarco stack.

Yet EPA officials are asking average citizens to do just that, at a Tacoma hearing scheduled Aug. 30. The agency says it wants the public to help EPA Administrator William Ruckelshaus decide whether the smelter should be shut down or compelled to install additional air-pollution-control equipment.

Since the early 1970s, Asarco has spent $38.5 million on anti-pollution devices and would have to spend — under one EPA proposal under consideration — some $3.5 million more to curb discharges further.

Obviously, there is some measure of risk in the plant's operation, else the government would not be considering additional regulations. At the moment, EPA says, there's no conclusive proof of danger.

Looking to the hearing next month, it's a fair guess that there will be highly charged testimony about Asarco's role in the local economy, including hundreds of jobs and a multimillion-dollar payroll.

The fundamental issue, however, is not related to socio-economics but to the specific question of how the smelter affects public health. Only the scientists can answer such a question definitively. That's what the EPA people are paid to do. The sooner they get the job done, the better.

OKLAHOMA CITY TIMES
Oklahoma City, Okla., July 18, 1983

CRITICS of the Environmental Protection Agency, as it has been operated under President Reagan, eventually achieved one of their goals with the resignation of its administrator, Anne Burford.

Her effort to run the agency along the philosophical lines expressed by Reagan in his presidential campaign got her in trouble with the environmental purists, and she finally had to resign. Her "crime" was in trying to carry out the Reagan directive of striking a balance between the sometimes conflicting aims of environmental protection and economic growth.

This gave rise to the perception, fed by certain special-interest groups, that the Reagan administration was prepared to give big industry and commercial developers a free rein in raping the environment.

The nomination of William D. Ruckelshaus to succeed her as head of the EPA was hailed by the Washington establishment generally and the agency's bureaucracy specifically. After all, he was "one of them," having set the EPA's course as its first administrator.

Some of the environmentalists were a little skeptical, suspecting Reagan's motives and fearing Ruckelshaus might not have a free hand in running the agency. They wanted to know more about his present views.

They didn't have long to wait. Ruckelshaus's first big test is looming as the EPA moves toward regulating arsenic air pollution. Pressed by a court order, the agency is proposing limits on copper smelters, glass makers and other plants that emit arsenic into the air. Critics say Ruckelshaus has opted for standards less stringent than had been urged by some in the EPA.

At stake, and the reason for Ruckelshaus's soul-searching on the new rules, is the fate of a major copper smelter at Tacoma, Wash., in the state where he lived for eight years before returning to the capital. The standards he prefers would save industry millions of dollars, avert closing of the plant and, thus, save about 600 jobs.

In short, Ruckelshaus has had to weigh the risks of a possible cancer-causing emission against the economic damage of strict regulation. That neatly delineates the whole environmental-economic dilemma.

Another problem is the reluctance of industry to get burned by environmental politics. An official of E.I. du Pont de Nemours has proposed that the chemical industry take a major role in the cleanup of toxic waste dumps by providing experts in planning, engineering and management. But he said it won't happen "unless the political interests are dealt with" so the EPA "can't be accused of being in bed with industry."

Protecting the nation's environment requires common sense, not extremism.

Chicago Tribune
Chicago, Ill., July 18, 1983

It is a cruel choice William Ruckelshaus is offering the people of Tacoma, Wash.: Accept a little arsenic in the air, with an increased risk of lung and skin cancer. Or face a shutdown of one of the area's big industries, with a loss of 800 jobs and consequent trickle-down economic problems.

But reality is often cruel when economic issues clash with environmental safety problems. And Ruckelshaus, as head of the Environmental Protection Agency, wants to turn the arsenic pollution situation into an instructive national lesson in risk/benefit analysis and environmental decision making.

Ruckelshaus, in fact, has chosen a relatively straightforward case in which to ask for public involvement. There are no doubts about arsenic's hazards to humans, unlike some other pollutants. The source of the arsenic is clear—a copper smelter. New national arsenic standards have just been put into effect which will cost the Tacoma smelting company $4.4 million and will reduce air pollution substantially, although some danger remains. Just how great the risks are has been calculated in terms understandable to the public: Tacoma residents face one additional case of lung cancer per year if the smelter remains open and operates under the new standards, down from four cases a year estimated under past pollution levels.

Environmental spokespersons call even the new, reduced level of arsenic emissions unacceptable. The EPA itself said in a statement that since "arsenic, like most carcinogens, is believed by most scientists to present risks at any level of exposure, any emissions will present some human health risk." But it calculates that any tougher standard will force the smelter to close with an annual loss to Tacoma of about $20 million. The smelter has been the largest source of arsenic pollution in the air in the nation; the new regulations also affect other smelting companies and glass manufacturing plants.

The EPA will give people in Tacoma a chance to have input into the agency's decisions when it holds a series of public hearings there next month. But Ruckelshaus says the final decision remains his.

Giving people a say in how much risk to health they are willing to run for economic benefits is a major improvement in EPA procedures. All of us repeatedly make such decisions in our own lives, in choosing, for example, whether to fly, drive, smoke, eat rich desserts or take medication that can have adverse side effects.

The issue is far less clear, however, when whole communities are involved and when benefits don't fall evenly or equitably on those who are forced to assume the risks. Not everyone in Tacoma will profit from keeping the smelter open, yet they may run risks as great as those who do benefit.

Americans face a "minefield of risks from hundreds, perhaps thousands, of substances," Ruckelshaus says. The nation is a long way from developing a workable consensus on how to deal with these dangers, many of which cannot even be evaluated in any useful way. But the EPA's candor in sharing the dilemma with the public and in spelling out the risks, the costs and the benefits to those involved is a welcome step.

The San Diego Union
San Diego, Calif., August 18, 1983

In deciding whether to use an automobile, the need for transportation outweighs the possibility of accident involving injury or perhaps even death. In deciding whether to smoke a cigarette, the addictive effect of nicotine may win out over the fear of contracting lung cancer or other often-fatal illness.

Now, the U.S. Environmental Protection Agency seeks a collective balancing of risks and benefits from Tacoma, Washington. The nearly 159,000 residents of Tacoma have been asked whether they prefer some risk of cancer from arsenic pollution or possible closure of the copper smelting plant that has been producing the carcinogenic emissions.

The perplexing question put to Tacoma by EPA Administrator William D. Ruckelshaus marks the first time a community has been asked to help make such a momentous decision about its environmental and economic future. Success or failure of this procedure in Tacoma may determine how officials deal with similar environmental problems elsewhere.

The stakes for Tacoma are highly significant. Operated by ARSARCO Inc., the smelting plant employs nearly 600 people and pumps an estimated $20 million a year into the local economy. It is only fair for the community to have some say in the matter.

Nevertheless, some residents of Tacoma and a number of zealous environmentalists contend that it is unfair for Mr. Ruckelshaus to seek a popular preference for jobs or health. They claim the decision should be left entirely to experts.

The experts, however, cannot agree. Although the law requires air-pollutant standards that provide "ample margins of safety to protect public health," many scientists believe there is no safe level of arsenic or any carcinogen. This means that merely decreasing arsenic emissions would provide inadequate health protection. Only closure of the plant would suffice.

On the other hand, some scientists contend that there is not a shred of evidence to indicate that arsenic emitted by the smelter ever has caused cancer.

Obviously, there is no easy answer for Tacoma. But easy answers are rare in an increasingly complex society. Industrial development provides economic benefits but environmental hazards. Environmental protection provides health benefits but at an economic cost.

In asking the residents of Tacoma how safe they want to be, Mr. Ruckelshaus is only following the democratic tradition of popular participation in government. The benefits of this tradition are many. But they never have come without risk.

ST. LOUIS POST-DISPATCH
St. Louis, Mo., September 24, 1983

When William D. Ruckelshaus took over as head of the scandal-ridden Environmental Protection Agency last spring, he was hailed as Mr. Clean. As far as personal integrity is concerned, Mr. Ruckelshaus seems still to deserve that reputation. But some newly proposed rules would allow the environment to be less clean than under past EPA standards and would permit far greater health risks.

The rules in question involve two serious industrial environmental hazards — radiation from active uranium mills and arsenic emissions at smelters. Under the proposed radiation standard, one death from cancer for every 1,000 people exposed near the mills could result. Under the proposed arsenic standard — which involves primarily a copper smelter in Tacoma, Wash. — the EPA estimates that two deaths could occur for each 100 people exposed to the highest amount of the pollutant for the most prolonged time.

Since it was formed in 1970, with Mr. Ruckelshaus as its first chief, the EPA, according to environmentalists, has generally adopted regulations that permitted risks to human life of no more than one in 1 million to one in 100,000. Under Anne Burford, Mr. Reagan's first EPA administrator, however, the agency began allowing higher risks by adopting less stringent rules on formaldehyde and several pesticides. Now Mr. Ruckelshaus seems to be extending that practice — one that, according to Robert Yuhnke of the Environmental Defense Fund, suggests "a pattern of sacrificing human health for the economic well-being of industry."

The guideline that governs present rule-making is what Mr. Ruckelshaus calls the most stringent standard achievable with practicable technology. But the Clean Air Act does not indicate that Congress intended to allow trade-offs between economic costs and human life and health. In earlier cases the law has been used to force regulated industries to develop effective pollution control technologies that were economical enough to allow them to continue operating. Some Tacoma residents say, for example, that arsenic emissions at the plant there could be reduced at a cost that would not bring about its closing.

But in any event, the environmental standards for an area should not be based on what an industry says it can afford. Nor should they be based on the number of people who would actually be affected. As one unnamed EPA official who was involved in the debate said, underpopulated areas, such as the West, could become a "risk dump" where higher risks were permitted because fewer people were exposed. Mr. Ruckelshaus has asserted that nothing has been finally decided on either the radiation or arsenic regulations. So there is still time for the EPA to revert to standards that show more regard for human health and less for industry profits.

THE DENVER POST
Denver, Colo., July 18, 1983

THE NEW Environmental Protection Agency chief has put the essentials of environmental decision-making on the line in Tacoma, Wash.

William Ruckelshaus is asking Tacoma's citizens to let him know what they want to do about a copper smelter in their town: Do they want to accept an increased risk of cancer from the plant's arsenic-laced emissions, or would they rather sacrifice the 800 jobs that would be lost if the plant were closed?

This is a fascinating experiment. Is the EPA abdicating its responsibility to prevent environmental health hazards? Does it point to a new style of federal regulation?

The issues are more complex than a simple question of putting decision-making at the most local level possible.

About a week ago, the EPA proposed a new standard for atmospheric emissions of arsenic. It conceded the new rule would entail some increased risks to health. A copper smelter operated by Asarco Inc. in Tacoma is the only one in the country that uses ore with high arsenic content, so it would be the one most directly affected.

The economic question is a blunt one: If the EPA orders Asarco to close its plant because of the emissions, the 800 people who work there will have to find other jobs or go on the public dole. The health questions are less distinct: Long-term risks from low-level exposure to hazardous substances are subject to considerable debate.

The EPA has begun a major effort to inform the people of Tacoma of what it knows about the health risks. About 80 of Tacoma's 160,000 residents die each year of lung cancer from all causes, the federal agency estimates, and about four of those are the result of the plant's emissions. It says the 282 million grams of arsenic now being emitted annually create about 9 chances in 100 that the people exposed will contract lung cancer sometime in their lifetimes.

The EPA says the new standard would lower the chances of contracting lung cancer from plant emissions to 2 in 100 — which is still a high risk by public health standards. Ruckelshaus said risk couldn't be eliminated entirely without eliminating the source — or imposing such severe restrictions that the economic viability of the plant would be destroyed.

Erling Mork, Tacoma's city manager, told The Post it's not clear exactly how the EPA chief proposes to gauge the city's feelings. "It's a highly honorable objective," Mork said, but it's difficult to determine a municipal consensus. There are no plans for a referendum.

The situation is in some ways similar to the government's attitude toward cigarette smoking. Rather than ban a known health hazard outright, the surgeon general's notice on every package advises smokers of the risks and lets them determine whether it's acceptable.

But smoking is a matter of individual choice; it's not so easy to avoid the widespread emissions of a major, polluting plant. And that raises a troubling long-range question: Who assumes the burden of caring for those who are made ill because the government would not shut down an acknowledged hazard?

Tacoma's responses — provided the EPA makes an effort to ensure that they are truly representative — should at least provide guidelines for future risk assessments.

THE ARIZONA REPUBLIC
Phoenix, Ariz., July 18, 1983

EVEN the most fervent environmentalists will agree that sometimes the cost of cleaning up the air and water can outweigh the benefits.

For example, the cost of removing 90 percent of the pollutants from a smokestack may be one the manufacturer can bear, while the cost of removing 95 percent may force him to shut down the plant, making hundreds unemployed.

Do the benefits of removing the extra 5 percent justify the price tag?

In the past, the Environmental Protection Agency and the courts have decided questions like that.

Now the EPA's new administrator, William S. Ruckelshaus, wants those directly affected to decide.

The case involves a copper smelter in Tacoma, Wash.

EPA recently issued a set of rules about how much arsenic a copper smelter may emit. EPA admits that even smelters that meet its standards would create a risk of cancer — a much lesser risk than now exists, but a risk nevertheless.

Should the standards be raised?

The answer will affect only the Tacoma smelter because it's the only one in the nation that uses ore with a high content of arsenic.

If the standards are raised, the Tacoma smelter will be forced to shut down, with a loss of 600 jobs.

Are the residents willing to accept even a higher risk of cancer to save those jobs?

Ruckelshaus has launched a campaign to inform people of the pros and cons.

At the end of the campaign, a series of public meetings will be held at which residents can state their views.

Ruckelshaus will make the final decision himself, but he says he will be guided by what the residents say. He does hope a consensus will emerge from the meetings because that would make his job easier.

This seems like an eminently sound and democratic way to reach decisions that directly affect thousands of people.

Assuming they have the opportunity to consider all the facts, they should have the right to decide for themselves which is preferable — accepting a slight risk of cancer or having a job.

Phoenix, Ariz., August 23, 1983

IT'S been more than three months since William Ruckelshaus took over as administrator of the besieged Environmental Protection Agency.

Under the new chief, EPA seems to have risen above past turmoil on a single promise — that EPA's affairs would be more open, and it would react faster to major environmental problems.

Ruckelshaus deserves praise for making EPA more accessible, appointing more experienced assistants, striving for more skilled technicians, working closer with Congress, and boosting agency morale.

However, the days of ready promises and euphoric expectations will soon end.

EPA now faces some tough decisions.

Ruckelshaus must soon announce an administration policy on acid rain, a decision that could cost Midwest industry millions of dollars in sulfur dioxide emission controls, to say nothing of thousands of jobs.

And this on the eve of election '84.

Rules on toxic wastes must be revised.

Some 14,000 disposal sites pose dangers. The extent of dioxin's threat to humans still raises major questions.

EPA also is considering stricter arsenic and pesticide standards.

Most of the major environmental laws enacted in the 1970s are being considered by Congress for reauthorization. EPA is expected to take a position on all of them.

These include clean air and water quality standards. Both bills have been in the midst of an intense environmentalist-industry debate.

Ruckelshaus believes environmental laws must be more reasonable while being more effective.

Environmental and industry groups are skeptical of such logic.

Presumably, more reasonable rules would help industry while more effective regulation would favor environmentalists.

Ruckelshaus has engaged in generalities to buy time.

One of his first acts was to appoint task forces to come up with answers to problems. They're still working.

He gave some insight into his future style in a decision involving Tacoma, Wash.

He asked Tacoma residents to help decide whether a local copper smelter should have stricter arsenic emission standards — perhaps forcing it to close — thus relying on a public referendum on a tough question.

Ruckelshaus has the advantage of conditional public support as he settles in to his new post. With public sentiment for environmental quality controls running so high, Ruckelshaus now can make policy decisions that will set a tone for environmental regulations for years to come.

THE ATLANTA CONSTITUTION
Atlanta, Ga., July 15, 1983

Now comes the classic environmental dilemma: Your money or your health?

In Tacoma, Wash., the Environmental Protection Agency has announced new emission-control standards for a copper smelter. As stringent as the new standards are, admits EPA Administrator William Ruckelshaus, they are not tough enough to ensure absolute protection from lung cancer caused by arsenic fumes emitted from the smelter's stacks. Standards that would be strict enough to guarantee safety from contamination probably would result in the forced closure of the smelter, thus costing the community many vitally needed jobs.

So the question: Do the citizens want to risk health hazards from the smelter in order to keep it open?

EPA has laid out the case clearly and simply. It will hold public hearings to gather citizen input on the crucial question. Although Ruckelshaus is reserving the right to make the final judgment on the new standard, he says public opinion will weigh heavily in his decision.

Putting this particular case this way unfortunately contributes to the broad misconception that environmental protection often is a choice between sweet air or jobs; the call is rarely that dire. But even so, this is a refreshing breath of clean air in EPA, whose recent history has been one of evasiveness and deception regarding hazardous wastes and emissions.

Ruckelshaus has chosen to make the public aware of the hazards and the economic trade-offs required to reduce them and then let the public play a major role in deciding what should be done.

William Ruckelshaus

This is obviously the kind of environmental policy that should have been adopted by EPA long ago. It will be instructive to see which door the public chooses: Prosperity or safety

Post-Tribune
Gary, Ind., July 23, 1983

If you were in the path of airborne arsenic coming from a plant and were asked how much cancer risk is acceptable, what would you say? If you

Our opinions

worked in the plant, or were otherwise affected by its economic health, what would you say to that question?

The Environmental Protection Agency plans to poll residents of the Tacoma, Wash., area, where a copper smelter plant (pictured here) operates.

It is a strange way to run an agency that is supposed to protect the environment. But EPA Director William Ruckelshaus said that "for me to sit here in Washington and tell the people of Tacoma what is an acceptable risk would be at best arrogant and at worst inexcusable."

We respect Ruckelshsaus and believe in his integrity, but why would he put this impossible choice on those people? He announced new emission standards for copper smelters and glass plants that produce arsenic. Deciding what is fair and what is economically feasible is difficult, but the agency should enforce its standards.

An EPA official said the poll "won't come down to jobs vs. arsenic." Ruckelshaus said it isn't really a referendum. Yet the director raised the question, perhaps only half seriously: "I don't know what we'll do if there is a 50-50 split." Good question.

This unusual approach should be monitored by the whole country to see what it suggests about future policies of the EPA under its new chief.

The Oregonian
Portland, Ore., July 21, 1983

The U.S. Environmental Protection Agency has asked the citizens of Tacoma, Wash., if they are willing to breathe in less arsenic than they have for past 90 years or see an aged copper smelter close that provides nearly 600 jobs in the suburban town of Ruston, less than four miles from the heart of Tacoma.

Whatever the outcome of Tacoma's decision, EPA cannot shed its responsibility for deciding how to comply with the spirit and letter of the Clean Air Act. The issue is important, not only as it affects the quality of air and the economics of the Puget Sound area, but also because of its national implications.

Being tested is a former Puget Sound resident, the new director of EPA, William D. Ruckelshaus, brought in to rescue a beleaguered agency from charges of sweetheart deals with industry. Closely watched in this case, he has been accused of giving the community an artificial choice of "your health or your jobs" after he made the proposal following the filing of a suit in New York state. It alleged the agency was not following the strictest possible standards in Tacoma.

Unfortunately, the situation does not have an early technological solution. Money won't solve the problem. If the smelter were rebuilt, new techniques would be used, but there has been no showing this would rid the air of substantially more arsenic or other contaminants. The smelter is operated by Asarco Inc., a New York firm, and in addition to producing copper and other metals, including gold, the plant reportedly is the nation's single source of arsenic, used extensively in agriculture for pesticides.

The company has spent $35 million to $40 million in recent years cleaning up its operation,

according to Tacoma Mayor Doug Sutherland, who reports the firm has been a good member of the community. Its payrolls mean "big bucks," he said.

Past air cleaners and proposed new equipment are expected to reduce estimated arsenic levels to about 1 to 4 percent of former heavy emission rates. Arsenic, a heavy metal, is carcinogenic, so "ample margins of safety" are required by law. The smelter also emits two other heavy metals, lead and cadmium, both potential health problems.

Proposed plans to put hoods over the smelting ovens are expected to reduce annual emissions from 313 tons to 189 tons. Technology, for the time being, then would have run its course. What this means, an EPA health authority has told Tacoma officials, is that the "levels of probability" of death from cancer, which are now at 20 percent from all sources for persons living outside the pollutant area, may be increased to 20.02 percent in a mile area around the smelter with the new equipment in place.

Public hearings are planned in August as a way of letting local people know the health risks and economic penalties. No referendum is now planned. Yet to be heard from are the courts, which may decide whether the EPA is abdicating its responsibility under the Clean Air Act of 1977.

We do not fault the EPA for seeking community opinions and technical information outside its agency, but in the end, EPA will have to decide, and that decision must comply with the intent of Congress on safe health standards, not with how community leaders, understandably concerned about job losses, feel about public risks.

Fluorocarbons:
Depletion of Ozone Layer Feared

Ozone is a bluish gas that, on Earth, is a toxic pollutant. A poisonous, unstable form of pure oxygen whose molecules consist of three atoms instead of two, it irritates the mucous membrane of the respiratory system and aggravates respiratory diseases. It is the major ingredient in smog, formed when hydrocarbons and nitrogen dioxide react. There is, however, an ozone layer in the Earth's stratosphere that is vital to the planet's ability to support life. This ozone layer, some 10 to 20 miles above the Earth's surface, protects the planet from solar ultraviolet rays. Its importance was suddenly brought into focus for the American public in the mid-1970's, when it was reported that the use of spray deodorants, air conditioners and other consumer products might be depleting the ozone layer. All of these products used fluorocarbons, compounds of carbon and fluorine that are useful as coolants in refrigeration systems because of their inert nature and low toxicity. Fluorocarbons were also used as propellants in aerosol spray cans. Scientists warned that the continued widespread use of fluorocarbons could result in a 15%–18% reduction of the ozone layer in a hundred years. Certain forms of skin cancer had been closely linked with ultraviolet exposure, and it was estimated by the National Research Council in 1982 that a 1% drop in the ozone layer could entail as much as a 10% increase in the frequency of these cancers. Researchers feared that there would be damage to plant life and even global climatic changes. Increasing ultraviolet exposure, the NRC reported, could also damage the human immunological system, which resisted diseases of all kinds. In its 1982 report, however, the NRC lowered to 5%–9% its estimate of the depletion of the ozone layer that would occur in a hundred years, at the 1977 rates of fluorocarbon usage. There is still wide disagreement among scientists concerning the extent of the damage that could occur, and the potential severity of the resultant health effects.

ST. LOUIS POST-DISPATCH
St. Louis, Mo., October 14, 1980

In an apparent effort to curb the resistance of the affected industry, the Environmental Protection Agency is considering methods other than mandatory production ceilings on fluorocarbons — the environmentally harmful gases that are used, among other things, as refrigerants and as a bubble-making agent for foam cushions. Scientists have concluded that continued worldwide use of these gases will deplete the ozone layer that protects the earth from excessive ultraviolet radiation from the sun. Increased ultraviolet radiation could cause a rise in skin cancer and perhaps genetic mutations and changes in climate and crop growth conditions.

Last year the EPA banned the further use of fluorocarbons as propellants for deodorants, hair sprays and other spray can products. Earlier this year the agency said it intended to order a ceiling and then a cutback in total U.S. production of fluorocarbons. Now the EPA says it is considering banning certain uses of these gases or possibly requiring certain users to recover and recycle them before they escape into the atmosphere. Yet the agency concedes that it still considers a flat production ceiling as the most "efficient method" of reducing environmental and human health risks.

Evidently the EPA is proceeding cautiously because of opposition from a domestic industry with $600 million in annual sales. Yet the problem is not just a national one. The agency will continue to encourage other countries, many of which have been unresponsive, to reduce global releases of fluorocarbons. The best way to do that may be to cut back U.S. production, since this country exports fluorcarbaons to foreign buyers.

The Knickerbocker News
Albany, N.Y., October 9, 1980

Back in 1974, scientists began to report evidence that chlorfluorocarbon, a manmade gas used in everything from aerosol deodorants to refrigeration units, might be depleting the layer of ozone in the earth's atmosphere.

This layer of unstable oxygen gas is vital, because it acts as a kind of filter to absorb ultraviolet rays in sunlight. Unfiltered ultraviolet light can cause skin cancer in humans and can harm plants, fish, everything that lives on the earth. There is no real way to calculate what harm the increasing damage to the ozone layer might cause because the effects are subtle and cumulative. We might know in 10 or 20 or 50 years, after it's too late to do anything about it. Ozone layer damage is irreversible, most scientists agree.

There is no absolutely conclusive evidence of the adverse effects of fluorocarbons on the ozone layer at the moment; but the scientific findings have held up under repeated testing, and the probability they are correct mounts each year.

The National Academy of Science issued a report last year that said an expected 16 to 30 percent depletion in the ozone layer by 1990, if fluorocarbon use continued and increased unabated, would produce very serious problems for health and agriculture.

The Environmental Protection Agency banned the use of aerosol cans with fluorocarbon propellants two years ago; the agency now wants to limit use of all fluorocarbons to 1979 levels.

Those industries using the chemical would like to see more tests before use is restricted.

The obvious problem with that is we have no way of knowing how much additional — and irreparable — damage to the ozone will be done in the next ten years, but scientists are afraid it will be substantial.

We are confident that, when forced to, many businesses and industries will find adequate alternatives to fluorocarbons, as the aerosol manufacturers did two years ago. We haven't noticed any dearth of hair spray or aerosol oven cleaner or anything else on the market since that ban.

And, after all, the substance is only being restricted, not banned — yet. Only if it proves as dangerous as suspected might its use be further curtailed.

It seems to us that the EPA has little choice but to restrict fluorocarbons. The inconvenience and extra expense to which manufacturers may be put now can't be compared with potential destruction of the ozone layer, and the pain and deprivation that could bring to the entire world.

The Providence Journal

Providence, R.I., January 2, 1980

The earth's ozone shield, that stratospheric gaseous layer that protects living things against excessive exposure to ultraviolet rays in sunlight, is disturbingly back in the news again.

In a 392-page report to the Environmental Protection Agency (EPA) last week, the National Research Council of the National Academy of Sciences (NAS) warned that unless the international community takes swift regulatory action relative to the use of fluorocarbons, the ozone layer will be reduced by 16 to 30 percent, endangering human health and the world's food supply.

An unusual note of urgency is evident in this latest report, unlike its predecessors. NAS concedes there are still unanswered questions about the fluorocarbon-ozone phenomenon. But it warns that known dangers are so great, there is no time to wait before taking concerted worldwide measures.

If other nations, which produce half the world's supply of fluorocarbons, had followed the example of the United States, which produces the other half, the global problem doubtless would be less compelling. In three stages this country has imposed a partial ban on the production, distribution and use of fluorocarbon propellants — the activating gas in aerosol containers. On Oct. 15, 1978, the manufacture of fluorocarbon propellants for non-essential uses was stopped. On April, 15 1979, all interstate shipments of such products were banned.

The gains thus achieved, however, appear only temporary. Other uses have rapidly increased abroad, as well as in the United States. These include air-conditioning, refrigeration, insulation, metal cleaning and drying and making furniture cushions. The NAS warns that if these uses continue to grow, U.S. gains from abandoning non-essential aerosols "will rapidly be offset" and along with several hundred thousand more cases of skin cancer each year, some warming of the earth's atmosphere is likely to occur.

For more than five years, scientists have been studying this problem, developing a considerable body of scientific data. The warnings become more and more strident, yet the world community appears unconcerned. As the search for fluorocarbon substitutes goes forward, the United States should heed the NAS call for this country's leadership in a global campaign to reduce the growing threat. A covenant banning non-essential uses ought to be drafted and circulated. Some means of establishing a quota system for more essential uses ought to be developed. And clearly in all this the World Health Organization of the United Nations should to play a pivotal role.

The scientific community has fulfilled its responsibility to identify the dangers. The world's political systems have a solemn obligation to respond.

The News Journal

Wilmington, Del., December 29, 1979

Fluorocarbons, those gaseous mixtures that enhance our quality of life through refrigeration and air conditioning, are under attack once again. The charge is that fluorocarbons deplete the ozone layer that shields mankind from excessive ultraviolet radiation, the kind of radiation that can cause serious forms of skin cancer.

Scientists from the prestigious National Academy of Sciences are so concerned over the hazardous long-range effects from ozone depletion that they are urging the United States to take the lead worldwide in curbing fluorocarbon usage and emissions. Conceding that the effects of ozone depletion are not fully understood, these scientists nevertheless say that fluorocarbon emissions are "clearly detrimental" to health. Ozone depletion will also contribute to a warming up of the Earth's surface and this may have as yet unknown effects on agriculture, say the scientists who conducted the fluorocarbon study for the academy.

Another group of scientists from the Du Pont Co. and from the Chemical Manufacturers Association is taking a different approach to the fluorocarbon issue. These scientists state that "No ozone depletion has ever been detected, despite the most sophisticated analyses" and they urge that no hasty fluorocarbon ban be imposed (fluorocarbon-powered aerosol sprays have been banned in the United States since April) until more research is done on the matter. Du Pont has a major share of the fluorocarbon market; other chemical companies represented by the association would also suffer from a cutback in fluorocarbon usage.

Is that commercial self-interest enough to make these companies toss aside all adverse indications and proceed in marketing a product that they know will in the long run harm people, including themselves and their families? Some detractors of industry would doubtless tar them with that brush, but we find it difficult to go along with that harsh judgment.

The controversy over fluorocarbons seems more like a classic case of scientists being unable to come up with "hard" facts, the kind of facts that naively we used to expect to get from the scientific community and that we since have learned rarely exist. Think of all the doubts that have arisen in just this decade over artificial sweeteners, risks from nuclear power, the pros and cons of flu vaccines.

Sometimes our nation has rushed into bans as with cyclamates only to learn later that they may not be as hazardous to our health as had been feared. At other times we have been too slow in uncovering the harm that certain hormone medications can cause. Cause and effect are far from easy to establish.

As science and technology continue to advance, old conclusions have to be tossed aside and new ones drawn. A certain flexibility has to be maintained in science as in other human endeavors and the advantages and disadvantages of questionable products have to be weighed.

Certainly, fluorocarbons when used in refrigeration contribute to the preservation of perishables, especially food. If fluorocarbons are replaced by another gas, what hazards could stem from that? Of what value do we consider air conditioning? Is it worth the possible risk of ozone depletion or would there have to be firmer proof of this happening before fluorocarbon use for air conditioning should be restricted? How about mechanical changes that would permit some form of recirculation of fluorocarbons? What price is that worth? Are the emission controls on automobiles worth the increased gas consumption they have caused?

In a matter where some of the best scientific minds disagree, a public policy decision becomes necessary. That's what happened on saccharin, where Congress overrode a ban based on test results showing a link between cancer and saccharin and listened instead to those experts who pointed to the decades of safe use of saccharin by thousands of diabetics.

Now before the public are two contradictory views concerning fluorocarbons. One argues that the potential hazard is so great, though it won't be fully felt for another century, that fluorocarbon usage should be restricted promptly. The other asserts that there is no need for haste and that a few more years should be spent on research and evaluation before a decision is reached. Is the Environmental Protection Agency equipped to make this far-reaching decision? Or does that matter belong more appropriately before Congress?

In this era of skepticism and uncertainty, we are not sure of the answers. But an immediate ban without fully exploring its consequences seems both inappropriate and unnecessary.

The Wichita
Eagle-Beacon
Wichita, Kans., December 21, 1981

Recent studies have produced new evidence that the ozone scare of a few years ago was indeed well-founded. Not only was the fragile atmospheric layer that protects the Earth from excessive solar ultraviolet radiation in danger then, but it still is, and damage already has been logged.

Citing findings based on satellite data, NASA's Goddard Space Flight Center reported last summer that the amount of ozone about 40 kilometers (24.8 miles) above Earth decreased at a rate of one-half percent a year between 1970 and 1979. Researchers point out the findings are somewhat preliminary and the indicated depletion rate is not yet alarming. But there is solid cause for concern: Loss of the ozone layer could result in much more skin cancer, damage to plant life and even global climatic changes.

The principal threat to ozone is the chlorofluorocarbons — also known as freon — once widely used in aerosol sprays, and also used as a common refrigerant for air conditioning and other purposes. Chemical industry arguments of a few years ago that the perceived threat was only theoretical, unsupported by substantiating evidence, now seem to have been disproved. Additional information about the situation is expected from a new research satellite launched Oct. 6, the Solar Mesosphere Explorer, and from an instrument-equipped balloon with which a Harvard University chemist, James G. Anderson, plans to probe the ozone layer at a slightly different altitude next summer.

Just last month, another report from Goddard indicated not only two types of freon but traces of two other Earth gases — methane, one of the types of "natural gas" commonly used for heating, and nitrous oxide, once called "laughing gas," have been found in our atmosphere.

Not only is Earth threatened with a possible excess of solar radiation, but the build-up of carbon dioxide in the atmosphere — due to the enormous consumption of fossil fuels in industry and in cars — threatens to trap the world in a gaseous greenhouse that could, some scientists fear, cause the planet's temperature to rise by as much as four or five degrees.

Already there is evidence of slight shrinking of the Antarctic ice pack. Other nations should follow the lead the United States has taken, through legislation and otherwise, and act swiftly to do their part to contain the chlorofluorocarbon threat before truly irreparable harm is done.

THE MILWAUKEE JOURNAL
Milwaukee, Wisc., August 18, 1981

This nation simply must resist temptations to back away from the environmental protection that Americans wisely embraced so strongly in the 1970s. One of the most troubling ironies of Ronald Reagan's ascent to power is that he and his backers — while calling themselves *conservative* — seem so heedless of life-endangering threats to the environment.

Take, as examples, two of the most recent findings of scientists regarding long-term atmospheric changes that apparently result from human activities and pose serious danger for the climate.

A team of physicists at the Goddard Space Flight Center laboratories has published a report saying that Earth's average temperature increased 0.2 degrees centigrade in about the last 15 years. That finding has not been confirmed by other scientists, yet. But it adds significantly to concern that the heavy use of fossil fuels, particularly coal, will cause a "greenhouse effect" that could turn lush US cropland into a dust bowl and melt polar icecaps, raising sea levels enough to endanger coastal cities. Some scientists talk of an ecological emergency as early as the year 2020.

Some of those dangers — and others — are cited in another recent finding, this one by scientists with the National Aeronautics and Space Administration. They say they have found the first real evidence that chlorofluorocarbons (once used in aerosol sprays and still widespread in industry) may be eating away the ozone layer that protects Earth from dangerous ultraviolet radiation. Further study on this potential danger will, of course, be necessary.

Our point is this: With scientific reason for concern on the rise, it would be foolish now to relax the prudent efforts of the past to avoid making those problems worse. Yet, the Reagan administration plans greatly to expand the nation's burning of coal, ease air quality standards and back away from tighter regulation of chlorofluorocarbons.

That is dangerously wrong-headed. Prudent Americans — real conservatives — would demand greater scientific evidence of *no risk* before plunging ahead on Reagan's course. Turning the concept of conservatism on its head, by avoiding caution until there is greater documentation of danger, could easily mean waiting until it's too late!

The Hartford Courant
Hartford, Conn., August 15, 1981

Scientists have long suspected that the layer of ozone, an atmospheric gas that helps protect the Earth from dangerous ultraviolet radiation, is being slowly destroyed by chlorofluorocarbons.

In fact, it was on the basis of their theorizing that CFCs, whose DuPont trademark is Freon, were banned from most aerosol sprays several years ago. They are still commonly used in many other appliances and products.

Evidence announced this week by three scientists with the National Aeronautics and Space Administration tends to corroborate the theory.

They have found that the most vulnerable area of the ozone layer seems to be undergoing depletion, although they are quick to point out that no firm cause-and-effect relationship between the CFCs and the depletion of ozone has been established.

Inconclusive though it is, the evidence should be enough to defeat bills now in Congress that would prevent the federal Environmental Protection Agency from setting new limits on CFCs. About 800 million pounds of the chemical are produced every year.

If the ozone layer were sufficiently damaged, the increased radiation that would hit the Earth could, it is believed, result in climate changes, damage to fish and crops, and an increase in the number of cases of human skin cancer.

The new evidence on the condition of the ozone layer is not enough to justify pushing the panic button and banning all CFCs.

But it is enough to make it unwise for Congress to render the EPA powerless, should a cutback on CPC production seem justified.

The Kansas City Times

Kansas City, Mo., April 2, 1982

Well, maybe after all, those aerosol spray cans were only half as dangerous as the scientists estimated three years ago. Whereas the National Research Council, an agency of the National Academy of Sciences, estimated in 1979 that chlorofluorocarbons used as propellants in the cans and in other products could erode the protective ozone layer in the Earth's atmosphere by 15 to 18 percent, the council now reduces that figure to 5 to 9 percent. That ozone layer 20 miles out in space shields us from the sun's ultraviolet rays which can increase the incidence of skin cancer.

No matter, the spray cans were outlawed in April 1979 and the container industry, despite some reservations, made the conversion to pump cans or those using such other propellants as hydrocarbons or freon. With alternatives to go to and high product turnover, the change was not as traumatic as some which have been inflicted on other industries by federal regulation. And with a question of public health involved, it was wisest to err on the side of caution and act on the best scientific data then available. The image of mankind unwittingly wafting upward a cloud of spray can effluvia to dissipate a safeguard against cancerous light rays from space was compelling.

The research council, while more sure than ever of a link between ozone depletion and skin cancer, now concedes that "recent advances in knowledge since the 1979 report . . . have clarified our view of the problem, but have also pointed out scientific areas not emphasized in earlier reports that confound the simple prediction of the effects" on living systems. So many of these potential environmental hazards are difficult to measure, both as to their toxicity and the provable effect on human health. Affected industries are thereby encouraged to challenge the assumptions leading to restrictions on them. But the regulators, dealing with an issue of possibly irretrievable human damage, have to move ahead, allowing for adjustments as more definitive knowledge of the problem is acquired.

THE SACRAMENTO BEE

Sacramento, Calif., July 1, 1981

One reason scientists are still debating the effect of spray-can propellants on the earth's stratosphere is that no one has been able to measure it. The critical ozone layer is too high to be reached by aircraft and too low for satellites; instruments dropped by parachute pass through the ozone layer too quickly, and instruments held aloft by balloon can't be exposed to enough different altitudes to be useful.

What seems to be needed is a way to suspend an instrument package from the right height, 25 miles above the earth, and slowly lower and raise it over a distance of about 12 miles. What seems to be needed, in fact, is a mammoth, floating yo-yo. And that is exactly what Dr. James G. Anderson, a Harvard professor of atmospheric chemistry, is building.

Anderson's yo-yo will be suspended from a balloon 100 times bigger than the Goodyear blimp and will consist of a 120-pound instrument package at the end of 12 miles of synthetic "string" that is stronger than steel and lighter than cotton. Just like a real yo-yo (hereafter known as a small yo-yo), the energy released as the instrument package descends will be used to bring it back up again. The role of the yo-yo hand and finger will be played by a winch and flywheel platform hanging below the balloon.

The first test flight of the world's largest yo-yo is expected this fall, and within a matter of months the missing data about the earth's ozone layer should begin pouring in. It will probably be years, however, before scientists learn all they want to know about around-the-world, walk-the-dog and loop-the-loop.

The Washington Post

Washington, D.C., August 25, 1981

THE TOUGHEST PROBLEM for health and environmental regulators is what to do about a particular risk when there are suggestions or theories of danger, but little proof is available. The risks posed by human meddling with the earth's atmosphere are among the hardest to judge and the most important to understand.

Relatively little is known about the atmosphere, which sets the biological and chemical conditions for life on this planet, though not for any lack of interest. The reasons for the modest knowledge are that anything so all-encompassing is exceptionally hard and expensive to study. The five thousand trillion tons of air that surround the earth represent not only a complex thing in itself; they also interact continuously with the oceans. Small changes in such huge realms are obviously difficult to predict or document.

Despite the difficulties, a great deal is being learned about one threat to the atmosphere—depletion of the stratospheric ozone layer caused by a class of chemicals called chlorofluorocarbons (CFCs). Depletion of this shielding layer would mean more skin cancers from higher levels of ultraviolet radiation and possibly also damage to vital crops and fish species. It might also affect the worldwide climate.

The hypothesis that CFCs could damage the ozone layer is very young: it was first proposed in 1974. Despite the lack of conclusive evidence that damage was actually occurring, the United States banned CFC use in aerosols just four years later. Canada, Norway and Sweden soon did the same. Last year the 10 members of the European Economic Community bound themselves to reduce CFC use in aerosols, and went a step further by agreeing not to increase production capacity for CFCs' many other uses. An international commission, under United Nations auspices, will meet in a few months to take the first steps toward a global convention to protect the ozone layer.

But this is only part of the picture. Concerted international action is still years in the future, and while scientific knowledge is growing much faster than most people had expected, the findings are still patchy. Recently the first direct evidence of depletion was gathered from satellite measurements, but a good deal of work remains before these findings can be confirmed and the extent of the damage calculated.

Meanwhile, American CFC producers, the world's largest, reportedly have persuaded the Environmental Protection Agency to abandon plans to limit future production levels. It would be profoundly wrong to break the momentum that has accomplished so much in so short a time. Here, as in all the unresolved questions of health and the environment, the only defensible rule is to err on the side of caution until the risks are more precisely known.

EVENING EXPRESS

Portland, Maine, April 5, 1982

Scientists now believe the Earth's protective layer of ozone is being depleted far more slowly than was supposed only a few years ago. But that news ought not to lessen efforts to reduce the amount of fluorocarbons released into the stratosphere.

The use of fluorocarbon aerosol spray cans in such products as hair sprays and deodorants has been banned in the United States since 1979. Fluorocarbons are stripping away the ozone layer which protects Earth from solar ultraviolet rays, vastly increasing the likelihood of skin cancer.

At the time the ban was approved by the Consumer Product Safety Commission, the National Academy of Sciences estimated that as much as 18 percent of the ozone layer would be destroyed by late in the 21st century. Now, however, the projected rate of ozone destruction has been cut in half.

That's hardly cause for celebration. First, any depletion of the ozone layer is irreversible. Second, for each 1 percent decrease in the ozone layer there will be an increase of up to 10 percent in some types of skin cancer.

Moreover, the national ban on aerosol sprays is only partially effective since it reduced worldwide emissions from that source by only 50 percent. Other nations continue to contribute to the problem. And large amounts of fluorocarbons continue to be released into the stratosphere from refrigerators, air conditioners and other home and industrial products.

Plainly, the problem of protecting the thin and fragile ozone layer is global. Efforts to reduce fluorocarbon emissions must be approached on a worldwide basis. The U.S. ban on aerosol sprays is only a tentative first step in coping with a threat to the lives of everyone on Earth.

The Morning News
Wilmington, Del., February 27, 1984

Opening [handwritten margin note]

PERHAPS we can learn a lesson from the ozone story.

For most of us, ozone became a key word in 1978. That's when the alarm went out about ozone depletion and the danger that that could pose to us earthlings.

Ozone, we learned, is a form of oxygen, found in the stratosphere. It helps shield the earth from harmful ultraviolet radiation from the sun. Excess of such radiation can cause human cancer, damage plant life and affect the climate.

But the ozone layer was in jeopardy, it was said six years ago. Gases released from aerosol spray cans, refrigerants and other sources were depleting the ozone layer at a frightening pace. A study from the National Academy of Sciences put the possible ozone reduction rate at 15 percent to 18 percent — an alarming figure.

What to do? The federal government promptly banned the use of the alleged offender — chlorofluorocarbons — in spray cans. Use of the chemical in refrigerants was permitted to continue, though there were arguments about that too. Scientists who contended that there was no imminent danger to the ozone layer were largely ignored.

In 1982 came another study sponsored by the National Academy of Sciences. It downgraded the estimate of ozone reduction to a range of 5 percent to 9 percent. Few listened.

Fortunately, the scientists kept on researching the matter. The latest findings from the academy are in. At the current rate of use of fluorocarbon gases, the ozone layer could be diminished by 2 percent to 4 percent by *late in the next century.*

That's a startling change of perspective in half a dozen years. During that time, data gathering techniques have improved, mathematical models have been refined, other variables such as projected climate changes have been taken into account.

By today's hindsight, the ozone depletion scare of 1978 was exaggerated. How could scientists of national renown have been so far off? one is tempted to ask. How should we view their latest findings?

With skepticism. It's time that the public learned that science does not equal certainty. Science is a continuing quest for answers, facts, new theories and proofs. We must learn to accept the limits of our knowledge and attempt to operate rationally within those limits.

This winter's hysteria over contamination from pesticides containing EDB (ethylene dibromide) is as unjustified as was overreaction to chlorofluorocarbons. There are dangers, but they must be defined, observed, studied, researched and remedies must be phased in as understanding increases. Total bans and rash substitutions are as fraught with danger as the substances they eliminate.

THE KANSAS CITY STAR
Kansas City, Mo., February 27, 1984

It's embarrassing to have gotten so exercised over another of those life-on-Earth-threatening disasters—which none of us would have lived to experience anyway—when science takes another look and concludes the problem isn't that bad after all. Such is the case with the danger that fluorocarbons released into the atmosphere might deplete our planet's ozone layer, admitting ultraviolet radiation from the sun that would give us all skin cancer, at the very least.

But now the National Academy of Sciences, through its National Research Council, has made another biennial review of the situation for the Environmental Protection Agency and decided that ozone depletion would be only 2 to 4 percent over the next century—if that—instead of their 15 to 18 percent forecast of five years ago. It has now been six years, however, since manufacturers of products marketed in aerosol spray cans spent millions in retooling to different types of containers in response to this supposed peril.

Members of the council study group said when new mathematical models were used, examining the effect of other trace gases, the threat of depletion of the protective ozone layer by chorine elements diminished until there was even the possibility of a 1 percent increase. Interestingly enough, another scare scenario, the "greenhouse effect," in which warming of the Earth's atmosphere by carbon dioxide would melt the polar ice caps and inundate the world's coastal areas, would at the same time cool the stratosphere and slow the ozone depletion process.

These scientific predictions of catastrophe usually rest on projections far into the future, projections vulnerable to drastic change or—in this case—to being overturned just by finding out more about the situation. With all the real and present dangers facing the residents of this globe, from starvation-causing drought in Africa to nuclear oblivion, we don't need to be scared out of our pants by theoretical and tenuous disasters, a century or more away, which might never occur.

DESERET NEWS
Salt Lake City, Utah, February 27-28, 1984

Remember the aerosol scare of a few years ago — the fear that gases from spray cans and other sources were depleting the earth's protective ozone layer? As a result, the U.S. banned almost all aerosol sprays in 1978.

Estimates of ozone loss have been lowered considerably, the National Research Council reported this week. Where a 1979 study estimated the ozone layer might be reduced 15% to 18% in a few years, the latest report figures it might be only 2% to 4% because of chlorofluorocarbons by late in the next century.

That's more because the original problem was overstated than because of the aerosol ban. By 1982, long before the ban would have made any significant difference, the council reduced the estimate to between 5% and 9%. The latest study revises that estimate even further.

If the "greenhouse effect" is considered, says the study, then total stratospheric ozone might *increase* by about 1% over the next century. Warming of the earth's atmosphere because of the heat-holding properties of carbon dioxide — the greenhouse effect — would produce a corresponding cooling of the stratosphere. That would slow down the rate of chemical reactions that destroy ozone.

Some 95% of ozone is in the stratosphere. Ozone is formed from sunlight reacting on normal oxygen atoms. It's destroyed in reactions with chlorine atoms contained in chlorofluorocarbons.

The optimistic ozone study is good news for a world that worries about such things as changes in the earth's atmosphere. And with good reason. Significant losses in the ozone layer would strip away much of the earth's protection from ultraviolet radiation from the sun. That would increase human skin cancers, damage plant life, and possibly affect the world's climate.

While mankind can breathe a little easier from the most recent study, scientists still need to keep a close watch. After all, there's only one atmosphere protecting all earth life.

Fossil Fuels: "Greenhouse Effect"

The Environmental Protection Agency and the National Research Council each issued reports in October, 1983 predicting that a build-up of carbon dioxide in the Earth's atmosphere during the coming decades would warm the global climate, with potentially severe consequences. This warming of the Earth, known as the "greenhouse effect," was expected to occur as a result of the burning of fossil fuels, which increased the carbon-dioxide concentration in the atmosphere. (Carbon dioxide allowed sunlight to enter the atmosphere, heating the Earth, but inhibited the escape of heat radiation into space.) Atmospheric carbon dioxide has already increased from an estimated 260 to 280 parts per million before the Industrial Revolution to almost 345 parts per million today.

By the year 2100, the EPA reported, average global temperatures could increase nine degrees Fahrenheit, causing such "catastrophic" changes as the disruption of food production, coastal flooding and the desertification of some already dry areas. One alarming possibility mentioned was the rapid melting of the polar ice caps. The report said that bans on the use of coal by the year 2000 and on the use of shale oil and synthetic oil would be the most effective means of delaying the warming process. By contrast, the NRC report, though it predicted a rise in average worldwide temperatures of two to eight degrees Fahrenheit by the year 2075, stated that farmers and other people would be able to adapt naturally to the climatic changes. The council reported: "We do not regard the hypothesized carbon dioxide-induced climate changes as a major direct threat to American agriculture over the next few decades." The report recommended only that research be continued, asserting that in any case "there is no politically or economically feasible way of heading off the greenhouse effect."

Recently, in fact, agricultural scientists have been encouraged by evidence that the increase in atmospheric carbon dioxide has spurred some kinds of plant growth, particularly at high altitudes. This is because the elevated level of CO_2 promotes photosynthesis. Environmentalists are concerned, however, that these effects will not be experienced equally by all types of plant growth, and that the resultant shifts in vegetative balance and wildlife populations could drastically alter or destroy many ecosystems.

The Cincinnati Post
Cincinnati, Ohio, September 18, 1981

Whenever we weary of contemplating the more immediate threats to the survival of civilization-as-we-know-it, there's always the "greenhouse effect."

This is the possibility that an accumulation of carbon dioxide in the earth's atmosphere could, by trapping solar radiation, raise temperatures, melt the ice caps and play general havoc with the climate.

Some scientists discount the threat, but researchers with NASA's Institute for Space Studies in New York City report they have definitely detected a global warming trend going back to 1880. Associated with this is a rise in the amount of carbon dioxide in the air from about 280 parts per million to 340 p.p.m.

If the burning of fossil fuels continues at the present rate, they say, the CO2 level could double in the next century and the global mean temperature rise as much as 9 degrees.

Before this happens, we trust the scientists will be decide whether the greenhouse effect really is something to worry about and, if so, what can be done about it.

Arkansas Gazette.
Little Rock, Ark., October 22, 1980

Carl Sagan, the articulate astronomer from Cornell, has been trying to get across several broad points about the planet earth on his *Cosmos* series appearing on public television, including the Arkansas Educational Television Network, this fall. One of these points seems to be that all humans need to become more aware of what is happening to our planet that threatens its very existence unless we alter some of our patterns of using and misusing natural resources.

The point keeps coming up in other places as well. One is a report on the "greenhouse effect" that appears in a recent publication of the Agency for International Development. Dr. Sagan also cites the "greenhouse effect" as a possible insidious threat. What is it? When fossil fuels (coal, oil, gasoline) and wood, etc. are burned, carbon dioxide is released into the atmosphere. In the past, a more or less natural balance has been maintained between the amount of carbon dioxide released and the amount used by plants for growth and absorbed into the earth's soil and water. In recent years, says AID, the burning of fossil fuels has increased rapidly and the practice has coincided with massive destruction of forests, most of them in the Third World, to provide firewood and more land for increasing populations. Since 1950, alas, the earth's forests have diminished from about 12 billion acres to about 7 billion acres and they are receding at a rate of 50 million acres a year. No matter how hard we try — and man isn't trying very hard in most places — to reforest, it is almost impossible to keep even.

The fewer trees, in any case, the less carbon dioxide can be absorbed and the excess carbon dioxide begins to build up in the atmosphere. At the same time, sunlight strikes the earth's surface and much of it becomes heat that rises but is trapped close to earth by the layer of carbon dioxide — thus making the earth's atmosphere something of a greenhouse and raising temperatures worldwide. Some scientists, it should be recognized, as has the Council on Environmental Quality, "believe that countervailing factors — such as increased dust particles in the air, which reflect sunlight away from the earth and have a cooling effect — could offset the greenhouse effect." But what worries others is that if humans worldwide wait for conclusive proof that a carbon dioxide buildup actually is causing a warming trend it might then be too late to do much about the condition.

What would be the effects? Probably an altering of growing seasons that would bring persistent drouth to the Midwest of the United States. The Interior Department, already worrying in 1970 about what the "greenhouse effect" might have, calculated that a five-degree temperature increase could melt the polar ice caps with disastrous consequences. Oceans would rise enough to inundate the whole of the Eastern Seaboard, including all of Florida, and move all the way up the Gulf Coastal Plain to the foot of Cantrell Hill at Little Rock, if we read the map correctly.

The larger point in all this is that the inhabitants of earth may be destroying their home without realizing fully what they are doing, in blissful ignorance of the earth's fragile ecology.

St. Petersburg Times
St. Petersburg, Fla., January 19, 1981

It sounds like a plot for a science-fiction movie: The year is 2050. A buildup of carbon dioxide in the atmosphere has raised the average world temperature by 5 degrees Fahrenheit. Rich farmlands are turning into deserts. The polar ice caps are melting. The sea level is rising. Floods are wiping out coastal cities throughout the world. More than 11-million people in the United States are left homeless. One-third of Florida is under water.

The title of the movie is *The Greenhouse Effect*. But it could be a documentary rather than fantasy.

As bizarre as the scenario sounds, the world's climate might be dramatically altered as early as the mid-21st century — if energy planners and government officials throughout the world do not act immediately to prevent it, warns a report released last week by the President's Council on Environmental Quality.

THE GREENHOUSE effect, which causes warmer temperatures, is created by a buildup of carbon dioxide in the atmosphere that traps more heat than normal in the lower atmosphere.

The 92-page report warns that in recent decades the concentration of carbon dioxide in the atmosphere has been increasing in a manner that corresponds closely with the increasing global use of fossil fuels. The burning of fossil fuels (oil, coal and natural gas) releases carbon dioxide, about half of which is retained in the atmosphere. The upshot is that a continued worldwide dependence on fossil fuels could bring about a global greenhouse effect.

A doubling of the amount of carbon dioxide in the atmosphere could raise the average world temperature by 5.4 degrees Fahrenheit, the report says. While raising the temperature by a few degrees may not sound disastrous, it could profoundly effect the lives of future generations and the economies of most nations.

Think about the flooding that would result from higher temperatures melting the polar ice caps. Scientists estimate that the sea level would rise an average of 16.5 feet, which would have a devastating effect on the shorelands of the world.

Much of St. Petersburg would be flooded. A 16.5-foot increase in sea level would flood a coastal area of the United States now occupied by 11-million people, the report says.

Scientists also believe that the productivity of different agricultural regions would change — some for the better, some for the worse.

The Corn Belt, for example, might shift toward the north where much of the soil is less suitable for agriculture. Wheat production

might be reduced by as much as 20 percent because of drier, warmer conditions in some of the present wheat producing areas of the world, including the United States and the Soviet Union. On the other hand, rice production might increase due to higher temperatures. Overall, though, scientists fear that drastic climatic changes would have an adverse effect on the world's ability to feed an increasingly large population.

This doomsday plot could be altered. The solution is obvious: Industrialized nations must increase energy efficiency and accelerate the development and use of renewable energy sources that do not emit carbon dioxide. In the short run, it may be necessary to convert some oil-burning electric plants to coal-burning units as a cost-reduction measure. However, to avoid the devastating consequences of the greenhouse effect, the long-range goal must be the development of alternative energy sources as part of a worldwide energy policy.

The time to act is now — when government officials in many countries are evaluating their energy policies and will be making decisions on what kind of facilities ought to be built to meet their energy needs for the next 30 to 40 years. That's why the report rings with urgency. If the world continues to rely on fossil fuels, the

likelihood of avoiding a global greenhouse effect will be much diminished.

THE SAD truth is that the international cooperation needed to address this worldwide threat has seldom, if ever, been achieved.

However, as the largest single consumer of energy in the world, the United States ought to exercise vigorous leadership in addressing the problem. As trustees of the earth for future generations, it is the responsibility of every nation to deal with the carbon dioxide menace. If the United States does not assume a leadership role by demonstrating a commitment to reducing the risks of a drastic global climatic change, which nation will?

Des Moines Tribune
Des Moines, Iowa, January 23, 1981

The President's Council on Environmental Quality last week warned that if the United States and other nations continue to place increased reliance on such fossil fuels as coal and oil and on synthetic fuels made from coal and oil-shale, a major change may be triggered in the world's climate that could cause such "widespread and pervasive changes" as a shift in the world's major crop areas, flooding of major coastal cities, and social and economic disruption.

The report concluded that burning fossil fuels leads to a net increase in the amount of carbon dioxide (CO_2) in the atmosphere. Since the Industrial Revolution, the concentration of carbon dioxide has risen 15 to 25 percent.

This buildup creates the so-called "greenhouse effect": Higher concentrations of carbon dioxide trap more heat in the lower atmosphere. The council's report estimated that, if the carbon dioxide is doubled — as it will be in 40 to 70 years at the recent rate of growth — surface temperatures might eventually increase by about 5 degrees Fahrenheit, while temperatures near the North Pole could rise by as much as 13 to 18 degrees during the winter.

Such a warm-up would likely lead to a shift in the areas best suited for agriculture. Corn yields might be cut in the Midwest, the report warned, and U.S. wheat production might drop.

A major warming trend might eventually cause disintegration of the West Antarctic ice sheet, which could cause ocean levels to rise gradually by 5 to 8 meters, enough to flood U.S. coastal areas where 11 million people now live.

Scientists do not fully understand the impact of a carbon-dioxide buildup on the world's weather, but the CEQ concluded that the problem is serious enough to require immediate action. A decision taken in the 1980s to greatly expand the use of fossil fuels would make it harder to avoid the dangerous worldwide warm-up described in the report.

The danger of a carbon-dioxide buildup strengthens arguments for energy conservation and the use of solar power.

The Reagan administration cannot responsibly ignore the findings of this Carter-administration report. As Gus Speth, chairman of the CEQ under Carter, warned in his introduction to the report: "The carbon-dioxide issue may present the ultimate environmental dilemma."

THE SUN
Baltimore, Md., October 29, 1981

As if mankind didn't have enough to worry about, there is the "greenhouse effect." Greenhouses stay warmer inside than out because their glass walls and roofs admit sunlight and hold in heat. Carbon dioxide gas creates a "greenhouse effect" in the entire atmosphere. As levels of the gas go up due to man's burning of carboniferous fuels (coal, oil, natural gas, oil shale, other "fossil" fuels) the atmosphere tends to trap more heat. Theorists wonder if the world may not get warm enough one day soon to melt polar icecaps, flood coastal cities, disrupt climate and agriculture and cause other major problems.

These omninous prospects edged a bit further beyond theory recently when Columbia University scientists reported that satellite observations tend to confirm that the melting may have started. Summer ice at sea near Antarctica is more extensive than it was earlier in the century (meaning that more ice is breaking away from Anarctica), and average air temperatures in certain parts of the Northern Hemisphere have risen almost 1 degree Celsius since 1938. While the scientists acknowledge the possibility of causes other than carbon dioxide increases, theirs isn't the only data. NASA reports, for instance, that the earth's tem-

perature has been on the increase since 100 years ago — roughly the time industry started burning fossil fuels in very large quantities.

Only continued research will prove beyond doubt that the greenhouse effect is at work. But the evidence is piling up fast. Moreover, there are many other reasons why it is in mankind's best interests to lessen reliance on fossil fuels — to prevent acid rain and the ravages of strip mining, for instance. The carbon dioxide danger may be the clincher.

Fossil fuels are, of course, exactly that: fuels which once were living plants and animals. These fuels were laid down during a geological period when the plants and animals flourished in a warm, moist, swampy, feculent world of a sort that hasn't been known since. Presumably the world was that way because carbon dioxide levels were higher than they are now — and the gas not only warmed the atmosphere but also was a major nutrient stimulating plant growth. Much of the carbon dioxide eventually was immobilized as plants died and were not recycled, and the world then cooled. Man is now recycling the ancient plants, thus ushering in a new warm, moist, swampy, feculent world—unless he takes firm steps to do otherwise.

DESERET NEWS
Salt Lake City, Utah, November 21, 1981

Sometime within the next century, the Council on Environmental Quality warned early this year, the concentration of carbon dioxide in the atmosphere could double from what it is today.

If that were to occur, the resultant heating of the earth from the "greenhouse" effect could melt the West Antarctic ice sheet, raising the sea levels and flooding vast land areas of the world.

Even more significantly, said CEQ, the warming effect might be considerably higher — perhaps as much as 40 percent — if increases in gases other than carbon dioxide disturb the existing chemical composition of the atmosphere. If that were to happen, the rise in ocean levels might begin as soon as the next 50 years.

A recent study by the National Aeronautics and Space Administration lends credence to those fears. NASA's Institute for Space Studies in New York has found that atmospheric carbon dioxide has increased from 280-300 parts per million in 1970 to 335-340 parts in 1980. The concentration of CO2 is projected to grow to 600 ppm by the end of the next century, even assuming that there's a slowdown in the rate at which fossil fuels are consumed. As a result,

the global temperature is expected to rise five degrees Fahrenheit, causing gradual melting of the polar ice caps.

That process would be speeded up by such factors as major deforestation, which is occurring right now; more rapid burning of coal as petroleum gives out; or disturbance in the chemical composition of the atmosphere.

Several unknowns could alter that future. For example, increases in global vegetation could keep CO2 levels down, the oceans could absorb more CO2 than expected, or dust in the atmosphere could reflect sunlight and tend to counteract the increased carbon dioxide by cooling the earth.

The bottom line is that we do not have adequate information yet as to the long-term effects of such atmospheric changes. The government agency charged with keeping a finger on such problems — the Council on Environmental Quality — was first reduced in size and then eliminated this year by the Reagan administration.

That assignment ought to be taken over by some other government agency or private group. If changes are needed to avert such long-term catastrophes, it's better to act early than wait until the process is irreversible.

The Evening Gazette
Worcester, Mass., May 19, 1981

Gloomy predictions have depressed human beings ever since the description of the Apocalypse in Revelations.

In our own day, the so-called "greenhouse effect" theory may be the most scary. It says that the increased use of fossil fuels, particularly coal, is gradually building up a layer of carbon dioxide in the atmosphere that eventually will raise the earth's temperature, melt the Antarctic and Greenland icecaps, flood the coasts of the continents and raise havoc with agricultural production.

But now comes Sherwood B. Idso, a scientist with the United States Department of Agriculture, to rebut the whole hypothesis. Idso, a physicist who has been measuring incoming solar and thermal radiation in the Science and Education Administration's Phoenix laboratory for the past dozen years, says temperatures will not increase nearly that much. Even a doubling of the carbon dioxide concentration in the atmosphere would be insignificant, he says.

What's more, Idso says, any increase in carbon dioxide would be beneficial, in that it would help increase food production. He claims that carbon dioxide in real greenhouses is often several times as concentrated as it is in the air.

"We must completely change the thinking of world leaders on this topic," says Idso. "We actually need more carbon dioxide in the atmosphere."

If Idso is right — and, of course, his prediction is only a theory, too — a lot of thinking could be changed. For one thing, one of the main arguments against burning coal will be refuted.

But it's a relief to have at least one authoritative voice speaking out confidently and optimistically on the future. We've had an awful lot of scientific pessimism lately.

The San Diego Union
San Diego, Calif., October 8, 1981

Scientists are warning that the "greenhouse effect" — a warming of worldwide temperatures as a result of the burning of fossil fuels — may be occurring faster than anyone expected. It is neither wise nor fair to assume that this problem is something for future generations to deal with.

The "greenhouse effect" has become a convenient term to describe the consequences of a build-up of carbon dioxide in the earth's atmosphere. This invisible blanket traps heat that would otherwise escape. The inevitable result is a subtle but persistent rise in temperatures.

There are frightening scenarios describing what could happen if the global temperature rises by only a few degrees. A melting of the polar ice cap would raise the level of the oceans and flood low-lying coastal areas. Some of the world's most abundant food-producing areas — the Western United States, Canada and parts of central Asia — would be stricken with drought.

Southern California would feel these effects acutely. Our coastline would be redrawn by the rising ocean levels. Declining rainfall in the West and Southwest

would aggravate our water-supply problems, which are formidable enough already.

Until recently computer projections were indicating that the continued burning of fossil fuels would begin producing significant effects on the world's climate sometime in the 21st century. Now that timetable is being reconsidered. A report published in the journal *Science* suggests that the phenomenon is occurring more rapidly, because the carbon dioxide build-up is accompanied by a build-up of "trace gases" — nitrous oxide, Freon and methane.

The study charts an increase of .4 degree Fahrenheit in the earth's average temperature between the mid-1960s and the present. At this rate we could see a rise of a full degree by 1990. A rise of four degrees would be enough to trigger major climatic changes.

Not all scientists agree on the meaning and implications of such atmospheric studies, but the warnings are consistent enough that we can no longer omit consideration of the "greenhouse effect" from decisions about fuel and energy we are making today.

Fossil fuels — oil, gas, and coal — are still the mainstay of energy production the world over. The best alternatives that do not contribute to the "greenhouse effect" are nuclear and solar power. But nuclear power development is currently stalled on issues of safety and cost, and the harnessing of sunlight as a basic energy source has many limitations.

The gradual warming of our atmosphere cannot go on indefinitely without cataclysmic results affecting millions of the earth's inhabitants. This cannot be ignored in debates about our energy options.

How do the risks associated with nuclear power stack up against the risks of not exploiting nuclear power? Are the nation's coal reserves going to be as useful as an energy source as we have always presumed? Will new energy technologies like fusion be available in time to keep the world from generating its own disaster with fossil fuels?

These are the kind of questions that the "greenhouse effect" is forcing us to answer, and perhaps sooner than we thought.

The Idaho STATESMAN
Boise, Idaho, March 30, 1982

New evidence supporting theories about the "greenhouse effect," which scientists say is causing the Earth to heat up, poses some interesting questions about how — and whether — society will act to control its future.

When climatologists bring up the greenhouse effect, they talk about farm regions becoming arid and deserts turning green as rainfall patterns shift. They also mention the possibility that coastal areas — up to 25 percent of places like Florida and Louisiana — could be flooded if rising temperatures melt part of the Antarctic ice cap, causing the oceans to rise as much as 15 feet.

The culprit in this scenario is carbon dioxide given off when fossil fuels are burned.

In the atmosphere, carbon dioxide works like the glass in a greenhouse. Transparent to visible light, it lets the sunshine in to warm the Earth. Because it is opaque to infrared radiation, however, carbon dioxide tends to keep the heat from being released back into space.

Scientists estimate that the level of carbon dioxide very likely will double within the next century, causing the Earth's average temperatures to rise by as much as 2 degrees Celsius. If that happens, we likely will begin to see significant climatic changes within 50 years.

The question is: What, if anything, can we do?

At one extreme, some might be tempted to suggest we quit using fossil fuels now. Practically and politically speaking, that's impossible, considering our energy needs and resources.

At the opposite extreme, others might say do nothing, that we will deal with the problems as they arise.

Where we end up between the two extremes depends on how seriously we take the scientific community.

Currently, scientists are certain climate will change significantly, but their research lacks the precision to say for sure how and where the changes will occur. The realistic course is to give researchers the money to improve computer modeling techniques and gather the massive climatic data needed to formulate predictions accurate enough for policy-makers to act upon.

President Reagan's 1983 budget has cut proposed funding for research in the area to $8 million from $12 million appropriated for 1982. Reagan proposed the cut even though the Energy Research Advisory Board, a panel of scientists and business representatives, recommended the funding be increased.

Congress would do well to ignore Reagan's recommendation and fund the program to at least the 1982 level. The difference of $4 million doesn't really sound too outrageous given the possible changes in climate.

Rockford Register Star
Rockford, Ill., March 30, 1982

Maybe you've heard of it? The "greenhouse effect"?

It's not science fiction. But it does have its horror side.

"What if . . . (as all horror stories begin). What if outside temperatures began to rise all over the earth? This would melt the polar ice caps. That would raise (the level) of the oceans. And global tragedy would result."

These were speculations when, in the early 1960s, Prof. Melvin Calvin, University of California-Berkeley, won his Nobel Prize for work on carbon dioxide collections in the atmosphere.

Such concentrations have increased 7 per cent in the last 25 years, causing scientists to fear that they will trap heat, making areas arid that were once fertile farms and causing coastlines to flood. The Antarctic ice cap already is diminishing.

Today, Professor Calvin sits on the Energy Research Advisory Board which has recommended that carbon dioxide research be this nation's second top priority this year, including the burning of fuels that produce this chemical.

The other shocker is that this is the year the Reagan administration decided to cut back carbon dioxide research by a third.

That's why two United States House sub-committee chairmen called a hearing recently to question this economy move. They are Rep. Albert Gore Jr., D-Tenn., and Rep. James Scheuer, D-N.Y.

Among their witnesses was Professor Calvin, testifying against a background in which it was established that the earth's atmospheric temperature will rise several 10ths of a Celsius degree in this decade alone.

Said Gore, "It is apparent that we are no longer dealing with merely a provocative theory."

What he and like-minded colleagues want is an "early warning" system to detect the onset of the "greenhouse effect." And that means more money in a harried budget year.

BUFFALO EVENING NEWS
Buffalo, N.Y., October 23, 1983

Every now and then, scientists come out with warnings of future catastrophes through global climatic changes. In the past, it has been difficult to get very excited about this, since the changes were far off in the future and the scientists, like economists, never seemed to agree. Some said temperatures were going to rise and melt the polar ice caps; others said temperatures were going to drop and create a new Ice Age.

Now, however, there seems to be something to get excited about.

A new report of the U.S. Environmental Protection Agency says that average global temperatures are rising, that the first effects will be noticeable in the 1990s and that the changes could be "catastrophic" by the year 2100. The rise in temperature is caused by the carbon dioxide poured into the atmosphere over the past century or more through the burning of coal, oil and natural gas. The carbon dioxide keeps the heat from radiating into space.

This is not just a theory, the EPA emphasizes, but something that is inevitable. If so, study and planning are needed to make any adjustments as painless as possible. A rise of several degrees may make some farmlands into deserts, but it may make possible the opening up of new regions. A rise of four to seven feet in the sea level by the year 2100 may cause flooding in many cities, but the rise will be gradual and population centers can be planned on higher ground.

There could be important strategic considerations. A warming trend would open up vast new areas of farmland in Canada and the Soviet Union. The mineral wealth of Siberia would be easier to tap. Similarly, Antarctica might be more attractive and, perhaps, the focus of international rivalry.

For Buffalo, the change might not be too bad. The EPA report says New York City could have a climate like that of Daytona Beach by the year 2100. If so, Buffalo would be only a bit cooler. With our beautiful Lake Erie, it could be a vacation paradise here.

Roanoke Times & World-News

Roanoke, Va., January 12, 1982

SCIENTISTS aren't yet unanimous in predicting the results of the carbon dioxide buildup in the atmosphere. But opinion now favors the "greenhouse effect," a global warming trend that could raise average temperatures more than 5 degrees in the next 50 years, setting off massive climate changes.

What's more, it seems the door to the greenhouse is even now swinging shut. If man wants to alter what is now happening to his atmosphere, there's only a decade or so to act. He'll live with the results for a thousand years.

Carbon dioxide (CO_2), released mainly through the burning of fossil fuels like oil and coal, has been increasing in the atmosphere since the Industrial Revolution. The buildup has accelerated in the last 20 years, to a rate of 1.5 percent a year.

These tiny particles let sunlight through but reflect heat and energy back to Earth, creating the greenhouse effect. The results, in the form of slowly rising average temperatures, may become apparent by the late 1980s. And once in the atmosphere, those particles are very slow to settle out. A report of the National Academy of Sciences has said:

"Unfortunately, it will take a millennium for the effects of a century of use of fossil fuels to dissipate. If the decision is postponed until the impact of man-made climate changes has been felt, then, for all practical purposes, the die will already have been cast."

The die could turn up a baleful snake-eye. The climate affects how and where everyone lives. Large-scale climate changes could lead to famine, economic upheaval, social disruption, mass migrations, even war. There could be long-term benefits to a warmer planet: e.g., longer and better growing seasons in some key agricultural areas. But those benefits would be generations away, and from here they hardly seem worth the trauma and suffering that would come along the way.

It has taken scientists about 25 years to near a consensus on the effects of CO_2 buildup; how long might it take for a majority of nations to agree? And since some parts of the world would fare better than others from the prospective change, could a new policy be forged and put into effect? If so, what needs to be done?

To slow or halt the increase of carbon dioxide in the atmosphere, man would have to substitute fuels that don't contribute to the CO_2 load: energy from the atom, wind, hydropower, biomass and the sun. He might also have to stop destroying the world's forests; this destruction, some scientists contend, releases CO_2 from biological matter.

Policies like these would require compromises and cooperation on a scale the family of nations has never achieved. The industrial nations are the big consumers of fossil fuels (the United States contributes an estimated 25 percent of the carbon dioxide added to the atmosphere); most deforestation goes on in the developing nations — which would like to be big consumers of fossil fuels.

Some believe there's no point in trying to roll back the CO_2 tide. Charles F. Cooper, an ecologist at San Diego State University, has written that "The only practical strategy is adjustment. Technology will help to ease that adjustment but the institutional rigidities of our advanced societies may correspondingly make the response to the changed climate stickier than in earlier, simpler ages."

That makes sense. It could be much easier to prepare for the future than to change its shape. But if realism is not to become fatalism, scientists must reach agreement on what the CO_2 situation demands, and then win over the policymakers. That's just the first step or two. As the doomsayers are fond of reminding us, time is short.

The News and Courier
CHARLESTON EVENING POST

Charleston, S.C., October 29, 1983

Two recent reports — one from the Environmental Protection Agency and one from the National Academy of Sciences — predict the average world temperatures will rise about 9 degrees within the next hundred years. Carbon dioxide from burning fossil fuels has made Earth a giant greenhouse as fewer of the sun's rays escape the atmosphere. If these predictions hold true, life as we know it will alter drastically.

More than shorter winters and extended summers are at stake here. Storm patterns and amounts of precipitation will shift causing major changes in agriculture. For example, the Midwest's Wheat Belt will move further north as the growing seasons in northern Canada lengthen.

But what's more worrying in the Lowcountry is the anticipated rise in the sea level brought about by heavier rainfalls and melting Arctic ice flows. It's possible that the Holy City, which has survived so much for over 300 years, won't make it to the next centennial.

The reports were not released to create panic but to urge planning. The challenges must be met by all nations, without regard to political persuasion. This is a tall order for a society that has not yet learned to cope with existing weather phenomena such as earthquakes, tidal waves and such. International committees should assemble to study alternatives in the areas of agriculture and coastal planning in light of what's to come. If definitive problems can be identified and isolated, then the next step would be for nations to work together for solutions starting perhaps with "mutual and verifiable" bans on burning fossil fuels. It's possible that, spurred on by Mother Nature, international agreements can be reached that have so far eluded the politicians.

Wisconsin State Journal

Madison, Wisc., October 24, 1983

At the end of the Carter administration, government experts warned in the controversial "Global 2000" report that carbon dioxide pollution could eventually produce a "greenhouse effect" — the warming of the earth's climate due to the trapping of the sun's heat in the atmosphere.

The experts said the change could alter global crop production and coastlines as icecaps melted somewhat.

That prediction was dismissed earlier this year in a counter-study financed by the Heritage Foundation. Researchers led by Herman Kahn of the Hudson Institute called it doomsday stuff put out by bureaucrats hoping to inspire greater government regulation.

Now the greenhouse effect is back in the news again.

The Reagan administration's Environmental Protection Agency warns the greenhouse effect will start changing life on Earth in the next decade. Not *might*, but *will*.

The EPA is paralleled by another study just released by the National Academy of Sciences.

The academy agrees the potential exists for the greenhouse effect, but is less alarmed than the EPA over the immediacy and extent of the threat.

The academy says the effects won't be felt until the beginning of the next century, and that problems of climatic changes and rising oceans will not be insurmountable.

Both groups acknowledge that readjustment on a global scale will be needed, and the root cause is human activity.

Although carbon dioxide enters the atmosphere as a natural by-product of respiration and decay, much of the greenhouse effect is the result of pollution from burning fossil fuels, combined with the widespread clearing of forests for agriculture. (Plants consume carbon dioxide in the atmosphere for use in photosynthesis.)

The two studies underscore that human activity can produce profound changes in the ecosystem. Some may be easy to adjust to. Others may not.

The reports remind Americans, and others, that we must take realistic stock of how human activity affects the environment, then do what is necessary and possible to prevent harm.

THE MILWAUKEE JOURNAL
Milwaukee, Wisc., November 4, 1983

Two eminent science groups recently issued separate reports that came to the same troubling conclusion: Carbon dioxide and other pollutants are building up in Earth's atmosphere. Humankind will pay a price for the inevitable consequences.

This buildup traps the sun's heat the way glass in a greenhouse does, and will warm the atmosphere. The warmup won't be more than a few degrees, but it is likely to produce big changes in the way we live, and the changes may come before the end of the century.

The reports were prepared by the Environmental Protection Agency and the National Academy of Sciences. The NAS study was less pessimistic than the EPA report, and far less urgent.

The EPA strongly suggested that plans be made now to deal with the changes. Its report bluntly said that "a global greenhouse warming is neither trivial nor just a long-range problem." The academy, by contrast, said the country could safely wait and study the problem for another 20 years before making drastic plans.

If nothing else, the two reports say what environmentalists have been saying for many years: that the earth is fragile as well as bountiful, that its resources are finite, and that its capacity for absorbing the punishment of pollution is limited.

The world cannot continue to burn fossil fuels and in other ways to exploit the earth without, sooner or later, facing the consequences.

Not all of the consequences will be dire, or even unpleasant. Increased amounts of carbon dioxide may increase agricultural production, for instance. But most of the consequences will be disruptive, some may be catastrophic, and all of them, as the scientists say, should be prepared for.

The Wichita
Eagle-Beacon
Wichita, Kans., October 23, 1983

Houston Chronicle
Houston, Texas, October 19, 1983

The "greenhouse effect" isn't new. For years, it has been discussed in abstract terms, something that might or might not happen sometime in the future.

Now, the Environmental Protection Agency says complacency had better be put aside and planning should begin. The temperatures are getting warmer, and by year 2000 there will be major changes. By 2100, New York City could have Florida-like weather.

The Earth will get warmer because increasing amounts of carbon dioxide in the atmosphere trap heat from the sun. The carbon dioxide is coming in a large part from the burning of fossil fuels. The result will affect crops, change rain and drought patterns and even raise the sea level because the polar icecaps will become smaller.

Consider what has struck our state this summer. As the western part suffered from lack of rain, the Gulf Coast had flash floods at frequent intervals. These weather disturbances have been considered within the normal range. If what the EPA says is correct, and the scientists say there is no longer any "if" about it, then there will be even more drastic weather patterns in the near future. Not only that, the scientists interviewed for the study claim that there is nothing that can now be done to prevent this change from taking place. Perhaps it can be delayed, but it cannot be stopped. Projected over centuries, the change could be catastrophic.

In schoolbooks, we read about ice ages and how seas covered what are now deserts. What was fascinating is now threatening.

It's 2040. New York City is renowned for its mild winters. The climates of Florida and Louisiana have become unliveably hot. Because the polar ice caps are melting, both states are a third smaller. Saskatchewan is the center of the wheat belt. Kansas still grows wheat, but far less than in the 20th century, and is fast becoming a desert. Everywhere in the United States winters are short and mild, summers long and hot . . . .

This isn't the setting for a "doomsday" movie, but, say the Environmental Protection Agency and the National Academy of Sciences, an accurate depiction of the 21st century American climate. Pollutants now accumulated in the atmosphere, most notably carbon dioxide, assure an irreversible warming trend on earth. The two organizations agree no amount of human intervention can forestall it.

These pollutants form a layer high in the atmosphere that prevents heat radiating from earth to escape into space. This is called the "greenhouse effect," for the layer is analagous to the glass roof of a greenhouse. Scientists estimate that since the dawning of the industrial age in about 1860, carbon dioxide and other pollutants in the atmosphere have increased more than 20 percent, trapping more and more heat. The primary source of the pollutants is carbon dioxide created by burning of fossil fuels — coal and oil. But the layer also includes woodsmoke, freon from leaking refrigeration systems, hydrocarbons from automobile exhaust, and methane gas emitted when animal and vegetable matter decompose.

It's too late for a worldwide moratorium on pollution-causing activity — which would be politically infeasible as well — to reverse the warming trend. The damage has been done, and, says the EPA, warming has begun already: In the past 100 years, the melting of the polar ice caps has raised ocean levels about one foot. But both organizations note that careful planning can avert some of the warming trend's effects. Residents of Kansas and other Great Plains states, in particular, have a high personal stake in preparing for the inevitable now.

The EPA advises that plant geneticists should begin developing strains of wheat that will withstand extreme heat and require little water. The presently available hardy strains of wheat likely won't be suitable to the region's new climate.

The politics of water, already controversy-ridden in this region, will become more so. Users of the Ogallala Aquifer should understand that that non-renewable source of water may give out just as the effects of the warming trend are beginning to be felt — and plan accordingly. The State Water Office should recognize that any water plan based on the assumption Kansas' climate is a fixed entity likely won't be adequate to water needs of the mid-21st century.

Perhaps more importantly, the efforts to conserve soil that grew out of the Dust Bowl, unfashionable in the present time of relatively plentiful water, again must be seen as imperative. "If the greenhouse warming produces the effect that I consider most probable, in spite of the uncertainties," writes climatologist Walter Orr Roberts, "the Dust Bowl of the '30s will seem like children's play in comparison with the Dust Bowl of the 2040s." He foresees a 40 percent decline in rainfall for the region, and adds, "The soils will dessicate, and the winds will lift them to the skies." The academy says humanity has perhaps 20 years to prepare for the changed climate.

None of this is to suggest that humans may as well continue to pollute the atmosphere. To do so would bring on the warming trend more quickly than now apparently is inevitable. And, considering that the warming trend isn't the only adverse effect of atmospheric pollution — acid rain is another that comes readily to mind — any reduction in pollution will make earth a more healthful planet on which to live.

The Washington Times

Washington, D.C., October 27, 1983

Don't jump! All is not lost! That's for all of you who read the EPA report about the dire advent of the greenhouse effect and were ready to take your chances with a possibly-cooler afterlife. Temperatures will rise, the report said, and by the year 2100, New York City might have weather found nowadays in Daytona Beach, Fla. Ice caps will do some melting, and coastal cities like Charlestown, S.C. could become bedroom communities for Charlie The Tuna and friends. Aieeeeee? Oh hush.

Without a doubt, the Earth's temperature will rise thanks to an upper-air buildup of carbon dioxide and other gasses which allow sunlight in, but keep heat from dissipating into space. The chief cause of the buildup is the burning of fossil fuels, augmented by the widening use of nitrogen fertilizers and the general decomposition of things. And even if we burned no more coal and let the land lie fallow, the heat will rise, and the surf will go up. As has been expected for years. There's no stopping it.

Just how hot things might get depends on the forecaster, however. The EPA warns of a fried egg on every sidewalk, saying that within 17 years, the worldwide temperature could rise by 1.1 degrees, then jump 3.6 degrees by 2040 and 9 more degrees by the start of the 22nd century.

Predictions of how high the sea will rise vary from 20 inches over the course of the next century to five feet or more by the year 2100. Some scientists say the figures are too conservative. Others say they are way too high. We're not going to commit, but we're not going to worry much either.

Certainly, let's not fear that the bread basket is about to burn up. Do look for wider use of corn hybrids that are less water-demanding — evaporation will be speedier in the greenhouse — and for an avoidance of water-intense technologies. And there are always the sorghums, which are adapted to hot and dry climates.

And not all of the changes will be adverse. More carbon dioxide in the atmosphere will spur many plants to grow faster with less water. Some areas that are now unproductive farming regions because they lack rainfall and enough heat will blossom.

In the northern states, expect an increase in the growing season of several weeks in both the spring and fall. Plus a scarcity of inland ice, which will make the Great Lakes more easily navigated in winter. And the wheat belt will probably move north: southern Ontario could become climatically what southern Illinois is today.

Who knows what else might be affected. Will Vail and Aspen close down? Will there ever again be a White Christmas after the year 2000? One scientist tells us that after a few centuries, once new energy sources are harnessed and fossil fuels deplete, things could get back to normal. And then we'll get reports about the return of the ice age.

Remember how hot it was last summer? That's probably going to be something of a summer norm. So what? This is the nation that swept the Nobels in science this year. We'll adapt, as acknowledged by a more thorough report on the greenhouse effect and its consequences released this week by the National Academy of Sciences. Could even prosper.

There's no sense in opening a vein because of a little more heat. Opening a window will do.

SYRACUSE
HERALD-JOURNAL

Syracuse, N.Y., October 19, 1983

"New York City could have a climate like Daytona Beach, Fla., by 2100."

One would think that statement — used in a new Environmental Protection Agency (EPA) study — would be reason for celebration.

But don't grab bathing suits and head for Manhattan yet. In fact, the EPA example is an indication of the immediacy of worldwide change with some ominous implications.

The EPA calls it the "greenhouse" effect — a warming trend brought on by a buildup of carbon dioxide in the atmosphere. "We are trying to get people to realize that changes are coming sooner than they expected. Major changes will be here by the years 1990 to 2000, and we have to learn how to live with them," says John S. Hoffman, the EPA's director of strategic studies.

Although the study concedes there is still considerable uncertainty about the rate of change, the EPA projects average worldwide temperatures could increase 2 degrees centigrade (3.6 degrees Fahrenheit) by the year 2040. That, the experts say, would lead to even greater temperature changes in the polar regions, which could cause polar ice caps to melt rapidly.

Changes is the earth's makeup have come slowly over the eons. Scientists believe it will be different this time.

Melting polar regions will produce higher coastal waters. Shoreline states cities will have their coastal boundries radically changed in a century or two; some may be under water. Altered weather patterns will also change inland climates.

Not all the alterations will be bad. We can expect more rainfall, which should be a benefit the worldwide production of food. But, the EPA report warns, changes — even ones to which humans can easily adjust — will bring economic disruptions costing billions of dollars.

The report is a startling reminder that what we do to the earth and its atmosphere has a direct impact on what future generations will be able to do. We have not yet come to grips with that aspect of the industrial-technological revolutions of the 19th and 20th centuries.

If we don't start doing a better job of thinking ahead and planning for consequences, we're apt to be washed aside by nature — which knows how to adjust.

The Star-Ledger

Newark, N.J., October 28, 1983

By the turn of the century, Florida no longer may have its main tourist staple for luring New Jerseyans who want to escape the rugged, chilling temperatures of Northeast winters for the soothing warm climate of the Sunshine State.

If the startling scientific findings of a study by the U.S. Environmental Protection Agency (EPA) have a valid basis, by the year 2000, the Garden State will have a balmy climate similar to the one now found in Florida during winter months. This radical climate transformation—a worldwide warming of the earth caused by a buildup of carbon dioxide in the atmosphere—is ahead of schedule.

The drastic change in weather patterns has been forecast for years by private scientists, but the EPA report is the first warning by the federal government that the "greenhouse effect" no longer is in the theoretical realm but a threat in the imminent future, a matter of only several years away.

However, there are skeptics in the scientific community. Other independent studies of atmospheric changes question the reliability of the EPA findings, which were based on research done by two laboratories. A change of potentially catastrophic implications should be a matter of far more extensive scientific investigation before any definitive conclusions can be reached.

A more conservative, cautionary approach has been taken by a Princeton University scientist, Joseph Smagorinsky, a highly regarded world authority on climate. He was sharply critical of the EPA study; he found it to be "unnecessarily alarmist... (using a) worst-case" scenario as a premise for its projections rather than an assessment based on some positive factors.

There is a scientific consensus that a worldwide warming of the earth is taking place, but there is wide divergence of thinking on the timing, a key element in the "greenhouse effect" meteorological phenomenon. The federal study projects an impending emergency which is disputed by Smagorinsky.

"Our conclusions," he said, "now show there will be a warming of the earth, but one that will not be dramatically discernible until the end of the 21st Century. It is impossible to know the exact impact of the temperature changes. We just don't know enough yet."

What has been firmly established is that this is a man-made problem, a contamination of the atmosphere by the emission of carbon monoxide that began with the advent of the Industrial Revolution in the mid-1800s. The United States has undergone a significant changeover from a basic smokestack industrial base, but there has been a serious deterioration of the atmosphere which appears to be irreversible.

Only now are we learning of the damaging global ramifications of a seriously polluted atmosphere, but there is still time to develop strategies to cope with the full impact of the warming trend, and to minimize its adverse effects.

The Salt Lake Tribune
Salt Lake City, Utah, October 24, 1983

The Environmental Protection Agency advances a mighty scary scenario; the world is going to get warmer, the polar ice caps are going to begin melting at an awesome rate, seaside cities will have to cope with massive elevations in the seas' levels, towns with temperate climates will become semi-tropical and areas now devoted to the growing of grains like wheat will no longer be suited for such enterprise.

This prognosis is not capriciously arrived at, but is the work of competent and highly qualified scientists who attribute the drastic climatic changes they foresee to the "greenhouse" effect. This is the name given to the buildup in the atmosphere of carbon dioxide, which will act like the glass in a greenhouse, allowing the sun's rays to warm the Earth and then trapping the heat. Carbon dioxide is produced primarily by the burning of fossil fuels — oil, coal and natural gas.

That climatic changes are possible and that they work significant and permanent change can't be denied. Reputable scientific studies of the Earth's geologic history have documented this. It was, after all, changes in climate that ended the multi-million year reign of the dinosaurs. And in more recent times, archaeologists have offered convincing hypotheses that the Anasazi culture of the Four Corners area was negatively impacted because of a prolonged drought in the late 13th century, resulting in the abandonment of large communities of cliff dwellings.

Press reports of the latest EPA predictions of a warming Earth, resulting from the fossil fuel induced accumulation of carbon dioxide in the atmosphere, make no mention of a natural process that might counteract, or at least inhibit, the "greenhouse" effect — green plant photosynthesis.

Photosynthesis works directly opposite to the combustion of fossil fuels. Instead of consuming oxygen to produce carbon dioxide and water, the consumption of carbon dioxide by plants, in the presence of light, yields oxygen and water. Thus, the metabolic processes of green plants might, by consuming large amounts of carbon dioxide, play an inhibiting role in the progress of the "greenhouse" effect on the Earth.

It is nearly impossible to conceive of competent scientists, like those at EPA, not taking photosynthesis into consideration when working up their postulates on the "greenhouse" effect. Still, we would like to know what role, if any, this basic, fundamental natural phenomenon will play in the forthcoming world catastrophe so graphically and urgently outlined by EPA.

The Birmingham News
Birmingham, Ala., October 23, 1983

Back in the 1950s, a number of Hollywood B-movies drew their inspiration from frightening portrayals of disasters brought about by dramatic changes in the world's climate patterns.

With the report of the Environmental Protection Agency this week on a peculiar atmospheric condition known as "the greenhouse effect," the stuff of Hollywood science fiction has now been annointed formally as the stuff of science.

The "greenhouse effect" is a long existing theory which says that a build-up of carbon dioxide in the earth's atmosphere — from burning fossil fuels and other sources — has created a condition in which the gases will allow sunlight through, but then trap heat like a greenhouse. The result, according to the theory, will be a gradual, but steady, heating up of the earth's atmosphere and severe alterations of many climate areas.

The EPA report, reviewed by 100 scientists, is the most pessimistic one to date on the phenomenon, with the agency warning that some changes could begin occurring within 20 years and that the problem has advanced to the point that little can be done except to prepare for it.

There are scientists who disagree with EPA's dire warnings (in fact a report on the same topic by the National Science Foundation expected to be released this Friday will probably paint a somewhat more optimist picture). But most of those dissenting scientists are disagreeing with specific figures used by the agency or the urgent nature of its report — not the conclusion that the "greenhouse effect" is in fact taking place.

The question now is what to do about it. EPA says its study shouldn't be taken as a "doomsday report" but rather as a vehicle to be used by planners to prepare for changes. That admonition in some ways is almost as scary, however. For, as things stand now, one gets a mental picture of politicians from coast to coast pinning their hopes for passage of pork barrel construction projects on public fears of the "greenhouse effect" (with the additional possiblity of EPA itself getting some stiff funding increases to help circle the wagons).

Obviously that is not called for. As odd as EPA's claims may seem, they have the support of a large body of scientists. But that doesn't call for for panic; it only means conditions should be studied and kept under close watch until there are clear signs available to determine if EPA's pessimistic findings are prophecy or pap.

The Burlington Free Press
Burlington, Vt., October 22, 1983

What has been termed the "greenhouse effect" by the Environmental Protection Agency in a report on serious climatic changes which could occur in the next century as a result of use of fossil fuels is not an accurate description of the phenomenon.

Greenhouse contents are warmed by reducing the heat losses that occur through convection, the process in which warm air rises and cold air sinks. But increased levels of carbon dioxide in the earth's atmosphere absorb morere of the planet's radiated heat, heating up the lower atmosphere. Since the beginning of the Industrial Revolution, carbon dioxide concentrations have gone up from 15 percent to 25 percent. Doubling of the amount which was in the atmosphere in pre-industrial days could produce dramatic changes in the world climate.

In its report, the agency predicted that average world temperatures could rise 3.6 degrees by 2040 and nine degrees by 2100. Temperatures in the northern part of the country could rise more than in the south, causing the wheat and corn belt to shift northward. Because temperature changes would not be uniform, the effects would be felt more perceptibly in the northern latitudes, particularly in the north polar region. Sea levels there may be boosted by as much as seven feet as the polar ice cap melted and causing erosion and flooding in coastal areas. Rainfall may increase 11 percent in some areas and there would be more hot days. Plants that need carbon dioxide will grow faster and need less water. At the same time, certain types of vegetation would not be able to survive in some changed environments. The climatic shift could have an impact on agricultural productivity in many parts of the world. Some areas would become more productive; others could become drier and less fertile. Developing nations might well have difficulty in feeding their people.

In a report to the White House in 1981, the President's Council on Environmental Quality called for the development of an American and global energy policy as part of an effort to cut down on the amount of carbon dioxide that is being fed into the atmosphere. The council recommended that consumption of fossil fuel be curtailed for the remainder of the century. Limitations on its use should be set by the federal government, according to the council. It called for greater stress on conservation and use of renewable energy sources.

But the Environmental Protection Agency report said that no strategy, even a ban on use of fossil fuels, could do more than delay the warming effect for a few years. It predicted that major changes will occur in the 1990s to 2000 and recommended that planning begin now to deal with the threat. "A soberness and sense of urgency should underlie our response to a greenhouse warming," the report said. Economic losses as a result of the changes could run in the billions.

But the National Research Council of the National Academy of Sciences issued a report that took a more optimistic view of the situation than the agency report. The report warned against panic and dramatic shifts in energy policy. "Our report is more conservative because we measure the effects of the problem against the real world and our ability to adjust to them," said Dr. William A. Nierenberg of the Scripps Institution of Oceanography, chairman of the study committee. "There are things we can do to mitigate these effects," he said. "We have a much more manageable future than some believe."

Because there will be no major weather or climatic effects as a result of increases in carbon dioxide levels in the atmosphere during the next two decades, he said there will be time to conduct more research on the problem and make realistic estimates of its impact.

However inclined they may be to respect the judgments of scientists, Americans must ask whether two decades are long enough to develop programs to offset the effects of the carbon dioxide buildup in the atmosphere.

The chances of coping with the changes successfully would appear to be better if the federal government took steps now to deal with the problem.

RAPID CITY JOURNAL—

Rapid City, S.D., October 21, 1983

A startling Environmental Protection Agency report issued this week warns that a "greenhouse effect" will, in years to come, warm the earth and gradually change the climate in many parts of the world.

The phenomenon is being caused by the increasing amounts of carbon dioxide in the air, produced by burning fossil fuels. The carbon dioxide allows sunlight to penetrate but keeps the heat trapped in our atmosphere.

Among other things, according to the report, temperatures will increase, the sun belt climate will gradually move north and plants, which need carbon dioxide to live, may need less water and could grow faster. That means plants that can't survive now in harsh northern climes could flourish.

Now this is important information, which the EPA hopes will be used to cope with the coming changes, especially the coastal cities which will have to adjust to a higher sea level because of melting polar ice.

But, somehow, it still brings to mind North Dakota.

Just think what a warmer climate could mean to our neighbors up there. The state's farmers might have to switch from growing the durum wheat for which they're famous to raising Bull Durham tobacco.

Residents wouldn't have to grin and bear all those terrible jokes about North Dakota winters. Like the one about the frozen columns of smoke above North Dakota chimneys, or the one that claims North Dakotans put runners instead of snow tires on their cars in winter.

And the campaign to drop the word "North" from the state's name could be abandoned. The word, residents felt, gave outsiders the image of a cold, frigid state. Well . . .

But that's enough fun poked at our much-maligned neighbor. This is serious stuff, and North Dakota isn't the only state that would face drastic changes should the predicted greenhouse effect occur.

Minnesota might become habitable enough that its residents and companies would quit crossing the border to South Dakota. Wisconsin's populace might switch from beer to mai-tais. Even Cleveland residents wouldn't have to set their rivers on fire to keep warm.

And how about South Dakota, which also isn't exactly a tropical paradise in winter? Well, maybe our ranchers could raise pineapples instead of Herefords. Sports enthusiasts could switch from snow skiing to taking jungle safaris. The vast sandy shorelines of the Missouri River reservoirs could rival the beaches of the Bahamas.

And you wouldn't have to burn this newspaper in your fireplace to keep warm.

The Providence Journal

Providence, R.I., October 24, 1983

They've been chattering for years, these scientists have, about something called a "greenhouse effect." Not many people paid attention. Now two high-powered Washington outfits, after much pondering and analysis, have concluded that yes, Virginia, there *is* a greenhouse effect silently at work warming the earth's climate. It's brought on by an increase in atmospheric levels of carbon dioxide, and (we are told, rather ominously) it may change crop patterns, melt some of the polar ice cap and raise sea levels all over the place.

Should we all flee for the hills? There was a twinge of such an alarmist reaction when the first such study, by the Environmental Protection Agency, was released last week. The warming trend, EPA intoned, is imminent, and mankind must start now to adapt as necessary. It sounded urgent, even epochal: One could conjure up visions of a Newport submerged, of waves lapping at the very edges of the Brown University campus. Was it, after all, to be The End?

Indeed (it now appears) not. For on Thursday the National Academy of Sciences (a rather more stately and learned body than EPA) came forth with *its* version of the greenhouse theory. Yes, said the NAS crowd, things look to be warming a bit (one almost could imagine them stifling a yawn), but it's not going to be drastic and, anyway, there's plenty of time for all of us to adjust.

Whew.

It seems that the "greenhouse effect" works this way. The burning of coal and oil, coupled with deforestation and altered land-use patterns, has been sending more carbon dioxide gas floating into the atmosphere. There it acts like a one-way valve, letting sunlight in but hindering the escape of heat radiation into space. Solar heat gets trapped (hence the term "greenhouse").

Even if carbon dioxide should double (and scientists think this likely), it still would constitute less than 1 percent of the atmosphere. This might — *might* — boost average temperatures as much as two degrees by the year 2040, and this *might* help raise sea levels by two feet over the next century.

Wisely, though, the NAS report stressed the vast uncertainty in its predictions and the need to avoid dramatic policy shifts based on what, as yet, is only a distant contingency. Climates have changed throughout history, it notes, and mankind, in its way, has managed to adapt. The most constructive approach for scientists, adds the NAS study, is to be flexible in analyzing the problem and debating its implications.

Cataclysmic effects? Don't bet on it; no one really knows

Future generations may want to avoid building on low-lying property, and way down the road it may be that farmers in Maine will have a longer season to harvest their potatoes. But cataclysmic effects? Don't bet on it. No one really *knows* what the weather is doing. In fact, another report last week suggested that the earth just now is in the midst of a long-term *cooling* trend. Who's right? It's hard to say. But if the earth wants to warm up, as the NAS study admits, there's nothing much we can do about it anyway — just learn to cope.

Fossil Fuels: Acid Precipitation

Acidic precipitation—including snow, hail and frost as well as rain—is believed to form when moisture in the atmosphere or on the Earth's surface combines with sulfur and nitrogen compounds. Power plants burning fossil fuels are the major source of sulfur emissions in the United States; midwestern states such as Ohio and Illinois, where there are large numbers of coal-fired plants and factories, are the worst offenders. Some of the areas in the U.S. that have suffered most from acid rain, due to prevailing winds from the industrialized Ohio River Valley, are the Northeast and the Mississippi Valley. When the resultant precipitation, sometimes as acidic as vinegar or lemon juice, falls to Earth, it can have far-reaching cumulative effects on forests, fish, soil, vegetation and water. The most dramatic effects are seen in freshwater lakes, where the acid can cause fish eggs to dissolve and the aquatic plant life to be destroyed; the end result is a lake which is crystal-clear and devoid of life. (In New York State, where many lakes have been badly affected by acid rain, legislation was signed in August, 1984 to require state reductions in sulfure dioxide emissions. The bill's requirements—a reduction of about 12% by 1988 and a total reduction of 30% by 1991—were nearly identical to those contained in stalled federal legislation.) These changes occur more quickly in lakes with little ability to neutralize the acids because of bedrock, such as granite or quartz, that is sensitive to acid. In addition, the acid pollutants react with toxic metals in the soil, releasing them into lakes or making them available to plant roots. Acid precipitation is also thought to affect the foliage of plants by extracting such nutrients as calcium and potassium from their tissues.

Since weather systems do not respect boundaries, acid pollutants travel from their tall smokestacks and other sources to wherever the wind carries them, into Canada as well as across state borders. Acid rain has become a major source of political tension between Canada and the United States; both nations contribute to the problem with smelters and power plants, but Canada claims it creates far less of the precipitation than it receives from the U.S. The Reagan Administration has so far refused to take steps to control sulfur emissions, instead calling for further research. (The two currently available methods of "cleaning" coal—either by preparation to remove sulfur before burning, or "scrubbing" to desulfurize the emissions after combustion—are both expensive and could result in raised electric rates.) After years of attempts to come to a mutual agreement, Canada decided in February of 1984 to act unilaterally, announcing that Canada's sulfur dioxide emissions would be reduced by 50% within a decade. At the same time, Canadian officials stressed that U.S. cooperation was necessary to prevent the further destruction of lakes and forests on both sides of the border.

One of the least understood aspects of acid precipitation is its direct effect on human health. Some of the sulfates and nitrates released into the atmosphere, instead of combining with water, return to the Earth as "acid dust." Scientists fear that these particles lodge in the lungs and contribute to respiratory illnesses. It is also suspected that the toxic metals leached by acid rain may enter the food chain and lead to disorders of the nervous system, kidney, liver and heart.

The Montreal Star

Montreal, Que., Canada, July 25, 1979

IT IS A SAD IRONY that the latest world energy crisis struck just at the moment when governments in Canada and the United States were gearing themselves to do something about one of our most dangerous pollution problems — acid rain. Because the measures being proposed in both the United States and Canada to reduce reliance on imported petroleum are guaranteed to make the problem worse.

It is already critical, especially in areas of Ontario. Dilute sulfuric acid, caused by air pollution in the United States, has already killed hundreds of lakes, leaving them empty of fish and other life.

The situation is sufficiently alarming that Ontario's Environment Minister Harry Parrott warned that the province would be devastated unless the United States could be persuaded to do a massive cleanup. And Canada's new environment minister, John Fraser, declaring acid rain Canada's most serious environmental problem, will be holding talks with U.S. officials in search of an anti-pollution treaty.

The determination to clean up the air, however, runs counter to President Carter's proposal to deal with his most serious problem — over-reliance on imported oil — by making more use of coal and of U.S. heavy oil resources. Those alternatives will add significantly to the amount of sulfuric acid in the rainfall. So will the development of heavy oil deposits in Alberta.

Governments in Canada and the United States, then, are faced with an agonizing choice — either remain dangerously dependent on OPEC imports or sacrifice the environment. President Carter has already indicated that, to some extent, he is ready to sacrifice the environment.

Environmental damage will be most acute if we insist on solving our energy problem by spending more and more money to develop ever more expensive sources of energy — sources where each new barrel of oil requires the investment of more energy. That process produces ever-increasing amounts of pollutants.

The more we depend on conservation rather than new production, the better it will be both for the environment and the economy. That is the message Mr. Fraser should take to his colleagues in the government and his counterpart in the United States.

Oregon Journal

Portland, Ore., December 27, 1980

A new study of the impact on acid rain on Adirondack Mountain lakes shows the destruction of fish life is even more widespread than believed earlier. The New York Department of Environmental Conservation has found that 212 lakes and ponds were too acidic for fish to survive in and that an additional 256 were approaching that condition.

The culprit is sulfur-dioxide pollution from coal-burning generating plants in Illinois, Indiana, Ohio, Michigan, Missouri, Tennessee and West Virginia. Sulfur dioxide and other chemicals are carried by rain and snow clouds eastward and deposited through the Northeast United States.

Concern about acid rain began in 1974. The new research found that 55 percent of 849 New York state lakes and ponds higher than 1,000 feet in elevation were either critical — without life — or endangered. Many of these lakes once were excellent brook-trout fisheries. Now they are dead or dying.

As the chemicals, carried by prevailing winds, fall to the ground, they leach into lakes and ponds and increase the acidity of the water.

Scientists count 2,877 lakes and ponds covering 282,000 acres in the Adirondacks. They say many more may be badly polluted. They haven't got around to studying all of them.

One answer is low-sulfur coal, such as that used in Western coal plants, including Portland General Electric's new Boardman plant. That coal is shipped to Oregon from Wyoming. Midwest and Northeastern coal-burning plants use coal with a higher sulfur content.

Logistics may keep these plants on high-sulfur coal, however, which leaves installation of emission control devices on the plants as the main alternative. Apparently adequate federal regulations exist that would curb emissions that result in acid rain, and the need is for intense enforcement of clean-air standards to halt this destruction of marine life.

The Toronto Star

Toronto, Ont., July 15, 1980

Within a few weeks, Washington and Ottawa expect to finish drafting a memorandum of intent which will set out the general scope and goals of their efforts to conclude a full-scale agreement to reduce acid rain. Such a memorandum is only a preliminary stage in the often cumbersome process of negotiating international accords. But it is nonetheless crucial. What it contains will, in all probability, determine the eventual success or failure of the agreement that follows.

The memorandum is being worked out by External Affairs Minister Mark MacGuigan, federal Environment Minister John Roberts, U.S. Secretary of State Edmund Muskie, and Doug Costle, the administrator of the U.S. Environmental Protection Agency (EPA). It will establish a timetable for the talks as well as commission further research on the industrial sources of acid rain and on the technology needed to reduce it.

But to be truly effective, the memorandum should also stress the need for strict anti-pollution controls and for some kind of international body that can help monitor and enforce them.

The memorandum should make it clear that the final agreement must pinpoint the particular industries most responsible for the acid rain fallout over the U.S. and Canada, and it should contain regulations that are specifically tailored to those industries or the regions they occupy.

If, for example, scientific evidence points to the Ohio Valley as a major source of the acid rain falling on Canada, then the factories in that area must be required to install the necessary anti-pollution devices to bring their noxious emissions down to levels that are acceptable to both Ottawa and Washington. And of course, the same should go for Inco, the Nanticoke power station or other Canadian installations if they are shown to contribute substantially to the U.S. acid rain problem.

Coal-burning power plants and smelters are known to be the principal contributors to acid rain. They emit vast amounts of sulphur dioxide into the air. This then mixes with oxygen to form acids that fall to earth with rain and snow, killing aquatic life and damaging crops, forests and possibly even human health. Recent studies have established that sulphur dioxide often travels hundreds of miles before turning into acid rain.

The exact routes followed by sulphur dioxide emissions have yet to be charted, but it is believed that more than half the acid rain falling on Canada originates in the United States, and that a significant share of America's acid rain originates in Canada. It is this interaction that makes an enforceable international agreement so imperative — neither country can truly solve its own acid rain problem without the help of the other.

Translating that knowledge into reality, however, is going to require great political will, perhaps greater than now exists.

Both Canada and the U.S. are reluctant to yield any of their sovereign powers by allowing the other a say in formulating and enforcing anti-pollution regulations. And this reluctance is heightened by political considerations. The U.S. coal mining industry, smelters and power companies have already begun to pressure politicians, particularly in the U.S., to oppose any strict measures to fight acid rain.

The EPA's Costle is as unyielding to this pressure as is Environment Minister Roberts. Both men have publicly called for an effective agreement to reduce the destruction caused by acid rain.

But their words and their sense of urgency would have a far greater impact on the public, and on their respective governments, if they were clearly and forcefully enunciated in the memorandum of intent that they are now helping to draft.

THE WALL STREET JOURNAL

New York, N.Y., June 30, 1980

Environmentalists are abuzz about acid rain. Ecology groups are lobbying for tight pollution restrictions in hopes of curbing acidic rainfall, and the Environmental Protection Agency took its first policy steps in this area last week by requiring two Ohio power plants to cut their output of certain emissions. But the problem in all this is that the nature of acid rain is unclear. No one yet knows precisely what causes it or what effects it has on the environment.

Rain and snow are normally slightly acidic due to the reaction of evaporated water droplets with carbon dioxide in the air. But rain falling on much of the Northeast and Scandinavia is now more acidic than that expected from natural causes. To date, the scientific explanation for this is sketchy.

Besides carbon dioxide, acidity in rain is usually derived from sulfuric and nitric acids. The exact composition, however, varies depending on locality, season, meteorological conditions and other factors, says the Electric Power Research Institute, which is conducting a major study of acid rain. These two acids are products of reactions involving sulfur dioxide, nitrogen oxides and water, and these gases are normally associated with the combustion of fossil fuels. For this reason, EPA is targeting its regulatory efforts at coal- and oil-fired power plants and other industrial users of fossil fuels. But anomalies in this thesis abound.

Acids are also produced by wood and refuse burning, automobiles, the rotting of plants, ore smelting, dust and debris swept into the air, sea spray and volcanic activity. No one has yet made a guess, for instance, about what effects the eruption of Mount St. Helens will have on rainfall acidity.

The National Coal Association says it is difficult to find a direct link between increased coal burning and higher rainfall acidity. Little, if any, more sulfur from coal is being placed into the atmosphere now than in the late 1940s since there now is greater use of low-sulfur coal and scrubbers in power plants. It also notes that sulfur dioxide emissions in Europe have risen by about 35% since 1965, but the rainfall in Sweden, which is downwind of major power plants, has shown a fairly level trend in acidity. Moreover, EPRI notes, sulfur dioxide emissions over the Adirondacks may vary by only 10% over a given period while acid concentrations may change by a factor of 10 from one rain to the next.

Higher levels of rainfall acidity are being found in quite unexpected areas. The Lower Mississippi Valley has experienced a rise in rainfall acidity that cannot be linked with the burning of fossil fuels because the predominant winds come off the Gulf of Mexico. Colorado also has highly acidic rain, though the prevailing winds come in over uninhabited areas devoid of power plants. Even the island of Samoa reports higher levels of rainfall acidity.

In the U.S., data on acid rain measurements over an extended period at the same sites are in short supply. From the mid-1950s to the early 1970s, only two such stations were maintained. At one site, rainfall acidity increased; at the other, it decreased. Europe has conducted more extensive studies into acid rain over a lengthy period, but even these data are not conclusive and more studies are needed.

It is known that sulfur and nitrogen oxides travel over long distances. But what winds are needed to transport the pollutants that far and what chemical changes they undergo during the journey remain unknown, EPRI states. Another important factor to consider, it says, is the probability that only a limited amount of air pollutants can be incorporated into a water droplet; thus, once exceeding some minimum level, rain acidity may be unaffected by the atmospheric concentration of pollutants.

As to the effects of acid rain on the environment there is also insufficient information. Some scientists blame acid rain for the disappearance of fish from some lakes. But, EPRI says, fish have disappeared from some lakes receiving acid rain, though in other lakes receiving similar rainfall fish have thrived. Several factors could be at work such as seepage from surrounding forests and soils, the reaction from fertilizers or even the composition of the lake-bed.

"There's still a lot we don't understand" about acid rain, EPA administrator Douglas Costle conceded last April. But he said EPA was still determined to move ahead with regulations in this area. EPRI says that at least five more years of study is required to identify correctly the causes and effects of acidic rainfall. Precipitous regulatory action by EPA could cost utilities and other industries billions of dollars. Until more is genuinely known about acid rain, these expenditures may end up only going down the drain.

Democrat Chronicle
Rochester, N.Y., September 22, 1981

IF IT hadn't been read into the record of a congressional committee, it would be hard to believe that a high-ranking federal official could put such callous words on paper.

The federal official is David Stockman, President Reagan's budget director. The words are about acid rain, the chemical showers that are dumped on the Adirondack lakes and forests by coal-burning power plants in Ohio.

New York Assembly Speaker Stanley Fink read a letter from Stockman to an anti-acid rain group into the record at a hearing held by the House Natural Resources, Agricultural Research and Environmental subcommittee. This is what Stockman wrote:

"I KEPT reading these stories that there are 170 lakes dead in New York that will no longer carry any fish or aquatic wildlife. And it occurred to me to ask the question . . . how much are the fish worth in these 170 lakes that account for 4 percent of the lake area of New York?

"Does it make sense to spend billions of dollars controlling emissions from sources in Ohio and elsewhere if you're talking about a very marginal volume of dollar value, either in recreational terms or commercial terms?"

FINK was right in telling the hearing that New York "did not set aside the Adirondack Preserve and protect it with the 'Forever Wild' clause in the state Constitution only to have David Stockman tell us that it is not cost-beneficial to the people of Ohio to pay for the cleanup of their own pollution."

The acid rain problem, as Stockman must well know, goes far beyond the wilderness beauty of the Adirondacks. Other eastern states and Canada have protested. The Canadian lakes and forests across the border from these pollution-pumping plants are being ruined and officials there have demanded action.

They haven't been successful. The federal government's only action so far was to form a study group to see how serious the problem is.

That could be, if past study groups or commissions are any indication, a way of forgetting the problem or putting it off.

It's a federal problem that can't be dealt with by individual states. As Fink said, facetiously, New York can't "send out its state militia to close coal-burning plants in Ohio."

Fink suggests an immediate low-cost plan be taken until final solutions can be found to stop the acid rain destruction.

The plan includes washing of high-sulfur coal, setting upper limits for emissions from plants and using lower pollution-producing plants when there is a lower demand for power.

There would seem to be little hope for federal action if Stockman's words are an indication.

TAKING STOCKMAN a step further, one could say that it would be better economics to bulldoze down the forests of the Adirondacks, level the mountains and drain the lakes to make way for more plants with their smoke-belching furnaces.

Who can place a value on the clear still water of a mountain lake or the whispering hush in a forest? Stockman ought to get up in the morning and go to one of those Adirondack lakes, look himself in the face on its mirror-like surface and try to say the furnaces of Ohio are more important. If he can do that, the Adirondacks are in real trouble.

The Philadelphia Inquirer
Philadelphia, Pa., January 11, 1981

In the next few months, Congress will review the Clean Air Act which expires this year. Debate over extending the controversial act is expected to be intensive. One of the most bitterly contested sections undoubtedly will be that dealing with regulations on acid precipitation.

A number of scientists have described acid rain as among the most critical environmental problems of the coming decade. In the eastern half of the United States, the average pH level of rainfall is now between 4.0 and 4.5, with normal being 5.7. Some rainfall has been measured at 3.0. That is as acidic as lemon juice. The ecological impact of the acidic precipitation is considerable.

The first to feel the effects are the microscopic organisms, then algae and tiny insects. Soon the trout, salmon and other species of fish die. Frogs and diving birds disappear, followed by otters, beavers and other animals once dependent upon the lakes and streams for food. Left behind is an eerie sight — startlingly clear water, devoid of indigenous life.

Hundreds, possibly thousands, of lakes in the Northeast and upper Midwest, and thousands of Canadian lakes have been turned acidic in the last two decades. Tens of thousands of others in both nations are threatened with a similar fate. While the lakes are not "dead," as many might believe, they have undergone such a dramatic qualitative change that they have become anomalies amid their natural surroundings.

The problem goes far beyond losing a recreational lake or favorite trout stream, however. When the ecological balance is upset, changes occur in crop productivity (as the surrounding soil becomes acidic); forest yields decline, and drinking water supplies are jeopardized.

In the eastern half of the United States, the acidity of rainfall appears to have increased about 50-fold during the past 25 years, according to federal studies. Similar sharp increases have been reported in Canada. The areas affected are those underlain with granite, as opposed to limestone (which tends to chemically buffer increases in acidity). Among the regions most susceptible are the forests and lakes of the Adirondack Mountains of New York, the northern wilderness of Minnesota and Wisconsin, and vast tracts in Quebec and Ontario.

•

The source of acid rain is not yet fully understood — a fact that is, in part, responsible for delays in dealing with the problem. Political and economic considerations, both national and international, also are at fault. The potential for permanent damage to people and the environment, however, is so great that these obstacles must be surmounted.

Two-thirds of the chemical composition of acid rain is sulfur dioxide — the emissions from fossil-fired combustion by utilities, automobiles and industrial processes. The remaining third is nitrogen oxides, also a byproduct of industry. As these pollutants are emitted into the air, they combine with water molecules to become sulfuric and nitric acids. Once aloft, they are borne by prevailing westerly winds, falling back to earth sometimes hundreds of miles from their point of origin.

The amount of raw materials for the acid brew is enormous. The U.S. and Canada discharge 50 million tons of sulfur and nitrogen oxides into the air each year. According to a joint study conducted in 1979 by the two nations, the U.S. emits five times more sulfur dioxide than Canada and 11 times more nitrogen oxide. However, the precipitation honors no national boundaries and both nations are experiencing severe damage.

Why, one might ask, is acid rainfall a problem now, as opposed to decades ago, when far more coal was burned? The answer, or at least a widely supported hypothesis, lies in the high temperatures at which coal and oil is burned today, and the unexpected side effect of federal air-pollution regulations.

When the Clean Air Act was passed in 1970, an acceptable means of compliance was to increase the height of smokestacks so that pollutants were emitted high into the air where they were carried off by winds. This concept became known by its critics as "air mailing" away pollution. Industry and utilities, faced with the option of installing expensive emission control units on their existing plants or building higher stacks, frequently opted for the latter.

•

The fate of the Clean Air Act in the Congress is uncertain. President-elect Reagan, who said during the campaign that he believes air pollution has been "substantially controlled," has pledged to reorient the act to aid the coal and steel industries.

There is considerable pressure to do nothing about the acid rain problem. The utility industry, which is responsible for 65 percent of the sulfur dioxide emissions in this country, has argued that no new emission standards should be enacted until more is known about the problem.

Inaction, however, is a guarantee of continued destruction. The Congress must extend the Clean Air Act with provisions requiring pollution controls on stacks now emitting sulfur and nitrogen oxides. The Reagan administration should act quickly to open discussions with the Canadian government with a goal of establishing an international treaty to deal with the problem, similar to the pact achieved to clean up Lake Erie. Provisions of both the Canadian and U.S. clean air acts authorize federal action to tackle transboundary pollution.

If there is any doubt in Washington about the need for extraordinary measures, a visit to the eerily quiet, lifeless lakes of the nation's north country should be arranged.

What exactly is this acid rain problem and what effects does it have on Canada, Pierre?...

CORKY -81
STAR-BULLETIN

The Register

Santa Ana, Calif., March 13, 1981

Two or three words strung together that for whatever reason work on the emotions are about all that's necessary in this crisis-oriented world to mobilize normally rational people into unthinking activists.

Remember the reaction to "Black Power" in the 1960s? Others come easily to mind: "ban the bomb," "better red than dead," "free speech," "flower power," "no more nukes."

When President Reagan went to Canada Tuesday the hecklers came armed with placards trumpeting the latest No. 1 slogan on the environmental scene:

"STOP ACID RAIN"

Heaven knows that if Ronald Reagan truly had the power to stop acid rain, he could have flown to Ottawa without Air Force One or walked across Lake Erie. But, in fact, when Adam and Eve first met in the Garden of Eden, the first rain doubtless contained "acid" and it's been that way ever since. It is one of the few things you can't blame on a politician.

Common lightning, for instance, produces nitrogen oxides in rainfall. Nitrogen oxide is acidic but plants can't live without it.

While "Stop Acid Rain" looks good on television and in newspaper photos, and although it produces sure-fire uncertainty in those who read the phrase and seems to demand that governmental action be taken, the fact is that no one knows precisely what causes acid rain or what effect it has on the environment.

It seems to be worst right now in the Northeast, in southern Ontario province, the backwoods of Maine and the mountains of northern New York. Thus it might be deduced that the cause is industrial smoke from some up-wind source — for instance the big power plants in the Ohio Valley, or autos in Detroit or smelters in Ontario itself.

For instance — and ironically in this case — the superstack at the immense Inco nickel-smelting complex in Sudbury, Ontario, has purportedly produced more sulfur dioxide in 10 years than all the volcanic eruptions in history.

Obvious conclusion: Industrial pollution causes acid rain. The 300 or so "dead lakes" in the Adirondacks and Ontario can be blamed on industry. The solution: shut down industry or force it to the wall with pollution controls.

The problem is that it's not that open-and-shut.

For totally unknown reasons, higher levels of acid rain are being reported in such places as Samoa (!), Scandinavia, Colorado and the Lower Mississippi Valley. None of those areas is downwind of industry or areas linked with the burning of fossil fuels.

Just as interesting is the fact that lakes declared "dead" of fishlife from concentrations of acid rain in the Northeast lie in the same downwind path as lakes not remotely harmed. An oddity is that lakes surrounded by foliage seem to escape the impact of acid rain, while rock-bound lakes seem most affected.

In short, "acid rain" is still a relatively new area of research, first viewed with interest in the mid-1950s, with a still unknown cause and largely unknown impact. If the thrust of the "stop acid rains" message is simply to impede industry, impose costly control measures and promote still another governmental activity, it could have the desired effect. Government is not known for high intelligence action.

But until more is known about acid rain — and major studies are now in the works — whatever is done at this stage is just guesswork. We hope the President didn't get any environmental ideas Tuesday.

Des Moines Tribune

Des Moines, Iowa, January 26, 1981

Acid rain was a subject that got attention at the recent meeting of the American Association for the Advancement of Science. It is a subject deserving greater attention from politicians (who usually move like snails on pollution issues) and industrialists (who see anti-pollution measures as brakes on economic growth).

Among the troubling reports heard by the scientific group was one from a Minnesota legislator who said that 60 to 80 percent of the lakes in the northeastern part of his state, including the Boundary Waters Canoe Area, are susceptible to damage from acid rain.

More alarming was a report from a University of Toronto botanist who said that several hundred Canadian lakes already have been "killed" by acid rain. Thousands more, especially in Ontario and Quebec, are in immediate danger. Up to 50 percent of the acid rain falling in Ontario is traceable to U.S. industrial plants.

Acid rain is the name given to sulfuric and nitric acids formed when atmospheric moisture mixes with the sulfur and nitrogen pollutants resulting from the burning of coal and other fossil fuels. Falling as rain, the acids accumulate in lakes, killing fish and eventually turning the lakes lifeless.

Growing concern has been expressed about the potential threat to agriculture and forestry. A recent report from the Argonne National Laboratory near Chicago indicated that sulfur levels within permissible federal standards have damaged soybeans. Minnesota scientists have expressed concern about their state's forests and farmlands.

Acid rain isn't an exclusive North American problem. Thousands of Scandinavian lakes are becoming acidified and devoid of fish, largely because of pollution from the West European industrial heartland.

Although the Canadian minister of environment has called acid rain the "most important issue" between his country and the United States, U.S. officials don't see it that way, and manufacturers, led by the electric-power industry, are urging modification or deletion of rules inhibiting wider use of coal.

The gravity of the situations described for scientists should not be ignored. Government leaders had better face up to the grim prognosis.

THE ANN ARBOR NEWS
Ann Arbor, Mich., October 12, 1981

ACID RAIN. Its immediate effect is not that of melting your umbrella, horror-movie style. Rather, acid rain's effect is cumulative on lakes and waters of eastern Canada and the northeastern U.S.

The National Wildlife Federation recently analyzed data on the acidity of rainfall in the eastern U.S. This group's researchers tagged 15 states as being extremely vulnerable to the harmful effects of acid rain.

Michigan was one of those states. And since our lakes are state tourism's calling cards, doing nothing about acid rain will hurt us economically.

(The Wildlife Federation's findings are at variance with the views of James Merna, research biologist in the U-M's Institute for Fisheries Research. Merna says the hardness of water in Michigan's lakes, caused by high calcium carbonate levels, acts as a buffer against the effects of acid rain).

ACID RAIN occurs when sulphur dioxide and nitrogen oxides oxidize and then combine with cloud moisture to form mild solutions of sulphuric and nitric acids.

The sulphur dioxide comes from the stacks of utility plants or smelters, the nitrogen oxides primarily from smokestacks and vehicle exhausts. Sulphur oxides are currently the main cause of acid rain.

These emissions travel with the winds. Acid rain has virtually killed animal and plant life in hundreds of lakes in eastern Canada and the northeastern U.S., according to "Canada Today/D'Aujourd'hui.''

To keep acid levels low, low sulphur fuels should be burned. Barring that, expensive scrubbers applied to gases in the smokestacks help keep acidity down.

IT WOULD certainly appear from all this that relaxing air pollution control standards, as envisioned by the Reagan administration, is not the way to fight acid rain.

Clearly, it's to the two nations' mutual advantage to eliminate the sources of acid rain pollution. Here is what Roger Simmons, member of Parliament and secretary to the Minister of the Environment, said recently:

"We cannot wait for a perfect understanding of the acid rain phenomenon before moving to control it. If we had waited back in 1972 for a complete understanding of the effects of phosphorus on the Great Lakes before starting our joint clean-up program, we would still be waiting and Lake Erie would be irreversibly dead....How many more lakes have to die before we get the message?"

And it's not just lakes. Acid rain may be reducing forest productivity. Tourism in Canada is a $5 billion-a-year industry, much of it based on fishing.

Michigan's stake ought to be apparent in our own vulnerability to acid rain. Tighter, not more liberalized, emission standards as applied to the sources of acid rain would go a long way toward control.

Portland Press Herald
Portland, Ore., October 22, 1981

The Reagan administration's attitude toward the problem of acid rain is clear: Millions for research but not one cent for controls.

Congress should press ahead in fighting this potentially disastrous environmental problem with or without administration support.

A spokesman for the Environmental Protection Agency told a House committee this week that the administration is prepared to spend more than $9 million to continue studying the problem, but it won't support a proposal by Maine's Sen. George J. Mitchell aimed at reducing acid rain beginning immediately.

Despite the EPA's contention that more study is needed, there is abundant evidence already to indicate prompt action is needed.

Acid rain has been traced principally to coal-burning factories of the Midwest. It is carried by high winds to the Northeast and Canada, where air and water quality is adversely affected. Hundreds of lakes have become so acidic that they no longer support fish life. In time, it is expected that acid rain will also cause harm to forest growth, erode stone buildings and affect human health.

Mitchell's proposal calls for industrial emission levels to be held at current levels to prevent the acid rain problem from worsening. Next, factories now venting high sulfur pollutants would be required to reduce emission levels by 40 percent before 1991.

It would be a costly program, but the price already being paid in terms of environmental pollution is potentially greater. The time to begin reversing that trend is now.

THE INDIANAPOLIS NEWS
Indianapolis, Ind., October 7, 1981

A new element of environmental concern has arisen in the form of "acid rain."

Acid rain occurs when air pollutants — generally identified with the burning of coal and oil — combine with water to form sulfuric and nitric acid in the atmosphere, which falls as rain or snow.

A few Canadian lakes and U.S. lakes in the New England area have already felt the effects of the acid rain, nearby residents claiming that fish have died from the pollutants. Indiana fishermen who have fished Canadian lakes for years have reported these conditions, too. Because of the widespread reports and complaints from Canadian authorities, the General Accounting Office in Washington (GAO) has been engaged in a comprehensive survey of the conditions supposedly causing acid rain and the damage it is alleged to have caused.

The GAO has not completed its full report, but has issued an interim document that tends to take some of the steam out of the negative claims. The damage to the lakes is still not fully assessed, but the report says there is "no firm evidence of harm to health."

The report emphasizes wide disagreement among "authorities" who have been writing and speaking on the problem. The environmentalists have contended that the acid rain argument is overwhelmingly weighted in favor of radical controls of oil and coal burning facilities and that the damage in terms of human health is significant.

The GAO says it isn't quite so cut and dried. "Summing up the evidence," says the report, "on the acid precipitation debate, even the most conciliatory representatives of the opposite sides arrive at different conclusions."

It agreed that some streams and lakes in the Northeast and Midwest are acidic and "there is substantial consensus among scientists, although not unanimity, that acid precipitation is the primary cause of this condition."

Environmentalists claim that the damage from the rain is not necessarily local; a smoke plume from a factory may degrade the atmosphere 500 to 1,000 miles from the source. This is the primary contention from the Canadians — that distant American coal-fired factories are the primary causes of the contamination of Canadian lakes.

The industrial companies — many of them utilities — say that their pollutants may be a factor, but are not the sole cause. The interim report appears to confirm their position.

One factor in the argument is the increased emphasis on the industrial use of coal in an effort to de-emphasize the use of expensive oil. Some industries have converted from gas to coal in their quest for cheap and available energy. Now, in retrospect, with the abundance of comparatively cheap gas, these conversions may have been a mistake. This problem is in the early stages of its development. It has the potential for affecting large areas of the Midwest, including Southern and Central Indiana.

Acid rain may pose no significant health risks except as it affects water supplies and some soils, which do not possess sufficient alkali to neutralize the acidity. But if more industries convert to coal, the problem may become more intense.

The full GAO report is awaited with rising interest. Meantime, the once-touted conversions to coal should be examined more carefully and with a view toward what the long range — and international effects may be.

Winnipeg Free Press

Winnipeg, Man., October 13, 1981

A Canadian delegation led by Environment Minister John Roberts ran into heavy criticism last week at an appearance in Washington to lobby against relaxation of U.S. clean air standards. The Canadian fear is that the Reagan administration and Congress are about to allow increased sulphur emissions by the coal-fired power plants in Ohio and other industrial states which contribute massively to the volume of sulphur blowing about the continent and hence to acid rain in Canada.

The contention of Ohio congressman Clarence Brown that Canada wants to interfere with his state's coal-fired power plants so as to increase Canadian exports of electric power scarcely deserves attention. The Canadian concern about sulphur emissions is real and legitimate on its own terms and can be met by U.S. power utilities in any one of a variety of ways, the most promising of which involve eliminating the sulphur on the way up the stack. Where and how the utilities obtain or generate power is of no concern to the Canadian minister of the environment so long as it does not involve polluting Canada.

But the other principal congressional objection to the Canadian stand struck closer to the bone. Canada, Mr. Brown and others pointed out, is a source as well as a victim of acid rain. That is a stock argument polluters' advocates always use — why should one polluter reform while others still offend? — and it is an argument that leads nowhere, since it is open to any polluter. So long as all sit tight waiting for the others, nobody moves. The argument can be met only by the polluter who moves first.

In that context, the advice offered by a parliamentary subcommittee on acid rain last week is sound. The panel, chaired by Sault Ste. Marie Liberal Ronald Irwin, urged that Canada should set about reducing sulphur emissions in this country without waiting for U.S. action. The alternative is to maintain Canadian sulphur emissions as a bargaining chip to give away in return for clean-up moves in the U.S. That strategy would be folly since the emissions from Ontario Hydro coal-fired plants, from Inco smelters at Sudbury and Thompson, from the Hudson Bay Mining and Smelting plant at Flin Flon and from other Canadian sources do at least as much harm in Canada as in the U.S.

The subcommittee has not provided the precise clean-up program the country needs, with price tags attached, for systematic and permanent reduction of sulphur emissions. But it has rightly warned against letting the Reagan administration's shortsightedness dictate Canadian policy.

The Globe and Mail

Toronto, Ont., October 16, 1981

Acid rain is the enemy. It has killed fish in hundreds of Ontario lakes, and stands to kill those in 48,000 more. It threatens a forest industry in Canada worth $20-billion a year. It has affected the Atlantic salmon in Nova Scotia and the freshwater fish populations of Newfoundland. The damage it does to automobiles, buildings and statuary on this continent costs us hundreds of millions of dollars a year.

This is the cutting edge of the report Still Waters, a stunning 150-page document compiled over the past 14 months by a subcommittee of the federal Standing Committee on Fisheries and Forestry. The report wastes no words; it is clear and authoritative. For once we have a paper which merits the expensive production it has received; it is being distributed to senior politicians across Canada and the United States, and its impact depends on how many of them read it.

The figures are staggering. Each year the U.S. emits 31.7 million tonnes (35 million tons) of sulphur oxides, the source of sulphuric acid rain; Canada emits 4.8 million. The U.S. emits 20.2 million tons of nitrogen oxides, which produce nitric acid; Canada emits 2 million. Three times more sulphur moves across the border from the U.S. to Canada than moves the other way; the U.S. contributes half of Canada's sulphur-sourced acid rain, and as much as 70 per cent in the Muskoka-Haliburton region of Ontario.

The report plays no favorites; it finds villains on both sides of the border. In both countries, motor vehicles top the list of sources for nitrogen oxides; but the U.S. has far stricter air pollution regulations on its automobiles than Canada. In Canada, the major source for sulphur oxides is non-ferrous smelters, which process high-sulphide ores to produce metals like nickel and copper; the main offenders are Inco Ltd. (866,000 tonnes a year at Copper Cliff, Ont. and 359,000 tonnes at Thompson, Man.), Noranda Mines Ltd. (604,000 tonnes at its two Quebec smelters) and Ontario Hydro (410,000 tonnes). The report criticizes federal and provincial Governments for acting "only reluctantly and cautiously" to reduce emissions.

In the U.S., the major sources of sulphur oxides are coal-fired power plants. The subcommittee commends the principle of energy self-sufficiency behind the conversion of oil-fired plants to coal, but stresses this is no excuse for lax controls: "The long-term environmental and economic costs of uncontrolled air pollution will far exceed the illusory short-term gains deriving from ecologically unsound conversions to coal by U.S. industry." According to the U.S. Environmental Protection Agency, 20 U.S. thermal power plants are already exceeding legal emissions of sulphur dioxide "by wide margins, some by factors of two or more"; the report urges the EPA to disallow proposals to substantially increase those emissions, a move which would "have the effect of sacrificing North American environmental quality on the altar of industrial convenience".

The subcommittee warns Canada against being smug in this game; the argument by its industries that their contribution is negligible in the North American context is a "spurious philosophy" which "will effectively doom the Canadian environment". We cannot hope for any real influence on U.S. officials until we put our own house in order, install "scrubbers" to remove sulphur from plant emissions, and set tough and effective regulations for pollution control.

Times-Colonist

Victoria, B.C., October 11, 1981

In its efforts to persuade the United States to reduce the acid rain fall-out from American industry, Canada resembles a phoney evangelist who urges sinners to repent — and then promptly adjourns to the local red-light district for a bout of heavy drinking, gambling and lechery.

It's a campaign flawed by hypocrisy and it deserved the derision it received recently when environment officials from the federal and Ontario governments took their crusade to Washington.

When the Canadians urged the congressional sub-committee on health and the environment to include tougher anti-pollution standards in a new U.S. Clean Air Act, congressmen from the Midwestern coal states claimed that Canada is not doing as much as the U.S. to curb air pollution.

With such crass Canadian models as the Inco Ltd. nickel smelter at Sudbury, Ontario, belching upwards of 3,000 tons of sulphur-dioxide emissions *per day* into the atmosphere, who can blame the Americans for adopting such a stance?

However, an awareness of the need for Canada to set an example finally seems to be dawning on Canadian legislators. After a one-year study a special Commons committee composed of MPs from all parties has concluded that tough measures must be taken to combat what it describes as "the greatest threat to the North American environment in the recorded history of this continent."

Among other actions, the committee's report calls for stronger controls in the federal Clean Air Act, national emission guides and fines heavy enough to force compliance by Canadian industry.

Federal and provincial governments must heed this advice and display more interest and resolve than they have done to date. Until we clean up our own act, we can hardly expect the Americans to take our preaching seriously.

Incredibly, the B.C. government did not even bother to make representations before the Commons sub-committee, apparently smug in the belief that acid rain is almost exclusively an eastern Canadian problem. But the report warns that the problem is growing in Western Canada, particularly in northern Saskatchewan and B.C.'s Lower Mainland, and that it threatens the multi-billion-dollar forest industry because acidic forest soils may be "uniquely vulnerable" to acid rain.

How many more warnings are needed before action is taken to avert this looming disaster?

San Francisco Chronicle
San Francisco, Calif., February 15, 1982

THOMAS NILES, a deputy assistant secretary of state, testified before a subcommittee of the Senate Foreign Relations Committee the other day that it would be "premature" to begin a program to clean up smoke emissions from Ohio Valley coal-burning power plants and smelters because of "the uncertainty of the scientific data currently available."

The gentleman, we must say, has either not done his homework or is attempting to negate many years of careful scientific study performed by both Canadian and American researchers working both together and separately. He is also, we must add, running roughshod over some deep and just Canadian sensibilities in proposing this American position.

The fact of the matter is that the Canadian and American governments and the Parliament and Congress have long ago pledged to undertake the necessary, costly cleanup. The Canadians, in our judgment, are entitled to substantial bitterness at the result to date.

A LARGE PORTION of Eastern Canada, New England and Upper New York suffer from a problem which should alarm any government. They are afflicted by "acid rain," a weak mixture of nitric acid and sulfuric acid which is produced when emissions from coal-burning industries mixes with atmospheric moisture and then descends as rain, sleet, snow or even fog. Some modern smoke stacks are built 1000 and more feet high; this effort to diminish local pollution means that pollutants are carried farther away by upper altitude winds. While the acids are weak, progressive accumulation in forests, lakes and rivers eventually destroys them. In water, algae dies, fish disappear, and lakes can become useless to beast or man.

Almost a year ago, on his successful visit to Ottawa, President Reagan addressed protection of shared environment on three separate occasions, promising cooperation. During the same visit, Secretary of State Alexander Haig told a news conference that budget cuts did not dampen his optimism that a program to abate "acid rain" would succeed. Canadians remember those words even if Deputy Assistant Secretary Niles does not.

Wisconsin State Journal
Madison, Wisc., February 23, 1982

U.S. Energy Secretary James Edwards has pooh-poohed the idea that acid rain is an environmental problem, saying he has a "hard time buying the theory that rain is killing our lakes."

"All rain is acid," Edwards noted.

Many folks in Canada might disagree. Acid rain is made acidic by atmospheric pollutants and widely believed to be responsible for the death of freshwater lakes in the northeastern United States and Canada.

"Acid rain is the most serious air-pollution problem facing our two countries today," Canadian Minister of the Environment John Roberts told a U.S. audience recently. "The situation is already intolerable. Unless we take swift action, it's going to get worse instead of better in the years ahead."

A Canadian government study committee concluded that acid rain "poses the greatest threat to the North American environment in the recorded history of this continent."

Normal rain is slightly acidic, with a PH of about 5.6. However, Canadian studies claim rainfalls in east-central Canada and the northeast United States typically measure between 4.1 and 4.3, with some individual rainfalls measuring 3.0 or lower. All rain is acid, but acid rain is much worse.

U.S. power plants are believed to be major contributors to acid rain, which could explain Edwards' tendency to discount the importance of the problem.

Until conclusive findings are in, Edwards would be wise to mute his speculations, lest he needlessly aggravate the issue by giving Canadians the impression that top U.S. officials don't take the problem seriously.

AKRON BEACON JOURNAL
Akron, Ohio. February 14, 1982

THE GREENPEACE oddballs climbing power-plant smokestacks these days are not winning a lot of friends in this part of the country.

The attack on acid rain they want would add dollars to already climbing electric bills in the Midwest and might cost some jobs.

But they have a point, and that point is not getting the serious attention it should in either Washington or Midwestern state capitals now: When we have ways to prevent it, people in one area should not dump their corrosive garbage on people in another.

Sen. Mitchell

People in Canada and the U. S. Northeast have been complaining at a climbing decibel level that this is what we in the Midwest are doing.

The rain that kills

Tons of sulfur oxides pouring from the stacks of Midwestern coal-fired power plants, they say, are combining with water droplets high in the air to form acid — which, in turn, comes down on their lakes and forests and fields and cities as rain considerably more acid than it would be without these man-made contaminants.

The most noticeable damage is the effect on lakes: They die. Rising acidity kills the plant and animal life on which fish depend for food, and the fish die. Eventually it can also dissolve large enough amounts of toxic metals to make the lake water and surrounding ground water unfit for human use.

There are also destructive effects, more subtle but still serious, on land plant life and the animal life dependent on it, and even upon buildings.

The linkage between such degeneration and acid rain, hardly more than a suspicion held by a few as recently as a decade ago, is no longer a matter of dispute. The extent to which power-plant flue gases are a proven villain in the piece, however, is hotly disputed by some.

Study, study, study

In Ohio — where Gov. James Rhodes' response long remained a dismissal of all this as fanciful nonsense — the official reaction was to authorize a study, from which nothing has been heard for some months.

And the federal Environmental Protection Agency is now taking the same path. During the Carter administration, the EPA accepted the power plant-acid rain linkage and started vigorous action to reduce sulfur emissions. By some it was accused of excesses and mistakes, and it probably made some. But now the EPA is calling for "accelerated research" over the next four to five years, and resisting congressional pressures for action until the research is complete.

Non-scientific Canadians, Northeasterners and "environuts," however, are not alone in their view that enough is already known to justify action now.

"Although claims have been made that direct evidence linking power-plant emissions to the production of acid rain is inconclusive," the research council of the National Academy of Sciences concluded in a report published last September, "we find the circumstantial evidence for their role overwhelming."

The report urged action "whenever possible."

Cures exist

Corrective action isn't easy, isn't technically perfect and isn't cheap, but it *is* possible — now. To some extent it is being taken, but not enough.

As was pointed out to the Greenpeace demonstrators down in Conesville last week, for example, the power company is working to clean up emissions. It has sulfur-removing gear on two of its six generating units there and in three others will soon be burning coal washed free of most of its sulfur; the one with the 805-foot stack two Greenpeace men scaled is to be burning "compliance" coal two years from now.

Such measures are costly and technically cranky; they add appreciably to electric bills. And if instead of "scrubbing" emissions, the plants shift to low-sulfur coal from the West, this will shrink the market for high-sulfur coal from Ohio and other Midwestern and Eastern mine areas where unemployment is already high.

But failure to clean up the emissions is also costly. The difference is that we're exporting those costs — some easily computable and some, perhaps the heaviest and the longest-enduring, beyond calculation — to be borne by "somebody else, not me."

This is unconscionable. And it won't stop until the federal EPA is once again pressing us firmly to accept our responsibility, and the state EPA falls in line behind the federal enforcers.

Maine's Sen. George Mitchell, co-sponsor of a measure calling for major reduction of sulfur emissions from plants east of the Mississippi, has a lot of company in seeing the administration's "study it" approach as "an excuse for inaction."

"If they won't accept scientific evidence from the National Academy of Sciences, what scientific evidence will they accept?" he asked. "The answer is clear: None."

RAPID CITY *JOURNAL*—

Rapid City, S.D., May 4, 1982

Before the federal government launches a massive and costly regulatory program aimed at curbing acid rain, there should be more convincing scientific evidence of the causes and effects of acidic precipitation.

Environmental groups in the United States, and, Canadian government officials maintain that acid rain is getting worse and it is caused by sulfur dioxide emissions from coal-burning power plants. Spurred by pressure from those groups, Congress is considering requiring power plants in the eastern half of the country to reduce emissions by 55 percent.

There is no assurance, however, that additional air quality controls on power plants will reduce acid raid, nor is much consideration being given to the costs involved or who will pay the bill.

Because the data base for monitoring acid rain is sparse and inconsistent, there is no sound evidence that acid rain is increasing or decreasing. According to the Environmental Protection Agency, atmospheric levels of sulfur dioxide are 67 per cent lower now than they were 15 years ago, and sulfur dioxide emissions from the burning of fossil fuels has gone down seven percent from 1970 to 1979. At the same time sulfur dioxide emissions were dropping, coal burning by electric utilities, which use four-fifths of America's coal, was rising 65 percent.

Compliance with the additional reduction in emissions proposed in bills under consideration in Congress would cost between $4 billion and $7 billion more per year than present costs of complying with existing air quality requirements. Residential customers served by utilities in affected states could see their electric bills rise from $200 to $500 a year. These costs would have the most severe impact on minorities, the elderly and the unemployed.

In 1970, a massive effort to clean up the nation's waters was launched. An investigation by the Washington Post recently concluded that despite the $30 billion spent in that program, there was no reliable evidence that our water is any cleaner.

The experience with the water pollution program indicates the need to understand the dimensions of a pollution problem before adopting and promulgating regulations to attack it.

There's no doubt that acid rain has had an effect on the environment. Funds have been appropriated to learn more about the phenomenon. The studies should be accelerated to develop a better scientific understanding of acid rain and to insure that the solutions proposed will be effective before another massive regulatory program is launched which will result in financial hardships on people.

'THOUGH ACID SHOWERS MAY COME YOUR WAY, THEY KILL THE FLOWERS THAT BLOOM IN MAY...'

The Houston Post

Houston, Texas, December 1, 1982

Acid rain is rapidly souring relations between the United States and Canada. Just as few householders would welcome lye thrown over the fence by a neighbor, so does the Canadian government resent the pollution loosed across the border by American smokestacks. From the Adirondacks to the Appalachians, polluted rainfall is damaging lakes and rivers. But this shared misery does nothing to temper the impatience of the Canadian government.

Findings on acid rain are often contradictory. It is true that any kind of water or rain may be naturally acid or naturally alkaline. But when the acidity surpasses the normal average rate, it must be assumed that the additions are man-made. Some scientists blame acid rain on the interaction of natural and industrial materials. The Department of Energy cites its own study to claim that to reduce air pollution in the Midwest would do little to moderate acid rain on the East Coast, Canada and Scandinavia. Farmers, foresters and fishermen of the East Coast, Canada and Scandinavia are unconvinced.

Early this month the Environmental Protection Agency released the 1,200-page draft of a report prepared by a team of 54 scientists. It said that sulfur dioxide pollution east of the Mississippi River "doubled from 1950 to 1978" and that "electric utility contributions tripled over this period and represent the major sulfur dioxide source." The EPA warned that much of North America has lakes and streams sensitive to the acid rain, including "much of eastern Canada; New England; the Allegheny, Smoky and Rocky mountains; and the Northwest and North Central United States." That leaves free the West where, the EPA team said, acid rain is not yet a problem because of substantially lower levels of sulfur dioxide and nitrogen oxides.

Kathleen Bennett, assistant EPA administrator, then reiterated the Reagan administration stand that more research must be done before the nation spends money to reduce sulfur dioxide emissions from coal-burning plants. But might it not be possible to make a start? Perhaps plants not yet built could be designed and equipped with pollution control devices to forestall costly revisions and adjustments later. Canada might be appeased at this evidence of good faith.

THE ATLANTA CONSTITUTION
Atlanta, Ga., October 1, 1982

Evidence of the illness and death caused by sulfur pollution such as that in acid rain continues to flow in, but the Reagan administration looks the other way, even slighting technology that could mitigate the problem significantly.

A draft report from Congress' Office of Technology Assessment says as many as 51,000 people in North America may have died in 1980 from illnesses caused by sulfur pollution — about 2 percent of the total deaths in Canada and the United States. Sulfur pollution could be responsible for as many as 57,000 deaths by the year 2000, just 18 years away.

Acid precipitation (rain or snow) contains significant amounts of sulfuric or nitric acid, formed when sulfur dioxide or nitrogen-oxide gases from industry or vehicle exhaust undergo a chemical transformation in the air. The Environmental Protection Agency says 65 percent of man-made sulfur dioxide comes from electric utilities burning coal, the rest from other industries. Most of the nitrogen oxide comes from automobiles.

No area escapes it. A special federal task force has found that 11 times more acid rain has been falling on Atlanta and other parts of north Georgia than is considered normal. Some farmers have noted a growing acidity in the soil. Fish from Lake Chatuge show the same bone disorder found in Northeast and Canadian lakes, in which the acidity of the water is growing.

The indifference of airborne pollution to state and international boundaries requires action at the federal level. The alarmed Canadian government has made numerous pleas for cooperation from Washington, but has been turned away by Reagan's wait-and-see policy.

Current technology could keep most of the pollutants from getting into the air. Yes, the retrofit would cost industry, which would pass the cost on to the public in the higher utility bills or product prices. But there is a cost benefit, of the sort Reaganism normally champions. The congressional study found that if sulfur-dioxide emissions are reduced by 20 percent by the end of the century, the number of deaths can be reduced by 40,000.

Instead of asking how much it will cost to rid the air of acid rain, the administration and Congress should ask: How much — and how many — can be saved?

THE ⬛ SUN
Baltimore, Md., February 9, 1983

Most of the acid in the acid rain that has destroyed much of the life in lakes in Eastern Canada and the Adirondacks probably comes from Midwestern power plants which burn high-sulfur coal. Some of the acid that ends up in Canada may also come from automobiles in U.S. and Canadian cities. But the source of Maryland's acid rain is still largely a matter of guesswork.

It is, in fact, an unknown quantity in more ways than one. Although it is quite highly acidic and may cause serious harm to fisheries, groundwater reservoirs and forests, clear proof of damage is still to be produced. Acid rain has been identified at six Maryland sites, ranging from Western Maryland to the Chesapeake Bay. Baltimore and Washington metropolitan areas have the highest levels, perhaps due to auto exhausts. Given prevailing westerly winds, Western Maryland's acid rain may come from power plants in West Virginia. The three Maryland power plants which burn high-sulfur coal — in Montgomery, Charles and Prince Georges counties — may or may not contribute to acid rain in the Bay and tributaries.

Two or three possible kinds of damage cause the most concern. One scientist fears that acid (or aluminum, leached from soil by acid) in streams will prevent spawning of commercially valuable fish that migrate from the oceans to reproduce. He also worries about contamination of underground reservoirs which supply drinking water. Another scientist thinks soil nutrients essential to plant growth will be leached away and forests will be harmed. Problems may be less serious in Western Maryland, where Deep Creek Lake, the major body of water there, contains chemicals that "buffer," or neutralize, acid. Even here, however, ignorance prevails: No one knows what these chemicals are, or their source.

Maryland state health and natural resources officials say they will have a clearer view of acid rain problems in Maryland after a preliminary study is finished in September. They suggest that may be an appropriate time to ask for more money for more detailed studies. It is possible their pace may be too leisurely. A Smithsonian Institution scientist says his measurements show a five-fold increase in acidity in the Rhode River watershed south of Annapolis since 1974. The problem is serious, and its causes need to be identified soon so steps can be taken to abate them.

WORCESTER TELEGRAM.
Worcester, Mass., June 29, 1983

The Reagan administration at last is about to grapple with the vexing issue of acid rain. A panel of scientists appointed by the White House has recommended immediate and substantial reductions in sulfur emissions from coal-burning factories and power plants. The National Academy of Sciences is releasing a report today that comes to the same conclusion.

The White House panel, headed by William Nierenberg, does not claim that the complex acid rain problem is completely understood. But it says that the government can't afford to wait for definitive evidence on all aspects of the matter. The damage to lakes, forests and farmland is progressing too rapidly.

Something must be done, and as soon as possible.

Nierenberg, director of the Scripps Institution of Oceanography in La Jolla, Calif., says that enough is known about acid rain to justify immediate action. No one doubts that reducing sulfur in the atmosphere will ease the danger, although no one is sure to what extent.

Acid rain is a special problem for New England because of the prevailing west wind. The sulfur from factory smokestacks in the Midwest is quickly dumped over our region, killing our fish and damaging our forests and soil.

Acid rain will take years to treat effectively, but the sooner we get started, the better.

Detroit Free Press
Detroit, Mich., June 18, 1983

IN ANOTHER move to soften its anti-environment reputation, the Reagan administration concluded last week that man-made pollution causes acid rain, and thus cleared the way for action at last by the federal Environmental Protection Agency and Congress to reduce the culprit industrial emissions.

A consensus has been growing among scientists that pollution from industrial plants in the Midwest causes high acidity in rain and snow that damages lakes and streams in New England and parts of Canada. Acid-damaged lakes also were recently discovered in Michigan's Upper Peninsula.

The commonly accepted solution for dealing with acid rain is to reduce sulfur emissions from coal-burning power plants and factories, especially those in the industrial Midwest. Until last week, the administration had argued that more research was needed to identify the causes and effects of acid rain before any action could be taken. That sounded suspiciously like a tactic to stall any regulation at all.

Although the president's Task Force on Acid Precipitation made no recommendations for action, it appears that Mr. Reagan is going to give William Ruckelshaus, the new director of the Environmental Protection Agency, enough room to come up with his own plan of action against acid rain. Mr. Ruckelshaus and the EPA are developing a plan that would initially require a modest reduction of pollutants from Midwest power plants and factories and gradually impose tougher standards.

That would be a compromise with proposed bills in Congress that call for a 50 percent reduction of sulfur dioxide emissions by the early 1990s. But it is encouraging to see that the appointment of Mr. Ruckelshaus, who has a credible reputation in environmental affairs, may turn out to be more than just a cosmetic effort to satisfy criticism of the EPA in Congress and among environmentalists, and may indeed have some substantive results.

The acid rain task force maintains that more research is still needed to identify exactly how much pollution causes what amount of acid rain damage. Such research should be continued, but as long as we have made the indisputable connection between man-made pollutants and acid rain, some steps must be taken to save our forests and lakes from continued acid rain damage.

ST. LOUIS POST-DISPATCH
St. Louis, Mo., May 20, 1983

The coal industry and the Reagan administration appear to be orchestrating a movement to kill any legislation aimed at dealing with the serious problem of acid rain, which has been linked to emissions from coal burning plants. One day the Peabody Holding Co., parent company of the St. Louis-based Peabody Coal Co., releases a study purporting to show that if Congress passes a bill requiring emission reductions, thousands of coal mining and related jobs will be lost, electric bills will soar and the Midwest in particular will suffer a severe economic impact. Then U.S. Energy Secretary Donald Hodel makes a speech in St. Louis warning that environmentalists, who want to control emissions, are trying to wreck the coal industry.

Both Mr. Hodel and a Peabody executive belittled the evidence linking acid rain to sulfur and nitrogen oxides from power plant stacks. Their stance is like that of the tobacco industry, which refuses to accept evidence that smoking causes lung cancer. But the National Academy of Sciences has reported: "Although claims have been made that direct evidence linking power plant emissions to the production of acid rain is inconclusive, we find the circumstantial evidence of their role overwhelming. Many thousands of lakes have already been affected."

The NAS also found evidence of a serious hazard to human health and recommended two years ago a prompt reduction in oxide emissions.

Acid rain is the name given to the depositing of atmospheric pollutants by rain, snow or fog. Such precipitation is destroying life in lakes and streams, corroding buildings and monuments and very likely causing extensive damage to forests and crops. Although the damage in North America is most noticeable in the northeastern United States and eastern Canada, the effects of acid rain are appearing elsewhere too. Arguments persist over whether Midwest pollutants are carried by the wind as far as New England and Canada. But the damage is nevertheless real, and efforts to control it should be national.

That is why Congress should adopt pending amendments to the Clean Air Act that would require a phased reduction by a stipulated number of tons per year of sulfur dioxide and nitrogen oxide emissions. Instead of accepting the need for legislation, the coal industry, abetted by the Reagan administration, is issuing dire warnings of economic collapse and is seeking to stir regional divisions by saying the Midwest will have to sacrifice more than the Northeast. The same unwarranted doom-crying tactics were used a generation ago to try to stop smoke control laws.

Missouri's Sen. Danforth has a far more constructive approach. Recognizing that acid rain is a serious problem that must be addressed, he favors legislation to mandate a reduction of emissions. But to avoid a heavier burden on states like Missouri where pollution control would be costly, he plans to offer an amendment that would spread the cost of control by creating an "acid deposition reduction trust fund" derived from a levy on every utility in a 22-state region of a 3-mill charge per kilowatt hour sold. States would receive grants from the fund based on the actual expense that utilities within their borders incurred for the installation of scrubbers or other measures to control pollution. Such an approach, costing an estimated 5 percent rise in utility bills, would help solve the problem fairly without causing the calamity being predicted by the industry and the administration.

THE ARIZONA REPUBLIC
Phoenix, Ariz., October 4, 1983

WEST Germans are in a state of anxiety about damage to their prized forests by acid rain and motor vehicle pollution.

The suddenness of the damage has been startling — it has mostly appeared in the last 18 months. Agriculture and Forestry Ministry experts cite decades of acid rain and vehicle pollution.

An estimated 3.5 million tons of sulfur dioxide are spewed into the air annually, to say nothing of nitric oxide from motor vehicles. German damage apparently has come more swiftly than in the Northeastern United States and nearby areas of Canada.

However, West Germany has made a decision that the Reagan administration and Congress have been unwilling to make. The government has accepted the view of its scientists that nitric oxide and sulfur dioxide are killing the forests.

Congress has refused to enact a revised Clean Air Act because acid rain's impact hasn't been unconditionally proven.

The West Germans obviously believe there are grounds to avoid such caution and hesitation. The country's 18 million acres of forest cover more than 30 percent of the land. Thus, the problem is much greater than in the United States where 482 million acres are capable of producing commercial lumber.

Some 40 percent of all West Germans vacation in their forests, especially the Black Forest, which also attracts foreign tourists. This has added to the emotionalism of the issue.

The German decision undoubtedly will set a precedent for action that the United States will find difficult to ignore.

The Evening Telegram

St. John's, Nfld., February 28, 1983

In the past couple of years, most countries of the "civilized" world have been deeply disturbed by the awesome threat of nuclear holocaust, made all the more pertinent by the casual way in which many military chiefs, especially in the United States, are looking at the possibility of fighting a "limited nuclear war." However, while the ordinary citizens are now beginning to look with horrified gaze at the frightening prospects from that source, they are not so concerned about another potential source of great danger to the human race, the new silent menace -- acid rain.

The Canadian government, especially the Department of Environment, has been making a notable effort to convince the government of the United States of the need for the two countries to cooperate to fight this menace as it affects their particular territory. Almost a year ago Minister of the Environment John Roberts offered to cut Canada's emissions of acid rain by 50 per cent, at a cost of $1 billion, provided the U.S. would do the same. At that time, the Reagan administration replied that the proposal was "premature." last week they indirectly confirmed that belief.

This came about as both governments unveiled the final report of a joint Canada-U.S. scientific study on the causes and effects of acid rain. The release was a year late, because the Canadian and American scientists disagreed on the conclusions that could be drawn from the facts presented.

Both groups agreed that the sulphuric acid, created when sulphur dioxide combines with moisture in the atmosphere and falls to earth as "acid" rain, are endangering lakes, rivers, forests and farmland; they also agreed that a reduction in the emissions from the tall chimneys of industrial plants in Canada and the U.S. could stop the trend, even reverse it by enabling some of the damage to be repaired. But after that they agreed to disagree.

The Canadian scientists concluded that the present levels of the sulphur deposits should be reduced to less than 20 kilograms a hectare, annually, in order to protect all but the most sensitive lakes and rivers; but the American scientist group wouldn't buy that. The combined team had been established in August, 1980, to work out a plan to cut back acid rain.

It was a mutual effort, because most of the poison comes from the industrial heartland of the continent i.e. central Canada and the United States. The Canadians admitted their industries accounted for about 25 per cent of the acid rain, while the United States was responsible for about 50 per cent. We would say the Canadians were too "generous" in taking the blame for such a large proportion of the silent menace.

The basic problem is that the Reagan administration, or at least some powerful figures in it, don't want the government to launch out on a vigorous program to combat the precipitation that is destroying so much of nature's wealth and beauty. The American leaders fear that the steps proposed to reduce the acid rain would put a heavy cost on industry and cause the loss of thousands of jobs; and they are not prepared to accept that kind of "short term pain" for "long term gain" in the restoration of life and grandeur to the desolate lands and waters of the two countries.

As for Canada, this country will go ahead with its promise to try to reach the 25 per cent reduction mark. Roberts balks at going for the 50 per cent as that take away American incentive to get involved, However, there are signs that a new groundswell of opinion is building up south of the border, with many prominent Americans in all walks of life waiting to stand and be counted. It's up to them to go after President Reagan and his cabinet, and also the Senate and House of Representatives, to try to get across to see what's going on in the minds of many Americans, who don't want to see the northern U.S. turned into a massive "Love Canal."

St. Petersburg Times

St. Petersburg, Fla., July 2, 1983

Former Energy Secretary James Edwards' comments on acid rain reflected a peculiar perspective on pollution that is killing fish in lakes and damaging forests. "I don't want to stop acid rain," he once said. "A little acid rain helps neutralize the soil," and is "good for crops" in many areas, he claimed.

Scientists much more knowledgeable about acid rain than Edwards have arrived at an entirely different conclusion. The prestigious National Academy of Sciences said this week there is a direct link between sulfur dioxide spewed from coal-fired boilers and the death of lakes in the United States and Canada.

The academy said the amount of acid rain in the Northeast is roughly proportional to the amount of sulfur dioxide and nitrogen oxides emitted by power plants, factories and other polluters in the eastern half of North America. The academy's study adds to the mounting scientific evidence that acid rain is caused primarily by power plants.

THE EVIDENCE is so convincing that even scientists who work for the Reagan administration are acknowledging it. A panel of nine advisers to the President's own science office also said this week that acid rain is a manmade phenomenon that could be dealt with by curbing the sources of pollution.

Undaunted by the facts, the administration still insists that more research is needed to determine whether to start a federal program to reduce acid rain by controlling smokestack emissions.

William Ruckelshaus, head of the Environmental Protection Agency (EPA), said the recent findings do not merit immediate action. "These scientific reports are further evidence that the problem is there and what the major cause of it is," he commented. "But precisely what you do about it involves a very great mix of complicated problems and people and interests, and we have to sort through all these in deciding what the public interest dictates."

Ruckelshaus is making it much more complicated than it really is. The solution is simply to require utilities to reduce the sulfur dioxide emissions from coal-burning power plants. Jack Calvert, who headed the academy's study, told reporters that the findings should show members of Congress considering legislation that "for the first time ... they are going to get something for their bucks that will pay off."

Several bills have been introduced that would require up to 50 percent reduction in sulfur dioxide emissions, primarily from coal-fired power plants over the next decade or longer. Any reservations Congress might have had about imposing tougher controls should be eliminated by the academy's report. There no longer is any justification for waiting to impose regulations that would reduce acid rain.

In the Adirondack Mountains of New York, there already are at least 170 high-altitude lakes barren of fish life. In Vermont, a mountaintop forest once deep green and dense is now scarred by "gray skeletons of trees," and the most likely cause is acid rain, according to a botanist who said that half the trees on the mountain had died. Ponds are dying in New Hampshire — "not even a weed will grow" in them, one expert said.

How many more lakes, forests, ponds and rivers have to die or be damaged before the administration is convinced that acid rain must be reduced? Congress should not wait.

The pressure from business lobbies to hold off imposing curbs on acid rain is tremendous. A new business-oriented group called the Alliance for Balanced Environmental Solutions was formed in February to counter efforts by environmentalists to persuade Congress to enact tough regulations. But, on this issue, Congress cannot afford to be bought off by special interests. Most Americans strongly support efforts to clean up air pollution even though it is expensive to comply with some regulations.

FLORIDA UTILITIES also are fighting legislation that would require them to add antipollution equipment to their power plants. Spokesmen argue that Florida utilities are not contributing to acidity in lakes in the Northeast and that acidity is not a problem in Florida lakes, so Florida utilities should not have to pay for more pollution control.

But some scientists disagree with the rosy picture painted by the Florida utilities. Dr. Thomas Chrisman, a professor in the University of Florida's Environmental Engineering Sciences Department has said: "We know the water is becoming more acidic. We know there have been changes in the lake communities." Victoria Tschinkel, secretary of the state Department of Environmental Regulation, said last September that acid rain was a growing problem in Florida.

Congress should listen to the scientists and turn a deaf ear to the special interests.

The Honolulu Advertiser
Honolulu, Ha., August 11, 1983

The Reagan administration has been slow to act against acid rain, even though scientific evidence links its damaging effects clearly to the sulfur dioxide and nitrogen oxide emissions from power plants, ore smelters, factories and automobiles.

This obvious lack of concern thus far, it has been speculated, may be due to the fact that acid rain's damage has been evident mostly on the East Coast and in parts of Canada.

BUT NOW, acid rain's corrosive consequences are spreading to the West, Reagan's beloved home base. Recently, Los Angeles fog has had the acidity of lemon juice. Two-thirds of Colorado regularly gets rain of abnormal acidity, in some places four times more than eight years ago.

"Across large parts of eastern North America, acidification has leached toxic metals from soils into lakes, streams and drinking water, killing fish, salamanders and other plant and animal life, robbing soils of nutrients and possibly killing entire forests.

"Acid rain. . .(has) corroded buildings and monuments and caused reduced yields of certain crops. These same processes are beginning to eat away at the Western U.S.," writes Michael Oppenheimer, senior scientist with the Environmental Defense Fund, in the Los Angeles Times.

Environmental Protection Agency studies indicate that large parts of the West — especially the Colorado and Wyoming Rockies, the Cascades in Washington and Oregon and California's Sierra Nevada — are extremely vulnerable to acid rain and may face the same irreversible damage that has killed lakes in the Adironacks.

Even when no rain is falling, undetermined amounts of dry acid are continually settling onto the ground, vegetation, lakes, buildings — and people.

CONTROLLING oxide emissions from existing and new smelters, plants and autos will be costly. And the difficulty is compounded because the places where acid rain does its damage often lie far from where it is "created" and must be regulated.

But having seen the effects in the East (and in Europe) of acid rain, there really can be no choice about whether steps should be taken against it. The only question is how much of the West will have to be eaten away before President Reagan gives control of acid rain the priority it requires.

ARGUS-LEADER
Sioux Falls, S.D., September 6, 1983

There's welcome new cooperation between the United States and Canada in tests to be undertaken this month to determine movement of air pollutants between the two countries that cause acid rain.

The experiment was approved in August in Ottawa by U.S. Ambassador Paul Robinson and Canadian Environment Minister Charles Caccia. The bilateral agreement calls for scientists to release three bursts of a tracer gas in mid-September from Sudbury, Ontario and Dayton, Ohio, and track it for at least 600 miles using seven aircraft and 85 ground stations.

The experiment is expected to show how pollutants are carried over long distances by wind currents in the atmosphere. Until now, the direction and distance that emissions travel have been estimated with complex computer programs.

Both Robinson and Caccia have called the agreement a significant improvement in relations between the two countries. Robinson said further study was needed before action could be taken to reduce sulfur dioxide emissions, a major source of the mild acid that causes damage to waterways and buildings and threatens drinking water.

Robinson stopped short of committing the U.S. government to Canada's proposal to cut emissions in half on both sides of the border by 1990. But he said both countries are not only talking to each other more ... we're listening to each other. "One of the concerns we're listening to is Canadian concerns about acid rain, as we are also listening to our American constituents on the very same subject."

Both countries share many natural assets on the North American continent. Both countries face the same problems in preserving lakes and forest lands from the damaging effects of acid rain.

The joint project is a helpful step toward some future solutions.

The Star-Ledger
Newark, N.J., September 25, 1983

Two independent scientific studies have confirmed that the deepening presence of acid rain represents a serious threat to the ecology of the Northeast, including New Jersey. These disturbing findings, based on extensive research, underscore a sense of urgency in beginning to abate the contamination caused by the emission of gas wastes from coal-burning power plants in the Midwest.

The remedy is relatively uncomplicated in technical terms. The emissions can be reduced by switching to low sulfur coal, washing coal before burning, or installing scrubbers that remove sulfur from gases emitted by industrial plants. These alternatives, however, have negative economic and social implications—involving matters of high costs and potential job losses in the coal industry.

There are, too, concomitant troubling issues: The assessing of the huge conversion costs (an estimated $2.2 billion), and which emissions should be reduced. The principal source of sulfur emissions is the Midwestern utilities that spew waste gases into the atmosphere which acidify rain that is wafted as far as the Northeast.

Closer to home, the problem was dealt with in an affirmative manner when Con Edison, the New York utility, was put on official notice by that state's environmental conservation commissioner that it would have to install scrubbers if it converted two generating plants from oil to coal. The decision was highly reassuring for New Jersey conservation aides who strongly opposed the Con Ed conversion plan because it did not provide for scrubbers.

* * *

At the federal level, the Reagan Administration, regrettably, has been laggard about attacking the acid rain problem. But the new studies by a White House advisory panel of scientists and the National Academy of Sciences recommended immediate remedial action by the government. A similar strong position has been taken by William Ruckelshaus, the new head of the U.S. Environmental Protection Agency.

Under these insistent pressures, the Administration is expected to draft an acid rain policy, albeit on a limited basis. A broader based legislative response is emerging in Congress, a proposal that would underwrite conversion costs by levying a nationwide tax on consumer utility bills, about 50 cents more on monthly charges.

While the House bill has some flaws, it would deal with the principal problem of conversion costs. But it should include a flexible feature contained in a Senate measure that would mandate a reduction in sulfur emissions, but would let utilities decide on the alternatives required to achieve it.

The scientific evidence developed in the independent evaluations validates a sense of urgency for a rigorous government policy to deal with this serious pollution peril. Congress should address the problem in an expeditious manner before it further worsens.

The Burlington Free Press
Burlington, Vt., October 30, 1983

Houston Chronicle

Houston, Texas,
November 29, 1983

Rain — acid rain, to be specific — is falling on William D. Ruckelshaus' parade. His appointment as chief of the Environmental Protection Agency was greeted favorably by most environmental groups, and he has enjoyed a sort of honeymoon for the last six months. But the acid rain issue is now dampening that relationship.

Ruckelshaus' problem is so complex that most people prefer to deal with it in generalities. Who wants to argue about pH factors and sulfur dioxide when it is simpler to say the fish are dying?

His task is to come up with a policy to deal with acid rain. He said he would recommend a plan before the end of September, but two months later, no plan is in sight.

Acid rain is caused by air pollution, principally sulfur dioxide and oxides of nitrogen that are put in the air by many sources, with coal-burning power plants getting most of the blame. There is evidence that acid rain is killing fish in the Northeast and Canada. There are claims that it is damaging forests. However, there is also evidence that acid leached from the soil is more of a problem than rain, that evergreen trees contribute heavily to acidity in lakes and that acid rain is a natural phenomenon caused by volcanic emissions and lightning.

A major government study was conducted to produce the facts. The report came out in June, and, unfortunately for Ruckelshaus, was inconclusive. The report called acid rain a threat and said it caused damage in sensitive areas. It also called for more research, said there were many uncertainties and said problems arise when scientists try to pinpoint exactly which sources of pollution affect which sites.

This leaves Ruckelshaus caught between political, environmental, industrial, scientific and, because Canada is involved, foreign policy pressures. Billions of dollars are at stake, depending on the type of controls imposed on power plants. The cost, of course, would eventually be passed on to consumers. Ruckelshaus has indicated he would like a "flexible" or "evolutionary" approach, but even that might require spending $20 billion or more with uncertain results.

The logical thing to do is delay a decision until there is sufficient data available. But that won't happen. Acid rain is a scientific problem that is headed for a political solution.

Because acid rain dissolves significant amounts of aluminum in the soil and the mineral then runs off into lakes and streams, there may be a link between acid rain and Alzheimer's Disease — which mainly afflicts elderly people, causing progressive mental deterioration, loss of memory and inability to carry on the routine activities of daily living.

Most victims of the disease ultimately become immobile and die. Occasionally it is contracted by young people who first lose their memories and eventually go through the same process as the elderly.

Dr. Daniel P. Perl, professor of pathology in the University of Vermont's College of Medicine, last summer discussed the possibility of the relationship between acid rain and the disease in testimony before the subcommittee on human services of the House Committee on Aging. Stressing that further research is needed on the subject, Perl said his studies still do not permit him to conclude that aluminum is the culprit in the disease, even though there are indications from research on Guam, southern Japan and southwest New Guinea that environmental factors might be involved in the disease.

In Guam, for instance, the brains of native Chamorros showed a tendency at an early age to develop "massive numbers of neurofibrillary tangles," according to Perl. Neurofibrillary tangles, he explained, are accumulations of abnormal fibers within certain nerve cells. As a result, the natives develop two severe disorders: amyotrophic lateral sclerosis (Lou Gehrig's disease) and a form of Parkinson's disease.

The incidence is 100 times that seen anywhere else in the world. The causes have been the subject of intensive research for the past 35 years and the answers have not yet been found, according to Perl.

In recent years, however, there has been a "dramatic decrease" in the incidence of the two diseases on Guam. "This marked change in the extent of the problem, according to experts who have studied it, suggests that environmental factors play a major etiologic role within Guam and that changes in environmental coditions, likely relate to increasing introduction of aspects of Western culture to the island, have modulated the underlaying causative factors," he told the subcommittee.

In his studies of Chamorro brain specimens, he said the neurofibrillary tangles are virtually identical to those encountered in Alzheimer's disease. "We have found evidence of dramatic accumulations of aluminum ..." he said.

The soil and water supply on Guam are rich in aluminum, Perl said. At the same time, there is a "marked deficiency" of calcium and magnesium in the water supply.

"It is important to emphasize that our studies still do not permit us to conclude that aluminum plays an active role in the mechanism by which neurofibrillary tangles occur," he said. The accumulations of the mineral might be the result of a malfunctioning, patially-damaged cell, he explained. "To determine whether aluminum plays a causal role in these phenomena will require further research," he said. Even so, the findings suggest that the possible role of environmental factors should be considered in connection with the problem, according to Perl.

"Most of the aluminum encountered in nature is present in an insoluble form," he said. "It is therefore with some concern that I view the increasing number of studies showing that a major byproduct of acid rain is the leaching out into solution of significant amounts of aluminum from the soil into streams and lakes. That this dissolved aluminum may be taken up by biologic species is adequately demonstrated from studies of fish and trees in northern New England. The eventual toxic effect of increasing concentrations of aluminum on fish and plant life is also well documented in the literature ... My concern is that these studies of acid rain environments have shown dramatic shifts in the nature and availability of large amounts of aluminum-containing compounds in our environment. It is clear that the aluminum can then be taken up by a variety of biologic species and exert long-term toxic effects."

Aluminum is apparently dissolved by the sulphuric acid in the rain and carried off to lakes and streams.

Even though there may only be a possibility that acid rain is linked to Alzheimer's disease, the suspicion should lead lawmakers from the Northeast to redouble efforts to pass legislation regulating the sulfur emissions in the Midwest that are believed to be responsible for the acid rain problem.

The Times-Picayune
The States-Item

New Orleans, La., July 4, 1983

The National Academy of Sciences has provided the most convincing testimony to date that the burning of coal in industrial and utility plants produces the acid rain that kills lakes in the United States and Canada.

The academy's report, made public this week, rebuts the claims of spokesmen for the coal and electrical industries that there is no scientific basis for linking acid rain to the burning of coal in industrial boilers.

While the academy's study undoubtedly will not be the last word on the subject, the prestige of the scientific institution means that its findings must be taken very seriously. Until and unless further scientific study disputes, contradicts or disproves its findings, the academy's document must figure prominently in the drafting of long-term federal and industrial policies on the control of emissions from coal-fueled plants.

The study, conducted by the academy's National Research Council, cites measurements taken over the last 18 years at the Hubbard Brook Experimental Forest in the White Mountains of New Hampshire. The academy reports that the tests revealed that the amount of acid rain falling in the forest was related to changes in the emissions of sulfur dioxide in local regions and others far upwind of the forest.

Scientists say sulfur dioxide reacts with moisture in the atmosphere to make sulfuric acid. It becomes acid rain when it falls with rain or snow. As the rainwater runs into lakes it raises their acidity to a point where they can no longer support life.

Until recently, the Reagan administration had taken the position that nothing should be done to control acid rain until more scientific evidence is in. That evidence seems to be in.

While the academy's conclusions might not merit immediate or hasty action, as EPA Administrator William D. Ruckelshaus says, they do provide a basis for policies designed to reduce and maybe one day eliminate the acid-rain problem. Mr. Ruckelshaus is right that these policies must take into account various factors, especially the economic impact of tougher regulatory requirements on industries and consumers of electric power.

The Washington Post
Times Herald
Washington, D.C., July 1, 1983

THE SCIENTIFIC case for controlling acid rain grows increasingly strong. That means further reductions in the air pollution generated by coal smoke. The costs will be substantial, and they will fall chiefly on people's electricity bills, since most of that smoke comes from utilities' power plants. Those costs, and the uncertainty over some of the scientific issues, had made the Reagan administration reluctant to do anything about acid rain. But two reports this week give the administration little choice but to begin moving toward much more stringent controls.

A special committee of scientists appointed by the White House warned it on Monday that waiting for further evidence would risk irreversible damage. Two days later the National Research Council concluded that there is a direct relationship between emissions of sulfur dioxide—which comes mainly from burning coal—and the acidity of the rain that falls in the northeastern United States and southeastern Canada. To cut the acidity of that rain in half, as the NRC urged two years ago, requires cutting sulfur dioxide emissions in half.

Nothing is simple in the chemistry of air pollution, and by no means all of the mysteries of acid rain have been resolved. Why, for example, is the damage from acid rain increasing while sulfur dioxide emissions have been decreasing since the early 1970s? One possibility is that other pollutants from other sources affect the process. Another is that there may be a loading process in the chain of causation, in which pollution's effects become cumulative.

The National Research Council's report draws attention to the disparity between the importance of the scientific choices to be made and the paucity of reliable scientific data on which to make them. For all the controversy over acid rain in the past decade, there is only one place in North America—at the Hubbard Brook Experimental Forest in New Hampshire—where anyone has actually taken accurate measurements over a period of time to track acidity. It is worth noting that recent cuts in the budget of the Environmental Protection Agency have resulted in discontinuing many pollution indicators. Once again the country is about to spend billions of dollars to combat pollution that it has been unwilling to spend a few millions to investigate systematically. That is a familiar theme in the politics of environmental improvement.

Another familiar theme also turns up in these recent reports. Americans tend to regard coal as familiar and therefore safe. But the pollutants in coal smoke carry many risks to human health and to the country's ecology. As Americans burn more of it, they are going to have to do it under increasingly tight rules. Acid rain is only one example of the costs of handling coal carelessly.

Los Angeles Times
Los Angeles, Calif., April 9, 1984

America is killing many of its forests not with chain saws but with smokestacks and tailpipes. The destruction—most evident in the red-spruce, balsam-fir and white-birch forests at higher elevations of New York, Vermont and New Hampshire—is caused by sulfur dioxide and, in the pines of California's San Bernardino Mountains, by ozone produced by industrial and automobile pollutants. The issue is clear enough for action; the question is whether enough people in government have the vision to take the first steps.

A grim future for both the nation's and the world's forests is outlined in a new report by Worldwatch Institute, a nonprofit research organization that studies global problems. "Air pollutants and acids generated by industrial activities are now entering forests at an unprecedented scale and rate," the report says. "Many forests in Europe and North America now receive as much as 30 times more acidity than they would if rain and snow were falling through a pristine atmosphere."

Acid rain—that is, rain, snow or fog carrying sulfur and nitrogen oxides from power plants that burn coal or from automobile exhaust—seems to do its poisonous work by leaching from the soil the calcium and magnesium that trees need for nutrition. Heavy metals that enter forests in smoke from smelters or the fumes of leaded gasoline also damage forests by attacking root systems through which trees absorb moisture and nutrients.

What do fewer forests mean for the world? Less oxygen, for one thing. Less lumber, for another. According to one recent study, a 5% decline in the growth of Southern softwoods "would translate into an annual loss of timber sufficient for about one-tenth of the new U.S. homes built each year."

While there are a number of elements that harm forests, most discussion now centers on how to deal with acid rain. Before that problem can be tackled, Worldwatch believes that government must change its philosophy. It "must begin to recognize that damage to forests, soils and lakes is an added cost of fossil-fuel combustion that is not now taken into account in the prices consumers pay," the report says. "Society is in effect subsidizing fossil-fuel-generated electricity, motor-vehicle use and metals production by allowing free use of the environment to absorb the resulting pollution . . . . Correcting this inherent market failure requires that those fostering the pollution begin paying for it."

The Reagan Administration so far has chosen to budget money for the coming year only for research. Congress has before it several bills that require cleanup efforts either financed directly by ratepayers who rely on the utilities that are the major source of sulfur dioxide or through a nationwide tax.

Legislators divide according to region on plans to pay for acid-rain cleanup: Northeastern legislators, whose districts are downwind from the major sources in the industrial Midwest, want the polluters to pay. Midwestern legislators understandably want the burden shared. Appalachian legislators have their coal-mining constituents to think about; so do Western members of Congress, although those mines produce coal with lower sulfur content than that from the East. Those whose regions don't contribute to the problem don't think that they should help pay for its solution.

We prefer the idea of a national tax to finance scrubbers to help reduce the sulfur-dioxide emissions, as embodied in legislation sponsored by Reps. Henry A. Waxman (D-Calif.) and Gerry Sikorski (D-Minn.). The decline of the nation's forests is a national problem that demands a national solution.

But the problem demands starting on *some* solution sooner rather than later, as the Reagan Administration would do. "Without efforts to brake these rising emissions," the Worldwatch report says, "destruction caused by acid rain and air pollution two decades from now may dwarf that evident today." The destruction of forests has unfolded with rapidity; the human response is unfortunately not as swift.

The Tennessean
Nashville, Tenn., January 16, 1984

AS sure as spring follows winter, the presidential primary season is accompanied by a plethora of rhetorical commitments. The latest bunch of campaign promises focused on an environmental problem — acid rain.

A recent conference — Acid Rain '84 — was fertile soil for campaign verbiage since the conference was held in New Hampshire, a state that will also host the first presidential primary late in February.

Five of the democratic presidential contenders were in attendance, and they agreed that acid rain is a serious environmental problem that is threatening a growing portion of this continent. They did not, however, agree on an optimum solution. More specifically, they disagreed on who should foot the clean-up bill.

The primary cause of acid rain is sulfur dioxide emissions of coal-burning factories. When mixed with water in the atmosphere, the emissions are chemically altered and fall to the earth as acidic rain, snow or dry deposits.

There is proof that the acidic precipitation is killing freshwater life in the northeastern states and Canada. Additionally, it destroys forest and plant life, and therefore threatens human health.

Each candidate has a plan on acid rain clean-up. For example, Mr. Walter Mondale wants to reduce the emissions by 50%, financed by a utility tax on the 48 contiguous states. Mr. John Glenn, whose home state of Ohio has done more than its share of polluting, would be satisfied with a one-third reduction in emissions, financed by a tax on fossil-burning plants in 31 states in the Northeast and Midwest.

But the Democratic candidates agree on one thing — any of their solutions to acid rain are preferable to the present administration's negligence of the problem. In fact, President Ronald Reagan has not proposed any program on acid rain control, even though he has been advised to do so by the EPA director, Mr. William Ruckelshaus.

The President's continuing neglect of the acid rain problem is threatening the plant and animal life of this continent. Perhaps the challenge of the November election will spur him to action.

THE CHRONICLE·HERALD
Halifax, N.S., April 5, 1984

A FAVOURITE instrument of government when it wishes to avoid taking action or when it desires to evade responsibility is to hide behind the unctuous announcement, "More research is necessary." With that excuse, a matter of grave importance can be delayed indefinitely.

A current example of that approach to a problem is being demonstrated by the United States in its response to the problems accruing from acid rain. Others of the industrialized nations are showing a conscientious concern toward a modern-day hazard while our neighbours to the South continue to dump tons of toxic emissions not only on Canadian territory but also on their own.

In an effort to institute a program which would reduce the air pollution causing acid rain, nine European nations have joined with Canada in the signing of a covenant. The agreement, which followed a two-day conference, requires the signatory nations to reduce emissions of sulphur dioxide by at least 30 per cent by 1993. Reductions also will be made in the outpourings of nitrogen oxides, another of the key ingredients in acid rain. The United States refrained from joining in the fight against acid rain, hiding behind the excuse that more research is necessary.

How much research is required? Conclusive evidence readily is available concerning the damage which acid rain inflicts on forests, lakes and streams. Eastern Canada and the northeastern states of the United States are feeling the damaging impact of the emissions from the industrial heartland of the continent. The environmental authorities of the American government are aware of the seriousness of the problem.

They are knowledgable, too, of the two-pronged approach to the problem that is necessary: one is to clean up damaged areas, the other to halt, or at least reduce, the volume of poisonous emissions.

The problem is discernible, the answers are clear. Why, then, must people living in eastern Canada continue to suffer while the United States does more research?

It appears that the foot-dragging in Washington is due to lobbying on the part of the industrial giants which do not wish to become involved in costly remedial measures.

The American attitude is deplorable. Ottawa must continue to exert every possible pressure on Washington to awaken a conscience in this serious matter.

THE PLAIN DEALER
Cleveland, Ohio, March 26, 1984

In the Netherlands, where the tulips grow, roughly 75% of the pollution that accounts for acid rain is exported by neighboring European countries. The rest is homemade and results from some industry, lots of cars and the penchant of the Dutch for storing and using manure as fertilizer, which produces ammonia.

The Dutch deserve both sympathy and credit. Although they also export air pollutants to other European nations, they are a net importer, or so it seems, and therefore suffer unduly from the prosperity and industry of their neighbors. As a result, they have instituted, along with a number of other European nations, extensive and laudatory efforts to control the emission of air pollutants generally and those that cause acidic fallout specifically.

Contrary to what the Canadian government would like us to believe, however, there is no compelling political lesson to be learned by Washington from Amsterdam's acid angst. American control of automobile emissions, for example, is far ahead of any European initiative, and developing disincentives with regard to the storage and spreading of manure is more appropriate to Washington bureaucrats than to the nation in general.

The failure

That, then, was the failure of the Canadian/European Ministerial Conference on Acid Rain, which concluded last week in Ottawa with a call for the reduction of annual sulfur emissions by 30% within all nations that are party to the Convention on Long-Range Transboundary Air Pollution. The conference had been held in Ottawa with the express purpose of applying pressure on the United States to take action on acid rain, and yet the majority of statements were so inapplicable to the American situation that the pressure became pious scolding. Not surprisingly, the United States responds to sanctimony about the same as any other nation.

As evidence of Washington's truculence, look no further than the American conference activity. There was none, except for the occasionally stormy press conference held the night before in the office of the American ambassador. Fitzhugh Green, an associate administrator of the Environmental Protection Agency, relayed Washington's "delight" at being invited as an observer to the conference, but refused to go any further. Time and again, he asserted the administration's position that more research is needed. And time and again, he pointedly remarked that Washington was not "susceptible to this kind of pressure at all." The Canadian press responded badly to the lack of an admission of guilt, and misrepresented some of his remarks in the next day's stories.

In any event, Green was so swamped by reporters during the conference that, after the opening day luncheon, he was not seen again, his place reportedly occupied by a low-ranking embassy functionary.

All of which would be just gossip, if it weren't for the deep feelings in Ottawa that the United States is fouling the Canadian nest, and were it not for compelling evidence that Ottawa is fundamentally right. Canadian estimates suggest that 50% of that nation's acid deposition results from transboundary pollution. The fact that Canada has tried just about everything else to get Washington to act against acid rain mitigates the conference's geo-political clumsiness.

The administration is wrong to pursue "perfect knowledge" (proof positive of the relationship between sulfur emissions, acid precipitation and environmental damage). Statements made at the conference by European nations, where the effects of acid deposition are acute, clearly indicate the link between sulfur dioxide and nitrogen oxide emissions and environmental crisis. Monuments, buildings, works of art, surface waters, aquatic life—all key components of the term "quality of life"—have been damaged, some perhaps irreparably. There is, as well, new evidence linking acid depositions with ground contamination and a threat to forest industries.

The administration is not wrong to imply that the acid rain problem in other countries cannot be compared to America's. The United States lacks a clear consensus on how to address the problem, which is not the case within the governments of most European nations. Further, the European crisis is much more readily addressed—even among the rightly anxious Nordic nations. Only West Germany faces the incredible expense of mandating technology on numerous stationary sources of air pollutants.

Canada, also, has a clean-up commitment (50% reductions by 1994) that is markedly easier and less expensive to achieve than anything to which the United States can look forward. Indeed, Ottawa and the provincial governments are naive to think that, with so much at stake, Washington would move with alacrity ... especially when the demonstrable decline in the American environment is still a relatively remote threat to most Americans.

Potent, potent, potent

If the lessons for Washington were not political, however, they were nonetheless important: Acid rain is a potent international issue; acid rain is a potent environmental poison; acid rain will become a potent political topic. The broad consensus among the Europeans that acid rain amounts to environmental death is evidence enough of the need to regulate it, and the administration therefore is playing fast and loose with the wilderness whenever it makes an egregious call for "perfect knowledge."

More important than such perfection is the two-fold understanding that, A) acid rain threatens the United States just as it now attacks the Europeans and Canadians and B) that it must be regulated and reduced with speed and efficiency. The risk to our environment, the risk to our health and quality of life, is clear and present. That was the point in Ottawa, and the administration is wrong not to pay it heed.

The Hartford Courant
Hartford, Conn., April 28, 1984

The new National Wildlife Federation report that acid rain afflicts the Sun Belt and other parts of the United States besides the Northeast merely confirms the findings of other studies — among them a U.S. Geological Survey report issued earlier this year.

Sulfur and other atmospheric pollutants are contributing to acidity in the rain that is damaging the environment — particularly lakes and forests — everywhere.

The problem, further, is not limited to the United States, or even to the North American continent. It is global in scope, prompting nine European nations and Canada in March to sign an agreement committing their governments to reduce sulfur emissions by at least 30 percent over the next decade. The nations call themselves the "30 percent club."

The international agreement was intended to goad nations that have not acted to reduce sulfur emissions, chiefly the United States and Britain, to finally do so.

The White House continues to oppose sulfur reductions in industry and power plant emissions, at least until more research is conducted. But perhaps the growing accumulation of studies and international sentiment will shame Congress to take action. Committee work is continuing on proposed amendments to strengthen the Clean Air Act.

Let's join the club.

The Washington Times
Washington, D.C., April 25, 1984

Every time we hear another sky-is-falling story about the menace of acid rain, we gag. Six states recently filed suit against the federal Environmental Protection Agency for allegedly failing to curb the sulfuric and nitric acid fallout from factory and car emissions. Canada has joined nine European nations, including West Germany, to fight the "menace," which is, they say, "exported" in part by the United States. Bonn officials allege that a third of West Germany's forests are being eaten away.

In this country, some environmentalists say our drinking water is poisoned, causing our brain cells to degenerate. In fact, though, clinical tests show no demonstrable health effects until sulfate concentrations in the air reach 10 times the highest concentrations so far measured.

As for those Canadian lakes — ruined by bad rain coming from the U.S., so the story goes — evidence suggests that the acid mostly comes from Ontario's large smelting complex. And in Minnesota, Wisconsin, and Michigan, scientists find no trend toward acid damage to lake and ponds over the past 50 years. In New York, sampling shows no acidity changes for the past 15 years.

New studies in West Germany implicate ozone and pollutants other than acid rain. In this country, laboratory and field experiments render inconclusive findings as to whether soil conditions, climate, or chemical rain is the culprit. In the case of red spruce, drought followed by infection appears to be the cause.

Acid rain may be an environmental problem, but a crisis it is not. National Academcy of Science records show that the pollution is actually getting better, not worse. Congress nevertheless is under pressure to require industries in 31 states to spend billions of dollars switching fuels or installing scrubbers to reduce coal emissions. The costs would be borne by the public; the benefits — well, environmental activists and certain politicians would *feel* better.

The Pittsburgh
PRESS
Pittsburgh, Pa., April 24, 1984

If it was typical, the spring rain that fell throughout Pennsylvania Sunday was heavily acidic — at least 40 to 50 times more so than normal rain — and inched the environmental calendar ever closer to a wintry darkness.

As the rain fell, it dispersed tons of acids over the landscape and into the waters. It ate away at the normal life chains of all — from humans to the lowest form of life.

And it fell even as more pollutants were discharged into the atmosphere to await their turn to fall back to earth to further accelerate the decay.

It fell, too, as the National Wildlife Federation was on the eve of releasing a report that supported one filed three months ago by the National Geological Survey. Pennsylvania, both reports said, routinely receives the most acidic precipitation in the nation.

Acid rain is created when sulfur dioxide and nitrogen oxides combine with water vapor in the atmosphere.

Pennsylvania is second only to Ohio in sulfur dioxide emissions, spewing more than 2 million tons each year into the skies. The state reaps its cancerous reward when that, and more, returns to gnaw silently at the environment.

And because the nation seems assured, at least in the foreseeable future, of an energy system fueled by coal, the identified source of the problem, there is every indication that the problem will only grow worse.

Nuclear power is, for all practical purposes, headed, if not for extinction, at least for a straight-line existence with no increase. Solar power has yet to see more than the glint of a sunrise.

There is no fork in the road, only a road with signs that warn of impending disaster should present see-no-evil policies continue.

While the federal government has continued its cool attitude to the problem — President Reagan has asked Congress for $55 million for research but has not asked for a cleanup — Canada has decided to go it alone. There, acid rain is to be cut by 50 percent in the next 10 years.

In the United States, there are 11 different proposals in Congress to reduce acid rain. The least of them is better than the Reagan administration's policy of more study and no action.

The Union Leader
Manchester, N.H.,
January 10, 1984

However difficult it may prove to be to eliminate or substantially reduce that part of the acid rain problem that is caused by industry, there is a far more difficult task at hand: how to keep presidential politicians from exploiting the issue for their own purposes and then discarding it when the campaign is over. Given the heavy political flavor of last weekend's Acid Rain '84 conference in Bedford, which assorted Democratic presidential candidates used as a political forum to attack President Reagan and-or each other, the latter task may prove impossible of accomplishment.

Lest anyone wonder why the administration didn't bother to send a representative to the conference, at which assorted speakers told all they know or think they know about acid rain, that answer was given many times. As when one Democratic candidate actually scored points with the audience for raging, "The most important thing we can do to stop acid rain is to help elect a new President."

He might have said, with far greater logic: "The most important thing we can do to stop acid rain is to stuff the tops of all the world's volcanos with politicians."

It was what was *not* said at the conference that seems noteworthy. It was *not* said that there is no guarantee that the expenditure of multibillions of dollars on scrubbers, coal-washing and other techniques of reducing sulfur dioxide emissions resulting from our present combustion of fossil fuels will solve or even make a dent in the problem.

It was *not* said that while coal-burning plants most assuredly pollute the atmosphere, and always have, their emissions are infinitesimal when compared to what is disgorged into the atmosphere by even one volcano. (This is not an argument for doing nothing about the problem of industrial pollution; it's an argument for perspective on the full extent of the problem.)

It was *not* said that the replacement of much of our present combustion of fossil fuels with nuclear energy could resolve at least that part of the acid rain problem that is not due to natural causes.

And it *was* said, but by only one presidential candidate, Senator Ernest Hollings, that blame for inaction cannot be placed totally on the White House when Congress itself has failed to act.

Lincoln Journal
Lincoln, Neb., May 13, 1984

It is incredible, the equanimity with which people greet the news that forests and groves and isolated trees in Europe and the Eastern United States are possibly being destroyed by acid rain, a result of air pollution.

Perhaps humans do not truly understand the importance of wooded areas in their lives, and the life of the entire world.

The people of Indonesia could tell them something about what the loss of forests mean. While differing in some respects from the situation now facing the industrialized world, Indonesia's experience still offers a lesson.

What caused the loss of a forest in East Kalimantan Province, formerly Borneo, last year was not acid rain but lack of any rain. Drought completely dried out the island's rain forest. Somehow a fire started, and months later — actually, it is still burning in some areas — 3.5 million hectares of forest had been destroyed, or a region about one-fifth the size of West Germany.

Obviously there was a devastating economic impact because of the loss of timber. But because so much wildlife lost its home, insects are now threatening wide stretches of the province. Flash flooding is a recurring danger. An Indonesian ecologist warns that vegetation and climate will be affected far beyond the burned-out area. In 70 years the forest might be restored ... might. Probably the area will never be the same.

Acid rain works more slowly than fire. But the result can be similar.

Rockford Register Star
Rockford, Ill., April 6, 1984

Six northeastern states acted responsibly in filing a federal suit to force the Environmental Protection Agency to enact tougher restrictions on sulfur-dioxide emissions in industrial smoke, a major component of acid rain.

The lives of our forests — one of our most essential and difficult to replace natural resources — is at stake.

We may be too late to save the first casualties of acid rain's ravages. From the Appalachians of Virginia and West Virginia, northward into the Green Mountains and White Mountains, red spruce shows serious "dieback." Damage is most severe in the high elevation forests of New York, Vermont and New Hampshire, on peaks forested mainly with red spruce, balsam, fir and white birch.

As in Central Europe — where devastation of forests has been massive — the deposit of acid by rain has been linked to the dieback of spruce.

A likely theory is that acid rain (sulfur and nitrogen oxides are chemically transformed in the atmosphere and fall to earth as acids in rain, snow or fog), in conjunction with other pollutants (such as ozone from auto exhausts) weaken the trees and make them more susceptible to other natural ailments; insects, drought, excess cold and wind.

Concerned Canadians have wisely signed an agreement with nine European nations to reduce sulfur dioxide emissions by 30 percent by 1993. Canadians are understandably frustrated at the lack of similar concern in the United States, a neighbor so close we share some forest ranges.

We learned too late the lesson of Love Canal, N.Y., to save Times Beach, Mo., from deadly dioxin. We can't afford to lose more precious forests before we demand that EPA get tough about acid rain.

The Courier-Journal
Louisville, Ky., May 5, 1984

IT SEEMS doubtful that any member of Congress — or at least any member who has spent much time examining the evidence — believes that the eastern half of the United States can indefinitely continue to pour 22 million tons of gasified sulfur into the atmosphere every year. Almost every month, there's new evidence of harm as pollution from stack gases in coal-burning plants returns to earth in the form of diluted sulphuric acid.

Yet what may have been the last chance for action on acid rain in the current session of Congress went down the drain this week. A House subcommittee voted 10-9 to kill a Clean Air Act amendment that would have required coal-burning power plants, the chief source of this pollution, to reduce emissions by 10 million tons over the next 10 years. That's close to the 50 percent cutback the National Academy of Sciences thinks is needed to bring acid emissions to a tolerable level.

The proposed legislation may or may not have been the best way to get the job done. It would have required the use of scrubbers to clean stack gases. That's admittedly a well-proven technology. Japan, through the use of scrubbers, made a 50 percent reduction in sulfur dioxide emissions between 1970 and 1975. It has since continued this improvement and pioneered new sulfur-removal techniques.

Shift to low-sulfur coal?

Nevertheless, many argue that the same result could be obtained at less cost by giving utilities the alternative of fuel switching — burning of low-sulfur coal rather than the high-sulfur fuel commonly used in the Midwest.

But fuel-switching, in human and political terms, is a more difficult option. Shifting to low-sulfur fuel would aid some mines in Eastern Kentucky and West Virginia. And it would especially benefit low-sulfur producers in the West, at the expense of high-sulfur mines such as those in Western Kentucky, Indiana and Ohio. Bad employment situations would become worse.

Moreover, it seems unlikely that even low-sulfur coal, in the long run, can be burned satisfactorily without emission controls. The U. S. mines more than 800 million tons of coal a year. Even a sulfur content as low as one percent would put a horrendous amount of this pollutant into the atmosphere. The future of coal depends on the use of technologies that make it environmentally acceptable.

Admittedly, better technologies than now exist will be available someday. Given the coal industry's reluctance to help clean up its own act, the technology probably will come from Japan or some other nation that is more concerned about the environment.

In any event, the effort to do something this year in Congress wasn't defeated on its merits. The swing vote on the House subcommittee was a young Ohio congressman who tacitly acknowledged that the bill was in the national interest. But the congressman, Dennis Eckert, represents a district near Cleveland that has an unemployment rate of 12 percent, two of the nation's worst-polluting utility plants, and an understandable fear of higher utility rates.

Mr. Eckert can console himself that his "no" vote probably didn't kill the measure single-handedly. The acid-rain measure would have faced an uphill battle on both the Senate and House floors, and President Reagan's almost total indifference to the environment.

The national interest will win out someday as the environmental damage increases and the scientific evidence mounts. But the delay, with likely irreversible damage to lakes, streams, forests, soil and human health, will be costly. The longer Congress keeps refusing to address the issue of acid rain, the more it adds to doubts about its ability to put the national interest ahead of special interests.

THE TENNESSEAN
Nashville, Tenn., June 15, 1984

THE 50th birthday celebration of the Great Smoky Mountains National Park is being dampened by reports of extensive pollution damage. The suspected culprit is acid rain.

The plant and animal life on the 52,000-acre mountain preserve is particularly susceptible to pollution because of the park's high altitude. Plant pathologists explain that trees on the highest peaks comb out the heavy pollution directly from passing clouds.

The results are ghastly on even the heartiest plant life. Red spruce and Fraser fir trees that normally live up to 400 years appear to be dying after just 45 years. And an estimated 100,000 trout have died since 1981 on trout farms near the park's borders.

Dr. Robert Buck, a plant pathologist investigating the destruction, has found acid precipitation in the park that is 100 times the acidity of pristine rainwater. Dr. Buck said the continued degradation of the park would "be the ecological disaster of the century."

The reported damage to the park is alarming — but it should not be surprising. Acid rain has been killing plants and animals in the Northeast and Canada for years. The fact that the plants and animals are on the nation's most popular wildlife reserve offers no protection from poisonous pollution.

Congressional attempts to curb acid rain — caused primarily by sulfur dioxide emissions of coal-burning power plants — have been unsuccessful. Bills now in Congress would mandate reductions of high-sulfur coal emissions. But the coal industry is afraid those measures would create additional unemployment, and the utilities are afraid they would cause rate increases.

It now seems likely that the congressional debate on acid rain may very well outlive many stately spruce and fir trees. That is indeed an ecological — and a political — misfortune.

ACID REIGN

THE CHRISTIAN SCIENCE MONITOR
Boston, Mass., June 28, 1984

NO one disputes today that rain laced with acid is inflicting serious damage on lakes, streams, forests, crops, buildings, and wildlife over much of the Northern Hemisphere.

But lack of agreement on solutions has so far stymied attempts to take specific remedial steps in the United States.

The scope and seriousness of the acid-rain issue were made clear earlier this week in Munich when representatives of 31 Western and Soviet-bloc nations urged immediate measures to reduce acid rain. Most of the delegates, including those from the Soviet Union and East Germany, said their governments were ready to commit themselves to a 30 percent reduction by 1993 of the level of sulfur dioxide emissions as of 1980. The US and Britain held back, citing need for further study.

American officials also feel that, since pollution levels in this country were already considerably reduced by 1980, the 30 percent commitment would be much more expensive and difficult to achieve in the US.

Neither the White House nor the Environmental Protection Agency is ready to press for specific emissions controls aimed at diminishing acid rain. Reagan administration reluctance to impose new controls on industrial plant emissions is due in part to the administration's "deregulation" drive and concern over the effect expensive pollution-control measures might have on a rebounding industrial sector.

A US Supreme Court decision Monday upholding the EPA's so-called "bubble" concept of air-pollution control appears to legitimize a policy that permits industrial and utility plants responsible for a great deal of the acid in the atmosphere to continue to emit a level of pollution well above standards set in the Clean Air Act. Environmentalists argue that, in fact, the agency's method of applying the "bubble" policy will actually increase the level of pollutants released, including acid-forming sulfur dioxide.

The same environmentalists admit that imposing a "bubble" on a particular region where stacks are emitting concentrations of sulfur dioxide could be an efficient means for controlling acid rain.

Experience indicates that waiting for indisputable findings by physical scientists is likely to be fruitless. Two recent reports illustrate the point:

On June 11 a group of US government scientists with 12 federal agencies declared that more study is needed to determine the role of industrial plant emissions and questioned whether controls on Midwestern smokestack emissions would result in significant reduction in acid deposits in the Northeast.

On June 20 the Office of Technology Assessment, after conducting a four-year study, told Congress in effect that the time has come to quit studying the problem and mount a serious effort to reach a political decision on what to do about it.

And earlier this year a National Academy of Sciences report stated that reduction of sulfur dioxide emissions from smokestacks would result in a proportionate reduction in acid rain.

When a House subcommittee killed, May 2, a bill calling for sharp reductions in sulfur dioxide emissions, it was generally acknowledged that that was the end of acid-rain legislation for 1984.

Many members of Congress are probably awaiting clear indications that their constituents want action on acid rain — the kind of grass-roots support that resulted earlier in the passage of landmark environmental laws.

Congress and the White House must give top priority to renewal of the Clean Air Act, which officially "expired" on Sept. 30, 1981. A key part of the drive to renew and strengthen that landmark law should be enactment of an acid-rain control program.

EVENING EXPRESS
Portland, Maine, July 7, 1984

It won't be long before "studies" of acid rain turn into seven-figure junkets for those doing the studying—if the studies spread as fast as the effects of damaging sulfur dioxide emissions in our continental air.

New England has long since evidenced mounting concern for the cause-and-effect relationship between sulfur dioxide emissions from coal-burning utilities and industries, particularly in the Midwest, and the damaging effects of acid rain. So have New York and Canada. Still, Congress has persisted in regarding acid rain as the Northeast's regional problem.

No more. Startling evidence of potential acid rain damage surfaced this week in Florida. Environmentalists reported that unless emissions levels of industrial pollutants are reduced, acid rain could kill the fish in 2,600 Florida lakes over the next two decades.

"I was stunned when I realized how bad it was," researcher Paul Hansen acknowledged. "We'd always known that Florida was receiving acid rain, but we didn't know it was this bad."

More than that, researchers with the National Clean Air Coalition and the Friends of the Earth Foundation have stated flatly that acid rain poses a "clear and present threat" to the South just as it does to the Northeast, where its damaging effects were detected earlier.

We point out Florida's plight not because misery loves company, but because geography can encourage a consensus to develop in Congress broad enough and strong enough to limit acid rain.

U.S. Sen. George J. Mitchell knows how to do it. Mitchell would require coal-burning industries and utilities to reduce their sulfur dioxide emissions by 10 million tons annually over 10 years.

Not long ago, a bill like Mitchell's would have had only limited appeal for Southerners who could regard acid rain as another region's problem.

But times change. And when enough states become aware of the damaging effects of acid rain in their own backyards, so too will the law that governs industrial pollutants.

Minneapolis Star and Tribune
Minneapolis, Minn., May 13, 1984

As the truth about acid rain becomes clearer, the culprits loom larger. They are the coal-burning utilities of the Ohio Valley, whose tall smokestacks spew out millions of tons of sulfur dioxide every year. The windblown result is acid rainfall in the North and Northeast, which has wiped out aquatic life in hundreds of lakes and threatens fish in many more. This weekend's opening of the fishing season should remind Minnesotans of the risk to their own lakes. The obvious solution is to force the polluters to stop polluting. But how? And who should pay?

That kind of question makes life difficult for presidents and lawmakers. So far, neither the Reagan administration nor Congress has mustered the political will to settle upon an answer. The White House refuses to consider any kind of acid-rain control, calling instead for more study of a problem scientists already know how to solve. Congress, at least, is trying to write an acid-rain law. But Congress seems unable to reconcile ideal legislation with political reality to reach a compromise.

A bill sponsored by Sens. Robert Stafford, R-Vt., and George Mitchell, D-Maine, is the most idealistic. It would force polluters in 31 states east of the Mississippi to reduce sulfur emissions by a total of 10 million tons a year. It embraces the time-honored "polluter pays" principle, awards credit to "clean" states like Minnesota and gives states that must clean up freedom to choose how they will achieve the goal.

But forcing nearly the entire cost of acid-rain control on big Midwestern polluters could spur a dramatic rise in Midwestern electricity bills. Allowing utilities to control emissions by switching to low-sulfur Western coal could induce severe unemployment in the Midwest's high-sulfur coal mines. Midwestern lawmakers insist that their economically depressed region cannot endure the Senate bill's consequences. Whether their economic judgment is right or wrong, their political judgment is probably right: Such a bill wouldn't stand a chance in the House.

Congress must search for a more saleable approach that includes some kind of cost-sharing. Minnesota Sen. Dave Durenberger's proposal for a nationwide tax on sulfur emissions, languishing in a Senate subcommittee, is one good idea. But critics worry about the high cost of administering such a cumbersome tax. They also worry about provisions that would give wide discretion for ordering emissions reductions to a still-shaky Environmental Protection Agency.

A more promising approach is a bill sponsored by Minnesota's Gerry Sikorski and California's Henry Waxman. Their proposal would impose a small tax on all electricity users — 50 cents to $1 a month for the average consumer — to help pay for smokestack scrubbers on the nation's 50 dirtiest plants. Installing scrubbers would achieve a speedy 6.5 million-ton reduction in sulfur emissions. Another 3.5 million tons in emission reductions would come in the program's second phase, when states would start new cleanup programs. Through its two-step strategy, the bill would prevent both "rate shock" and immediate job loss in vulnerable industrial states.

The Sikorski-Waxman bill, which suffered a subcommittee defeat earlier this month, remains the best hope for acid-rain control. Even though the bill is unfair to states like Minnesota, it ought to be passed. The alternative appears to be no legislation and dying lakes. The bill is a pragmatic national response to a nationwide problem.

DESERET NEWS
Salt Lake City, Utah, February 9-10, 1984

Why should Utahns have to pay for cleaning up Midwest coal-fired plants when industry here has already invested a billion dollars in scrubbers and is already burning coal much cleaner than that used in the East and Midwest?

It shouldn't, of course. Just ask Governor Matheson and other Western chief executives, who started mounting a fight this week against such an injustice. Their constituents certainly ought to support them on this issue.

At stake is a 3% tax the Senate Committee on Public Works and the Environment wants to put on electricity to curb acid rain damage in the East. Environmentalists claim much of the damage is coming from Midwestern coal-fired electrical plants.

The figures tell the story:

While the Western region has equipped 31% of its coal-fired electrical capacity with scrubbers, the East figure is only 7%. That difference in effort and money expended means the conscientious pollution-fighters would have to pay twice to clean up pollution sources — once to clean up ours and once to clean up theirs.

In addition, the West is burning coal that has only .64% sulfur content, considered a "clean" fuel. By contrast, Eastern coal has 2.0% sulfur, more than three times the Western content. The East, which needs the scrubbers far more than the West to control emissions, is putting far less effort into its controls effort.

That's still only part of the problem on controlling acid rain. There's still no solid evidence that the Midwestern plants are contributing that much to the East's — and Canada's — problems with acid rain. In his State of the Union message, President Reagan dodged the issue of a massive cleanup program for acid rain, much to Canada's chagrin. He did, however, call for a doubling of research funds to probe the issue.

But whether Midwest plants are or aren't the cause, Utahs who are burning clean coal in clean plants shouldn't also have to pay for the dereliction of duty that apparently abounds in the East.

LEXINGTON HERALD-LEADER
Lexington, Ky., June 25, 1984

While Congress and the Reagan administration continue to hem and haw while doing nothing about the acid rain problem, scientific evidence linking sulfur emissions from coal-burning utilities to this dangerous phenomenon keeps piling up on the bookshelves.

The latest report comes from the congressional Office of Technology Assessment. Its analysis of the cause of acid rain is in line with most other scientific studies. Sulfur dioxide in the atmosphere — 70 percent of which comes from coal-burning utilities — is the primary cause of the acid rain that has already killed aquatic life in some rivers and lakes in Canada and the Northeastern United States and that is now threatening water supplies in the Southeastern United States.

But more is at stake here than just the fish and plant life in a few rivers and lakes. The report issued last week says sulfates and other fine particles in the atmosphere could be responsible for 51,000 early deaths each year in the United States and Canada. That's far too great a price to pay for dawdling by our elected officials in Washington, who seem to worry more about the coal and utility industries than about saving human lives and the environment.

Something needs to be done, and done now, to reduce sulfur dioxide emissions. Of course, that something — as the Office of Technology Assessment points out — need not fall into the traditional pattern of rigid federal rules by specifying down to the last nut and bolt what utilities must do to clean up their emissions. On the contrary, the report suggests that a better approach would be to establish a specific level of permissable sulfur dioxide emissions and leave it up to the utilities to meet that goal as they see fit.

Letting the utilities choose their own method of reducing emissions would have a definite economic benefit.

For instance, mandating the use of wet scrubbers would cost the utilities as much as $1 billion more than other methods, such as switching to low-sulfur coal or using a mixture of low-sulfur and high-sulfur coal in combination with coal washing and smokestack scrubbers. Since any cost to the utilities is going to be passed along to consumers, that's a $1 billion savings for the American public.

Letting the utilities choose a less expensive, though equally effective, method of cutting back on sulfur dioxide emissions apparently would reduce the adverse impact of acid rain legislation on the coal industry. In addition to the economic benefit of keeping miners at work, that might have the political benefit of reducing the industry's opposition to the legislation.

Too often in its attempts to solve the nation's problems, the federal government has been overly rigid in its rules and regulations. In this instance, and probably in others as well, it would seem better to establish the goal and monitor compliance with that goal, while leaving the method of achieving that goal up to the ingenuity of private industry. After all, it is the end result, rather than the specific nuts and bolts, that is important.

Arkansas ⚜ Gazette.

Little Rock, Ark., June 9, 1984

Acid rain is a blot on the environment that the administration and Congress cannot ignore, although both have been doing a pretty good job of excusing themselves from taking remedial action. As a way to increase the pressure in Washington for corrective legislation, the National Clean Air Coalition, in association with Friends of the Earth Foundation, has released a summary of 140 studies related to acid rain in 13 Southern states, including Arkansas, to demonstrate that the problem is more widespread than many Americans may believe.

This compilation of studies comes two months after a similar comprehensive compilation of research studies by the National Wildlife Federation, the nation's largest conservation organization, covering 21 scattered states, and showing that problems of acid rain are national in scope, although they vary by state and region.

The overwhelming evidence of the scientific studies in recent years is that the major problem resides in the Northeast, which receives the fallout from the massive smokestack industries, principally power plants in the Ohio River Valley, that burn high-sulphur coal. Although nitrogens are one culprit, the principal contributor to the acid rain is sulphur dioxide emitted by coal-burning plants. Once in the atmosphere, the pollutants combine with moisture to form an acidic solution that falls to earth as rain or snow or fog, often killing aquatic life, damaging vegetation and eroding structures. A conclusion of the National Wildlife Federation report is that 212 lakes and ponds in the Adirondacks are devoid of fish and that 250 additional bodies of water are in danger.

The National Clean Air Coalition's new report, while recognizing that the problem in the Northeast is broad and compelling, turns attention to the "likelihood" that the South is standing at the edge of a similar problem with acid rain. Already, it notes, the rainfall over much of the South averages 10 or 20 times more acidic than normal, "with occasional storms more than 100 times more acidic."

Some of the acid rain is produced in the region — the South produces 28 per cent of the nation's total emissions of sulphur dioxide — and some of it drifts in from other domestic locations, including the same coal-burning plants of the Midwest that cause so much pain in the Northeast. Emissions from regional sources are projected to increase by 8 per cent between 1980 and 1995.

Soil and water conditions vary greatly throughout the Southern region and these conditions can have a strong bearing on the environmental impact of all the gook falling from the sky. In the generality, however, the coalition's report finds that acid rain "poses a clear and present threat to the South's lakes, streams, and abundant estuaries." What is more, says the report, "damage to the South's forests, crops, and soils may prove to be even more serious than the damage to its lakes and streams." Timber and agriculture make up major components of the South's economic structure, and it is easy to imagine that the effects of acid rain could prove to be more costly than the efforts that have been proposed in Congress for bringing it under control.

The National Clean Air Coalition report did some figuring on this point. Taking one approach that has been in legislation before Congress, aimed at reducing sulphur dioxide emissions from coal-burning power plants by 10 million tons a year, the report finds that the cost on the consumer's electric bill would vary from a low of 45 cents a month in Arkansas to a high of between $3.29 and $3.88 a month in Kentucky, which is a large coal-producing state.

Not as many references to Arkansas are found in the report as there are to other states, principally because fewer studies applying to Arkansas have been done. Acidity is measured on what is called a pH scale in which 7.0 is neutral and 5.6 is considered normal. In 1981-82, the coalition reports, the average pH for rainfall in Arkansas was 4.74, which is over eight times more acidic than normal.

Certainly the coalition's report, while falling short on some details, makes the valid general point that the South has no ground to feel that it is immune from the damaging effects of acid rain and has a direct stake in whether adequate acid rain legislation ever makes its way past Congress and the White House into the law of the land.

The Wichita Eagle-Beacon

Wichita, Kans., June 19, 1984

One reason the federal government continues to study instead of treat acid rain is that the problem is perceived as regional: Acid rain, it's said, threatens only the northeast United States and parts of eastern Canada. Politicians often find problems confined to one region uncompelling.

That's why it's good — in a way — that acid rain now has been cited as a threat to 13 southern states as well. The National Clean Air Coalition's recent announcement that rainfall in much of the area from Maryland southwestward to Arkansas has become 10 to 20 times more acidic should help many more Americans — and their elected representatives — to achieve a necessary perception: Acid rain is a national problem demanding a timely national solution.

Predictably, the coalition's announcement, based on an analysis of 140 scientific studies, drew heavy criticism. The Edison Electric Institute — which represents a significant number of the smokestack industries often blamed for acid rain — deemed the report "another example of the politics of fear." And the U.S. Environmental Protection Agency, which continues its arguably redundant studies of the problem, deemed it "speculation." But neither refuted the report's central finding, that southern forests, crops, human health and fresh water already suffer from acid rain, and stand to suffer far more if the nation's nitrogen and sulfur emissions — more than a quarter of which originate in the South — aren't significantly reduced.

Only increased political pressure can change the Reagan administration's counterproductive stance on legislation to control such emissions. One hopes news that a third of the nation's land area — and nearly half its population — are exposed to acid rain now will bring such pressure about.

THE KANSAS CITY STAR

Kansas City, Mo., June 15, 1984

Slowly and not at all eagerly the administration is being dragged down the road to judgment on the acid rain problem. In its second report, the multi-agency task force of government scientists appointed by the president to study this phenomenon concedes that forests in the northeastern U.S. have shown a systematic and sustained deterioration for more than two decades, a decline not attributable to natural causes.

Mind you, the panel didn't go so far as to pinpoint acid rain as the cause, but rather various pollutants, possibly including acid rain. Still, it was a step forward for this commission to recognize that plant growth, as well as fish and marine life in lakes, may be suffering damage from sulfur dioxide emissions from coal-burning industry and vehicle exhausts.

The task force also accepted the National Academy of Sciences' conclusion that a direct, one-to-one relationship exists between the amount of sulfur dioxide generated and the acidity of subsequent rainfall. But the administration agency, in its first report last year, said it is not yet possible to connect a given volume of pollutants from one source to specific damage in a certain distant area.

This circumstance is what makes acid rain such a political hot potato when Midwestern coal-fired electric utilities in Missouri and Ohio, for example, are ordered to spend millions suppressing sulfur dioxide emissions for the benefit of upstate New York, New England and eastern Canada. Some experts believe most of Canada's problem, despite bitter protests on supposed U.S. foot-dragging, originates at the giant smelter complex in Ontario. Western states are even more upset by the proposal for a national tax on power plants without coal scrubbers to finance an acid rain clean-up.

Canada and this country have been conducting joint experiments to trace the path of acid rain pollution from source to deposition site, and the results of these tests may push the administration yet closer to moving on this touchy environmental problem, even in an election year.

Clean Air Act:
Industry, Environmentalists At Odds

The 1970 Clean Air Act, passed at the beginning of the "environmental decade," is the most ambitious of the nation's environmental laws. Its purpose is to regulate emissions from factories and vehicles of the seven most common air pollutants—carbon monoxide, hydrocarbons, lead, nitrogen dioxide, ozone, particulates and sulfur dioxide—and to require states to meet national standards for the maximum allowable concentration of each of these pollutants. The deadline for state compliance was originally 1975, but was extended in the 1977 amendments of the Act to 1982, or, for areas with exceptionally high automobile pollution levels, until 1987. The act established two levels of "national ambient air quality standards"—one a primary standard to protect the health of all the population including those most vulnerable to pollution, and the other a secondary one to prevent damage to crops, buildings, water and other items. The Environmental Protection Agency was required to set both standards without considering the cost of their achievement. No deadline was set for compliance with secondary standards. Among the many abbreviations commonly used to refer to key parts of the Clean Air Act are the following:

NSPS—The EPA was required to set maximum emission levels for new or expanding individual plants and factories within each industry; these were called *new source performing standards.*

SIP—States were required to develop clean-up programs, *or state implementation plans,* in order that the air within their borders might meet the national standards by the deadline.

PSD—Regions that met national air quality standards (called attainment areas) for a pollutant were required to keep the concentration of that pollutant from rising above certain levels. The guidelines established for this purpose were called *prevention of significant deterioration* rules. (The act divided the country into 247 sections, many of them congruent with counties, known as air quality control regions; if the air within such a region failed to meet federal standards for two of the seven regulated pollutants, it was designated a "nonattainment area.")

The Clean Air Act was set to expire in 1981, but has remained in effect through annual appropriations by Congress. All attempts to revise it have so far ended in stalemate. In 1982, as the December deadline for meeting ambient air quality standards approached, a battle in Congress over whether to adopt industry-backed amendments to the act stalled over such issues as relaxed emission standards for both autos and factories, extended deadlines for clean-air standards, higher maximum air pollution levels for such areas as national parks, and sulfur-scrubbing requirements for all coal-burning utilities. (See pp. 102–121.) In November, the EPA released a list of 472 counties that faced possible sanctions as "nonattainment areas" in violation of the national standards; the move was viewed by many environmentalists as a tactic to push through the controversial Clean Air Act revisions. Potential sanctions included a ban on construction permits in the region, or the cutoff of federal grants for highways, sewers and clean air programs. The sanctions were blocked by the House in 1983, but no further progress was made on revising the act. More recent attempts at compromise have foundered over the question of new acid rain provisions.

The Reagan Administration won a major victory in July, 1984 when the Supreme Court ruled that the EPA could expand the so-called "bubble policy" established by the Carter Administration. The Carter policy had amended the Clean Air Act to allow leeway for industrial expansion in "attainment" areas, by permitting the measurement of air pollution over an entire industrial region rather than at specific locations. Thus new boilers, smokestacks or other sources of emissions could be established so long as the overall level of pollution within the bubble did not increase. Under the interpretation given it by the Reagan Administration, the "bubble policy" could also apply to areas of the nation that have not yet met federal clean-air standards. In supporting that interpretation, the Supreme Court noted that the original provisions establishing the policy were ambiguous, leaving it unclear whether an emissions "source" referred to a single smokestack or an entire plant.

The Washington Post

Washington, D.C., January 24, 1981

THE COMING MONTHS will be filled with debate over the merits of environmental regulation, headed by the controversial reauthorization of the Clean Air Act. Changes in many of the act's specific provisions will be proposed, but affecting all of them will be the proposal to require that henceforth clean air controls must first be subjected to a formal cost-benefit analysis showing that they provide benefits of greater monetary value than the costs of achieving them. Cost-benefit analysis is not, however, the hoped-for answer that will automatically produce more sensible and effective government regulation.

Such accounting requires, first of all, that dollar values be assigned to intangible benefits. For example, the medical costs incurred by exposure to unhealthy air can be measured, as can lost wages and lowered productivity, but pain and suffering, psychological trauma and premature death are far harder, if not impossible, to gauge. Attaching specific values to aesthetic improvements is almost as difficult.

Another complicating factor is that costs and benefits arise at different times, with the costs generally arising immediately and the benefits spread over years or generations. Traditional economic methods for treating such problems do not work well. Cost-benefit analyzers must also make assumptions about the value society attaches to lowering a particular risk, yet society's choices in this realm tend to be irrational and unpredictable. Americans are willing to pay a great deal of attention to some risks (a nuclear accident) and very little to lower the risk of a more familiar but much more certain danger (cigarette smoking). Finally, cost-benefit analysis cannot accommodate the fact that costs and benefits are usually felt by different people.

Though costs are usually easier to measure than benefits, they also raise enormous uncertainties. Unpredictable technological innovation frequently invalidates the most careful estimates of the costs of meeting new controls. Productivity may be lowered by the diversion of scarce research funds, or raised by new processes or equipment developed to meet environmental requirements.

Putting all these uncertainties together means that the final answer is likely to mean little or nothing. Applying hard numbers to soft assumptions does not hide the fact that the method is highly subjective—not the hard-headed objective analysis it might seem. The Clean Air Act, for example, is the most exhaustively studied of all environmental laws, yet recent, reputable estimates of its annual benefits differ by tens of *billions* of dollars. Its estimated costs are equally vague.

Eventually, cost-benefit analysis should develop into a more useful technique, but at this point it is too crude to be made the sole basis for deciding on a particular regulatory goal. The most likely outcome of such a requirement will be years of expensive delay while courts struggle to decide among wildly conflicting estimates. For the present, the provisions of the Clean Air Act need to be individually examined and weakened or strengthened on the basis of all of the best available evidence.

THE MILWAUKEE JOURNAL
Milwaukee, Wisc., July 31, 1981

Air pollution is not one problem, but many problems — all of them complex. Thus, everyone knew from the start that the national effort to control air pollution would be neither simple nor cheap. And that's why we tend to dismiss most complaints about the complexity and cost of the provisions of the Clean Air Act, which expires Sept. 30 and is before Congress for reauthorization.

We think the approaches of the act are basically sound. The billions of dollars spent so far have been of great value — slowing, halting or, in some cases, even reversing trends toward unhealthful air. While fine-tuning is always in order, there is no need, as some in the Reagan administration insist, to radically change the direction of this hard-won, evolving, useful body of law. Particularly objectionable is the notion — still under debate in the administration — of "simplifying" the law by putting clean air on a cost-benefit basis.

Try as we might, we can't see the sense in applying to clean air a rigid economic technique that expresses everything in monetary value. For one thing, both costs and benefits of clean air are largely intangible — involving the multi-layered physical health of the entire populace. For another, health effects of bad air may be delayed for years after exposure. When they do appear, they may be compounded by other factors such as smoking.

Furthermore, the cost-benefit approach would not simplify the law; it would add to the complexity because costs always would be arguable. Finally, in a more practical, informal way, the law already provides for cost considerations on a case-by-case basis. That occurs when obviously a deadline can't be met or the costs of the next increment of improvement may well force the shutdown of needed industries.

Originally, the administration intended to reveal its air pollution proposals by June 30. The date has slipped to mid-August, leading some environmental groups to believe that the heat is temporarily off, that the present Clean Air Act may be extended for a year and the debate delayed. Maybe so. But whenever the administration's panel considering air pollution laws finally issues its report, we hope that the unworkable cost-benefit scheme will already have been debated, and properly junked.

The Seattle Times
Seattle, Wash., March 20, 1981

THE federal Clean Air Act has generally "worked well." That was the most important conclusion of a comprehensive, congressionally mandated, two-year study of the act released this month by the National Commission on Air Quality.

But that finding was overshadowed, in some reports, by suggestions for changes in the landmark law, which expires this year unless Congress opts for revision.

The commission recommended, among other things:

— That national 1982 and 1987 deadlines for achieving air-quality standards be dropped in favor of review every three years.

— That current restrictions on industrial development in some pristine, pollution-free areas be loosened a bit.

Environmentalists, including some members of the commission, cried foul, fearing that the act might be gutted. We, too, oppose any wholesale changes in the law, but it may be ready for some fine-tuning.

Actually, the law already has been altered considerably since its passage in 1970, in response to increased knowledge, new data and changing conditions. The sky did not fall as a result; on the contrary:

"Overall, the nation's air quality is continuing to improve, although serious problems exist in many areas," the Council on Environmental Quality said in its latest annual report.

A majority of the commission (whose chairman was Senator Gary Hart, a Colorado Democrat of sound environmental credentials), found the present law's deadlines unrealistic, ineffective and unlikely to be met no matter what the government does.

Abandoning them would not bring an end to progress, however. The commission stressed greater emphasis on installing new equipment at old, polluting plants; significantly reducing sulfur-dioxide emissions (linked to "acid rain") by 1990; and continuing to set standards that safeguard public health.

This is at least the second major study in two years to suggest some changes in the Clean Air Act. In 1979, Resources for the Future concluded: "Air-pollution control is probably the only area of U.S. public policy that uses brute-force regulation to try to solve a problem of such magnitude and complexity. The result has not been remarkably successful, even in terms of air - pollution control itself . . ."

That study, together with a Harvard University report released last month, recommended more reliance on market forces to fight air pollution, instead of bureaucratic regulations and arbitrary deadlines. Pollution taxes, "emission rights," excess-pollution fines and other economic incentives should be considered carefully by Congress during its upcoming review.

Those who argue that any changes will mean cutting the heart out of the act must realize that there's more than one good way to protect people's lungs.

The Idaho STATESMAN
Boise, Idaho, March 8, 1981

New federal recommendations to ease portions of the Clean Air Act include at least one major pitfall. Congress would make a mistake if it accepted the Commission on Air Quality's proposal to do away with 1982 and 1987 deadlines for meeting clean air standards.

The deadlines send a clear-cut message. They say to the citizen, here is a date by which you and your elected officials reasonably should be able to clean up the air you breathe. That way, if no progress is made, citizens can see they should start getting alarmed.

Ada County provides a perfect example. For several years, resistance to regulation has stymied efforts to improve the county's air. Subsequently, public concern has grown as the deadlines continued to draw nearer with little or no visible progress being made.

Now, the Legislature appears to be responding to public concern. On Wednesday, the House passed a bill that would allow counties to set up inspection and maintenance programs to test motor vehicle emissions systems. Ironically, this bill designed to help Ada County meet the 1987 federal deadline may pass just in time for the Congress to do away with the deadline.

Clearly, deadlines perform a potent political and technical function, allowing the public to make sure its representatives keep to a rigid schedule for cleaning up the air. Where the deadlines are realistic, as in Ada County, they should be maintained. Where they are unrealistic, as in severe problem areas like Denver and Los Angeles, there should be extensions.

In another recommendation that relates to Boise's carbon monoxide problem, the commission proposes to relax limits on allowable emissions from new cars, lifting the standard to 7 grams per mile from 3.4 grams. The commission found that devices designed to meet the tougher standard broke down so often the long-term effect on the air would be worse than using less-complicated devices that currently meet the 7 gram standard.

If that finding is accurate, keeping the less-restrictive standard only makes sense. However, Congress should keep the pressure on automakers to find more workable ways to make emissions cleaner.

The only way the air in places like Boise will be cleaned up is if the citizens are determined to do it. Still, the federal government must play a major role, keeping an eye on those responsible for technological advance and making sure citizens keep posted on the problems that may or may not be getting solved.

The Dispatch

Columbus, Ohio, August 7, 1981

HALLS OF CONGRESS will be the scene of a battle this fall between environmentalists and the Reagan administration regarding the renewal of the costly and complex Clean Air Act of 1970.

The issue has been raised with a proposal by the administration to lower pollution standards for cars and to reduce "unnecesssary restrictions."

Environmental Protection Administrator Anne M. Gorsuch explained the draft handed Congress is broad but says the basics are there to protect the public and the environment. Improvements will be at a more "reasoned pace."

Liberalizing current auto emission standards could save $100 on the price of a new car, a boost to the industry, she says.

But the nation's EPA chief emphasizes the national air standards will continue to be based primarily on public health considerations rather than potential costs to industry, thus reducing some opposition.

States, she notes, would have more of a voice in establishing EPA standards and the proposal retains exacting standards for national parks and wilderness areas.

One area of strong conflict will involve acid rain disagreements. The new proposals do not call for a program to curb acid rain but do recommend accelerated research on the program.

More effort is advised to control toxic pollutants while the coal industry should be happy with a section calling for reduced restrictions on new coal-fired plants.

Manufacturers already have lauded the proposals but environmentalists are looking with alarm.

The present law has come under fire from such officials as Gov. James A. Rhodes who argues the Clean Air Act, in part, is unrealistic and is too much of a burden on industry. The result, he says, is a loss of jobs and industry.

The proposed revision of this controversial law will be debated at length and should be. But its approach appears more practical and sensible than the current restrictive act.

SAN JOSE NEWS

San Jose, Calif., July 6, 1981

THE language of federal air pollution regulation is a jungle of exotic acronyms all but impenetrable to the unaided human intellect. Strangers who venture into it without a skilled guide have been known to vanish without a trace amid the trackless wilds of RACT, BACT and LAER, SIPs and NSPS and CASAC.

For general purposes, though, only one acronym is crucial to understand: NAAQS. NAAQS stands for National Ambient Air Quality Standards, which are the heart of the Clean Air Act and the federal clean air effort.

The Clean Air Act requires the Environmental Protection Agency to set two types of air quality standards: primary, relating to effects of pollution on human health, and secondary, relating to effects on buildings, crops, wildlife and so on. In setting the primary standards, the act directs EPA to protect against pollution levels that would "endanger public health, allowing an adequate margin of safety."

Today the "adequate margin of safety" concept is under heavy attack. Rep. James T. Broyhill, R-N.C., and some business leaders say it should be replaced with more lenient language and that EPA should have to prove that the benefits of any pollution standard would outweigh its costs.

Such changes would be dangerous. In our present state of knowledge, cost-benefit analysis is too crude a tool to use in setting air pollution standards; the effects of pollution are too hard to gauge, the benefit of controls is too hard to measure precisely. And the "margin of safety" concept protects those who are most vulnerable to pollution — the very young, the elderly, the sick.

We agree with the National Commission on Air Quality, the national Clean Air Coalition, the National League of Cities and many responsible business leaders that:

• *Primary air quality standards must be based on considerations of health, not cost, and must provide an adequate margin of safety for all citizens.*

Some business groups have also suggested that air quality standards — at least the secondary ones — should be set by the states or by regions rather than the federal government. The idea is that local officials are best able to balance local environmental concerns against the need for jobs, tax revenues and other benefits of development.

The theory sounds plausible, but the results would be disastrous. As they did in the days before federal regulation, industries would be able to engage in environmental blackmail, playing state against state, region against region, threatening to pull their factories — and jobs — out of any area that enacted high air quality standards.

• *Air quality standards should continue to be set on the federal level.*

However, as the National Commission on Air Quality suggested, Congress might consider directing EPA to set different secondary standards for different regions, depending on local problems and needs.

In setting air quality standards, EPA consults the Clean Air Science Advisory Committee, whose members are appointed by the EPA administrator. Business is skeptical about the panel's objectivity and has suggested creating a more independent commission and giving it virtual veto power over EPA's decisions. We think that would be going too far. But we agree with the U.S. Chamber of Commerce that:

• *An independent scientific advisory committee should be created to review air quality standards on a regular periodic basis.*

For at least eight years EPA has acknowledged the need for an air quality standard covering fine particles, which can be inhaled deeply into lung tissue, but has not yet established one. The agency has been equally sluggish in identifying and setting standards for toxic pollutants; in the past 10 years it has established levels for only four such substances.

• *Congress should direct EPA to establish standards for fine particles within two years, and set a deadline for a decision on the 40 suspected toxic pollutants now under evaluation.*

Four years ago, recognizing the need to protect the clean air we have as well as to clean up the air that's already polluted, Congress wrote the Prevention of Significant Deterioration (PSD) program into the Clean Air Act. The program envisions three classes of air quality regions. In Class I areas — national parks and monuments, wilderness and other pristine sites — virtually no air quality deterioration is allowed; in Classes II and III, progressively larger pollution increases ("increments") are permitted. Since no Class II area has ever been shifted into Class III, in effect there are only two classes.

The PSD program, with its complicated system of "increments," monitoring and review provisions, and requirements for sophisticated and costly pollution control technology, has been a headache for many industries, and some of them are out to kill or at least cripple it in this Congress.

The basic PSD program is worth keeping. Preventing pollution is easier, cheaper and healthier than trying to cure it; besides, the PSD restrictions discourage industries from relocating to clean-air areas to escape pollution controls. But we agree that the program as it stands is unnecessarily complicated and cumbersome.

• *Congress should eliminate Class III (which exists only in theory anyway). EPA should replace the elaborate increment system with special air quality standards for Class I and Class II areas, and require industries locating or expanding in such areas to adopt the best available pollution control technology. Further,*

• *The permit process for industries seeking to locate or expand in clean-air areas should be simplified, streamlined and speeded up. However,*

• *EPA should continue its policy of protecting visibility in Class I areas and adjacent landscapes.*

In sum, we believe the national air quality standards and the Prevention of Significant Deterioration programs can and should be stripped of a lot of red tape, but their basic premises are sound. The radical changes proposed by certain business interests and lawmakers would compromise our environmental quality and health for the sake of marginal and uncertain economic benefits.

Tomorrow, we will look at the Clean Air Act's enforcement provisions and proposals for changing them.

RAPID CITY JOURNAL—
Rapid City, S.D., August 4, 1981

The Clean Air Act which expires on Sept. 30 can be made more efficient without sacrificing air-quality gains and the result could be a savings of billions of dollars.

Many requirements of the act have little or no impact on air quality but impose substantial costs on industries and consumers. The legislation has delayed industrial expansion generally and has also slowed development of domestic energy sources.

An analysis of reports from companies of all sizes made by Environmental Research and Technology, Inc., an independent consulting firm, found 92 cases of adverse impact from the Clean Air Act. The 92 cases represent only the proverbial tip of the iceberg in terms of the total number of facilities affected by the law.

The study found that requirements of the statute often create a severe burden for small business. One foundry, for instance, faces a 23 percent increase in unit costs of production because of energy expenses required to control emissions.

Cost of pollution controls at another foundry was found to be two to three times higher than the cost of the firm's production facilities

In some cases, the cost of enforcing the Clean Air Act included loss of employment opportunities and local tax revenue.

A ban on new construction imposed in one area under provisions of the act is preventing an oil refinery

modernization that would reduce emissions, improve energy efficiency, increase local tax revenues by $7 million and provide 1,200 jobs during construction and 50 permanent jobs thereafter.

The report also cited the case of another refinery that planned to install equipment to reduce sulfur-dioxide emissions but was forced to delay the addition for a year pending a study required under a provision of the law designed to assure that new equipment does not worsen air pollution.

In reviewing experiences of individual companies with the act, the survey found four basic problems: Excessive and/or unnecessary costs; undue uncertainty in project planning; avoidable delays in decision making by review agencies and unjustifiably stringent control-technology requirements.

Congress has begun a broad review of the Clean Air Act. In view of the short time remaining before it is due to expire, it's probable the statute will be extended while major revisions are considered.

We all want the air to be as pollution free as possible. But in achieving that goal, a balance must be struck between the benefits and the cost.

As it considers revisions in the Clean Air Act, Congress should eliminate or amend its counterproductive requirements that don't do much to improve air quality but have an adverse impact on the economy.

The San Diego Union
San Diego, Calif., August 15, 1981

The federal Clean Air Act is 11 years old, and due for the overhaul proposed earlier this month by the Reagan administration. While the law has done much to arrest a decline in national air quality, standards and rules that seemed reasonable a decade ago have become unreasonably costly, unduly restrictive, or simply obsolete.

The world energy shortage, unimagined when the anti-pollution law was enacted in 1970, has brought hard times upon the automobile industry, which was expected to bear much of the cost of cleaning up the nation's air. Inflation has made economic nonsense of factory emission standards that once seemed attainable.

These economic facts of life, as well as advances in scientific knowledge and anti-pollution technology, need to be reflected in a revision of the clean air laws.

The problem facing Congress this year is to write Clean Air

Act amendments embodying principles enunciated Aug. 5 by Environmental Protection Agency Administrator Anne M. Gorsuch without nullifying essential provisions of present law. The health standards applied to air pollution, for example, can be based upon current scientific data, as the administration recommends. But they cannot be diluted solely because they are not cost-effective.

Some secondary standards protecting crops, visibility, and buildings, should however, be put to the cost-effectiveness test. Thus, it may prove wasteful to forbid the burning of high sulphur oil for power generation if the rule achieves a negligible improvement in air quality at extremely high cost to utility customers.

The administration is also on firm ground in urging uniform pollution technology for factories in clean-air areas. There can be no basis for requiring industry to

install costly anti-pollution equipment in areas where their emissions would not reduce air quality below national standards.

Southern California's continuing battle against air pollution would be set back, on the other hand, by the proposal to roll back new car emission standards for 1982 to the 1977 levels — twice the current limits in some cases — even though most new cars are already meeting the present tougher standards.

And while it makes sense to extend compliance deadlines for such problem areas as the Los Angeles basin, as the administration wishes, the new deadlines should reflect what is practically attainable rather than a full-scale retreat from enforcement.

Applying cost-effectiveness tests to pollution standards, simplifying compliance rules, and relaxing rigid EPA enforcement procedures can lift a needless burden from the nation's economy. Healthful, breathable air must remain the goal, however.

The Philadelphia Inquirer
Philadelphia, Pa.,
November 6, 1981

Round one of the Clean Air Act reauthorization fight was waged Tuesday in the Senate Environment and Public Works Committee, and those who thought they might see some bare-knuckles fisticuffs walked away disappointed. By a 13-0 vote, the committee decided to keep one controversial part of the law precisely as it is.

That decision handed the Reagan administration a defeat, along with lobbyists for business and industry that had argued that the so-called margin of safety provision of the act should be eliminated. The provision requires that clean-air standards be set stringently enough to protect the especially sensitive, including children and the elderly.

The Clean Air Act, which expired this year and must be reauthorized by Congress, has become the prime target of those seeking to reduce regulation and federal involvement in what the act's opponents claim to be duties better delegated to the states. Although there initially existed in Congress a strong sentiment to strip the 1970 act of many of its provisions, that movement appears to have waned in recent months.

The change of heart may be due to widespread support by many Americans for the act and strong opposition to efforts to dismantle it. A recent Louis Harris public opinion survey found that "the desire on the part of the American people to battle pollution is one of the most overwhelming and clearest we have ever recorded in our 25 years of surveying public opinion in this country."

The vote in the Environment and Public Works Committee indicates that that public sentiment has made itself felt in the Congress.

The Morning News

Wilmington, Del., November 30, 1981

Everybody wants clean air.

The National Clean Air Coalition, it goes without saying, wants it, and that includes such varied interests as the League of Women Voters, the Izaak Walton League, the United Steelworkers, the Sierra Club and the National Audubon Society.

Mobil Oil Co. wants clean air. The secretary-treasurer of the Amalgamated Clothing and Textile Workers Union, writing in the New York Times of Friday, spoke fervently in its behalf. The American Petroleum Institute wants it. The Motor Vehicle Manufacturers Association wants clean air.

How do we know they all want clean air? Because they all say so.

Perhaps most to the point, the U.S. Senate's Environment Committee this month voted for clean air. Sure it did. Would anyone go on the record against motherhood? Well, they won't go on the record against clean air, either.

A Louis Harris Poll last month found Americans by a margin of 80 to 17 against any relaxation of federal regulation of air pollution. It did not give a clear answer on how much those same Americans are willing to pay to get the air how much cleaner.

The National Academy of Sciences says, ringingly, that action on combating the problem of acid rain — just one element in clearing the air — is needed, "Now!"

Certainly action to clean the air is needed in Delaware's part of the country. In the East, for instance, we would all be part of a 31-state Acid Deposition Control Area under legislation proposed in Congress this month. Senate Bill 1706, one of the two measures introduced in this field, would demand that the states in this area reduce the air's sulfur dioxide (SO2) content from a present 23 million tons to 10 million tons during the next decade. (A vastly complicated system of actions by the individual states, including "tradeoffs" would be invoked.)

The vehicle manufacturers tell us that under the past decade of federal regulation, the amounts of hydrocarbons (HC) and of carbon monoxide (CO) in the air have been reduced by 96 percent. Oxides of nitrogen (NOx) have been cut 76 percent. The association would have us think that this is sufficient, that retaining, not increasing, present standards while increasing the percentage of vehicles that meet those present standards will bring us optimum ambient air quality in the next two years (as far as "mobile sources" are concerned). Meanwhile, they argue, this approach will not create a burdensome increase in vehicle prices.

The informal Senate committee vote early this month was fairly unequivocal: Air pollution standards, said senators, should continue to be set *without regard* to cost.

That's easy to say. There is a point, somewhere, however, at which further improvement in air quality will not be cost effective. There is a point somewhere, at which eliminating the last fraction of the last percentage point of pollution will cost more in other socially desirable goals than it is worth.

It isn't likely that we are getting dangerously near to that point. That seems especially so when we hear such a preponderance of the argument against higher standards from sources which needed the imposition of the present act 10 years ago to get them even to start cleaning up their act.

Government and industry sources said last week that the federal Environmental Protection Agency is considering a major relaxation of auto emission standards in an attempt to settle eight lawsuits filed by the auto industry. That is not the sort of action that will help the Reagan administration's growing reputation in some quarters of being peopled by "environmental illiterates," as they were called by Russell W. Peterson, a former governor of Delaware who now heads the National Audubon Society.

THE SUN

Baltimore, Md., December 22, 1981

A recent Brookings Institution study of the Clean Air Act may be one of the more stimulating studies of federal regulation to come along in years. It is bound to be influential—in ways that will both please and displease environmentalists—when Congress begins seriously to revise the act next year.

The study's prime thesis is that cleaner air enjoyed recently owes little to the act and much to reduced coal burning and economic decline. It contends the act does not face up to real problems, two of which are the tendency of auto-pollution controls not to work in actuality and the outlook for severe pollution problems from increased coal burning. These problems would be exacerbated in a prosperous economy.

Both emphases please environmentalists, many of whom for years have shared the authors' contention that the act doesn't do enough in these areas. Environmentalists also applaud the study's criticism of the act's failure to attack problems caused by various exotic pollutants—some of which may be potentially more dangerous than the seven common ones with which it does deal.

But other aspects of the study displease environmentalists, especially sections they regard as placing too much stress on cost-effectiveness. One suggestion is for scrapping, as too costly, the part of the act which protects certain highly sensitive asthma and emphysema sufferers who are a mere 0.05 percent of the population. While the environmentalist contention that these people deserve protection is persuasive, we wonder if there might not be cheaper ways to attack the problem than to embark on a massive campaign to try to make urban air pure enough for them.

The study is not the last word on the subject, and it has detractors. One, a former federal environmental enforcer, says it deals too much in averages and misses significant clean air act accomplishments, such as cleanup of some severely polluted areas. He adds that the emphasis on economic incentives, as opposed to law enforcement, too often amounts to proposing "licenses to pollute." He thinks monitoring these would require a far larger bureaucracy than the law-enforcement approach. The study certainly introduces fascinating elements for next year's debate.

Arkansas Gazette.

Little Rock, Ark., December 14, 1981

That public opinion strongly supports the Clean Air Act and opposes any attempts to weaken it has been demonstrated repeatedly in many ways, the most telling of which may be the reluctance of the Reagan administration to follow through on its earlier intentions to offer a new and weaker version to Congress.

Now comes fresh evidence of support for the act in a national survey conducted for the United States Chamber of Commerce by Opinion Research Corporation, whose president says flatly that "There is no support by the general public for any changes in the Clean Air Act that would lower the standards of air quality."

There are some parallels between the ORC survey and recent surveys by Louis Harris. ORC found that the public does not oppose all revisions of the act, and a Harris survey recently found that any attempt to weaken the law would be unacceptable to the public.

OCR finds, and it is easy to sense that the finding is on the mark, that the American public will welcome administrative changes that will save industry money and red tape without sacrificing air quality. In a way, this observation strikes at the heart of the issue between Congress and the administration on a renewal of the Clean Air Act, which was supposed to come this year but almost certainly will not because time is running out.

Just about everyone who is familiar with the act will recognize that it needs some regulatory streamlining, but the trouble is that those in industry who oppose the law in its entirety have been trying to eviscerate the act by offering amendments they say would simply ease unreasonable regulations. Stalemate means they haven't succeeded.

The Houston Post

Houston, Texas, August 10, 1982

Under the Clean Air Act, now being studied for revision, the Environmental Protection Agency is required to protect the public from hazardous air pollutants — substances that cause cancer, birth defects or other serious diseases. But in the 12 years since the law was passed, the EPA has identified only seven pollutants as hazardous and has regulated only four. For the past four years the agency has continued to study 37 chemicals from a list of more than 600 as being likely sources of human ills.

Republican Sen. Bob Stafford of Vermont apparently feels that the EPA has not sufficiently applied itself to the task of evaluating the 37 that it has pinpointed for priority study. Some 3 billion pounds of these 37 chemicals are released into the air annually. If they do hold dangers, they should be regulated. He has introduced an amendment to the Clean Air Act that would give the EPA three more years to determine which ones meet the statutory definition of hazardous air pollutants. After three years, any chemical on the list not yet ruled upon would be automatically listed as hazardous until proven otherwise. Stafford makes no attempt to change the law's definition of hazard. Nothing in his amendment would enlarge the number of chemicals for consideration or regulation. It is designed merely to spur action.

Scientific research takes time. Statistics have to amass before they begin to trace a trend or reveal a weakness. But much of the data on these chemicals has been under study for many years by other agencies of government and of science. The National Cancer Institute, for example, has recently finished a study showing that between 3,500 and 19,000 cases of lung cancer could be prevented if air pollution were better controlled. After adjusting the data for age, smoking, occupational exposure and X-rays, the institute scientists concluded that between 11 and 21 percent of current lung cancer cases were caused by air pollution.

Sen. Stafford is aware of economic and practical realities. His draft would let the EPA set standards below those needed for ample margins of safety if the standard should be economically or technologically impractical. This should ease reasonable fears on the part of industry that his amendment would require uneconomical solutions. But the goal of protecting the public health with ample margins of safety must continue to be maintained as a factor in further technological development.

Detroit Free Press

Detroit, Mich., March 1, 1982

THE DEBATE over the Clean Air Act is churning up more dust than a windstorm over Texas, but it could well end in stalemate this year. There may be general agreement that the act is as cumbersome as it has been successful, but there is notable lack of accord on how to revise it. If a compromise isn't struck soon, Congress is expected to abandon the effort until after the election.

Much of the debate centers on House Bill 5252, a package of revisions that has been, on the one hand, assailed as a gift of dirty air for the nation and, on the other, defended as a reasonable, necessary streamlining of the law. The bill is in some part the handiwork of Michigan Democrats Bob Traxler and John Dingell, whose special interest is in rolling back auto emissions standards to last year's levels.

It is not the emissions question that is the chief worry about HB 5252, however, but the cumulative impact of the many changes it proposes and the amount of discretion that would, as a result, be left to Anne Gorsuch's Environmental Protection Agency. The EPA has been so decimated and demoralized in recent months that it may be simply incapable of carrying out its responsibilities, even if its top officials were committed to strong enforcement — a point on which there are very serious doubts.

It is not very comforting to sit here in Michigan, where clean air laws have generally been enforced, and ponder what will happen if deadlines and mandatory penalties and strong federal oversight are removed from the Clean Air Act, and the EPA is too limp to enforce what is left. It would

be difficult to resist the clamor for weaker regulation, especially if other states were willing to attract industry by loosening their own standards.

There are ways to improve the Clean Air Act without retreating from its basic principles, and some of them are incorporated in HB 5252. We need to make the act more efficient. We need to retain the present air quality standards and to set standards for pollutants not now covered by the act. We need to set tight limits on those pollutants, but to give industry the flexibility to decide how to meet them and the encouragement to innovate. We do not need to give up clear skies, breathable air and the breathtaking vistas from national parks, wilderness and other places with pristine air quality.

As for the emissions rollback, it is not by itself going to save the auto companies, but neither is it going to poison the air in most of the country. Uniquely affected areas such as Denver and southern California should be free to impose their own tighter standards on auto emissions, but elsewhere a return to last year's standards will have insignificant health effects.

A rollback accompanied by a guarantee that emissions controls will work for several years in actual driving conditions would even improve air quality. And if a reduction in the cost of producing cars led to a modest increase in sales, the improvement could be more rapid, as newer, cleaner models replaced old, polluting junkers.

That is going to be a tough argument to sell outside Detroit, however. It will be even tougher if the rollbacks are perceived to be part of a package that looks like a full-scale assault on a popular and effective Clean Air Act.

Richmond Times-Dispatch
Richmond, Va., April 7, 1982

To hear the die-hard environmentalists and the die-hard industrialists tell it, the nation's choice in revising the Clean Air Act is between (a) polluting Yellowstone and (b) torpedoing the economy. Some choice.

It's one the public does not want made. There's bound to be a happy medium. Whether Congress will find it in the current climate of debate is questionable.

The House Subcommittee on Health and Environment has reported out a bill that has its shortcomings. It is silent, for example, on acid rain and toxic air pollutants, two subjects that at the least require study. And it relaxes auto emission standards that were set in 1977 and met, contrary to the industry's predictions, on schedule in 1981. The auto makers' argument that removing $100 to $350 worth of anti-pollution equipment will get Detroit moving again without adversely affecting air quality is not overwhelmingly persuasive.

But to inveigh against the bill as "a collection of polluter's proposals" and the congressmen who put it together as a "dirty dozen" who have "marched through the Clean Air Act like Sherman marched through Georgia" — as environmentalist groups have done — serves no one. Nor does descending on congressional committee rooms with "Wanted" posters bearing the committee chairman's picture or disseminating suspect data about alleged polluters. They've done that too.

The fact is that unnecessary regulation has hamstrung economic growth. The people who are trying to redress the balance hardly deserve impugnation.

The tactics of some in industry are more subtle, but still discomfitting. In a sagging economy, some seem to think, the bottom line on the P&L statement should be the top priority of the Clean Air Act. They charge that standards based on only circumstantial evidence will cost them billions in unnecessary equipment. That shouldn't happen, though it has: EPA has been sloppy in setting standards for some pollutants, remiss in setting others and arbitrary in specifications for compliance. Some of the computer models to which environmentalists point may not be definitive.

But it also happens that the nation can't hold its breath waiting for 100 percent proof of the connection between every pollutant and ill health effects. Common sense — and a sense of public opinion — require that when a link is strongly suspected, remedial action should be pursued.

Nobody disputes that the Clean Air Act has worked. Few people dispute that it is cumbersome and costly or that it must be made to work more efficiently and economically. Deciding who's crying wolf and who's legitimately crying foul is Congress' job. An increasingly acrid debate won't help.

ALBUQUERQUE JOURNAL
Albuquerque, N.M., July 5, 1982

The Clean Air Act may need revision, but there is no objective way to know if changes would lead to air that is more or less healthy than it is now. Nowhere is this more evident than in the debate about percentage reduction.

Percentage reduction, or scrubbing, was written into the act's 1977 amendments. It requires that new power plants remove from 70 to 90 percent of their sulfur emissions. The law does not recognize that eastern coal has much more sulfur in it than western coal. It merely requires that whatever coal is used, 70 to 90 percent of its sulfur content be removed.

This is bad law, as Sen. Pete Domenici, R-N.M., has charged. It protects eastern coal suppliers. It also has meant higher coal sulfur emissions in the east and has contributed to ever-more-furious debate about acid rain which may be partly caused by these high sulfur emmissions.

Western coal suppliers claim they are shut off from the big eastern market . They reason that use of western coal would reduce sulfur emissions that may contribute to acid rain. Seeing an opportunity, Domenici and other western senators want to repeal percentage reduction to boost western coal interests.

But if that happens, the percentage reduction requirement that has protected western skies from a higher level of emissions will be gone. Visibility — and the quality of western air that is generally more healthy and clearer than eastern air — could deteriorate.

The western senators rewriting the act know this. Three of them, Baucus of Montana, Simpson of Wyoming and Hart of Colorado, proposed on June 21 a two-tiered "performance standard" to replace percentage reduction. Congress has taken no action yet on the proposal.

The performance standard would specify what level of emissions would be allowable for a given region. This is an attempt to write federal law to apply to regions. The challenge is to create performance standards would prevent deterioration of air quality. that

So far, clean-air advocates, some eastern politicians, and some coal suppliers have expressed interest in replacing percentage reduction with performance standards. If lawmakers can come up with a formula that preserves generally pristine western skies, it may be the best solution to one segment of the clean air problem

WINSTON-SALEM JOURNAL
Winston-Salem, N.C., July 28, 1982

For a classic study in congressional factionalism and impasse, it would be hard to match the current efforts to rewrite the 1970 Clean Air Act, which expired last September. One might suspect that such a bill would simply pit environmentalists on one side against industrialists on the other, but in fact the situation is far more complicated than that.

The clean air question affects particular industries and regions dramatically and directly. The auto industry, for example, has been lobbying hard for an easing of the exhaust emission standards, which it claims add substantially to the cost of a car without adding to its buyer appeal. Midwestern utilities complain that standards for sulfur dioxide stack emissions will cost billions to comply with. Energy companies argue that air quality standards for "pristine areas" whose air is already cleaner than air elsewhere obstruct efforts to develop new sources of synthetic and conventional fuels.

Environmentalists argue that acid rain and other forms of air pollution are destroying plant and animal life and creating health problems for millions of Americans.

What is making compromise so difficult is that each faction in the debate, each region, views the rewriting of the act as its best, perhaps its only, chance for legislation with which it can live. Everyone wants clean air, of course, but two questions defy consensus answers: How clean is clean, and at what point do costs outweigh benefits?

Some members of Congress also want the Environmental Protection Agency to hurry up and finish studies it has been conducting for four years now on 37 air pollutants. They argue that if those studies are not complete within a year, the substances should all be declared hazardous. Industry counters that such a requirement would force EPA to investigate "totally unsupported, frivolous allegations" that a particular emission is harmful.

Not the least of the problems confronting Congress is that evidence about air pollution is expensive to gather and difficult to evaluate. Often sources of pollution are far removed from areas that pollution affects, causing jurisdictional disputes that question not only regulation authority but also the credibility of the evidence of pollution and its source. It is not easy to convince Gulf Coast industries that they are causing acid rain in North Carolina.

The current logjam in Congress over clean air may not be broken at least until after the November elections. That means the 1970 act as amended in past years will remain in effect. Predictably enough, that situation pleases some people and frustrates many others. For the most part, environmentalists are pleased. They realize that the troubled economy is putting heavy pressure on Congress to ease air standards so that American industry can again become cost-competitive, or, in some cases, simply survive. But delay would also mean postponement of tougher standards governing acid rain.

That pleases Midwestern utilities in particular, and although features of the 1970 act do not make them happy, some industry representatives now prefer putting off the whole debate about clean air until next year to risking acid rain legislation. The auto and steel industries contend the current regulations are killing them with only minimal improvements expected for air quality if emissions are further restricted as called for by the 1970 act. The coal industry is divided. Western coal interests say scrubbing requirements for their coal discriminate against Western coal, which is already cleaner than Midwestern and Eastern coal. Midwestern and Eastern coal companies want to keep the scrubbing requirement, but they argue that acid rain regulation will cause them great hardship.

All these groups have well-financed, powerful lobbies in Washington. The elections in November complicate the issue by making congressmen reluctant to deal with any controversial subject that they can avoid.

Perhaps the worst idea by way of compromise that has surfaced during the debate so far is one that would leave the easing of standards in "pristine areas" up to the states involved. The clean air regulations are complicated enough already without adding new layers to them.

But clearly, compromise is necessary. Clean air and a healthy economy are not unequivocally incompatible. One can only hope that special interests, politics and reason are not, either.

DAYTON DAILY NEWS
Dayton, Ohio, March 28, 1982

The answer is not blowing in the wind. What's blowing in the wind is the problem, smoke — car smoke, factory smoke and the political smoke of two sides in pitched battle.

The answer is in the middle if someone can get there long enough to run up a truce flag over renewal of the Clean Air Act.

The debate has been waged recently in the trenches of a House subcommittee, which released House Resolution 5252. Industries and the Reagan administration regard H.R. 5252 as the fairest creation since medieval maidens. Environmentalists see the bill as the Typhoid Mary of the regulatory kissing booth.

Both sides do agree that the Clean Air Act should be purged of the regulatory tapeworms thriving in the belly of its good intentions. The act is not sacrosanct. It should be efficient enough to deserve the goodwill, or at least general acceptance, of industries as well as environmentalists.

The issues are technical and many in areas where there is a lot of slack between causes and effects. The measurements in between are uncertain.

As a result, the issue and the legislative proposals are subject to a tangle of interpretations by both sides. But there are some fundamental issues, too. One, for example, is whether the cost of controlling pollution should be a consideration in setting standards.

Environmentalists say no; health standards should be absolute. But in fact, cost is a consideration, as when a deadline for meeting a standard is delayed. The real issue is how to use any cost-benefit consideration in a way that it doesn't become, as the Reagan administration seems to want, a loophole to blow a sky full of poison through.

Environmentalists accuse H.R. 5252 of "doubling" permissible emissions of carbon monoxide and nitrogen oxides. That's misleading. It would double the current auto emission standards of 3.4 grams per mile of carbon monoxide and 1 gpm of nitrogen oxide. Opponents of the bill don't want to change the car emission standard at all.

The dug-in opponents miss the point. Though a car may theoretically meet, say, the 3.4 gpm standard, because the auto maker's prototype car did, EPA studies show the same car on the road is coughing out 19 gpms of the poison. Why? Because the anti-pollution system has to be so refined to meet the high standard it deteriorates later, or because it was tampered with or somebody messed it up with leaded gasoline it wasn't designed to take.

What is a solution? Instead of freezing or doubling the standard, increase it to say, 5 or 6 gpms so auto makers don't have to make such a demanding and thus delicate system. And require inspection and maintenance as H.R. 5252 does not do.

Though there are a lot more issues involved, the central point is that the combatants have set up such a heavy crossfire the sensible middle can't raise its head or be heard. This is one case in which both sides can have what they want if they can get out of their bunkers.

THE ANN ARBOR NEWS
Ann Arbor, Mich., September 7, 1982

Take a deep breath and ho-o-o-o-ld it; that air in your lungs is the cleanest Michigan has had in a decade.

The air quality in Michigan, reports the Department of Natural Resources, is the best since the state began monitoring it 10 years ago.

(Of course it is, says the cynic in the back row, the factories are shut down and nobody's working).

The enactment of air quality standards and the installation of up-to-date pollution control equipment by Michigan industry are largely responsible for that improved state of aff-airs. The gains noted by the DNR are especially timely with respect to the Clean Air Act which Congress has been trying to revise.

THE COMMITMENT DURING the early seventies to clean air and water was well and wisely taken. The outlays of federal dollars for this purpose have served national needs and promoted better public health.

But times change and so do national administrations. The present climate of de-regulation and relaxed environmental safeguards does not augur well for such landmark legislation as the Clean Air Act.

The shift in sentiment showed in the House version of the Clean Air Act rewrite. The House Energy and Commerce Committee, chaired by Michigan's Rep. John Dingell, has, especially in the area of auto emissions, shown an inclination to weaken standards.

Now there clearly is a point of diminishing returns in auto emissions. It's impossible to get to a zero-risk situation in reducing air pollution. As Rep. Bud Hillis of Indiana said earlier this year, "a point of diminishing returns is reached when the cost to society in the loss of jobs (and) industrial capacity...is greater than the benefit of ever increasing air standards."

Michigan Living, the AAA monthly magazine, put it this way:

"If $300 can be taken off the price of a GM car (through eased emission standards), and this will not impair health, it should be done."

BUT WHERE A CASE can be made for easing up on auto emission standards, the same can't be said for air quality in other respects.

The Senate version of the Clean Air Act rewrite showed a commendable concern for preserving existing standards. The acid rain phenomenon for example, which has so exacerbated relations between the U.S. and Canada, is one which is downplayed by the president as only meriting "further study."

The Senate rewrite, according to *Congressional Quarterly*, added provisions designed to combat acid rain and to force the Environmental Protection Agency to regulate hazardous air pollutants.

The upper house's is the preferred version. Let's not throw away the gains of a decade in a burst of de-regulation and on the grounds of short-term gains, i.e., helping recession-struck industries over the hurdle of anti-pollution costs.

Our basic environmental laws have served us well. That is something to keep in mind as Big Industry cozies up to government and Interior Secretary Watt is unleashed to spread the message of strip, drill, mine and cut.

THE INDIANAPOLIS NEWS
Indianapolis, Ind., December 22, 1982

Midnight, Dec. 31 — deadline time for the Clean Air Act.

By the stroke of midnight most counties and municipalities throughout the country are required by law to meet national air quality standards for several pollutants. Many won't make it for every pollutant. Marion County is one of 13 Indiana counties that may not.

But not to fear — yet. Come the 12th chime on Dec. 31 the county won't be changed into a toad or, worse yet, automatically lose its Federal monies. As is the case in most governmental matters, it takes time to learn if, indeed, a deadline has actually been met and what punishments will be imposed if it hasn't.

And then there are exceptions which may be cited or extensions given or changes that Congress could make in the law. Many things could happen before the Environmental Protection Agency imposed any of three basic penalties available: Cut off Federal money for highways; withhold Federal money for cleaning the air or block permits for construction or modification of facilities which would add more of the pollutant in question.

EPA says it will work quickly to determine which counties have not met the deadline. Right now the agency is supposed to be checking the list issued Nov. 15 of 472 counties once declared in "non-attainment." Those which have come into compliance since will be struck from the list. According to a memo drafted by EPA Assistant Administrator Kathleen Bennett and reported by Congressional Quarterly, the remaining counties will be presumed to be out of compliance unless a state can show that it does meet the standards.

Then the agency will divide those on the list into two groups: Those that appear able to achieve compliance quickly and those that cannot. Those in the first group will not be punished "until the supporting data can be gathered and evaluated and until the area can be formally redesignated." To do so, EPA needs air quality data, which is collected on a quarterly basis and submitted the following quarter to EPA.

S. William Baker, executive secretary of the State and Territorial Air Pollution Program Administrators, told Congressional Quarterly, "I don't think the deadlines will work fast. It could take six months to two years."

Marion County can show it has met standards for ozone levels, but probably not for total suspended particulates (TSP), according to Harinder Kaur, legal counsel for the Air Pollution Control Division of the Indiana State Board of Health. Most of these particulates come from open burning, according to one study. The county is under some restrictions on industrial growth from EPA because of TSP pollution. Seven other Indiana counties are under sanctions, too.

The Clean Air Act enforcement machinery could change, of course, if Congress would change the law, which is due for revision and extension. Congress worked on the bill periodically throughout the last session, but like many important matters, the bill was not settled before adjournment — a casualty of the elections, misleading rhetoric and muddled political priorities.

The law still should be reviewed. The national air quality standards, as well as other pollution standards in the bill, should be re-examined to see if they are appropriate and reasonably attainable throughout the country. If they are, then they should be extended. If not — and there are several reputable groups who suggest that some may not be — then the standards should be revised.

It's two years until the next election, plenty of time to consider this tough issue before campaigning resumes in all its bluster and hype. But time is running out on the act and on those who must comply with it, built-in delays notwithstanding. When Congress opens its new session in a few weeks, it should begin work on revising the Clean Air Act — and begin work soon.

THE BLADE
Toledo, Ohio, February 12, 1983

IT is time for Congress to wake up and do something about the Clean Air Act of 1970, provisions of which could clobber 144 counties in 31 states — including nine in Ohio — many of them already reeling from recession-induced economic shocks.

The need for cleaner air in many of the nation's cities has been apparent, and a lot of progress has been achieved in improving air quality over the past 13 years. But to insist on letter-of-the-law compliance with unnecessarily high standards at a time of near-depression in many cities is economically unsound and destructive. Certainly extending pollution-control deadlines for five years, as the Reagan administration is proposing, is not an unreasonable step.

As matters stand, counties that do not measure up stand to lose millions of dollars in federal highway funds and face bans on construction of new factories or additions to existing ones. How better to put a stop to any thoughts of reviving depressed industrial areas? In Ohio some of the hardest-hit regions of the nation are involved in this exercise in futility, including Youngstown, Cleveland, Akron, Sandusky County, and several of the severely depressed counties along the Ohio and Mahoning valleys in the eastern part of the state.

The Environmental Protection Agency, often a whipping boy for imposing what critics regard as unrealistic standards, in this case is locked into enforcing the law as it is written. To her credit, EPA Administrator Anne Gorsuch concedes that "the imposition of these sanctions will be incredibly disruptive, both economically and environmentally." Her assistant, Kathleen Bennett, adds that the penalties are "arbitrary at best and have little relation to the attainment of cleaner air for the 1980s."

Given those assessments, it is understandable that both officials took Congress to task for failure to rewrite the Clean Air Act and to approve the five-year extension of anti-pollution deadlines. What more do the legislators need to ease the requirements, if they are as concerned as they profess to be about helping their hard-pressed constituencies?

It surely makes sense to do a little fine tuning at this stage to avoid kicking industrial communities when they are down. To do nothing about the regulations is a little like hitting oneself over the head with a club.

Chicago Tribune
Chicago, Ill., February 9, 1983

Under Director Anne Gorsuch the federal Environmental Protection Agency's enforcement of pollution control laws has been about as vigorous as the enforcement of vice laws in Chicago's old South Side Levee.

Aside from slashing the budgets for toxic waste removal, clean air, clean water and noise control by as much as 35 percent, she decreased the agency enforcement staff from 1,677 to 1,447 and the number of cases from 273 in 1980 to 107 last year.

Now, she's suddenly turned into an anti-pollution Carrie Nation, lashing out at 213 counties (including Cook and eight others in Illinois) for violating a 1982 air pollution control deadline. She's threatening to withhold federal funds for highway construction and other projects, which in Illinois' case alone come to $780 million.

Local governments, hard pressed by the recession, find this puzzling. So does the state EPA, which according to a spokesman knows of no air pollution problem in at least three of the counties. So does Rep. Henry Waxman (D., Calif.), chairman of the Health and Environment Subcommittee of the House Energy and Commerce Committee, who said that only about 50 of the target counties should be liable to federal sanctions, which are discretionary.

But there's usually a method in Mrs. Gorsuch's madness. For the last two years, the administration has been trying to weaken the Clean Air Act to double auto emissions, eliminate the requirement for coal smoke scrubbers and postpone action on acid rain. The Senate Environment Committee last year blocked that effort and came up with its own measure to strengthen the act. The issue is once again before the Congress.

Could it be that she's trying to cause as much pain and confusion as possible to put political pressure on members of Congress to ease the law? Said one county board, "They figure we'll do a lot of hollering. We're being used as a pawn."

Reno Gazette-Journal
Reno, Nev., February 8, 1983

Maybe the federal Environmental Protection Agency's air pollution crackdown is a political ploy, as some Reagan administration critics claim.

Maybe the Reagan people hope the reaction to the crackdown will be so violent that the administration can proceed with plans to weaken the clean air law.

Or maybe the EPA is simply enforcing the law because it is the law. Maybe that is why 217 counties were told to bring air pollution programs up to snuff in 45 days or risk severe penalties.

Either way, the message is clear: A great many counties are not making progress in fighting pollution. Not surprisingly, Washoe is one of those on the EPA hit list.

Washoe is also one of 102 counties denied an extension of auto pollution requirements.

Washoe's prominence in this regard is easy to understand. Air pollution is not bad all the time, but when it is bad, it is very bad. And the bad days seem to be getting progressively worse year by year.

Washoe faces sanctions for two primary reasons: automobile emissions and wood-burning.

Of the two problems, auto pollution might be closer to a solution. Mandatory engine emission inspections are scheduled to go into effect in Washoe and Clark counties on July 1, as mandated by state law. However, these inspections were originally to begin in 1979, and have been twice delayed by the Legislature. Another delay is possible, especially if the Legislature does not propose a better system than the present one.

The main problem with the emission inspection program is that the state designates certain private facilities as official inspection stations. But when this was done in Clark County in 1979, an uproar ensued. Motorists complained about the cost of having engines tuned, and they claimed that service stations were ripping them off. The program was abandoned after only a few months.

If the public is to accept auto inspections, it must have a guarantee that tuneups are necessary and that only needed work is done. This will require either extremely close supervision by the state and/or county, or government must operate the inspection stations itself.

But however they are handled, these inspections are vital to improving air quality. Even with them, the county might not be able to meet federal requirements; but definitely it cannot meet the requirements without them.

Under existing law, Washoe County commissioners can require auto inspections any time they wish, but no board has had the courage to do so. If the state extends the deadline again, the commission should gather that courage and move ahead by itself.

As for wood-burning, the main impetus at present is an educational program aimed at reducing burning. The success of such a program seems problematical, if the reaction to the December smog alert is any indication.

During the alert, people were asked to stop burning wood. Some did, but many did not, including wood-stove owners. These people rely on the wood stoves for cheap heat, and they were not about to use other sources because of the expense. Also, some of them blamed Sierra Pacific Power Co.'s high gas rates for the smog, and not themselves. In a sense this is true, but it does not indicate a high degree of civic responsibility on the part of the wood burners.

Perhaps an education program will work, but it is more likely the county will need a law restricting wood burning at certain times.

However, more will be needed than that. Wood burning creates pollution at all times, and helps to create the severe smog that leads to alerts. Logic suggests that the county control burning all the time. It would not be fair — or politically feasible — to ban wood stoves. But the county could require catalytic converters when these devices are considered reliable.

In addition, the county and Reno and Sparks should study methods of making homes more energy efficient.

In short, a good deal of positive action is needed from both state and local governments. Polite requests are nice, but hard-nosed regulations are much more likely to succeed.

St. Petersburg Times

St. Petersburg, Fla., February 5, 1983

It's so unusual for the Reagan administration to enforce any antipollution law vigorously that you'd expect environmentalists to cheer the crackdown on clean-air violators. They're jeering, instead. And their scorn is justified.

The Environmental Protection Agency (EPA) put 144 counties in 31 states, including Florida, on notice that they could lose millions of dollars in federal highway funds and face a ban on certain types of construction because their air is dirtier than government rules allow.

USING SUCH SANCTIONS to prod local officials to do a better job of cleaning up the air is good policy. When the Clean Air Act was passed in 1970, the nation made a commitment to protect Americans from unhealthy air. No American should be denied that protection simply because he or she happens to live in a city or county that is not as committed to reaching this goal as the law requires.

In this case, however, there's reason to question the administration's motives. Under Mr. Reagan's leadership, the EPA has adopted a policy of lax enforcement and breaks for polluting industries. It seems curious that the agency, all of a sudden, has gotten serious about enforcing an antipollution law.

The curiosity can be explained, however.

Since Mr. Reagan took office, he has tried to convince Congress to weaken the Clean Air Act and extend all deadlines for five years. So far, the President has had no success. By threatening construction bans and loss of federal funds, the administration apparently is hoping to create a backlash against the Clean Air Act so that Congress will weaken it.

Congress should not be influenced by such a ploy. The American people overwhelmingly support a strong Clean Air Act. In a recent poll conducted for *Business Week* by Lou Harris & Associates, more than 95 percent of those surveyed said they favored keeping the act as strict as it is now or making it even stricter.

IF THE ADMINISTRATION were listening to the people, it would abandon its efforts to weaken the Clean Air Act and work with counties to revise their compliance plans so that they result in cleaner air. Sanctions should be used, but with reason. Counties that can demonstrate progress toward meeting the standards should be granted short extensions. Counties that are making no effort to clean up the air should face cutoff of federal funds.

The people have spoken loudly and clearly on this issue. If the President won't listen, Congress should. No one wants to live in towns and cities choked by air pollution.

THE SACRAMENTO BEE

Sacramento, Calif., July 4, 1983

As William Ruckelshaus warned in his first major policy address since taking over the Environmental Protection Agency (EPA), "We are now in a troubled and emotional period for pollution control." Scientists, the public and the nation's policy-makers have reached a particularly awkward stage: Their ability to detect health risks in the environment is far more advanced than their ability to reduce or even evaluate these risks. It is a stage inherently conducive to public fear and mistrust, and that tendency was only aggravated by two years of Reagan administration support for the scandal-ridden and environmentally negligent regime of Ruckelshaus' predecessor at the EPA, Anne Burford.

Ruckelshaus, as he acknowledged in his speech to the National Academy of Sciences, has his work cut out for him. But there are already signs he may succeed. For one thing, he has not underestimated the complexity of environmental policy making. Nor did he try, in his speech, to turn the nation's unresolved questions about the costs and benefits of pollution control into an excuse for government inaction, as others in the Reagan administration have done. Just because there are other "societal goals" to be weighed against the goal of protecting the public from risks, he said, does not mean that federal air or water quality goals must be lowered or that the public's health concerns should be belittled.

The speech was a notice of Ruckelshaus' intention to solve problems, rather than use them as springboards for scoring ideological points. And that attitude in itself could do much to improve the EPA's regulatory activities and the public's faith in them.

Ruckelshaus' first major policy decision reflected the same sensibility. Last week, he undid months of unnecessary fear and deceptive politicking by simply rescinding Burford's plan to impose economic sanctions on 213 counties in 17 states that, despite their best efforts, have not yet met federal air quality standards.

In an attempt to scare Congress into amending — and seriously weakening — the Clean Air Act, Burford had claimed that the act gave her no choice but to punish the 213 counties (including Sacramento). But Ruckelshaus, eschewing her ulterior purposes, was able to interpret the law more rationally. Economic sanctions, he announced, are required when a community has failed to enact reasonable pollution control measures; they should not be imposed on a county that is doing all it reasonably can to achieve the clean air goals set by law.

Where Burford was hoping to force a choice between economic well-being and clean air, Ruckelshaus is attempting to accommodate both. It's an approach aimed at making pollution control work, rather than at making it fail. And while that seems little enough to ask of an EPA head, it's more than the Reagan administration has offered before.

It's too early to guess if Ruckelshaus will succeed in creating the consistent environmental policy for this administration that he told the National Academy of Sciences he is aiming for: a policy that is economically reasonable and scientifically valid and that engenders the trust of the public. But the first steps he has taken are certainly in the right direction.

THE KANSAS CITY STAR
Kansas City, Mo., July 5, 1983

The new head of the Environmental Protection Agency, William D. Ruckelshaus, has moved promptly to repair one piece of silly mischief at the agency. He has dropped the threat of his predecessor, Anne Gorsuch Burford, to impose sanctions against 213 counties in 33 states which failed to meet a deadline last Dec. 31 for compliance with clean air standards on all five major pollutants. The sanctions would have been in the form of withholding of federal highway aid and air pollution grants, and a ban on new factory construction.

The deadline was impossibly demanding in the first place, but Mr. Ruckelhaus' objection was that the imposition of sanctions would have failed to distinguish between those states and local governments which had EPA-approved compliance plans on file—and were making a reasonable cleanup effort—and those which were not even trying. It would have been manifestly unfair for EPA to endorse a compliance plan, then turn around and levy sanctions for missing the deadline.

The shortcomings were marginal in many cases. Wyandotte County made the dirty list in one category—particulates, or soot and dust—because one of its six monitoring stations was picking up dust from adjacent unpaved streets and parking lots. EPA officials conceded that some counties on the list might be in compliance through new clean air readings not yet reported or processed.

The sanction plan was suspect from the first, because Mrs. Burford and her assistant, Kathleen Bennett, both said the penalties were a bad idea and wouldn't improve matters, even as they announced them. Members of Congress and environmental groups joined in denouncing the move as an administration ploy to create so much misery for businesses and communities that Congress would be under pressure to soften the Clean Air Act, which has been up for revision since October 1981.

Mr. Ruckelshaus made it clear he means to review the noncompliance list and sanctions still are a possibility against those communities making no cleanup effort. They might also be used against urban centers with dirty air which refuse to implement inspection programs for motor vehicle emission controls, to assure they are working properly. But the blanket sanction threat was an unjust and foolish idea that should never have seen the light of day.

THE DENVER POST

Denver, Colo., March 11, 1984

THE ENVIRONMENTAL Protection Agency has just proposed that the federal law controlling airborne dust and smoke be changed to focus on the small particles that pose the greatest threat to people's health.

But while the revision in the rules may continue to safeguard the health of millions of vulnerable Americans — mainly children, the elderly and people with respiratory ailments — it won't necessarily improve the quality of life for the millions who live in Brown Cloud country.

Depending on what specific ceilings actually become law, the air in some Colorado locales that meet the current standard might be allowed to become as filthy as the air in downtown Denver.

The health-related aspects of particulate pollution are of primary concern in the East, where emissions from power plants and motor vehicles are heaviest. In the wide-open spaces of the West, where it takes much less pollution to create a problem, the threat to visibility is a greater worry.

The political implications are obvious. Once the Reagan administration has satisfied the East that it is dealing with the health problems associated with dirty air, it can safely ignore its less powerful constituents in the West.

State governments would be free to address the problems of visibility, of course.

But the states' power would be limited by the fact that the prevailing challenge — regional haze — extends across state lines.

In announcing the proposed change, EPA Administrator William Ruckelshaus said it would be designed to limit people's exposure to the tiny particles that can be inhaled deep into the lungs — generally, those with a diameter smaller than 10 microns, or one-tenth the thickness of a human hair.

But at the same time, the EPA chief said he wanted to postpone any action to control the even-finer particles — less than 2.5 microns in size — that cause the greatest impairment of visibility. (Paradoxically, it is not the largest particles that block Denver's view of the mountains, but the smallest. That is because the closer the size of a particle gets to the wavelength of light, the more light it refracts.)

What Ruckelshaus actually said was that the agency wanted to "defer" a decision on standards for fine particles "so that it can consider such a standard as part of a more detailed look at regional air pollution problems such as visibility degradation and acid rain." The

EPA, he added, expects to notify the public of a proposed rule on this matter "in the near future."

It should be remembered that this is the same official who said when he was appointed nearly a year ago that his top priority would be to come up with a program to control acid rain. No such strategy has yet been seen.

And this is the same EPA that said four years ago that the particulate standard proposed last week would be issued within six months.

Perhaps the most ominous aspect of the EPA administrator's announcement was his suggestion that the agency be allowed to take the "practical problems" of compliance into account in setting the primary air-quality standards — something not permitted under current law.

It may sometimes be justifiable to delay the enforcement of a standard because the technology is not yet available to enable polluters to comply with it.

In fact, the EPA decided to do just that in January, when it announced a two-year postponement of more stringent emissions-control limits for diesel cars and light trucks.

But weakening a primary health standard, just because it would be inconvenient or cost a lot to achieve it, makes no sense at all.

Minneapolis Star and Tribune

Minneapolis, Minn., February 23, 1984

The Clean Air Act is the crowning achievement of the environmental movement. Yet the 1970 law fails to regulate many toxic pollutants that should be strongly curbed. A bill sponsored by Minnesota Sen. Dave Durenberger would plug that loophole. The proposal is a necessary response to a pressing public health problem.

Billions of pounds of potentially toxic air pollutants are dumped into the air every year. Long-term exposure can cause cancer and other severe illnesses, even at relatively low dose levels. The health threat is particularly acute in cities, where the air contains up to 30 times the level of cancer-causing pollutants found in rural areas. Scientific studies have estimated that such pollutants cause 10,000 to 20,000 lung cancer deaths annually.

Although the Clean Air Act gives the Environmental Protection Agency the power to regulate toxic pollutants, in 13 years the agency has set standards only for asbestos, beryllium, mercury and vinyl chloride. Three others — arsenic, benzene and radionuclides — were listed as "hazardous" in the late 1970s, but all three are still unregulated. Dozens more have remained "under review" for more than six years — even though many of them are already acknowledged as carcinogens by government scientists.

No good reason exists for such stalling: According to a recent report by the General Accounting Office, the EPA simply hasn't been doing its job.

Durenberger wants to amend the Clean Air Act to break the bureaucratic logjam. His bill, cosponsored by Sens. Max Baucus of Montana and Gary Hart of Colorado, would force the EPA to decide by the end of 1986 whether to regulate approximately 40 suspected hazardous chemicals.

In cases where regulation is warranted, the bill would require that the agency promptly issue strong standards to protect public health. It would also force industries that emit hazardous pollutants to obtain permits for continued operation and to install the best available pollution-control technology. Newly discovered hazardous pollutants would have to be regulated within a year of detection. Whenever the EPA fails to meet the bill's deadlines, pollutants would automatically be treated as hazardous.

Durenberger argues that every year the EPA delays in controlling toxic pollutants means thousands of endangered lives. His bill would be a major stride toward the Clean Air Act's basic goal: for all citizens, air safe to breathe.

THE CHRISTIAN SCIENCE MONITOR

Boston, Mass., March 13, 1984

ONE part of the proposed changes in federal clean air standards deservedly is winning applause. But there are two reasons why other aspects of the changes are cause for concern.

The changes in the public health aspect of clean air standards were proposed over the weekend by William Ruckelshaus, administrator of the Environmental Protection Agency. No decision will be reached for several months, while all sides have the opportunity to comment on the proposals.

Even then, if the past is a guide, there could be several years of court suits before it is certain whether the new standards will be judged legal.

What is being generally applauded is the proposed change from measuring particles of all sizes in the air to concentrating on the very small ones, with a diameter much less than that of a human hair. Environmentalists are pleased with this change, noting that many scientists conclude small particles, not large ones, constitute a possible threat to health. In addition, removing the strict standard on large particles would ease a cleanup burden now seen as unnecessary for some industries, such as mining firms, which produce relatively large particles, but not small ones.

However, there is concern that, after all public com-

ment is complete, the EPA could decide to weaken the public health standards, a move that environmentalists would consider unwise. Rather than offer a specific standard for the smaller particles at this time, Mr. Ruckelshaus at the moment has proposed a sliding scale, from which the precise standard later will be selected. This scale ranges from, at the stricter end, an approximate continuation of the current level of stringency, to a considerable relaxing of that standard.

At the weekend Mr. Ruckelshaus said he was "leaning" toward coming down on the stiffer end of that scale; there is no certainty that this will be the final decision.

There is one other area of concern — that there may be no effective federal health standard for the amount of particles for several years, until the new standard becomes fully operational.

Keeping existing standards on all airborne particles should be considered until the new standard not only is decided upon, but also has met all court challenges — and until states have come up with their own regulations for putting it into effect.

As things now stand, once the EPA decides on a new standard the old one will cease to have effect. As a consequence there could be an effective hiatus in health standards of four to five years while the legal testing proceeds.

THE LOUISVILLE TIMES
Louisville, Ky., June 28, 1984

The U. S. Supreme Court's support of the administration's "bubble" approach to air pollution control has been denounced by some environmentalists, including members of Congress, as a near disaster and hailed as a victory by businessmen and the Reagan administration. But the court's message is not that the bubble is good or bad but that Congress can preserve, change or kill the policy by writing its intentions more clearly into the Clean Air Act.

The legal debate turned on the definition of a pollution "source." The administration argued that an entire building or factory, enclosed in an imaginary bubble, could be treated as a single source so long as total pollution remained within legal limits. A plant within a "bubble" can therefore expand, even though it is in an already polluted area, so long as older sources of pollution are reduced enough to compensate for the new.

The groups that challenged the EPA policy insisted that every boiler or other component of a plant must be treated as an individual source of pollution under the law. When an existing factory enlarges or changes its operations, according to this view, which was the prevailing one until three years ago, each new source must undergo a complicated review process and meet strict pollution limits.

The court decided that when Congress does not make its wishes clear — in this case by failing to define "source" in amendments to the Clean Air Act — then policy-making agencies have considerable leeway in interpreting the law. Congress can decree exactly what it wants the policy to be, in other words, if it will ever get around to finishing its review of the Clean Air Act.

While some polluters may have abused the bubble as the Reagan administration loosened environmental regulations, the concept makes considerable sense, especially since it is a way to permit economic growth without further damage to air quality. The Jefferson County Air Pollution Control District was a pioneer in the use of the bubble and officials believe it has worked well.

So Congress will make a serious mistake if it decides to junk the policy entirely. However, bubble arrangements need to be closely supervised and may often be inappropriate in communities with dirty air. And the ultimate objective — a net *improvement* in air quality — must not be compromised.

When the "bubble" first entered environmental jargon in the Carter administration, it was seen as an important step toward regulatory reform. The flexibility it offers can be a boon both to regulators and to regulated companies. Some corporate officials say the bubble has made plant modernization simpler and less expensive. Cleaner air is the payoff when dirty old factories are brought up to date.

The idea, therefore, is by no means faulty. But Congress should listen carefully to those who see risks in letting bubbles float too freely across the industrial landscape.

THE ATLANTA CONSTITUTION
Atlanta, Ga., June 29, 1984

The victory cheers emanating from the White House, following the U.S. Supreme Court's ruling on administration attempts to weaken federal air-pollution standards, might lead one to suppose the administration's standards had been declared sound. Not so.

The high court has determined only that the administration has the authority to set this policy.

The *wisdom* of the policy wasn't addressed in the opinion, written by Justice John Paul Stevens, except as a factor of no consequence as far as the law is concerned. The question before the court was, simply, whether certain acts of Congress were sufficiently "ambiguous" to permit different interpretations by different administrations. It decided they were.

In the Carter administration, the Environmental Protection Agency required industrial plants in regions of the country that had not yet met Clean Air Act standards to undergo environmental reviews before expanding. The Reagan administration in 1981 adopted instead a "bubble" policy. It permits a plant to increase pollutant emissions in one facet of its operations so long as it decreases them in another.

The policy is, in practice, a declaration that where the air is already bad, it will not be made better. The National Resources Defense Council, if a bit overwrought, is close to the mark in complaining that the ruling "completely guts" Clean Air Act programs in regions not yet in compliance.

The court affirmed the administration's right to set regulatory policy, right or wrong, but in no way quarreled with Congress' authority to rewrite and tighten up the laws if needed, to safeguard health and environmental interests.

Indeed, Congress has its work cut out for it. Its task will be to rewrite portions of the Clean Air Act, so as to prevent bureaucrats from caving in to special interests, and industrial polluters from jeopardizing the public health — quickly, and with a nod to the high court for showing the way.

Los Angeles Times
Los Angeles, Calif., June 27, 1984

A U.S. Supreme Court decision on Monday relaxed air-pollution controls in ways that Congress seems not to have sanctioned, even though the law's language may have been as vague as the court said it was. How serious the consequences will be in some states depends on how fast Congress can move to make the law perfectly clear.

The court's ruling involved the quite sensible "bubble concept" of smog control that allows industries to pay more attention to the total emissions under some imaginary "bubble" over a factory than to the smoke from any one stack.

The court not only extended the bubble principle from areas where the air is clean enough to meet federal standards to areas where it is not, it seems to have judged that what goes on under the bubble is nobody's business as long as total emissions do not grow. An owner, under that reasoning, could replace old equipment without installing tight pollution controls on the new source as long as total pollution levels did not grow.

It may be enough to stay even on pollution in cities where skies are relatively clear. But that approach would be disastrous in regions like Southern California where it may never be possible to meet federal air-quality standards and the best that citizens can expect is some improvement each year.

Fortunately, the court ruling will have little if any effect on California. For one thing, the state's pollution-control districts draw their legal authority from Sacramento, where the law says clearly that they may impose tougher control laws than the federal Environmental Protection Agency requires. For another thing, the South Coast Air Quality Management District follows procedures that the court ruling seems to say will not be necessary in other states. Factories have most of the freedoms provided by the bubble approach except when they replace a boiler or other piece of equipment. In those cases, regardless of the amount of pollution in the bubble, industries must install "best-available" pollution controls on new equipment. That usually results in a reduction of total emissions in the bubble, although the district says that it has not yet collected enough data to tell how large the reduction is.

A bill to require all states to follow procedures similar to those in California has been approved by the Senate Committee on the Environment and Public Works, and is awaiting action on the Senate floor. That may require a green light from the Reagan Administration. Judging from the Administration's crowing over the court decision, it may be a very long light.

Houston Chronicle
Houston, Texas, June 29, 1984

On the one hand, there is the desire for clean air. On the other hand, there is a need for more jobs and new products. The question is how to accommodate the two objectives. Naturally, there is controversy. Naturally, it wound up before the U.S. Supreme Court. We like what the court decided.

The Environmental Protection Agency came up with the "bubble concept." The idea is to allow industrial expansion as long as any additional pollution is offset by improvements within the plant. The "bubble of air" must not deteriorate.

The Supreme Court said this policy is a reasonable one. The court went even further. It said that the controversy is "over wisdom of the agency's policy" and should not be decided by the judges. Congress or the administration can change the policy, the court said, but the court can't.

The court made way for economic growth and deferred to the other branches of government on a policy matter. Not bad for a single ruling.

Part III: Water

Water, vital to all life, is perhaps our most valuable natural resource. Unfortunately, most chemicals introduced into the environment eventually make their way into drinking supplies. Despite its own built-in purifying and replenishing processes, there is thought to be little truly "clean" water left. Much progress has been made, through enforcement of the Clean Water Act, toward cleaning up the most obvious sources of pollution in surface waters. Thus, it is no longer common to find huge deposits of oil or raw sewage in lakes and rivers. But scientists fear that many tasteless, odorless chemicals may be taking the place of such gross pollution in the nation's waters. These chemical pollutants may come from municipal sewage plants, industrial discharges, runoff from agricultural areas, or even from additives used in water treatment plants to lower bacterial counts or remove brown coloring. Many of them are either known or suspected to be carcinogenic or mutagenic. Among the most dangerous ingredients found in drinking water are nitrates, asbestos, lead, chlorine, and trace minerals such as cadmium, mercury or zinc.

Especially worrisome is the growing contamination of ground water in the United States. Since surface and ground water are intimately connected, this is a problem that requires joint management of the two inseparable resources. Often, large aquifers lie beneath several states, making the coordination of any pollution control effort more difficult. But once the water contained in the soil and in aquifers is polluted, it is nearly impossible to cleanse; most owners of polluted wells drawn from ground water simply choose to dig new ones. This is because it is an expensive and tedious process to monitor aquifers; since the water moves so slowly, a polluted area within an aquifer may lie only yards away from unaffected water. The future direction and speed of the flow of water within an aquifer is also unpredictable. All ground water is vulnerable to chemical contamination because the natural filtering action of the soil is not sufficient to remove many toxic organic chemicals. A 1980 study by the Environmental Protection Agency of the ground water at 50 industrial waste disposal sites found the following hazardous contaminants: selenium, arsenic, chromium, lead, cyanide, copper, nickel, barium, polychorinated biphenyls (PCBs), chlorinated phenols and benzene.

Surface Water:
Rivers, Creeks, Lakes, Streams

It is surface water that usually comes to mind when the problem of water pollution is mentioned. The source of drinking supplies for about half of the nation's population, surface water includes rivers, lakes, streams and even puddles. Its pollution does not pose as great a potential health threat as the contamination of ground water, since it can be submitted to treatment at processing plants before being used as drinking water; pollution control efforts have at least partly succeeded in bringing back to life rivers and lakes that had been declared "dead." But the pollution of these bodies of water greatly diminishes their usefulness, as the level of pollution becomes too great to allow swimming or fishing in them. A report by the Environmental Protection Agency in July, 1981 listed 34 "hot spots" that it considered the most polluted water sites in the nation, including large parts of the Arkansas, Delaware, James, Ohio, Hudson and Mississippi Rivers. The dozen most pervasive pollutants listed in the report were arsenic, cadmium, chromium, copper, cyanide, lead, silver, mercury, polychlorinated biphenyls (PCBs), phenols and cresols, and phthalate esters. Many of these chemicals and heavy metals are released by industries. When the sources of such pollutants can be pinpointed—for instance, a particular factory or sewage treatment plant—they are much easier to control than "non-point" sources such as runoff from farms and construction sites. Pollution by pesticides is ranked second only to metals among the toxic substances most widely reported to be contaminating rivers. One of the largest and most widely publicized surface water pollution problems is that suffered by the Great Lakes, where the seepage of agricultural fertilizer has resulted in heavy phosphorus contamination, particularly in Lakes Erie and Ontario.

A growing, widespread pollution problem in lakes is related to the over-supply of plant nutrients provided by some forms of waste. These discharges promote unnaturally fast growth of aquatic plant life, which settles to the bottom of the lakes and decays. The resultant depletion of the lake's oxygen and its choking with algae, killing off fish and invertebrates, is referred to as "eutrophication."

The Providence Journal
Providence, R.I., June 20, 1981

It is not generally known, but the Atlantic salmon, which in centuries past flourished in New England rivers, was wiped out by pollution before the War of 1812. This bit of lore came to light in one of those encouraging (if all too rare) news reports on efforts to bring pollution under control. If experiments in the Connecticut River are any guide, the Atlantic salmon may be on its way back to some New England waters.

Thanks to years of anti-pollution efforts, the Connecticut is cleaner than it has been in years. Thanks also to a 15-year program of stocking the river with young salmon and building fish ladders, the fish are beginning to thrive. This year, they are returning to the Connecticut from the Atlantic Ocean in record numbers. Some of the fish have returned from as far away as Greenland, and have been seen as far upstream on the Connecticut as Turners Falls, Mass., and Brattleboro, Vt.

While no one is willing to predict that a permanent salmon population can be re-established in the Connecticut, the signs look promising. More than twice as many of the fish have been counted this year as in 1980. Unlike the Pacific salmon, which return from the ocean but once to spawn, and then die, the Atlantic salmon can return to fresh water for several years to spawn, given satisfactory conditions. This habit makes the fish especially suited for restocking efforts.

This program is run jointly by federal agencies and by the states that share the Connecticut and Merrimac Rivers. It is a worthwhile effort that deserves the continuing cooperation of all the states involved. And, quite apart from the satisfaction in having this fine fish again running in nearby waters, the patient efforts show convincingly the good things that can happen when man undoes the pollution that his forebears allowed to develop.

Reno Evening Gazette
Reno, Nev., February 13, 1980

The discovery of the harmful chemical PCB in the Truckee River is a stark reminder of the fragility of our major water supply.

It is unknown at this moment how serious the contamination is. It appears that the amount of PCB is below dangerous levels, but further tests are needed.

So, for the present, residents of the Truckee Meadows cannot use the river's water, and are even being warned not to eat fish from the river.

That means the entire Sierra Pacific Power Co. water service area is dependent upon a limited number of wells and Hunter Creek for all its needs. If does not appear this situation will continue into the summer, when water needs increase dramatically, but if it should, the Meadows would be in for a difficult time. Even during the winter, Sierra Pacific has asked residents to reduce water use 10 percent and not to water outdoors at all.

These reminders of the Truckee's importance in our lives should renew our determination to keep the river as pure as humanly possible.

In particular, state and local agencies must be careful to forbid industries from locating near the river if there is any possibility of contamination. This was not done in the case of Viking Metallurgical, the Verdi jet engine plant which uses PCB-laden oil and has an oil dump 300 feet from the river. It was at first thought the PCB contamination originated at Viking, but preliminary tests indicate Viking is not at fault. Also, Viking is to be commended for initating the testing by immediately informing authorities when it appeared there might be a high concentration of PCB in its ponds. In an age when many companies — Three Mile Island comes to mind — seek to hide public dangers to protect themselves, Viking has demonstrated responibility and good citizenship.

Still, the point remains: Local and state officials must not permit operations which use potentially dangerous substances to locate near the river.

In addition, communication must be improved between public agencies. State and county officials were surprised to learn this week that the U.S. Fish and Wildlife Department found PCB in Truckee fish going back at least 10 years.

A federal spokesman said the PCB level was not hazardous, but he also said the tests were not completely valid because so few fish were tested. In these circumstances, Nevada officials should have been made aware of possible trouble long ago.

The PCB problem also underscores Nevada's concern about a proposal to reopen three North Tahoe sewage plants which could damage Truckee River quality. The proposal originally surfaced in a committee of the California State Senate, then died in another Senate committee. But it resurfaced at a Jan. 30-31 public hearing when Dart Resorts proposed the reopening to the Lahontan Regional Water Quality Control Board.

Lahontan executive officer Roy Hampson says the proposal has very little chance of approval. That is probably true, but we all know how developers continue to beat on government's door until they beat it down. So Sparks Mayor Ron Player deserves full support in his opposition to the proposal. He has urged the business community in particular to help out, and it should do so, as should all others interested in our water quality.

Players' criticism of the State of Nevada's watchdog efforts in this regard is perhaps a little overdone. The state did send a mailgram to the California Senate committee in which the sewage plant proposal first surfaced, and has expressed its concern in the past to the Lahontan board.

But the state apparently was not involved in the Lahontan public hearing, and a representative should have been there to protest with Reno and Sparks officials. The state, like the Truckee Meadows, must vigorously protest harmful upstream California projects continually and forcefully.

In both these areas — sewage pollution and chemical pollution — and in all other areas as well — an uninterrupted vigilance must be maintained to assure that our drinking water supply is neither degraded nor contaminated, now or ever. To fail in this effort could be disastrous.

Arkansas Gazette.
Little Rock, Ark., April 16, 1980

Arkansas's effort to clean up and to keep clean its rivers and streams has been moderately successful over the last decade, if new data from the state Pollution Control and Ecology Department provide an accurate gauge. As one state official interprets findings from computer computations, "We're holding our own."

This is better news than it might seem, for the pressures working against improvement in water quality have greatly intensified through the years. Arkansas's population increased in the '70s, as the 1980 Census figures are almost certain to show. With more people there is more demand for water for many purposes; the more use, the greater becomes the opportunity to abuse the quality of water that is available.

In addition to population growth, this state has continued to enjoy industrial growth and while much of the industry is considered to be the "clean" variety it still exacts a price on the environment. Demand for agricultural products from East Arkansas has soared in an expanding world market. As a result, more and more land has been cleared and more and more chemicals have been applied to the land. A good deal of land clearing also has been taking place in the Ozarks and Ouachitas, for a variety of reasons. Forestry has taken its toll, but the Ozarks, especially, have attracted many modern-day settlers looking for a simpler life.

All of these factors, and others, would surely have resulted in massive destruction of Arkansas's water quality had there been no effort, by local, state and federal authority, to clean and to keep as clean as possible the streams that in a real sense are the life blood of this state.

"Holding our own" under these circumstances is a commendable record and it should encourage all of us who care a great deal about the vital qualities of Arkansas to push toward an even better record in the '80s.

To be sure, the data compiled in the computer survey of water samplings are uneven. The indications are, for example, that streams in the Ouachita River and White River Basins have declined slightly in quality, for varying reasons. The St. Francis River in East Arkansas, which flows through a region of heavy agricultural use, has never been in good condition in recent times, but, says John Giese of the state Pollution Control and Ecology Department, "it's not getting any worse." Slight improvements are shown in water quality in the Arkansas River and Red River Basins.

We are especially pleased that improvement has been shown on the Arkansas River, which has massive unrealized potential for the betterment of this state and its people. Change on the Arkansas is surely rooted in the $1.2 billion navigation project of the 1960s. It made no sense for the federal government to be spending that much money if municipalities and industries were to continue using the stream for a place to dump their wastes, in many cases untreated wastes. New federal regulations and federal matching money for waste treatment in many cases have assured a higher quality of water even though substantial industrial growth has been concentrated along the Arkansas River Valley in the western part of the state.

A good bit of disappointment, therefore, attaches to a recent decision by the federal Environmental Protection Agency, which has contributed so much to the cleanup of the Arkansas, to reject a recommendation of the Statewide Policy Advisory Committee that the EPA fund a detailed study of water quality in the Arkansas. One of the committee's most valuable members, a knowledgeable water chemist, apparently was so frustrated by the EPA decision that he resigned from the committee.

There are persistent reports, as well as the encouragement to be found in the report from the state Pollution Control and Ecology Department, that the quality of the Arkansas's water has improved enough to make it a source for municipal supplies. Even the salt content, so the speculation goes, has declined enough to make the water potable after treatment. Several cities and towns in West Arkansas have been looking around for new sources of water supplies, even to the point of casting covetous glances on some of the free-flowing streams that would have to be dammed, and destroyed, in the process. The Arkansas River, in any case, should be the first place to look. A great deal of money was spent to channelize the river and to clean its waters. It is only sensible policy to make the maximum use of these waters.

Democrat Chronicle
Rochester, N.Y., May 23, 1981

AS ROCHESTER area fishermen and swimmers can testify, Lake Ontario is making a dramatic comeback from the polluted conditions that closed so many beaches not so long ago.

Lake Erie, once dismissed as a "dead lake," now boasts some of the best walleyed pike fishing in the world.

When the International Association for Great Lakes Research met recently in Columbus, Ohio, it was agreed that excellent progress had been made.

Most researchers at the meeting gave primary credit to effective government regulation (including federal bans on DDT and manufactured PCBs), bans in most Great Lakes states on phosphate detergents, the updated 1972 Water Quality agreement between the U.S. and Canada, and Washington's dollar encouragement of the building of secondary sewage treatment plants.

But the job isn't finished. Reporting this important conference for *The Christian Science Monitor,* Lucia Mouat noted the warning sounded by Dr. Wayland Swain, director of the Environmental Protection Agency's Large Lakes Laboratory at Grosse Ile, Michigan.

"We've done all the easy things — we've taken out the big chunks such as the acids, the greases, and the oils. But there is this whole spectrum of compounds that tend to be terribly persistent... Toxics are probably the hardest challenge we've ever faced."

THAT CHALLENGE has now to be met with less money than before. How much less is hard to say until the budget cuts are worked out.

But almost certainly new sources of funding will have to be found, and perhaps new priorities set. The public may also have to become better informed about the Great Lakes.

Swain, who has traveled widely in the Soviet Union, says that youngsters there tend to refer to Siberia's Lake Baikal, the deepest and largest freshwater lake in the world, as a priceless national resource.

By contrast, he says, few American children coming through his lab in Michigan can name all five lakes, "let alone refer to them as a national treasure." Yet, he points out, the shortage of water in the West could one day make Great Lakes water as "valuable as OPEC oil."

Progress in cleaning up the Great Lakes has been good, but it's going to be harder to maintain without fresh infusions of money and ideas.

The Hartford Courant
Hartford, Conn., August 9, 1981

Using a pretty esoteric methodology, the federal Environmental Protection Agency discovered some pretty obscure streams for inclusion in its preliminary list of the most polluted waterways in the nation.

It included the Loosahatchie River near Memphis, Tenn., Irondequoit Creek near Rochester, N.Y., and Chicopee Brook near Springfield, Mass.

Hartford was included among those 34 areas with waters so polluted by toxic wastes that they require special attention. It won that dubious distinction because of Willow Brook, a narrow stream that stretches a mile or so through East Hartford before emptying into the Connecticut River. So obscure is the brook that some state environmental officials had never even heard of it before the EPA released the list.

The culprit is Pratt & Whitney Aircraft, which dumps treated industrial wastewater into the stream. P&WA is apparently violating no laws, but the EPA says the water in Willow Brook is so contaminated that further action — beyond current legal requirements — may be needed to clean it up. The brook is so small that the wastewater stays highly concentrated.

The problem in Willow Brook might be solved if P&WA would bypass the brook and pipe its wastes, containing heavy metals and chemicals, the relatively short distance to the Connecticut River, where they are more greatly diluted.

Connecticut environmental officials are also concerned about those waterways polluted by, not one, but dozens of industries dumping wastewater containing toxic contaminants.

Those include, for example, the Naugatuck, Hockanum and Quinnipiac rivers. The state Department of Environmental Protection is now trying to put together an equitable strategy to reduce the flow of toxic pollutants into those and other streams.

In spite of advances over the past decade in cleaning up Connecticut's waterways, there's still plenty to do. Willow Brook got the most recent EPA attention, but the cleanup effort has to be extended, to further reduce toxic threats to human and aquatic life in all the state's waterways.

The Kansas City Times

Kansas City, Mo., April 15, 1980

A decade ago there was justified national alarm over the pollution threat to the Great Lakes, the world's largest reservoir of fresh water. Even with good will and intentions it was difficult to get a large-scale abatement effort going as the various affected states and the United States and Canada negotiated between themselves. But in the period since then the federal Environmental Protection Agency has spent $5 billion on the cleanup work and additional billions have been committed by the states, local governments and industry.

Now the EPA, in the first status report in several years on the Great Lakes, can speak of "cautious optimism" and markedly improved conditions.

Lake Erie, once mourned as a dead sea saturated with sewage and industrial wastes, its beaches closed, has been turned around. Most noticeable are the shrinking blankets of green algae whose decay once soaked up the oxygen in the water to the detriment of all marine life. Lake Ontario, the next most damaged lake, has seen a huge shoreline cleanup, including new sewage treatment plants and a Canadian-New York ban on phosphate detergents.

Lake Michigan, once in serious trouble, has witnessed the wastes from a dozen sewage treatment plants and one industry diverted from the lake. The 1972 ban on DDT has cut DDT levels in the fish by 90 percent. Lake Superior, the least hurt by pollution, this year will see an end to dumping of industrial taconite tailings at Silver Bay, a source of asbestos-like particles in drinking water systems, after lengthy litigation. Lake Huron, the second least-polluted, has recorded improvements in its Saginaw Bay trouble spot.

The huge cleanup investment is yielding tangible returns in renewed outdoor recreation — boating, sport fishing, vacation resorts and re-opened beaches — that boost shoreline property values. But the basic incentive for attacking pollution in the Great Lakes was inescapable moral responsibility not to allow a natural asset of this magnitude to be destroyed through careless neglect.

AKRON BEACON JOURNAL

Akron, Ohio, November 21, 1980

RECENT studies done for the International Joint Commission on Great Lakes Water Quality indicate that the phosphate pollution levels in Lake Erie continue to be discouragingly high.

Yet except for Pennsylvania, which has only about 45 miles of lakefront and very little Lakes drainage, Ohio, the state with the biggest stake of all in Lake Erie, remains the only Great Lakes state or province without some form of limitation on the phosphorus content of laundry detergents sold in its Lakes watershed.

Akron and neighboring communities using Akron's disposal facilities have banned sale of high-phosphate detergents for almost eight years, without evident ill effects and at considerable savings in sewage treatment costs. But the state isn't getting the message.

Six Cleveland-area disposal plants were cited in the studies as among the 20 worst phosphate pollution sources on the Lakes, and Cleveland's and Toledo's plants were listed as special problems for Lake Erie in this and other ways.

Bills to ban sales of high-phosphate detergents have lain untouched in the Ohio Legislature for years, and the Ohio Environmental Protection Agency continues to oppose a ban — echoing the detergent-makers' argument that it would do no good because most phosphate pollution comes from other sources.

Maybe so. But this is like arguing that since you can't stop all dandelions it is useless to fight them in your own front lawn.

Useful action about the other sources is technically possible, but it is at best difficult.

And phosphate removal at treatment plants is of course possible — and still necessary even with a phosphate ban, as Akron has discovered. But the more phosphate the process must remove, the more it costs.

A ban on high-phosphate detergents, by contrast, is cheap and easy. Few in the Akron area would maintain that it has increased their home laundry costs appreciably, and it has saved money at the treatment plant. And while by itself it cannot be expected to solve all of Lake Erie's problems, it obviously removes at least one of their causes.

The discouraging new reports underscore once again that laggard Ohio should do what almost all the rest of the Lakes community has already done — ban high-phosphate detergents in its Lakes watershed.

Akron, Ohio, June 14, 1981

IT'S ONLY one more tiny item in the long, long list of programs up for lopping by the budgetary axes of the Reagan administration, but it is one that should concern the tens of millions of Americans whose lives are directly or indirectly affected by the condition of the Great Lakes.

The budget-cutters want to slash two-thirds from the money the federal government has been spending on efforts to clean up the lakes.

The whole cost of this federal effort at the current level is hardly a droplet in the vast sea of the federal budget: $30 million, less than half a 100th of 1 percent of the total budget. The $20 million contemplated saving is correspondingly insignificant; it's no trick these days to spend that much on a single military airplane.

But the effect on the lakes effort is likely to be drastic.

It would grossly reduce current efforts to monitor contaminants in the lakes, trace them to their sources and fashion means to abate the nuisances where possible. It would virtually wipe out any further federal effort in restocking game fish.

And it would come close to ending federal support for all research into lakes problems beyond the minimum to which this government is obligated by a treaty with Canada.

And this would come just as some progress is beginning to appear in the work to clean up the lakes, some of which were regarded only a few years ago as in imminent danger of "death" from pollutants and dropping levels of oxygen.

Under the "New Federalism" theory, it is presumably supposed that the affected states will pick up at least some of the work that has been federally sponsored. But there are eight states with frontage on the lakes, plus the province of Ontario; all are in tight financial binds, and in every case there are political and economic preoccupations that put the matter of the lakes' condition far down their lists of priorities.

The Great Lakes, as the world's largest reservoir of fresh water, are an international asset of huge proportions. Their health has some bearing on the lives of millions upon millions beyond the 22 million people who get their drinking water from them.

Leadership, coordination and stimulation of effort toward cleaning up the lakes and keeping them clean is a truly international problem that cries for involvement at the top levels of the governments of both the United States and Canada.

These are matters that cannot safely or wisely be simply turned over to the states involved.

It may indeed be a time for "reassessment" of federal funding for research on the lakes pollution problem, as a spokesman for the Reagan budget proposal put it. That funding has been going on for nearly 10 years, and every such program should be under constant scrutiny and re-evaluation to make sure it is well directed and likely to be effective.

But a "reassessment" and a two-thirds cut in funding are two quite different things. This one sounds like a new invitation for dirty, fishless lakes and irreversible damage to a great international treasure.

THE PLAIN DEALER

Cleveland, Ohio, June 15, 1981

The Great Lakes, beneficiaries of increased public environmental awareness for more than a decade, virtually are being stripped of that attention in the rush to cut the federal budget. We consider the cutbacks of Great Lakes projects to be shortsighted, but recognize that the programs to identify and correct abuses of the lakes could be victims of their own successes. When weighed against proposed reductions in other budgetary areas — aid to education, the elderly and the poor, for example — clean water programs have not excited much of a constituency to save them despite polls showing overwhelming public approval of such projects.

Federal support of Great Lakes management, surveillance and research programs this fiscal year amounts to just over $30 million. Target amounts for fiscal 1982 in preliminary congressional budget-making indicate cuts of more than 75%. In addition, the Reagan administration has not asked for any funds for sewage treatment improvements for the next fiscal year. Instead it sent Congress a list of changes it wishes to see made in current funding — it wants grants for separate sanitary sewers to be stopped, for example, and financing of reserve capacity at sewage treatment plants eliminated — and indicated that if it were satisfied with Congress' actions, it would recommend a $2.4 billion grants plan for fiscal 1982. That would represent a cutback of about 30% from this fiscal year's appropriation.

Great Lakes programs, which grew out of the ecology consciousness of the 1960s and early 1970s, have worked. Lake Erie is the best example of that fact. In the late 1960s it was held to be on death's doorstep due to abuse by municipal and industrial wastes. Scientists were enlisted to identify the problems. A permit program was initiated to squeeze off damaging discharges. Federal aid was provided both in capital funds and research capability. Frequent sampling of effluents and lake waters was conducted. As a result Lake Erie today is said to have been stabilized, although it still receives a higher load of phosphorus than is good for its health.

Persons who have been employed in the programs, aware of the near impossibility of saving them, seem to have chosen to switch rather than fight. Many already are lining up other jobs. One scientist at an agency to be abolished at the end of September told The Plain Dealer that the dismantlement of Great Lakes offices means it could take four to six years to re-establish such an effort with any degree of expertise.

What the practical effect will be on the lakes will only become evident with time, observers agree. If there is no sampling, no monitoring, no anticipating going on, problems will not become apparent until something out of the ordinary occurs — such as a fish kill, evidence of dangerous pollutants in water taken in by a distributing authority, or even, perish the thought, human illness or worse. Ingestion by fish of contaminants such as mercury and dioxin, which has been identified and controlled in the past, could escape detection in the future if testing is cut drastically, as it appears it will be. On the other hand, industries could be kept operating under restrictive waste discharge permits that fail to take into account improved conditions because inspection activities are spread too thinly.

A Senate subcommittee already has started rewriting the Clean Water Act, which it was not scheduled to do until next year. Its preliminary plan would reduce capital grants by 45% over four years and extend deadlines for secondary treatment of municipal wastes by five years to 1988.

The Plain Dealer would rather there were not such drastic cutbacks in the offing, with their potential for damage to the lakes, but we see no chance they can be avoided. The Reagan administration has said states and local governments should get together to fund the programs if they want them. There is practically no chance that will happen since states also are having to cut their budgets.

We support approval of House Resolution 3600, a bill by Rep. James J. Blanchard, D-Mich., that would declare the Great Lakes a national resource. Co-sponsored by Rep. Dennis E. Eckart, D-22, of Euclid, it would create a Great Lakes office to coordinate what is left of the programs — now scattered among more than 20 federal agencies and federally funded university research centers, overseen by a number of congressional committees. It could provide a base upon which to build a revived Great Lakes program if — when — the need again becomes obvious.

Cleveland, Ohio, August 17, 1981

The Plain Dealer supports Ohio Senate Bill 209, which would limit the phosphorus content of household detergents sold in the state to 2.2% by weight. State Sen. Ronald L. Nabakowski, D-13, of Lorain, sponsor of SB 209, is asking that hearings be scheduled for next month by State Sen. Sam Speck, R-20, of New Concord, chairman of the Senate Energy, Natural Resources and Environment Committee. We urge Speck to do so.

Until now we have recommended that Ohio enact a stricter limitation of 0.5% by weight. But legislation to accomplish that has been blocked consistently in the Ohio House. Although the 0.5% limit would be of greater benefit to Lake Erie and other Ohio water resources, we recognize that a 2.2% law would be better than no limitation at all. As it stands, Ohio is the only state with a significant portion of Great Lakes shoreline that does not limit the amount of phosphates in detergents sold within its boundaries.

In the 112th and 113th General Assemblies 0.5% bills were approved by the House Energy and Environment Committee only to die in the Rules Committee when Speaker Vernal G. Riffe Jr. did not send them to the House floor for a vote on their merits. In the 114th General Assembly the membership of the committee was changed and earlier this year the panel shunted aside House Bill 63, another 0.5% measure.

Detergent and appliance industry interests have fought diligently over the years against limiting phosphates in detergents. But phosphates are a nutrient, and when they reach Lake Erie and other Ohio waters they become food for algae, which proliferate into lush waterborne carpets. They use up oxygen, degrade conditions for fish and accelerate natural aging processes.

Although industry interests are expected to fight even a 2.2% limitation, it will be difficult for them to make any kind of a case for their position. In fact, detergent manufacturers have been producing a combination product of 2.2% phosphates and an agent known as NTA for years. They use that formulation to comply with Canadian laws. The product has been shown to be safe and as effective in cleaning laundry as the 6%-8% phosphate detergents sold in Ohio, and has not imposed other penalties on consumers such as higher energy costs or damage to washing machines.

A twin to Nabakowski's bill has been introduced in the Ohio House. House Bill 649, sponsored by Rep. Kevin M. Kapel, D-74, of Chesterland, has not been referred to a committee yet. But it has 36 co-sponsors, a fact that should indicate to House leaders there is widespread support for this compromise legislation. The Senate and House owe it to Ohioans and to the future well-being of the state to see that these bills become law.

The Toronto Star

Toronto, Ont., October 14, 1981

An American citizens' action group — the New York Public Interest Research Group — has characterized the scenic and spectacular Niagara River as "a sewer" that is "polluted . . . almost beyond belief."

Its recent report on the toxic chemical load of the Niagara — entitled The Ravaged River — echoes and underlines the dire warnings on the Niagara River sounded earlier this year by the International Joint Commission on Great Lakes Water Quality.

Quite simply, both groups state that the shared Canada-U.S. water course is staggering under a disastrous overload of industrial and biological wastes and, unless a concerted and immediate effort to halt the flow of chemicals into the system is undertaken on both sides of the border, hundreds of thousands of people who live on the banks of the Niagara will suffer the consequences in disease, death, and human misery.

It's a message that is chilling in its intensity and urgency. According to the research group, the industrial sites and dumps on the banks of the Niagara are the source of more than 500 million gallons of chemical wastes daily — a witch's brew of dioxin, mirex, lindane, phenols, phosphates, lead, arsenic, copper and mercury. For all its surface beauty, the Niagara is fast becoming a lethal cesspool. It already poses a potential long-term hazard to the people who rely on it for drinking water and crop irrigation.

And, disturbingly, the group alleges that U.S. environmental officials have deliberately kept Canada in the dark about the true nature, amounts, and dangers of the wastes going into the Niagara.

As the report also points out, the amounts of pollution now measurable in the river water are probably only a mere foretaste of a future in which — unless immediate and dramatic steps are taken — the Niagara, and Lake Ontario into which it flows, will be irreversibly contaminated.

The U.S. research group lays much of the blame for the continued torment of the Niagara River at the feet of American regulatory agencies that have, in the past, emphasized the cleanup of biological wastes such as sewage, and have largely ignored the less visible but more dangerous problem of industrial chemical pollution.

In fact, the group reports that 700 industries along the river neatly side-step American national and state pollution control regulations entirely by discharging their wastes through municipal waste centres.

It is distinctly uncomfortable to note that the municipal waste water treatment plant in Niagara Falls, N.Y., built specifically to handle a growing amount of toxic industrial waste from these sources, broke down under a corrosive burden of chemicals after only six weeks of operation. The substances that contributed to the ruin of the $48 million plant now flow, untreated, into the Niagara River.

It is a lamentable situation, and one that should be battled on all fronts by the combined efforts of the Ontario and federal governments. The wastes that seep or are dumped into the Niagara River end up in the drinking water taps of Canadians, as well as Americans.

And as the Public Interest Research Group points out, the Niagara is fast running out of time. Delays and evasions on both sides of the border have already allowed the negligent dumping of chemicals into the river to reach a flash-point of danger. Any further delay in halting further dumping and in cleaning up the dump sites from which the poisons leak could be no less than disastrous.

Detroit Free Press

Detroit, Mich., March 14, 1982

THE 1983 FEDERAL budget threatens to sink research efforts on the Great Lakes the way the iceberg sank the Titanic. The EPA Large Lakes Lab at Grosse Ile is to be abolished. The Great Lakes Environmental Research Laboratory at Ann Arbor will be no more. Sea Grant, trimmed in the 1982 budget, will be finished off in 1983. The EPA Great Lakes national program office in Chicago will have been cut back by 66 percent since the last Carter budget.

What is proposed is the virtual elimination of research on wind, waves, shoreline erosion, water currents, toxic contaminants, spills of oil and hazardous substances, phosphorous loading and water levels in the Lakes. These are not arcane studies, of interest only to hopelessly pettifogged academics. They affect health, lives and dollars as directly as any research can.

The abolition of the lab at Grosse Ile amounts to thumbing our nose at Canada, since it will make it impossible to meet our treaty obligations to monitor water quality. The shutdown of the Ann Arbor lab would end basic research into the circulation of toxic materials in the Lakes, the rate at which PCBs are breaking down and the effect of wave heights and water levels on shoreline erosion.

Researchers have just begun to track the long-range transport and buildup of organic pollutants in the Lakes — a phenomenon of intense interest to the 25 million people who depend upon them for pure drinking water. The travel and recreation industries, a revitalized sport and commercial fishery, shippers interested in reliable forecasts and navigation information, property owners and communities that want to know about shoreline protection — all benefit from the current research programs.

The Great Lakes are America's fourth coastline and the greatest freshwater reservoir on earth. Continued funding of research on the Lakes, particularly in the Ann Arbor and Grosse Ile labs, is vital, whether your concern is the short-term economic return from their research, or the assurance that your grandchildren will have drinking water free of carcinogens. Congress saved the Grosse Ile lab from extinction once before. If the budget-cutters still do not recognize the importance of the research effort here, Michigan's congressional delegation must spearhead the effort again.

DAYTON DAILY NEWS

Dayton, Ohio, March 18, 1982

With Reaganomics leading the way, we are returning to those thrilling days of yesteryear when U.S. industry wasn't burdened by lots expensive regulations and the federally employed inspectors who enforce them.

Canada, however, would prefer we didn't include the Great Lakes in any of these back-to-deregulation movements.

Canadians remember, even if Americans do not, those good old days as the time when toxic levels in fish were so high many commercial fishing operations closed, when beaches were littered with dead marine life and aquatic birds, when waterways were unsafe to swim in and when garbage on the Cuyahoga River near Lake Erie was so concentrated it caught fire — twice.

Canadian officials have been trying to get answers from Washington about how budget cuts will effect U.S. implementation of the 1978 Great Lakes Water Quality Agreement designed to eliminate toxic discharges, lower phosphate levels and set limits on radioactive water pollution from nuclear plants on the Great Lakes.

Ohioans should be asking for answers to the same questions since one of those lakes offers us an abundance of drinking water, commercial transportation, food and recreation.

If the administration follows through on its plan to halve the $7.9 million 1982 budget for the Environmental Protection Agency's Great Lakes office in 1983, it will severely curtail the work of the only U.S. agency which studies toxic chemical effects and monitors pollution sources on the Great Lakes.

Other proposed cuts would close the only research lab dealing solely with Great Lakes pollution; shut down our Great Lakes environmental research laboratory and end a grant program five universities participate in to study Great Lakes pollution.

Ohio sits on the edge of a natural resource the whole world envies — the largest freshwater supply on earth. In fact, the 65 trillion gallons in those five lakes are a fifth of this planet's fresh water and 95 percent of the drinkable and navigable water in this country.

Great Lakes states cannot complacently watch as programs designed to upgrade our water supply go by the boards in the name of economizing.

If we do, we will find ourselves subsidizing the western states which are lobbying hard for big federal bucks to underwrite their search for fresh water needed to attract even more northern businesses and industries.

THE BLADE
Toledo, Ohio, July 7, 1982

ALTHOUGH the U.S. House Science Committee recently expressed concern over funding cuts relating to Great Lakes research, two government reports indicate that money is not the basic question; the more immediate problem may lie in the federal bureaucracy itself.

The specific problem the House committee had in mind was that of water quality in the lakes, and its report stressed the need to meet U.S. commitments to Canada in cleaning them up. That is important, of course, but it appears to be more than just a matter of restoring money to government programs.

In 1980 the International Joint Commission, established to coordinate U.S. and Canadian efforts relating to the Great Lakes, cited a lack of coordination among federal agencies in meeting water-quality standards. And just last week the General Accounting Office released a report also blaming the Federal Government. It recommended that the IJC receive more support from Washington and that federal agencies be more responsive to the commission's proposals and requests for information.

A lot of money and work has gone into cleaning up the lakes over the past two decades, much of it right here in northwestern Ohio through the efforts of Clear Water, Inc., and the Ohio Water Development Authority. This valuable progress should not be permitted to be lost now.

Even under the restrictions of a tight federal budget, it appears that procedures and attitudes can be changed so as to maintain a credible lakes maintenance program in cooperation with our Canadian neighbors. It is well worth the attempt.

THE ANN ARBOR NEWS
Ann Arbor, Mich., June 15, 1982

Once again the issue is clean water vs. the threat of pollution from (once again) the detergent manufacturers.

The soapmakers try to come on as Mr. Clean in this battle but it doesn't wash. They say phosphates are needed to loosen dirt from clothing and that no satisfactory substitute has been found for phosphorous in cleaning agents.

Phosphorous in water speeds up the aging, or eutrophication, of the lakes. Weeds and algae thrive. Oxygen is choked off to fish populations.

That's why five years ago Michigan moved to improve water quality by reducing the phosphorous content.

Since that time, the soapmakers have not been idle. They got Rep. Connie Binsfeld, R-Maple City, to sponsor legislation raising the allowable limit on phosphorous in dishwater detergents.

That bill went nowhere. Now the soapmakers have taken to the Supreme Court a lawsuit questioning the authority of the state Natural Resources Commission to adopt a rule lowering the legal phosphate level in household cleaners.

GAINS HAVE BEEN MADE in the fight against phosphorous and the lakes are better for them. Passage of the Binsfeld bill, for example, would have set back years of effort to improve water quality.

And water quality — with all its implications for tourism and business resurgence — is the real issue here, not whether tableware comes out of the dishwasher with fewer water spots.

At last look, Procter and Gamble seemed to be doing okay in the competitive world of big business, so it isn't as though the phosphorous penalty has their executives jumping out of windows.

And as far as the "no satisfactory substitute" argument is concerned, the soapmakers should put as much effort into R & D in that area as they are finding work for their legal departments.

The continuing health of the lakes is the overriding concern. Legislators sniffed out the special interest aspect when they rejected the Binsfeld bill.

That precedent ought not be lost on the court justices who are reviewing the soapmakers' lawsuit — another "suit" which should be taken to the cleaners.

SYRACUSE
HERALD-JOURNAL
Syracuse, N.Y., June 15, 1982

When Ronald Reagan goes to bed at night, do you suppose he lies there wondering if becoming president was all it was cracked up to be?

As if he didn't have enough trouble with his critics at home, now he's got the Canadians on his back.

A congressman just back from Ottawa said officials there are "offended and outraged" over what they consider a backing down by this government from its agreement with Canada to clean up the Great Lakes.

According to Bruce Ingersoll of the Chicago Sun-Times, the Canadians are worried about cutbacks in water quality and surveillance programs and also in federal dollars for municipal sewage treatment plants, particularly in the Great Lakes area.

The fact is that the United States has spent $30 million on the Great Lakes problem since 1971 and, according to the State Department, will spend another $10 million in the next year or two.

▽   ▽

Reagan's problem, as it has been right along, is one of priority. It appears he wants to slow the program, not eliminate it as the Canadians seem to believe. In that way, he can spread the costs and not heavily damage the fragile budget of any one year.

What will be lost or at least slowed down?

The Environmental Protection Agency's Chicago office says its research vessel will have to remain tied up at its dock this season because the office's budget was cut from $10 million to $7.1 million.

Two labs, one a research facility run by the EPA at Grosse Ile, Mich., and the other doing environmental work at Ann Arbor, Mich., probably would be phased out.

The administration made the attempt this year, but the Michigan congressional delegation succeeded in restoring $6.1 million to keep both in business.

Six universities in the Great Lakes area share $5 million annually for research under the Sea Lab program. That would end next year.

▽   ▽

Several questions have to be asked.

How much money is enough for such programs?

How well is it being used? How much of it goes for bureaucratic costs?

How does the president take us out of inflation, balance the budget and do the other things he was elected to do if we don't cut back on programs like this?

Granted we have a huge stake in the purity of the Great Lakes. Our drinking water comes from one of them.

Billions upon billions of dollars will be spent by the government next year and we assume Reagan's people are looking very closely at priorities.

What's good for the United States, not Canada, should be the top priority on their list.

The Philadelphia Inquirer
Philadelphia, Pa., December 15, 1983

A three-day meeting on Chesapeake Bay pollution ended on a somewhat disappointing but not entirely unexpected note last week when William D. Ruckelshaus, head of the U.S. Environmental Protection Agency, failed to firmly commit federal funds for a cleanup. That was not the final word on the federal role, however.

At the meeting, held in Fairfax, Va., and attended by 700 public officials, environmentalists and other concerned parties, Gov. Harry Hughes of Maryland, the state principally affected, pledged $70 million in the first year of a cleanup that is expected to take a decade. Gov. Charles S. Robb of Virginia, the only other state bordering the bay, pledged $6 million. Lt. Gov. William W. Scranton, representing Pennsylvania (Gov. Thornburgh was in China) pledged $2 million — a substantial commitment that was fully justified considering that the Susquehanna River is a major source of pollution in the Chesapeake.

With those levels of state commitments the Maryland, Virginia and Pennsylvania congressional delegations can make a strong case for federal funding when the fiscal 1985 budget begins to take shape early next year. Mr. Ruckelshaus, who was careful not to rule out a future commitment of federal funds, should use the state pledges as an argument for including U.S. assistance for the Chesapeake in the budget message that President Reagan will submit to the Congress soon after it reconvenes in January.

A $28 million EPA-funded study, completed last summer, documented serious declines in the oyster, crab and fishing industries on the Chesapeake. The cause, the report said, was pollution originating mostly in more that 50 rivers (the Susquehanna is by far the largest) that flow into the bay. It estimated that it will cost $1 billion and 10 years to restore the bay to acceptable clean-water standards in which shellfish, striped bass and other aquatic life could thrive as they once did.

Though Pennsylvania does not touch the Chesapeake at any point and has no direct economic interest in communities on the bay, state officials are right to recognize that the state has a responsibility to help in the cleanup. Gov. Thornburgh hosted a meeting with Govs. Hughes and Robb in June that resulted in last week's conference. Details of Pennsylvania's $2 million pledge will be spelled out in the next budget but are expected to include programs to curtail discharges of sewage and industrial wastes into the Susquehanna basin and fertilizer runoffs from farms bordering the river and its tributaries.

Restoring the Chesapeake to environmental and economic health is attainable, but no time must be lost or effort spared. Federal participation with the states is essential considering the urgency of the situation.

The Virginian-Pilot
Norfolk, Va., August 12, 1983

Peaceful farms in Virginia, Maryland and Pennsylvania are some of the worst polluters of the Chesapeake Bay.

But state officials believe that persuasion, rather than coercion, should be used to reduce agricultural runoff. Fortunately, they have persuaded their counterparts in the U.S. Environmental Protection Agency to drop proposed mandatory controls to curb farm pollution.

Sometimes it's hard to believe that the bucolic, plowed fields can inflict such havoc upon the bay. The pollution isn't as obvious as a belching smokestack, or a drainage pipe spewing malodorous toxic wastes.

But fertilizers and pesticides washing off agricultural land are slowly killing the Chesapeake, according to a just-concluded, $28-million EPA study.

An EPA proposal would have forced farmers to limit runoff; in some cases, it would have meant substantial changes in the way they operate.

The EPA dropped the idea of compulsory controls after hearing objections from Virginia officials and bureaucrats in other states.

Instead, the EPA and the states agreed earlier this week to see how well the farm pollution problem responds to voluntary land management programs and financial incentives, at least for five years. After that, if improvements aren't seen, strict requirements can be enacted.

Environmentalists object that state officials are more concerned about offending politically powerful farmers and large land owners than quickly addressing one of the bay's largest sources of pollution.

Well, of course the state agency heads are being sensitive to political considerations. That's not necessarily a bad idea.

Cleaning up the bay will cost hundreds of millions of dollars, maybe even a billion or more. That kind of money will be appropriated in Congress and by the state legislatures only if public support for the Chesapeake cleanup remains strong.

Harsh mandates could shatter this consensus by making the farmers angry. The farmers can be formidable opponents when stirred.

The agricultural pollution compromise, hammered out Wednesday at an Annapolis, Md., meeting, seems reasonable.

Farm pollution control steps have been taken already in Virginia and other states in the bay's watershed. Farmers are being paid modest amounts to maintain a buffer strip between their fields and bay tributaries. And it's in the farmers' best interest to minimize the erosion of rich topsoil into neighboring streams, even without government payments.

It makes sense to try persuasion, education and financial incentives first. Restoring the bay to good health will be a sufficiently difficult job without the added political problems that will be created by heaping mandates upon the farmers.

THE CHRISTIAN SCIENCE MONITOR
Boston, Mass., October 18, 1983

A quiet bit of good news has crept out of Nova Scotia: The US and Canada have agreed to trim by 15 percent the phosphorus pollution in the Great Lakes. The agreement supplements the five-year-old Great Lakes Water Quality Pact between the two countries.

The new signing is particularly welcome at this time. For one thing, it shows the US remains committed to cleaning up the Great Lakes. Enormous progress toward ending its pollution has been made in the past 15 years, after decades of talk. But in recent months environmentalists have expressed concern as to whether the federal and state governments were still committed to continuing the cleanup. The pact just signed shows Washington's answer is affirmative. States abutting the Great Lakes are more likely to continue to do their part now that the Reagan administration has shown leadership.

The timing of the agreement is propitious for another reason. It reminds everyone involved in another thorny issue — acid rain — that reaching major international agreements often takes years, as in the case of the Great Lakes cleanup. In such instances, each nation first must reach its own consensus on problems and possible solutions; then comes a period of discussion to reach international approval. Still, the contrast between progress on the phosphate cleanup and a near-stall on acid rain should only quicken pressure for the latter's progress.

Concern has been expressed since early this century about pollution in the Great Lakes, but the major cleanup did not start until some 15 years ago. This was followed by stepped-up effort when the Clean Water Act was signed in 1972; in 1978 came the US-Canada accord.

All this should be remembered in the current thrashing about over acid rain, now properly of very major concern to the northeastern areas of Canada and the United States. Questions needing resolution include: what kind of cleanup action should be taken to achieve the greatest result with the least dislocation to others, such as coal miners; who should pay for it; and to what extent is US pollution responsible for acid rain in Canada?

In Washington, Environmental Protection Agency administrator William Ruckelshaus is having difficulty obtaining agreement within the Cabinet on what the US should do to curb acid rain; he notes the complexity of the issue. In Nova Scotia, he said he hoped but could not be certain that before next year's election President Reagan would decide what action the US should take.

That is a good deadline to aim for, provided research now going on in several areas yields wise answers to today's questions.

The Dispatch
Columbus, Ohio, March 15, 1984

Water is essential to the survival of man, fish and wildlife and the maintenance and preservation of usable freshwater supplies are the focus of National Wildlife Week, March 18 to 24.

"Water: We can't live without it," is the theme selected by the National Wildlife Federation, which one week each year lends added emphasis to the importance of preserving fish and wildlife for present and future generations.

The mismanagement and contamination of freshwater supplies is of special concern to Ohio as one of the Great Lakes states. The Great Lakes contain 20 percent of the world's freshwater supply.

Apart from supporting governmental efforts to preserve water supplies, individuals can help in refraining from introducing litter and other contaminants into our rivers, streams and lakes.

One opportunity to aid in preserving endangered native non-game wildlife species is the tax checkoff program available in Ohio and 30 other states. American families can contribute all or part of the income tax refunds in this manner.

The federation's continued vigilance in protecting our precious wildlife resources is worthy of continued support.

Newsday
Long Island, N.Y., November 30, 1983

The Environmental Protection Agency spent more than five years and $27 million studying Chesapeake Bay. The federal agency's conclusion: Pollution is choking the nation's most productive marine estuary, and it's getting worse.

Prompted largely by the EPA's report, which was released in September, legislation was introduced in Congress to provide $40 million in federal matching funds over the next four years to help states in the region begin cleaning up the bay.

Astonishingly, the EPA itself opposed the bill in testimony before a House water resources subcommittee earlier this month. The agency contended that the job could be done under the existing Clean Water Act.

Why, then, hasn't it been?

Despite the Clean Water Act, the oyster harvest has decreased by a third, blue crabs are far less plentiful and striped bass have declined by half since 1970. Long Island fishermen know the sad story well; many of the bass in the Island's waters were hatched in Chesapeake Bay.

EPA researchers traced the deterioration in fish and shellfish to the damage done by effluent from sewage treatment plants, to run-off from fertilized farmlands and livestock operations and to heavy metals and toxic chemicals from industries.

The only way to get at these problems is at the source. The EPA study called for "immediate steps" to halt further pollution. Clearly, much more money will be needed than the modest amount contained in the bill the EPA opposed, but it would at least help the states make a start.

Cleaning up the Chesapeake isn't going to get any cheaper or easier, and there's nothing unreasonable about federal aid to rescue a bay the EPA's administrator has called "a national treasure." Fish from the Chesapeake range far and wide on their own and grace tables across the country.

The administration should drop its shortsighted opposition, and Congress should put the matter high on next year's agenda.

THE SUN
Baltimore, Md., July 27, 1983

As symbolism, the sight of the nation's top environmental officer touring the Chesapeake Bay with two governors, a lieutenant governor, four senators, scores of underlings and dozens of reporters in a flotilla of seven boats was an immediate success. The message was clear: Both state and federal officials are concerned about the growing pollution of this country's largest and most bountiful estuary.

But William D. Ruckelshaus's recent day on the bay, spent watching crab potters, clam dredgers and oyster tongers in action, may prove of far more lasting value. For he will be a key player if there are to be any large-scale federal efforts to clean up the Chesapeake Bay. Judging from his reaction to his travels along the bay, the Environmental Protection Agency chief seems ready to help protect this region's most precious resource.

He called the bay "a very important national treasure." We already know that here, but it helps to hear it from an influential Washington figure. Sensitizing Mr. Ruckelshaus to the dangers of continued pollution to the rich bay harvests can only increase his appreciation of this unique estuary and the need to protect it.

Moreover, the EPA chief seems enthusiastic about the regional approach taken by officials in Virginia, Maryland and Pennsylvania to come up with a joint management plan for the bay. The key, though, will be the level of federal assistance to implement what are sure to be costly recommendations. Some of the proposals that may be adopted at a planned December "summit" of the three state governors include a ban on detergents containing phosphate, a major pollutant flowing into the bay, and vegetated buffer strips on bay farmland to prevent chemical runoffs.

There already is a provision in a bill before Congress extending the Clean Water Act that would authorize $150 million to control "nonpoint" pollution, such as farmland chemical runoff. But Mr. Ruckelshaus rightly points out that this may be far too little, given the dimensions of the problem nationwide. What level of funding the EPA does have in mind should be known next month. That may tell us much about its position on clean-up funds for the Chesapeake Bay.

The kind of regional cooperation being discussed, in which the states and the federal government would form a partnership to reduce sources of bay pollution, fits in nicely with President Reagan's "New Federalism" philosophy. Mr. Ruckelshaus said as much during his tour of the bay. The important thing is to agree on a management plan for the bay and get on with the clean-up before the damage to the Chesapeake becomes irreversible.

BUFFALO EVENING NEWS
Buffalo, N.Y., May 20, 1984

WHILE SOME progress has been made in cleaning up the Great Lakes, much remains to be done in the years ahead. This is a particular cause for concern in light of drastic cutbacks by the Reagan administration in support of environmental research and development in the lakes. Last year, American and Canadian officials warned that there had been little progress in reversing "serious degraduation of water quality" in the most seriously polluted areas of the lakes.

The critical need to advance cleanup measures would be well served by enactment of a measure sponsored by Rep. Henry J. Nowak, D-Buffalo, and approved by the House Public Works Committee. In contrast with the $1.7 million proposed by the administration next year for the Environmental Protection Agency's Great Lakes National Program Office, Mr. Nowak's bill would provide $50 million over the next five years for a renamed Great Lakes International Coordination Office.

Mr. Nowak sees the proposed agency as a "prerequisite for meeting our Great Lakes commitments." His bill would give the EPA and the Coordination Office a mandate to "take the lead in working with other federal agencies and state and local authorities in developing water quality strategies."

Under the bill, the EPA would create a Great Lakes systemwide surveillance network to monitor water quality. Federal funding would be increased for construction of municipal sewage-treatment plants, with an increase in the federal share of such projects from 55 percent to 65 percent. The legislation also would provide an annual $150 million program to control pollution at its source.

A key provision in the bill would require that the EPA work with the Canadian government and the International Joint Commission to carry out the goals of the 1978 Great Lakes Water Quality Agreement. This accord is due for revision by the United States and Canada in 1986, and scientists and researchers have urged a faster pace of funding and sterner enforcement to speed the cleanup of toxic chemicals. The Nowak bill would provide timely support for such improvements.

The cutbacks in research programs on the lakes for eradication of pollution — down to only $3 million currently from $33 million under the Carter administration — give reason for restoring funds and advancing the cleanup programs needed if the lakes are to remain a vital source for the economic, commercial, recreational and fishing needs of the Great Lakes region.

THE ATLANTA CONSTITUTION
Atlanta, Ga., April 5, 1984

Question: What substance is the No. 1 pollutant in Georgia's rivers and streams? It is neither sewage nor industrial waste nor any other chemical exotica, say the people who make it their business to know about such things. It is dirt — gritty, garden-variety dirt.

A key cause of excessive sedimentation in metro Atlanta's streams is erosion that results from careless construction. While the state seems to have adequate erosion-control laws, not every developer sees fit to comply with them. The consequent damage can create a considerable public expense.

Too much sediment can interfere with a city's water supply. It can create major expenses as waterworks equipment is cleaned and maintained. And it can cause environmental problems as it chokes life from ponds and eventually fills them in.

The Georgia Environmental Protection Division recently warned Gwinnett County that its power to issue building permits could vanish if the county doesn't force developers to toe the mark. The EPD found six projects that, it said, failed to comply with erosion regulations.

It is doubtful that Gwinnett stands on this shaky ground alone. Every county with significant new development (and that means every county in metro Atlanta) must take pains to ensure that its builders are being properly regulated.

The means to stop soil erosion are readily available. Often, hay bales will do it, as will sediment fences or sediment retention ponds. But these precautions cost a builder money. Because the regulations are relatively new, not all contracts cover such costs. That is not a legitimate excuse for disobeying the law.

It is imperative that, as the metro area prospers, it must keep one eye on the ledger sheet and the other on the environment. Encroaching civilization has its rewards for any county. But it is hardly a civilized practice to displace land and then allow it to clog up ponds and municipal waterworks downstream.

When it comes to the environment, we're all in it together.

The Evening Gazette
Worcester, Mass., June 18, 1984

Acid rain is the environmental catch phrase of the moment, but one survey of Connecticut lakes shows an even worse problem.

Scientists at the Connecticut Agricultural Experiment Station say the state should be more concerned about "galloping eutrophication," a buildup of phosphorus in the state's lakes, than about the effects of acid rain.

Sewage and construction of roads near lakes has increased the phosphorus content in lake water. The phosphorus encourages growth of algae and weeds, which in turn choke many of the lakes. Increased development during the past two or three decades is to blame.

A certain amount of phosphorus enters a lake regardless of land use, according to Charles R. Frink, a co-author of the station's report. He said the highest amount of the material originated from urban land. About one-tenth of a pound of phosphorus per acre per year comes from forested land, one-half a pound from agricultural areas and 1.5 pounds from urban land, he said. "Man's disturbance of the watershed, particularly as he paves roads and restricts the amount of water that goes into the soil, . . . increases phosphorus," Frink said.

A comparison of the alkalinity of some of the state's 1,000 lakes over the past couple of years with records that were kept in the 1930s showed little change, the scientists found, leading them to conclude that acid rain is not the major threat to Connecticut lakes it might be in upstate New York and northern New England.

Since development won't stop, the scientists said the eutrophication process can be slowed only by watershed management. Their warning is important to recreational users of lakes and water companies, both of whom suffer when lakes become victims of weeds and algae.

THE MILWAUKEE JOURNAL
Milwaukee, Wisc., February 8, 1984

It's almost a classic example of the squeeze that the Department of Natural Resources often gets into.

Researchers are concerned that a much higher than normal incidence of deformities in birds in the Green Bay area may result from toxic materials in the water and in fish eaten by the birds. PCBs and furans are a special cause of worry. Everybody agrees that more research is needed.

But Asst. Atty. Gen. Thomas Dawson, one of the state's public intervenors, wants the DNR to do more than study. He calls for immediate action to curb further pollution.

Not surprisingly, the DNR believes it will be accused of unfairness and will be thwarted in court if a crackdown on industrial discharges is begun before more scientific evidence is obtained.

No doubt that's right. History suggests that few polluters willingly do anything that costs much money.

In a sense, Dawson and the DNR are both right. The DNR has to operate in a political climate, and the department's governing Natural Resources Board is none too zealous about environmental protection. Thus unless Dawson and people like him keep pushing, the DNR is apt to take the politically cautious course.

That word was "cautious," not "conservative." The truly conservative approach is to err, if necessary, on the side of environmental safeguards. Curbing the *further* pollution of Green Bay with toxic substances should not have to await conclusive research on the full effects of pollution that has already occurred.

The Wichita
Eagle-Beacon
Wichita, Kans., February 18, 1984

The latest assessment of the quality of America's water shows that attention to yesterday's unaddressed problems is paying off today. The Environmental Protection Agency recently told Congress that significant reductions of key industrial pollutants — from 52 to as much as 80 percent in some cases — were recorded in the five-year span ending in 1977.

Additional improvements undoubtedly have occurred since then. Little, if any, of that progress would have occurred had it not been for a heightened awareness of the insidious problems of water pollution and a commitment by the federal government to set about correcting them.

The dilemma is far from resolved. Although sewage treatment plants now are able to remove nearly two-thirds more contaminants than they could a decade ago, the actual amount of sewage entering American waterways remains basically unchanged. That's because there are 18 million more Americans contributing to the daily outflow of sewage than there were then. In that context, it's debatable as to whether the nation's water quality actually is improving. The EPA findings indicate practically every state still has serious problems with industrial and municipal wastes contaminating downstream water resources.

It has been forecast that adequate water supplies may be the next major crisis facing America. That makes it all the more important that Americans not settle for maintaining current, less-than-desirable levels of water quality, but strive to continue reversing the processes that have degraded so many once-pure rivers, streams, lakes and ponds. It may not be possible to restore them to their pristine state — but, barring that, there's always room for improvement.

The Dispatch
Columbus, Ohio, February 6, 1984

Gov. Richard F. Celeste has lent his support to a misguided bill in the General Assembly that would reduce the amount of phosphorus permitted in laundry detergents sold and used in the state.

The object of the legislation is to reduce the amount of phosphorus that ultimately winds up in Lake Erie. Too much of the chemical can limit the lake's oxygen supply and promote the growth of bacteria and algae.

The irony of the bill is that phosphorus in detergents poses no danger to the lake, but the source of phosphorus that does threaten the lake's quality — fertilizer — would not be affected by the bill. Passage of the bill would result in higher laundry bills for consumers, headaches for detergent manufacturers, useless governmental interference in the private sector and a false impression that the state is acting forcibly to protect one of Ohio's finest resources. This hoax would, in the long run, damage the state's environmental protection efforts.

The bill would limit the phosphorus content of household detergents to no more than 2.2 percent. Granular detergents now being sold contain phosphorus at an average level of 6.9 percent. The curb would apply to 35 counties in northern Ohio which drain into Lake Erie and would be in effect until 1990.

The problem of phosphorus was recognized years ago by the state and was one reason why Ohio embarked on a large-scale effort to build waste water and sewage treatment facilities designed to filter out harmful substances in the water destined for Lake Erie. Thus, any phosphorus that finds its way from detergents to the treatment plants is filtered out of the water before it arrives at the lake.

What is not filtered out is the phosphorus that drains off farmlands into Lake Erie. The chemical is used in fertilizing farm acreage and there is no way that it can be captured and prevented from winding up in the lake. The effort to curb detergent phosphorus is targeted at the wrong source and is, therefore, a useless undertaking.

But it is more than just useless. It would hurt consumer interests by limiting a valuable cleaning agent. This would raise consumers' bills by requiring more frequent washing of clothes, hotter water temperatures and the use of more additives — additives which have been found to shorten the lives of washing machines.

The bill would also unnecessarily complicate the distribution of detergents in Ohio. Making life more difficult — for no good reason — for the businesses providing jobs to Ohioans is in no one's interest. The state government should, instead, seek to create a climate that is conducive to private sector growth. Enacting politically popular but unneeded legislation does not serve that goal.

The Seattle Times
Seattle, Wash., June 14, 1984

WHAT is in question about this week's announcement of a plan to clean up the disgraceful pollution in many parts of Puget Sound is not its necessity. That point is beyond dispute. Rather, the questions concern whether what is being proposed is enough; who should pay its costs, and whether the existing labyrinthine pollution-control apparatus can function quickly and effectively.

Regional officials of the Environmental Protection Agency have recommended $12 million in federally funded studies of, among other things, how to rid the Sound's urban bays of toxic metals and chemicals, and halt worsening contamination of shellfish populations.

Governor Spellman, meantime, moved to beef up the state Ecology Department's focus on Puget Sound water-quality problems and promised to ask the next Legislature for more pollution-control funds.

All well and good, albeit a belated and overdue response to circumstances that have been evident for a long time. While contamination can be traced to a variety of sources, experts such as William Ruckelshaus, EPA's top administrator, note the strong evidence that the principal problems are caused by industrial wastes. This creates a heavy responsibility for polluters to bear a fair share of cleanup costs.

In a three-part series of articles earlier this week, Times reporter Eric Pryne outlined the range and severity of contamination and the difficulties of controlling it. Not the least of those problems is the astonishing array of 18 federal, state, regional and local agencies responsible for water-quality control.

Despite this often uncoordinated maze of bureaucracy, pollution dangers have continued to deepen, reinforcing the call by several critics for a single agency with centralized authority to carry out remedies.

Thanks to earlier efforts such as the Metro cleanup of Lake Washington and Elliott Bay, and a halt to the discharge of certain untreated wastes into urban waterways, some of the region's most visible pollution has disappeared. But public support now is needed for an effort to attack the often unseen contamination that still lurks beneath the Sound's seemingly pristine waters.

Ground Water: Aquifers, the Purest Water

Ground water, contained in the soil and in aquifers, is an irreplaceable resource that is severely threatened by the seepage of toxic chemicals from abandoned dumps, pesticide runoff and other sources. Unlike surface water, it is nearly impossible to cleanse once contaminated. Aquifers, the underground reservoirs from which wells draw their water, are natural formations with porous walls of gravel or sand and a layer of bedrock at the bottom. Because of their construction, these cavities are relatively watertight; the water that has seeped into them through the decades, filtered by the layers of soil above, moves very slowly if at all. There are also two layers of ground water contained in the soil. Shallow ground water is held between soil particles and effectively filtered by them, while deeper down the water content of the soil increases to the saturation point. The upper limit of this saturated portion of the soil is known as the water table. This system is constantly in flux; underground waters flow into rivers and streams through springs, and every rainfall adds its contribution to the waters slowly making their way to aquifers. There is estimated to be about 50 times more ground water under the United States than is flowing in its surface lakes, rivers and streams. Ground water supplies about half of the drinking water for the American populace; rural areas are particularly dependent on untreated ground water from individual wells. Over 40% of the water used by farmers for irrigation purposes is also drawn from this subterranean supply.

While the pollution of aquifers is less visible than that of lakes or rivers, its effects are often more serious; because they are so difficult to monitor, the contamination of aquifers may not be discovered until illnesses traceable to infected well water develop in a local population. The most common cause of ground water contamination is from septic tanks or cesspools; other sources include garbage dumps, landfills, agricultural runoff, mines, and industrial liquid wastes stored in pits or ponds. Among the most dangerous contaminants are synthetic organic compounds produced by industrial processes. On Long Island, where toxic industrial wastes have regularly been dumped into sewage systems, it is suspected that the very high cancer mortality rate may be attributable to the resultant contamination of the water supply.

Houston Chronicle
Houston, Texas, May 24, 1982

Revisions of the Clean Air Act are causing a battle royal in Congress. There is also a small skirmish going on concerning clean drinking water, something most of us take for granted in this country.

The Safe Drinking Water Act was passed in 1974 and is up for review and funding. The debate centers around just how clean water should be. New techniques can discover very minute quantities of chemicals in water. Some of these chemicals are carcinogens, and the question is how dangerous they are in their very diluted state. Water company officials say they have the ability to remove all these chemicals — if people want to pay $5 for a glass of water. Congress, after hearing from the experts, is going to have to decide how to balance risk and cost in settling the debate over chemicals.

One disturbing fact has been revealed as the water act revision is considered. The General Accounting Office reports that in one year alone more than 146,000 violations for failure to test or meet water quality standards were found, but little action was taken. The trouble is primarily in the smaller water systems.

The Environmental Protection Agency announced last week it is going to put more emphasis on enforcement. Part of that effort should be devoted to getting every source of drinking water in compliance with the laws already on the books.

The Pittsburgh Press
Pittsburgh, Pa., April 11, 1980

The good news about the long-term seepage of chemical wastes on Neville Island into the water supplies of the West View and Robinson Township authorities is that this apparently poses no danger of immediate poisoning.

Indeed, the drinking water from both authorities continues to meet established federal guidelines.

Nevertheless, both state and federal environmental officials have expressed concern over the possible long-term risks associated with the drinking of water laced with trace amounts of chemical pollutants.

And this points up once again the hazards — some long known, some only now being recognized and others still to be discovered — that are part and parcel of modern man's need of industrial products.

★ ★ ★

Actually, it was back in 1972 that the Neville Chemical Co. was ordered by the state environmental protection officials to close a chemical-waste lagoon on Neville Island, which had been used for more than a half-century as a wastepit.

But it wasn't until last January that the company — again on state orders — finally finished pumping the lagoon dry.

In the meantime, chemicals matching those from the lagoon began to seep into the water table on Neville Island, forcing the West View Authority to close three of its radial wells there.

And when the West View Authority pumped the polluted water from these wells into the back channel of the Ohio River, this water mixed with the current and drifted into intakes of the Robinson Water Authority a few miles downstream.

★ ★ ★

As for the future, the State Department of Environmental Resources expects Neville Chemical to eliminate the risk of more contamination by removing some of the chemically saturated soil from under its now dry lagoon.

And the West View Water Authority is finishing construction of a new treatment facility designed to take water directly from the Ohio River instead of from its wells on Neville Island.

Meanwhile, however, the Robinson Authority remains dependent upon the diluted chemically-laced water coming from upsteam.

So it's imperative that local, state and federal officials continue to be on the alert to the long-term implications of drinking such water. And to seek new ways of making all water supplies absolutely safe to drink.

THE EMPORIA GAZETTE
Emporia, Kans., January 2, 1981

IT was bound to happen. First one thing and then another has been found to cause cancer — if consumed in large enough quantities by laboratory rats. The trend is familiar: "Government scientists today said that leafy spinach, consumed at the rate of 2,000 pounds per day over a period of 75 years, may lead to malignant tumors in white mice." Taken in such quantities, even marshmallows might cause cancer. Or carrot juice. Or Johnny Walker Red.

It was inevitable, then, that the scientists would finally get around to testing water. And what did they find? Trouble. Right here in River City.

Emporia is one of eight Kansas towns with water that contains strong traces of trihalomethanes — compounds created when chlorine, added to kill bacteria, reacts with organic matter such as leaves. Emporia ranked fourth of the eight Kansas cities listed, with 240 parts per billion of trihalomethanes. Over a period of 25 to 50 years, the Environmental Protection Agency says, this could be harmful.

Local and state officials do not agree. Jack Burris of the Kansas Department of Health and Environment said his staff has "serious doubts" about the Federal findings. City Engineer Lee Stolfus estimates that chlorine has been used to treat water here for 50 years and there has been no statistical evidence of an outbreak of the types of cancer supposedly caused by trihalomethanes. Besides, as City Manager Virgil Basgall pointed out, state law requires the city to use chlorine.

The E.P.A. was right to make its findings public, yet the warnings are no reason for panic. Until scientists can agree on the dangers of trihalomethanes, Emporians should feel relatively safe. — R.C.

The Hartford Courant

Hartford, Conn., December 28, 1982

The agreement under which the Solvents Recovery Service of New England will clean up contaminated ground water in Southington is a small bit of progress in dealing with the pervasive problem of hazardous wastes.

As part of a consent decree filed in U.S. District Court in Hartford, Solvents Recovery will act to prevent further pollution of town wells by its operations and pay the town for the cost of finding new water supplies.

The state Department of Environmental Protection took a back seat while the accord was negotiated among town, company and federal Environmental Protection Agency officials.

The state DEP had better get behind the wheel to solve many problems that remain. Solvents Recovery is not even the most serious source of hazardous waste pollution in Connecticut. It was only one of four dangerous toxic dumps in the state identified by the EPA as eligible for federal clean-up money.

Perhaps the most serious problem is at the Laurel Park Landfill in Naugatuck, where chemical wastes have polluted surface waters and created a plume of contaminated ground water. The other two named by the EPA are the Beacon Heights Landfill in Beacon Falls, where industrial wastes also are heading downhill through the ground toward drinking-water wells, and the Yaworski lagoon in Canterbury, which periodically spills contaminants into the Quinebaug River. There are many others.

The state must be more aggressive in getting responsible parties to own up to the problems they have created, and to pay to solve them, as the federal government did in Southington.

A more assertive DEP, with better funding to enforce hazardous waste regulations, is needed not only to ensure the cleanup of remaining trouble spots and to oversee existing off-site hazardous waste treatment facilities, but as a prerequisite to starting new waste sites in Connecticut.

The Capitol Region Legislative Caucus, an arm of the Capitol Region Council of Governments, has recommended the creation of an agency that would "plan, site, design, construct, finance, manage and maintain" a hazardous waste treatment facility. The agency was originally proposed by the Greater Hartford Chamber of Commerce.

There is neither an incinerator nor a so-called "secure" landfill site for hazardous wastes in all of New England, even though the region produces considerable waste. And the Connecticut Siting Council, established to license a private facility, has had no applicants.

But as the Legislature prepares to form a new state agency to authorize new facilities, it also should show more convincing evidence that the state is committed to cleaning up the mess that's already here.

TULSA WORLD

Tulsa, Okla., November 3, 1983

THE temptation is to criticize government officials for the delay in starting a program to attack pollution from water-filled mines in northeastern Oklahoma, pronounced by the Environmental Protection Agency to be the worst site of pollution in the U. S.

Since the EPA pronouncement a year ago and a promise of "emergency action," very little has been done, aside from continuing studies of the problem.

It appears that as much as $750,000 has been spent from various state and federal sources to study the Tar Creek area.

The pollution comes from water that has infiltrated the abandoned lead and zinc mines of the area and now is rising to the surface, complete with iron oxide, lead and other minerals and ores.

The Tar Creek name stuck because that stream first started showing the pollution from the mines. Tar Creek runs from the mine fields into the Neosho River and Grand Lake, perhaps the most valuable stream and lake in eastern Oklahoma.

Despite the obvious need for action, it is too early to start criticizing.

After all, this is not a matter that can be quickly fixed. Engineers have calculated the contaminated water contained in the underground shafts and caverns to be about half that in Grand Lake.

Early estimates are that it will take $25 million to successfully cope with the Tar Creek pollution, and committees of technical personnel disagree over the approach and the cost.

At present, it appears it might be another year before any actual work is done on Tar Creek.

State Sen. Bill Schuelein of Miami, who's close to the problem, is anxious that remedial work be started. But he cautions that it is better to go slow in the planning stage to be sure whatever solution is selected is the best one.

He's right, of course. Now is the time to patiently seek the full scope of the problem and the remedy dictated. Once that is done, the time for impatience will have arrived.

THE MILWAUKEE JOURNAL

Milwaukee, Wisc., September 28, 1981

Fifteen farm families in Lafayette County are suing the state, trying to force official action against a mining company that they believe has polluted their drinking water supplies.

* * *

In central Wisconsin's heavily irrigated agricultural area, excessive concentrations of a particular pesticide have shown up in some well water. The State Department of Health and Social Services wants the pesticide's manufacturer to withdraw the product voluntarily.

* * *

The Department of Natural Resources, along with university researchers, is studying whether nine other pesticides may have contaminated groundwater. Meanwhile, the Legislature and the DNR are trying to write better laws and rules for protection of groundwater from mining operations.

* * *

Such problems illustrate vividly the potential for conflict between separate interests that depend on the same natural resource base. There is no doubt that the heavy use of fertilizers and pesticides, along with heavy irrigation, has turned the so-called Central Sands area of the state into a lush cropland, producing unprecedented private and community wealth. Yet, there is also a dark side.

For example, taxpayers in Portage County's Village of Whiting will be paying until the year 2000 to retire bonds that financed a 1964 well system that can no longer be used because nitrates in the water exceed federal guidelines. "In spring, they fertilize like crazy and then irrigate like crazy," said one village official. Another explained: "There's no law against it. . . . What people want to hear is that laws are being created."

Some people, perhaps — but probably not those who profit from the agricultural or mining practices that may jeopardize everybody's groundwater. Such threats come from other sources as well, including animal waste storage areas, sewage treatment systems and solid waste disposal sites (the biggest one in Waukesha County is under orders to close because of pollution of nearby residential water supplies).

The DNR understandably proceeds with caution in such matters. It should try to halt any threat to groundwater, yet it sometimes lacks the specific legal authority that would be needed for effective action. (So the DNR should request such powers from the Legislature).

And when the DNR does have sufficient authority, it often comes under political pressure to go easy, particularly when agribusiness, chemical manufacturers or other commercial interests are concerned. Sometimes the DNR even is resented by the very people it should protect; residents whose water may be contaminated also might depend for their livelihoods on the polluters.

Thus, we think these principles should prevail:

— The primary effort should be to prevent pollution; cleaning it up later may be impossible.

— Nobody should be allowed seriously to degrade the water quality beneath someone else's property, or have an unlimited right to pollute the water beneath one's own property (since groundwater travels).

— Nobody should be permitted to pollute groundwater in ways that imperil lakes and streams.

The Legislature should make sure its laws and the DNR's new rules adhere to those principles, and that the agency has all the legal authority it needs to enforce them.

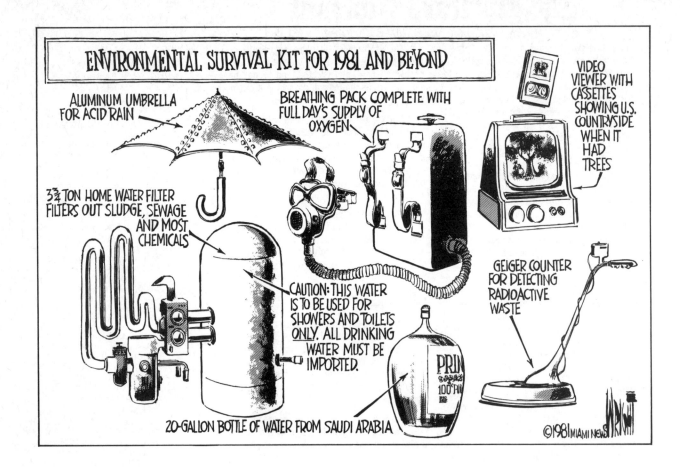

ENVIRONMENTAL SURVIVAL KIT FOR 1981 AND BEYOND

ALUMINUM UMBRELLA FOR ACID RAIN

BREATHING PACK COMPLETE WITH FULL DAY'S SUPPLY OF OXYGEN

VIDEO VIEWER WITH CASSETTES SHOWING U.S. COUNTRYSIDE WHEN IT HAD TREES

3¾ TON HOME WATER FILTER FILTERS OUT SLUDGE, SEWAGE AND MOST CHEMICALS

CAUTION: THIS WATER IS TO BE USED FOR SHOWERS AND TOILETS ONLY. ALL DRINKING WATER MUST BE IMPORTED.

GEIGER COUNTER FOR DETECTING RADIOACTIVE WASTE

20-GALLON BOTTLE OF WATER FROM SAUDI ARABIA

©1981 MIAMI NEWS WRIGHT

Arkansas Gazette.

Little Rock, Ark., December 20, 1982

Half of the nation's water for domestic use — for drinking, bathing and cooking — comes from wells that tap natural underground storage areas known as aquifers. In rural areas, about 90 per cent of the water for domestic use comes from these groundwater supplies.

So far, so good, but the worrisome fact is that the danger that this groundwater will become polluted, and therefore dangerous to use, has grown rapidly in recent years. Just exactly how bad the problem is may be outlined next month when an Environmental Protection Agency study group releases the findings of an extended effort. In the meantime, the EPA confirms that it plans to establish a special groundwater office to co-ordinate ways of dealing with groundwater contamination problems.

The EPA, says Jack Ravan, assistant administrator for water, has found "toxics * * * showing up in drinking water supplies." The concern should be obvious, and there is no assurance that adequate remedial or preventive measures can be expected in each state. That is, contaminated groundwater is principally a problem of national implications that requires national standards if it is to be addressed adequately.

The TENNESSEAN

Nashville, Tenn., January 10, 1983

THE Environmental Protection Agency has released drafts of a final report on groundwater pollution stemming from contaminated pools, pits, lagoons and ponds around the country. The conclusions are disturbing — but so is the fact that the study was finished more than two years ago, while the agency is only now on the verge of making it public.

According to the drafts, between 1978 and 1980, the EPA collaborated with various states and collected data at more than 80,000 sites where more than 180,000 contaminated pools — from cattle ponds to industrial waste lagoons — were located.

An agency spokesman has said that the drafts are "desktop" research resulting from states' having gathered information by on-site visits or aerial photography. However, the EPA is sticking by its assessment that more than 90% of the pools pose some threat to groundwater and thus to community drinking water supplies, since most are situated in permeable soils.

The question, then, is why did the agency wait until now to publicize the draft reports? Bits and pieces of research have been released in congressional testimony, but Rep. Toby Moffett, D-Conn., of the subcommittee on energy and natural resources, complains that a final report could have — and should have — been published back in the summer of 1980.

Considering the gravity of the water pollution threat, the Reagan administration should also endeavor to explain why it has sought to cut back by nearly 50% the agency's budget for groundwater inspection. Congress has appropriated for fiscal 1983 approximately $87 million to fund programs on water supply, water quality and underground injection of waste matter, yet the Office of Management and Budget has told the EPA to apply only $45 million of the total.

The EPA owes the public, as well as Mr. Moffett, an explanation. The agency should stop delaying and release the final report.

The Philadelphia Inquirer
Philadelphia, Pa., March 31, 1984

Believe it or not, Pennsylvania's drinking water is watched over by laws that have not been significantly overhauled since 1905. They worked fairly well back then. But they don't today. The state ranks second only to Colorado in the number of water-borne disease outbreaks.

That's not news to the hundreds of thousands of unfortunate souls in Clearfield County and McKeesport and suburban Wilkes-Barre and, most re-cently, Scranton. For weeks, in the case of Wilkes-Barre since Dec. 23, residents have dealt with contamination so severe they have had to boil drinking water or suffer the consequences.

Chief among those consequences — and nearly 300,000 consumers have been exposed recently — has been a nasty protozoan parasite carried in fecal matter called *Giarda lamblia*. Its symptoms are akin to violent stomach flu. In Wilkes-Barre alone 374 cases were confirmed.

While the people suffered and local economies wrestled with lost productivity — to say nothing of crippled public images — Pennsylvania's regulators were hamstrung. The state had a choice of taking over offending companies or begging them to issue warnings. There was not much in between.

"It was like we had a nuclear bomb, but nothing for the brushfires," said Bruce Dallas, a Department of Environmental Resources spokesman.

After 78 years, a bill that would revise the state's drinking water rules has finally emerged from the Senate. It is Senate Bill 201, the Safe Drinking Water Bill, and after amending it once, the House will vote on it again when it reconvenes next month.

It would bring Pennsylvania's drinking water standards into conformance with federal rules that regulate — as the state's do not — dangerous organics, toxic chemicals and pesticides of recent vintage. It would extend the state's reach to set and enforce standards for thousands of small water suppliers such as trailer parks and campgrounds.

And it would go a long way to moderating situations such as the *Giarda* outbreaks. For one thing, it would require emergency plans before permits were issued. In Wilkes-Barre, for instance, the water company was not even certain which pipelines carried water from which reservoirs.

The state's public notification authority also would be clarified. And so would its authority to order companies to come up with alternative supplies.

In the case of older open reservoirs in rural areas — those susceptible to *Giarda* contamination from untreated sewage or animal waste — the state's authority to order upgraded filtration techniques would be made more straightforward. DER has been operating under improvised legal strategies when the issue has been forced.

By finally adopting drinking water policy conforming with federal standards — the last Eastern state to come into line — Pennsylvania also would see up to $1 million more in federal funds for enforcement and inspection.

Federal drinking water laws have been relatively tough, but have had small staffs to enforce them. The state has had the opposite problem — a staff, but creaky, antiquated laws. The Safe Drinking Water Bill would beef up the staff and put teeth in the state law.

DER can hardly claim universally high marks for aggressiveness. New teeth do not a bulldog make. But with a $220 million low-interest loan program to offer to communities that need to upgrade water systems and the prospect of a modernized drinking water law, the agency cannot help but be better equipped to face — to prevent — future emergencies.

That's belated good news for the thousands of Pennsylvanians still boiling their water. And for the millions who aren't? Perhaps it will spare them ever having to tangle with invisible menaces such as *Giarda lamblia*.

THE ARIZONA REPUBLIC
Phoenix, Ariz., September 16, 1983

AFTER two years of study, the Academy of Natural Sciences has found that no one really knows to what extent ground water in the United States is contaminated.

The study also reveals how states approach the problem in varying ways and, thus, that there is no national strategy or coordination to deal with the wide range of toxic substances science identifies as ground-water pollutants.

Experts involved in the analysis — the most extensive ever attempted — found that ground-water contamination is increasing. Ground water provides about half the nation's drinking water, but it also is used for irrigation and industrial purposes.

The study singled out septic tanks as a major problem because their use is so widespread — about 30 percent of the U.S. population depends on them.

Nitrate from septic tanks, fertilizers, feedlot wastes, chloride, salt water intrusions, various industrial wastes and even household cleaning fluids were cited as top sources of contamination.

Less than 2 percent of U.S. ground water has been categorized as contaminated, but some scientists believe the actual total is much higher.

Federal laws affecting ground-water contamination are a hodgepodge that address individual parts of a large problem. The nation needs a clearer sense of how to deal with ground-water contamination.

At the very minimum, there should be a far more coordinated program of exchanging scientific findings so states can attack the problem.

A survey that is comprehensive seems to be a necessary first step.

Surely citizens need to know to what extent their sources of drinking water have been polluted so that proper safeguards can be developed to outlaw dumping and to protect public health.

Phoenix, Ariz., February 21, 1984

ENVIRONMENTALISTS and political leaders have been arguing about the extent and solutions to the country's ground-water contamination for nearly four years without agreement on an effective national policy.

The Reagan administration plans to announce its strategy to deal with such problems in May. Preliminary recommendations are already public. They suggest creating three classes of ground water, from the most serious to lesser problems, each requiring different protection.

The new plan sets priorities — from writing off very poisoned aquifers to cleaning up those that can be saved through Superfund grants. The Environmental Protection Agency also would do more research on general ground-water issues and look at local problems more closely.

Most environmentalists call the administration plan cosmetic. They say hard, clear solutions are needed now as a result of unchallenged pollution.

Yet, many members of Congress claim not enough is known about the effects of poisonous substances on ground water, particularly in the highly controversial area of industrial injection of hazardous wastes deep underground.

The White House Council on Environmental Quality estimated three years ago that toxic chemicals had seriously polluted ground water in 34 states. EPA estimated later that only 1 percent of the country's ground water is contaminated. However, the real question is how much potable water is polluted.

Pollution is difficult to monitor because it moves slowly. And much of it involves natural runoff from farms and cities.

However, thousands of wells in many states have been shut down because suspected cancer-causing agents were found in drinking water.

EDB, a soil fumigant suspected of causing cancer, has polluted 300 wells in Florida alone, some near cities. Unsafe levels of EDB also have been found in grain-based food products. The EPA has suspended use of EDB and may ban it.

These are a mere handful of many scary cases across the nation.

Americans are getting mixed signals from Congress, the administration, various experts and environmentalists on the real extent of the dangers involved. The public is perplexed, and some live in fear of cancer-causing agents in their drinking water.

This will continue because experts disagree on what constitutes a reasonable cause-and-effect relationship between contaminated ground water and toxic substances in the area.

Such identification will be a long, arduous process with thousands of site-specific studies.

However, there is no reason why Congress, the administration and others cannot make a start on the problem.

It certainly is not reassuring that they have not been able to do so. The American people may eventually pay a very high price for the delay.

Reasonable cause-and-effect relationships can be found by reasonable people genuinely seeking reasonable solutions.

More indefinite delays would be intolerable.

The Times-Picayune
The States-Item

New Orleans, La., February 16, 1984

Under pressure from two private organizations, the U.S. Environmental Protection Agency has finally agreed to a plan for controlling the injection of toxic wastes into underground water supplies.

The EPA has estimated that more than 10 billion gallons of wastes are injected into underground water supplies each year through special wells used by industries. The contaminants could pose a long-term major threat to aquifers that are the principal or sole source of water for many communities.

Some 88 billion gallons of water are pumped from the ground every day in the United States. The largest single use is for crop irrigation, but half of the drinking water in the United States comes from underground sources.

In an out-of-court agreement with the National Wildlife Federation and the Colorado Wildlife Federation, the EPA will require all states to have underground water control programs in place by next Oct. 29. The EPA will impose its own regulations for controlling well injection of toxic wastes on states that do not meet EPA standards.

The EPA program is required by the federal Safe Drinking Water Act. When the act's deadline for beginning state programs passed last year, only eight states, including Louisiana, had federally approved plans for controlling underground disposal of toxic wastes. The two conservation organizations subsequently filed suit against the EPA, charging it with failing to meet its responsibility under the terms of the act.

Approved plans call for governments or industries to obtain special permits to use wells for waste disposal. States can come up with plans stricter than EPA standards if they so choose.

Underground waste disposal is particularly hazardous because of the potential danger to water supplies for generations. Contaminants can migrate for miles through subterranean water tables.

Establishment and enforcement of basic national standards to protect underground water supplies is long overdue. The standards should be reviewed periodically and revised, if necessary, as experience dictates.

The Miami Herald
Miami, Fla., May 1, 1984

CAUTION: Florida's drinking water may be hazardous to your health — and don't blame germs. Medical science conquered typhoid, dysentery, and other water-borne diseases decades ago.

This time the danger is chemicals. Florida's groundwater has been contaminated by agricultural pesticides, industrial wastes, and other byproducts of our high-tech civilization. The result: More and more wells have been found to contain a witch's brew of chemicals.

Some of these chemicals are known to be harmful to human health. Others are suspect. Researchers hardly can keep up with the substances turning up in Florida's water supply.

The cost of neglect is high, as Alachua County Rep. Sid Martin pointed out the other day in Washington during testimony before the House Subcommittee on Environment, Energy, and Natural Resources.

Representative Martin cited two situations. First, the Florida Department of Transportation dumped wastes into a pit near the small town of Fairbanks for more than 20 years. The result: The wells became contaminated, the water unsafe to drink. A second situation: Ethylene dibromide (EDB) was used on citrus, peanuts, soybeans, and golf courses to control nematodes. The result has been the contamination of 618 of the 4,100 Florida wells tested for EDB.

So Mr. Martin was not exaggerating when he told Congress that preventing such occurrences is essential to the nation's physical and fiscal health. As he pointed out in his testimony, the Fairbanks cleanup will cost the state of Florida more than $3.4 million, and the EDB cleanup's costs will exceed $10.5 million.

In Mr. Martin's view as chairman of the House Natural Resources Committee, "The present policy of simply reacting to groundwater contamination will bankrupt both the state and Federal governments."

Mr. Martin and the Florida Department of Environmental Resources (DER) want to do more than merely react. They want Florida's groundwater tested for chemicals known or suspected to be hazardous so that the spread of contaminants can be prevented or halted sooner. His committee has proposed a farsighted package of legislation to test and cleanse Florida's waters.

Incredibly, the legislation has encountered opposition from penny-wise, pound-foolish lawmakers such as Sen. Tom McPherson of Broward County. Senator McPherson looked at the projected costs of safeguarding Florida's drinking-water supply and issued a totally off-the-wall forecast. Said he, "You're going to see water bills getting up there with electric bills."

Send Senator McPherson to remedial math. The highest cost projected thus far for testing and protecting Florida's drinking water amounts to less than $3 a year per Floridian. Many families spend more than that each week for water filters and bottled water. Surely it's a bargain compared to the cost of treating cancer, birth defects, and other conditions attributed to hazardous chemicals.

Lawmakers should ignore nay-sayers such as Senator McPherson and enact the House package of water-protection legislation in its pure, uncompromised form. Doing so would affirm everyone's basic right to safe drinking water.

Sunday Journal and Star
Lincoln, Neb., April 15, 1984

Every so often the international news report contains a story of someone in Europe digging up a bomb or an artillery shell dispatched with lethal intent during either of the World Wars. Occasionally, the story may tell of a delayed explosion, with fatal or crippling consequences.

Those brief dispatches usually also bring forth a fresh recall of William Shakespeare's insight as mouthed by Mark Antony over the perforated corpse of Julius Caeser: *The evil that men do lives after them. The good is oft interred with their bones.*

An area west of Grand Island has been having something akin to that experience recently. Traces of the explosive chemical RDX used years ago at the Cornhusker Army Ammunition Plant were found earlier this year in 246 private water wells east of the old plant. Some of those wells are as much as three miles from the plant, indicating considerable eastward groundwater migration. At two wells within a mile of the plant, TNT and DNT, other ordnance wastes, were detected.

More recently, Nebraska Health Department personnel discovered traces of trichloroethane in three wells already known to be RDX contaminated. Whether that chemical, whose cancer-causing properties are in dispute, came from the ammunition manufacture is uncertain.

In any event, the U.S. Army has been supplying families with bottled water for consumptive purposes since February. An extension of the Grand Island municipal water system may be the most likely permanent solution to the problem, although that could entail annexation, already a very hot subject around the Hall County seat.

The whole subject is raised this morning simply to polish up the point that in an industrial society such as ours, we frequently have not reckoned the through-time consequences of dumping chemicals and other materials in the air, earth and waters. Taking responsibility for polluting was not an idea which emerged coincident with the industrial revolution. The preference, even when someone reflected deeply about it, was to try and shunt the problem on others, either in time or space.

For the human race, however, space has finite dimensions. Perhaps the same thing can be said of time; certainly for individuals.

Ammunition contaminates in the Grand Island wells, like nitrate concentrations occuring artificially elsewhere, are signals we can run from, but finally we can't hide from irresponsibility.

Newsday

Long Island, N.Y., January 20, 1984

Since 1977, the federal Environmental Protection Agency has considered the Long Island aquifer important enough to qualify for the agency's highest level of protection. Unfortunately, even that isn't high enough.

Under federal law, the best the EPA can do is block any federally funded project that threatens to contaminate the underground water supply. There's no money available to buy crucial aquifer areas.

But there should be. Federal involvement in water projects has ample precedent. The river systems of the West, for example, have been extensively dammed to prevent flooding, provide hydroelectric power and store water for drinking and irrigation.

This year, the Reagan administration asked for $1 billion for western water projects. But efforts to expand the federal role in safeguarding underground water supplies have not fared well. An amendment to the Safe Drinking Water Act, first proposed in 1982, would have allocated federal money for the purchase of land in prime water-supply areas. This is particularly important to Long Island, where development is threatening the last major unpolluted aquifer, beneath the Pine Barrens of Suffolk County. The amendment emerged from committee in the Senate last year in the form of a $15-million suggested allocation.

But the measure still seems stalled in the House. Two Long Island congressmen — Reps. Thomas Downey (D-Amityville) and William Carney (R-Hauppauge) — are sponsoring different versions. The Downey amendment is broader in that it would permit federal court actions to be brought against polluters, but either version is acceptable.

According to testimony given when the Senate subcommittee on toxic substances and environmental oversight conducted hearings on the amendment, about half of the nation's population depends on underground sources for its drinking water. So this amendment should be able to generate broad support. Not only Long Island but the whole country needs it.

Portland Press Herald

Portland, Ore., March 28, 1984

As taxpayers well know by now, the cost of cleaning up the environment after years of heedless polluting can be enormous. Maine alone has invested millions upon millions.

The best way to avoid high pollution control costs is to take preventive action before a serious problem develops.

Right now, leaky oil tanks, particularly those buried in the ground, threaten to become a significant environmental hazard in Maine. The time to do something about it is now, before the problem gets completely out of hand.

It's already late in the game. According to a study conducted by the Department of Environmental Protection, as much as 11 million gallons of oil are leaking from underground fuel tanks. Reports of water supplies tainted by fuel leaks have increased sharply.

The Legislature is considering a bill to authorize the DEP to conduct an inventory of both aboveground and underground oil storage facilities, issue permits for new installations and set up a monitoring process to detect leaks early when they occur.

Currently there are few controls over the installation and maintenance of either commercial or private fuel tanks. Many of them are old and develop leaks either in the tanks themselves or in pipes connecting them. The DEP estimates that fully a third of oil tanks at gasoline stations and commercial buildings exceed the 15-year average life of underground storage tanks.

If the Legislature is interested in avoiding a potentially costly environmental problem in the future, the time to deal with it effectively is today.

The Wichita
Eagle-Beacon

Wichita, Kans., January 16, 1984

Part of the appeal of "high-tech" industries — especially microchip manufacturing — is that they are perceived as "clean" enterprises. And that's a generally accurate perception: industry minus the billowing smokestacks and massive factory buildings.

But, according to the publication New Scientist, the famed Silicon Valley in California is discovering that producing microelectronic components is not without environmental dangers. Acids are used in etching the miniaturized circuits, and solvents are used to cleanse the finished product. Wastes from those processes naturally accumulate and have been stored in underground tanks.

As all too often is the case, some of those tanks have leaked and allowed potentially harmful substances to seep into nearby groundwater. Although the leakage hasn't yet contaminated large amounts of drinking water, which usually is drawn from deeper strata, at least 10 wells in the San Jose area reportedly have been closed as a precautionary measure. There's valid concern the wastes eventually could percolate into needed water supplies.

New regulations have been drawn to address the dilemma, and it must be hoped they'll be effective in reining in what could become a serious problem. As unfortunate as this unforeseen development is, it should help in the long run by alerting microchip manufacturers to this danger.

The Courier-Journal

Louisville, Ky., February 3, 1984

FOR MUCH of the two years preceding the current Kentucky General Assembly session, state environmental officials and other water experts have wrestled with two tough pollution problems. One is how to protect invaluable ground water supplies; the other how to protect streams from brine generated by oil and gas wells. Their efforts were reward with a big fat zero in the executive budget presented by Governor Collins last week.

It's inevitable, no doubt, that many worthy things will be slighted when the state's revenue prospects are as dismal as they are this year. But a legislative budget hearing scheduled today should conclude that this omission is penny-wise and pound-foolish.

The state Division of Water needed everything it asked for and didn't get in the current budget, including $5.5 million a year to help fund local sewer projects. That and other desirable things may have to wait. But the $470,000 it sought to study Kentucky's underground water, to develop a state water management plan and to establish a registration program for water drillers is essential if water resources are to be mapped and protected.

One possible savings

The Natural Resources cabinet also was denied about $600,000 it sought to enforce new rules aimed at preventing water pollution by brine from oil wells. Perhaps this cost could be lessened by transferring the state Division of Oil and Gas from the Department of Mines and Minerals to the Natural Resources Department, where it logically belongs anyway. The oil and gas agency has personnel who could combine environmental protection functions with their present duties.

It simply makes no sense to have regulations controlling oil well pollution and hardly any means of enforcing them. It's particularly important along the Kentucky and Licking rivers, from which many Central Kentucky cities get their drinking water. Oil well brine produces chemical reactions that create cancer-causing chemicals in the water. The problem is persistent — and increases during drought periods, such as last summer.

This problem, however, is at least monitored by water companies. The 600,000 Kentuckians who use private water sources, mostly wells, lack even this protection. It's time to adjust the budget a bit and give them some.

Detroit Free Press
Detroit, Mich., April 23, 1984

MICHIGAN faces a mammoth groundwater contamination problem stemming from improper disposal of hazardous wastes. This state is one of the few that has actually catalogued the extent of its groundwater contamination: So far, 1,000 trouble sites have been identified, and about 100 more are added to the list every year. But identifying the problem, as Free Press reporters David Everett and Bob Campbell pointed out in a recent series of stories on groundwater contamination, is merely the beginning.

The contamination incidents are not confined to isolated private wells. Cities such as Battle Creek and Charlevoix, which rely on underground aquifers for their water supplies, are now suffering from tainted water supplies. The sources of contamination are widespread, and it is not always a mysterious, sinister chemical bubbling up out of the ground that causes the problems; the No. 1 source of groundwater problems in Michigan is gasoline seeping from leaky underground storage tanks.

The governor's Cabinet Council for Environmental Protection is expected to release a comprehensive plan next month for dealing with the problem. Money and understaffing remain major obstacles to the cleanup. The governor has proposed $21 million over the next two years to begin a cleanup of more than 100 sites and to provide alternate water supplies to those people whose water is undrinkable. Given the state's economic situation, that's not a bad beginning. But, as Mr. Everett wrote, it is just "a teardrop in a barrel" compared with what needs to be done.

The governor admits that to clean up the sites we already know about will take at least $3 billion over the next 20 years. If the federal Superfund program is reauthorized at no less than current funding levels, and if the state can identify the companies responsible for the contamination and wring some financial settlement from them, Michigan may have to come up with only half that $3 billion cost. Nonetheless, the governor's plan will have to include some provision for long-term funding, whether that means a cigaret tax, or groundwater user fees, or some other source of revenue.

The frustration in dealing with groundwater contamination, as in most cases of toxic waste management, is that not all the sources or sites of contamination are immediately apparent. There is pitifully little known about the effects of many contaminants on human health. Even when clusters of unusual ailments or diseases occur near a contaminated area, it is difficult, scientifically and epidemiologically, to link a specific malady to a particular instance of human exposure. We all live, breathe, drink and consume a soup of chemicals whose effects, after 20 or 30 years of exposure, or after dilution in the groundwater to parts per million or billion, are not fully known.

Of necessity, cleanup and management of the problem has to be spread over many years. The governor's funding proposals are a start. But the key is how soon and how willingly the Legislature will act. When the governor's comprehensive plan emerges, he and the Legislature will have their work cut out for them: to develop a strategy for cleaning up, for funding the cleanup and for convincing the public that the benefits of clean drinking water and toxic waste management are well worth the high cost.

THE SACRAMENTO BEE
Sacramento, Calif., May 2, 1984

Assemblyman Richard Katz's bill to prevent toxic contamination of the state's water supplies seems to have been born of sheer frustration, which would certainly be understandable. After years of lawmaking, California's hazardous-waste monitoring and licensing programs are still inconsistently enforced, when they're enforced at all. The state has yet to locate all its toxic waste ponds or impose safety measures on all those it has located. Its latest regulatory proposal (not yet even in effect) would merely require that operators of existing toxic waste ponds monitor nearby groundwater — and thus discover contamination after it has already occurred.

Although lately things seem to be getting a little better, they're still far from good enough. So Katz is proposing to replace the existing laws that have merely *allowed* state officials to enact strict water contamination regulations — and which these officials have so little used — with a new law that *requires* decisive, preventive action.

Under the Katz bill, no toxic waste ponds would be allowed within one-half mile of a potential drinking water source. (About half now are, even though so far it's been technically impossible to design a pond that is fully leakproof.) All other ponds would have to be monitored with state-of-the-art methods that permit detection of leaks before they reach underground water supplies. State officials would be required to inspect them annually and to shut down disposal facilities that don't have adequate monitoring plans or, in the case of leaking ponds, lack double linings and leak-detection equipment.

By thus eliminating regulatory discretion, Katz hopes to insure some regulatory action. And in that respect his bill is, indeed, radical. But it is, at the same time, appropriately limited in scope. It doesn't deal with toxic spill cleanups, with toxic chemical products or with many forms of toxic waste. It seeks only to prevent those liquid wastes now stored in hole-in-the-ground ponds, pits and lagoons — a form of toxic waste disposal that has proved exceedingly dangerous — from contaminating public water supplies. It is a goal that justifies strong measures.

The cost of the precautions Katz would require is high, but according to the Congressional Office of Technology Assessment, it would cost ten to a hundred times more to clean up contaminated soil and water after leaking toxic pits and ponds have done their damage. Indeed, the risks of such disposal facilities are so great that several chemical companies have lately decided to switch entirely to other methods of waste storage and disposal — to containerization, incineration or recycling of toxic wastes.

The chemical industry says Katz's bill is too rash. And in some respects, this is true. The bill needs to be amended to cut the unnecessary costs it would now impose. It should be rewritten to allow more reasonable timetables for the development of the required safety measures, to involve less red tape and to insure that the administrative costs of dealing with unsafe ponds are not passed on to the operators of safe ponds.

But with those changes, the Assembly Ways and Means Committee today and the full Assembly next week should have no qualms about passing the measure. It is suited to the drastic nature of the problem and the lack of effective regulation to date.

For reassurance that the Katz bill is the way to go, the Assembly need look no further than the state Senate, where the only alternative approach is sponsored by the chemical industry and carried by Sen. Ralph Dills. Under the Dills bill, each toxic pond operator would be responsible for adopting a self-monitoring plan, which only the industry — and not any government agency — would have to approve, which would not even be accessible to government officials except by specific request, and which could be kept from the public by calling its key aspects "trade secrets" (a designation to be determined by the operator himself).

The state, under the Dills bill, would not be required to shut down any ponds in any circumstances. The bill's language suggests California officials might even be prevented from enforcing any but the least stringent safety standards currently set by state or federal regulation. And the operator would not be required to take any of the more effective preventive actions the Katz bill mandates.

Between these two approaches — the Dills bill enshrining in law the current lack of regulatory oversight, and the Katz bill requiring by law that oversight be tough — the choice should hardly be difficult.

The Des Moines Register

Des Moines, Iowa, April 3, 1984

It took more than five years for the liquid wastes from the Diamond Vogel Paint Co. plant in Orange City to seep through the soils of a dump site in Sioux County. But when they finally reached an ancient riverbed covered over by glacial action, the underground waterway quickly spread the hydrocarbons to test wells at the site.

That's one theory as to why high readings of xylene and tolulene suddenly showed up in samples that Diamond Vogel sent to the Iowa Department of Water, Air and Waste Management.

Whatever the cause, the DWAWM has told Diamond Vogel to pump the chemicals out of the ground and dispose of them in some other fashion. The hydrocarbons can be recycled and used as solvents, or could be burned under controlled conditions.

The flammability of xylene and tolulene are part of the danger they pose, said Pete Hamlin, director of field services for DWAWM. The liquids just could seep into basements and explode.

The paint firm dumped the liquids for nine years, and the dump site has been monitored through seven water wells. Samples that Diamond Vogel took from two of those wells showed "incredible" concentrations of the chemicals, Hamlin said.

Potentially dangerous as the Sioux County contamination is, there is a bright side: This hazardous-waste escape was monitored with the cooperation of Diamond Vogel and can be corrected. What worries the DWAWM — and should worry all Iowans — are the unknown dumps, the illegal disposals of dangerous chemicals that may not be detected until it is too late.

The U.S. Environmental Protection Agency lists 156 uncontrolled hazardous-waste disposal sites in Iowa; the DWAWM puts the figure at more than 200.

The EPA estimates that more than 250 million tons of hazardous wastes are produced in the United States each year by more than 14,000 firms. Disposal of much of it is unregulated, and most of it goes into landfills, which engineers contend cannot be built to contain the wastes for as long a time as they threaten health.

The U.S. Office of Technology Assessment contends that cleaning up a leaky landfill and compensating victims of its pollution could cost 10 to 100 times more than the cost of safe disposal by other means in the first place. But there are not yet tough enough laws to discourage landfill disposal — and may not be, until safe drinking water becomes scarce.

St. Petersburg Times

St. Petersburg, Fla., April 16, 1984

About half the nation draws its drinking water from underground supplies. In Florida, subsurface sources provide 90 percent of the public water supply. More than 9-million Floridians have depended upon the state and federal governments to keep the water safe to drink.

Don't count on the feds to protect the public health by preventing hazardous wastes from contaminating the nation's precious water supplies. If Florida is to have safe drinking water, it will have to guard its own water resources a lot better than has been done in the past — including more stringent regulation of agricultural pesticides.

A report prepared for Congress includes some alarming assessments of the threat to the nation's water supplies. The study found that the Environmental Protection Agency's rules for monitoring and controlling hazardous wastes are inadequate to protect underground supplies. It says that many or most toxic waste disposal sites will have to be cleaned up in the future — at a far higher cost than if the ground water had been protected from the initial pollution.

INADEQUATE federal regulations, coupled with insufficiently developed techniques for preventing water pollution, "are likely to cause serious problems for future generations," said the researchers.

There is no nationwide system for monitoring underground water pollution — a reckless gamble, considering the warnings by experts that once ground water becomes contaminated, it is difficult if not impossible to reverse the process.

But the federal regulations on hazardous waste disposal are riddled with dangerous loopholes. Current EPA rules are not based on the effects of toxic wastes on human health and may not be sufficient to protect humans from contaminated water, the congressional report said. The rules were designed to keep testing costs low, which increases the risk that water contamination will go undetected.

WITH THAT KIND of federal negligence, Florida needs to continue the process started last year to protect ground water resources. A statewide monitoring system is being developed, and the Environmental Regulatory Commission has adopted a tough new rule that sets safe levels for eight dangerous chemicals, including ethylene dibromide, EDB, and requires that water be monitored for traces of 119 other toxins.

What Florida needs now is a permanent revenue source to enforce the water protection program. The money should come from new taxes on chemicals sold in Florida and on the hazardous wastes generated by industry.

That not only would make those who cause the problem pay for the public's protection, but would be an incentive for industry to reduce hazardous wastes.

St. Petersburg, Fla., April 28, 1984

No one really wants to guess how much it will cost to make Florida's poisoned water safe to drink again. State officials already have discovered hundreds of water wells contaminated by gasoline, pesticides and other toxic chemicals. Some 50,000 Floridians already have been warned not to drink or cook with the water that comes out of their kitchen taps. And health officials expect to find many more contaminated wells as they continue testing the hundreds of thousands of wells in the state.

The question now is not whether some of Florida's water is unsafe to drink, but how much it will cost to cleanse it and who will pay the bill.

AN INDICATION of how big the cleanup tab could be was provided this week by the State Department of Transportation (DOT). The water supply of the entire town of Fairbanks in Alachua County was contaminated by toxic chemicals that DOT buried in a nearby borrow pit. As part of the settlement for poisoning their water, the agency agreed to give the 238 households in Fairbanks $1,500 each to pay their water costs for the next decade. DOT also will spend $1.2-million to lay water pipes that will tie Fairbanks into Gainesville's unpolluted water supply. Total price tag for the clean-up operation: more than $3.5-million.

Another indication of how much it will cost to provide Floridians with safe drinking water comes from Lakes Wales. One of its municipal water wells was tainted with the cancer-causing pesticide ethylene dibromide (EDB). The State Department of Agriculture injected massive doses of EDB into the soil around citrus groves in Lake Wales to kill microscopic worms that destroy tree roots. The pesticide also seeped into the city's water supply. A House committee last month acknowledged state liability for the contamination and approved a $1.2-million payment for drilling a new well.

The tab for cleaning up the tainted water in these two small towns is almost $5-million. No wonder state officials shudder at the thought of estimating how much it will cost to clean up hundreds, perhaps thousands, of wells in Florida's other cities and towns.

Since state agencies were responsible for polluting the water in Fairbanks and Lake Wales, the state also should bear the responsibility for cleaning up their water. Government agencies, of course, are not the only contamination culprits. Many of Florida's tainted wells were poisoned by toxic chemicals carelessly dumped by private companies and gasoline leaking from rusting underground tanks. It still is not clear who is going to pay to clean up those wells.

State Sen. Karen Thurmond, D-Dunnellon, has one suggestion. Under a bill she is sponsoring, Floridians whose drinking water is tainted by pesticides or gasoline would be eligible for low-interest state loans. The bill would allow the Department of Environmental Regulation (DER) to make loans of up to $5,000 at 6 percent interest to pay for drilling new wells in uncontaminated areas, patching into clean wells or providing charcoal filtration of contaminated drinking water supplies.

THURMOND'S PROPOSAL is better than nothing. But it is not a fair solution. Homeowners should not have to borrow money, no matter how favorable the interest rate, to clean up wells they did not contaminate. The clean-up costs should be borne by the industries and agencies that created the problem.

Last year, the Florida Legislature rejected a proposal to tax chemicals sold in the state to pay for more stringent pesticide regulation. That was a mistake that should be corrected this year. And some of the money raised from a tax on chemicals and hazardous wastes generated by Florida companies should be set aside to compensate Floridians for the cost of cleansing poisoned water.

Until last summer when state officials first began notifying thousands of people that their wells were contaminated, Floridians took the purity of their drinking water for granted. Now, all Floridians live with the fear that they will get the same bad news. At the very least, the Legislature should assure the people that they will not be stuck with a large clean-up bill for a problem which they did not create and which could jeopardize their health.

Sewage:
Threatening Our Water Supply

The largest volume of waste water discharged directly into ground water—an estimated 1 trillion gallons per year—comes from septic tanks and cesspools. In rural areas, these discharges often contaminate private wells with fecal and toxic material. In towns, where domestic wastes are treated by municipal sewage plants, the contamination of drinking water supplies is often due to inadequate purification of the waste water, crossing of water and sewage lines, or a "backflow" of polluted water into the distribution system. The problem in rural areas is sometimes exacerbated by homeowners' use of septic tank cleaners, which not only release sludge into ground water but also spread chlorinated organic solvents, many of them carcinogenic, that are the active ingredients in the cleaners. Most municipal sewage plants treat waste water by filtering it through beds of sand, trapping particles of waste. This is ineffective for removing toxic organic compounds such as those contained in pesticides. A more effective method called activated carbon purification, in which many of the organic compounds adhere to the surface of a bed of adsorbent carbon granules, is more widely used for food and beverages than for waste water treatment. One source of organic compounds in drinking water is the decaying plant matter in lakes and reservoirs.

In 1972, with the passage of the Federal Water Pollution Control Act, a far-reaching program to build thousands of sewage treatment plants was established. The Clean Water Act, in 1977, amended that legislation and authorized $5 billion in federal grant money each year through fiscal 1982, to be used by states for sewer construction and other waste water treatment. This federal program was attacked from all sides as staggeringly expensive, unevenly enforced, ineffective and ill-equipped to suit the varied needs of different areas of the country. In 1981, President Reagan refused to request the sewer grant funds unless Congress passed a proposed revision of the Clean Water Act, reducing the annual grants to $2.4 billion. Reagan's plan would shift most of the allocations from the rural areas of the Sun Belt to cities, particularly in the Northeast, and would require that the funds be spent primarily in areas where there was an immediate health hazard associated with waste water. This would return the responsibility for funding new treatment methods, and most other sewer repair or construction costs, to the states. Congress passed the $2.4 billion annual authorization as a budget-cutting measure in 1981, but the revision of the Clean Water Act is still being negotiated. (See pp. 169–175.) In June, 1984, the House passed a version of the act that would increase federal aid by 38% for public sewer systems. It would also require the Environmental Protection Agency to develop regulation to control disposal of sewage sludge. In addition, New York City would be required to phase out by March, 1986 its current practice of dumping untreated sewage into the Hudson River. The version of the bill pending in the Senate, much preferred by the Reagan Administration, would provide considerably less in federal funds.

THE SAGINAW NEWS
Saginaw, Mich., July 7, 1981

Ronald Reagan of California was not a likely ally of the older, settled — and troubled — cities of the Midwest and Northeast.

But he has come up with a proposal that helps America's established urban areas, makes environmental and planning sense, and follows his frugal financial principles as well.

The president's changes in one of the most un-sexy federal programs of all, grants to help build sewage treatment plants, are expected to spark bitter congressional debate once again between the so-called Frostbelt and Sunbelt regions.

Under his plan, aid for treatment facilities would be sharply cut. The emphasis would shift from high-growth areas of the South and West to the older cities, on grounds that plants there would serve the most people per federal dollar.

Since the program dispenses about $5 billion a year — some $33 billion since its beginning under the Clean Air Act of 1972 — a great deal is immediately at stake. The impact would be even more dramatic as we approach the turn of the century. New plants would cost another $120 billion by then, with the federal share pegged at $90 billion. Reagan says the government just can't afford that.

But the aspect of good planning is just as important as the taxpayer cash involved.

Critics of the sewer-grant program complain that too often, the money has subsidized facilities in sparsely populated areas. More than that, it has encouraged residential sprawl where good public policy should discourage growth instead.

"All of us concerned with the environment have deplored the fact that farm land has disappeared, and urban sprawl has taken place," said Sen. John H. Chafee, R-Rhode Island, chairman of the committee working on the bill.

Chafee might have added that failure to use public dollars wisely has resulted in many more dollars being wasted on other new services demanded by residential growth. For instance, much of the Sunbelt is seriously, and expensively, short of water.

In contrast, the Northeast and Midwest, whatever their drawbacks, have well-established public services and, at least in the Great Lakes region, plenty of fresh water.

It only makes sense to concentrate improvements in other services in those areas. Reagan's plan would do just that.

It may be the first federal-policy tilt in years toward the struggling Frostbelt.

It also suggests that the president has better sense than the commission which earlier this year advised the government to waste its resources on developing the warm but empty parts of America, and virtually abandon the settled areas and their enormous public and private investment.

If there is to be a battle over this in Congress, it is one worth winning. Michigan's congressmen and senators should enlist under the president's flag.

The Providence Journal

Providence, R.I., September 21, 1981

Clean-water advocates, alarmed by the prospect of sharply curtailed federal funds for new sewer construction, may consider an alternative being tried in one Georgia county. There, sewage soon will be sprayed on trees.

It's not as weird as it sounds. Clayton County, south of Atlanta, was facing growth restrictions unless its water treatment system was expanded to meet federal standards. To handle the increased sewage flow from new subdivisions, the county tried an experiment: It would buy a 3,500-acre pine forest, lace it with sprinklers, hook them up to its sewage-treatment plant and spray the semi-treated, odorless sewage onto the trees.

Why? one may ask.

Forest experts who suggested the treatment technique say it: gets rid of sewage safely and inoffensively; helps raise depleted water tables; produces purified and drinkable runoff water; and, by fertilizing the trees, speeds their growth by as much as 20 percent. What's more, say the sewage-on-the-trees folks, the whole system (including the sprinkler installation and purchase of the forest land) costs no more to build than a typical sewage treatment plant — and can be operated for only about one-sixth the cost.

How well this technique works won't become clear until November, when the Clayton County sprinklers are started up. But alternative treatment techniques already are being used in hundreds of communities elsewhere in the country. A small town in Maine sprays the stuff on a stand of Christmas trees, a Michigan town puts it on cornfields, and a Texas town treats cotton crops with sewage.

These novel approaches grow more interesting as federal funding for new sewage systems shrinks. President Reagan, citing an existing backlog of $90 billion in sewer-construction projects, wants to trim federal aid for new projects. Congress appears likely to balk at this. Even so, the nation's mounting water-pollution problems are going to demand continuing attention. Since many conventional sewage treatment plants across the country still cannot meet federal environmental standards, new disposal methods are worth the serious attention of water-pollution officials everywhere.

The Boston Globe

Boston, Mass., October 15, 1983

Four of the Boston Harbor islands – Deer, Moon, Nut and Spectacle – have been ruined for our times, if not forever, because it was once convenient to turn them into places to dump trash or treat sewage. In a wrong-headed move, the US Environmental Protection Agency now suggests a similar fate might also be appropriate for Long Island.

Although the EPA has begun to back off a bit, its request that Long Island be included in a pending environmental impact study as a possible site for a new sewage treatment plant illustrates the environmental and political problems raised by the recently aroused concern over Boston Harbor pollution.

There is a range of sensible solutions under consideration by the harbor task force headed by former Gov. Sargent and by the parties to the suit filed by Quincy against the Metropolitan District Commission. The EPA, however, is pushing for a secondary treatment plant that would provide a higher level of treatment than now provided by the Deer and Nut Island plants.

There is considerable doubt that secondary treatment is needed. Given the nature of Boston Harbor, it is probable that acceptable water quality can be achieved by repairing the existing treatment plants, closing the combined storm and sanitary sewer outlets near beaches and shellfish flats, and by sharply reducing the "infiltration and inflow" of rain and groundwater which must then be "treated" as if it were sewage.

Officially, the EPA has refused to waive secondary treatment for Boston on the grounds that the MDC's initial request lacked sufficient scientific evidence. The Sargent committee has requested the MDC to submit a better-researched request.

If the EPA continues to insist on secondary treatment, then there is the political question of where to put the plant. There are active constituencies in Winthrop and Quincy that argue that their communities, already stuck with the existing treatment plants, have suffered enough. There are equally vocal constituencies in the farther suburbs who have been able to block serious consideration of satellite treatment plants that would treat sewage closer to its source.

Long Island thus sounds good to planners because there is nobody out there to complain, except for a declining number of hospital patients and witnesses being held for federal trials.

In terms of the future development of Boston Harbor, however, Long Island, with its commanding position at the head of President Roads, makes far more sense as an entry point for the Harbor Islands Park, a short ferry run from the downtown waterfront.

Fifty and 100 years ago, the harbor islands were the dumping grounds for unwanted people and projects – paraplegic children on Bumkin Island, Italian prisoners of war on Peddocks Island, smallpox victims and female paupers on Rainsford Island, as well as the garbage dumps and sewage treatment plants.

With the gathering of the islands into a state park a dozen years ago, those who saw them as an invaluable environmental and recreational resource had hopes that such policies had at last been abandoned.

It certainly makes little sense to propose solutions for the harbor's pollution that impose environmental costs that would themselves harm the harbor.

The Louisville Times

Louisville, Ky., June 6, 1983

"We all need to take the philosophy that if we err at all, we err on the side of environmental protection." The statement is reassuring since it came from a U. S. Environmental Protection Agency official in refusing the Metropolitan Sewer District permission to dump 26 million gallons of raw sewage a day into the Ohio River this summer while Louisville's Buchanan Street pumping station is repaired.

The EPA's decision has to be an encouraging one to anyone in this River City who has legitimately worried that the "P" has come to stand for pollution, not protection. But considering the very outrageousness of the proposal, and the opposition it provoked from neighboring Indiana, the EPA really had no other choice.

The alternative to bypassing the pumping station will be an expensive one, unfortunately. A temporary pumping station will be installed to ship the sewage to the Morris Forman treatment plant. The cost will be almost $600,000. Indiana's protest of the proposed dumping, though, serves to call attention to the ridiculous double standard applied on opposite sides of the Ohio River. After all, if the EPA concludes that the main waterway flowing through the region must be kept reasonably clean because a number of states share it, why not apply that logic to the air? The agency's decision early last year to allow power plants in Indiana to emit sulfur dioxide at five times the rate permissible in Jefferson County is clearly contradictory to its ban on sewage-dumping.

While the Louisville Gas & Electric Co. has installed sophisticated "scrubbers" on its smokestacks, Public Service Indiana's Gallagher plant just across from Louisville's West End operates as though it is in a different era. And all because the EPA has put Indiana in a different region. Such bureaucratic categorization may make sense when shuffling paper, but not when breathing polluted air. The agency should run the risk of "erring" again: this time in favor of the lungs of Louisvillians — and Hoosiers.

The Times-Picayune
The States-Item
New Orleans, La., February 5, 1984

Jefferson Parish residents soon will be asked for their cooperation as their public officials try to come up with the money to pay for a $350 million sewerage improvement program.

That better sewerage is needed throughout the parish has long been evident to public officials and many other residents. For years, parish officials have worked to produce the necessary technical planning and means of financing an updated system. Sewerage and drainage needs go hand in hand, as many residents have learned to their dismay in heavy rains when both systems became overloaded.

Lending urgency to the effort to raise revenues for improved sewerage are deadlines set by the U.S. Environmental Protection Agency. The agency is pressuring the parish to comply with federal guidelines governing sewage treatment by 1986. The parish must complete the diversion of treated sewage from wetlands to the Mississippi River by 1988.

To raise the necessary revenues, Parish President Joseph S. Yenni, who has displayed determined leadership on the matter, said he will submit revenue proposals to the Parish Council within two weeks. He will recommend either a sales tax increase, a sewerage service charge or a property tax increase.

The final revenue proposal will be submitted to parish voters on April 7.

Jefferson Parish's drainage and sewerage problems, which have resulted in large part from growth outstripping revenues and services, have festered for years. Now the day of reckoning is rapidly approaching. Bringing both of these services up to date is essential to the progress and well-being of the parish and its residents.

THE ☼ SUN
Baltimore, Md., February 10, 1984

A plan to triple the sewage capacity in the Carroll county town of Hampstead has been compromised to reduce by half all phosphorous content, and that's a relief. The town's sewage flows into the Loch Raven reservoir, one of the chief sources of water for the households of metropolitan Baltimore. Years of phosphorous build-up — not just from sewage, but from heavy storm-water run-off, too — has lowered water quality in the opinion of many experts.

By agreement, county officials will hold down phosphorous to one part for each million parts of Hampstead waste water headed for Loch Raven. Earlier, the state's environmental programs office was poised to approve an allowance of 2 parts phosphorous for each million of waste water discharged from the greatly expanded plant.

The reduction is not total, but it is one way of assuring healthier reservoir management until 1989 when the agreement expires. By then a new, far more comprehensive system to safeguard storm run-off and provide a filter for the Hampstead plant will reduce the phosphorous level to .03 part per million.

Protection against further phosphorous deposits in the reservoir will not slow Carroll's growth plans. Hampstead already has successfully attracted residential and industrial sewer users. Tripled capacity of the town's treatment plant is but one more step in preparation for increased population and production.

The Carroll county commissioners deserve credit for their willingness to work with state officers to seek sensible solutions that satisfy both environmental and economic priorities. But the parties involved in this compromise must realize that the solution is shortterm, and must be replaced in five or six years by a program that will cut phosphorous not only from Loch Raven, but from Liberty and Pretty Boy reservoirs as well.

The Hartford Courant
Hartford, Conn., February 24, 1983

There must be some reason why Gov. William A. O'Neill hasn't spoken publicly about the threat of sewage pollution in the Connecticut River.

Perhaps it is because Mr. O'Neill has been stunned into silence by some of the absurd commentary issuing from Springfield officials: for example, Mayor Theodore E. Dimauro's admission that he wasn't aware, until he read the newspaper, of his own city's plans to divert up to 15 million gallons of untreated sewage daily into the river during pump station repairs, or Public Works Director John J. Lyons' reasoning that the discharge could have "a positive impact on fish and plant life" because the river would have more "food" in it.

Notwithstanding the apparent indifference of the Connecticut governor and the Springfield mayor, and Mr. Lyons' exotic logic, some federal officials have responded appropriately.

The regional office of the federal Environmental Protection Agency has announced that it will hold an "informational" public meeting before acting on Springfield's request to divert the raw sewage into the river.

EPA's decision could delay the first of two sewage diversions, scheduled by Springfield for March and November. The delay would allow authorities to consider alternatives which might eliminate the need to dump sewage in the river, or at least minimize the impact.

Among the more promising solutions offered by Connecticut's congressional delegation, U.S. Reps. Samuel Gejdenson D-2nd District, and Barbara B. Kennelly, D-1st District, have called on the EPA to finance alternatives that would eliminate or reduce the sewage discharge while the repair work is done.

EPA has the discretionary authority to provide such grants. This approach would maintain water quality in the Connecticut River while lessening the burden on Springfield.

Hartford, Conn., March 19, 1984

After a year of fighting, Connecticut has apparently won the day and managed to stop the city of Springfield, Mass., from going through with its plans to dump millions of gallons of raw sewage into the Connecticut River while repairing a pumping station.

The Massachusetts Legislature came through with $1.3 million last month, which will presumably enable the city to conduct its repairs with a minimum of discharge — perhaps only a few hours' worth later this month. The money was allocated just in time to avoid a long-threatened lawsuit by Connecticut.

Springfield had originally planned to dump up to 10 million gallons of untreated sewage a day for two four-week periods while conducting the repairs. The federal Environmental Protection Agency whittled the plan down to eight days, but even that was unacceptable to Connecticut, only 4.5 miles downstream from the discharge point.

From both political and ecological points of view, it would have been unconscionable to have allowed a massive discharge of raw sewage into a river that has cost hundreds of millions of dollars to get as clean as it is today.

In the euphoria of this victory, however, let's not lose sight of the formidable clean-up job that remains. ◂

Springfield, like many other Northeast-ern cities, has a sewer system that combines sewage and street runoff — millions of gallons of which overflow into the Connecticut with every heavy rain. Since industries aren't required to treat their wastes before discharging them to the sewer system in Springfield, that overflow also contains high levels of toxic chemicals and metals.

Connecticut does require industries to pre-treat their wastes, but many of its cities and towns also have decrepit combined sewer systems that pollute the rivers during rainfall. Some communities, like Enfield, are already correcting the problem, but it will take hundreds of millions of dollars more to eliminate it altogether — an estimated $100 million just for Hartford.

The Metropolitan District Commission, the regional sewer and water district for the Hartford area, is planning a new system under order from the state Department of Environmental Protection.

Alleviating all pollution problems from combined sewers would take a financial commitment from the federal, state and local governments that makes the money that will be spent to avoid the massive dumping from Springfield seem insignificant — and poses a real test of determination to clean up our rivers and streams.

The San Diego Union

San Diego, Calif., November 21, 1983

A Lowery and Associates study confirms that the United States and Mexico face a monumental problem in dealing with sewage disposal at the border between San Diego and Tijuana. The study estimates that a sewage treatment plant of capacity to be sufficient for the next 20 years would cost $729.4 million, a figure removing any doubt about the need for major federal participation in the financing.

The plant envisioned by the study would process 100 million gallons of sewage from Tijuana and 30 million gallons expected to be generated soon from industrial development on Otay Mesa and other growth in the South Bay area on this side of the border. This would serve until population growth requires an additional investment of perhaps another $100 million after the turn of the century.

The study recommends that the plant include not only conventional sewage treatment facilities but also an "aquaculture" system using water hyacinths to purify water to the point that it can be used for irrigation and a recreational lake. The experiments with this type of water reclamation in Mission Valley have been promising enough to warrant its inclusion in any future sewage treatment plan.

As for the site of the plant, we see no reason to quarrel with the recommendation in the study that it be located four miles inland along the Tia Juana River, on this side of the border just across from the present main pumping station of the Tijuana sewage system. From the standpoint of environmental impact, it would be hard to find a better site. Like the San Diego treatment plant near Lindbergh Field, this one would discharge its treated effluent through a pipeline reaching several miles offshore.

Two conclusions leap out from the Lowery report. One is that the U.S. government must bear a substantial part of the financial burden for the plant, even if the need for it arises mainly from the inadequacy of sewage disposal measures on the Mexican side of the border. Ideally, Mexico would contribute a significant share of the cost, but given the fiscal problems already saddling the Mexican government, its contribution doubtless will have to be deferred under some arrangement with our federal government.

San Diegans should not assume, however, that a new sewage plant is going to be handed to them as a gift from Washington. The plant will be accommodating our own population growth and industrial development. To that extent, a portion of the cost must be underwritten locally.

The other conclusion is that whatever international negotiations must be carried out, whatever clearances must be obtained from federal, state and local regulatory agencies, whatever steps must be taken to arrange financing — these preliminaries must be put on a fast track.

The Lowery report says the earliest we could expect the new plant to be completed is 1988, although construction might take until 1991. The target date of 1988 must be adopted and pursued by all agencies concerned. Even if work were started on the plant tomorrow, we face a period of increasing contamination of our beaches, as far north as Coronado or beyond, because of the discharge into the ocean of raw sewage from Tijuana. That situation should not be tolerated one day longer than is absolutely necessary.

San Diego, Calif., March 16, 1984

U.S. and Mexican negotiators have finally achieved some progress in staunching the flow of Mexican sewage pouring into the United States.

The Mexicans have pledged to complete construction of a new sewage pumping plant by the end of this year that will enable them to handle 30 million gallons of sewage a day — 10 million gallons a day more than Tijuana is currently producing. The sewage will be dumped in the ocean at least 5.6 miles south of the border. And Mexico is considering treating the sewage so that it doesn't flow in a raw state into the ocean. Some of it might even be diverted for agricultural use.

In addition, the Mexicans have completed repairs on their broken sewage pipeline that was dumping 3 million gallons of sewage a day into a South Bay holding pond. The pond, according to

Mexican authorities, may not be needed by the end of next week.

To guard against further environmental crises, the United States and Mexico have agreed to form three study groups to investigate problems of air pollution, soil pollution, and water pollution afflicting both sides of the border.

Despite this welcomed progress, however, serious issues remain unresolved. Only half of Tijuana's residents are connected to a sewer system — meaning that a million gallons of raw sewage continues to gush through the Tijuana River into the United States every day. And even with its new pumping plant, Mexico will be unable to deal with the expected increase in Tijuana sewage by the end of the century. By 1990 the sewage flows from Tijuana are expected to double. And by the year 2000 they are

expected to double again. Moreover, American officials are worried that Mexican sewage dumped in the ocean will be swept back upon San Diego's beaches.

The health hazard presented by untreated sewage is frightening. Indeed, the county Department of Health Services fears an outbreak of encephalitis — a brain infection carried by mosquitoes — this summer because of the raw sewage in the Tijuana River Valley.

We are reassured by the strides the Mexicans are now taking to regain control of their sewage. But cooperative steps, as urgent and welcomed as they are, can only bring temporary relief. A long-term solution must be found, and soon. Nothing less than the health, well-being, and the very future of San Diego are at hazard.

THE KANSAS CITY STAR
Kansas City, Mo., April 4, 1984

In the early years of the 1970s the federal government, swept along by a tidal wave of environmental alarm, plunged vigorously into cleaning up the nation's air and water. A major element of that effort was a federal grant program to enable local governments to build sewage treatment plants. In the last 12 years $40 billion has been poured into this program. It is second only to highways in federal public works spending.

Much good has been accomplished in improving the water quality of streams and lakes which once served as dumps for raw sewage. But the plants could treat wastes only from specific discharge pipes, while random runoff from the landscape and farm fields or barnyards, including insecticide and herbicide residue and rotting vegetation, continues to taint our waterways.

As with the federal interstate highway program, on which most of the mileage has been completed, the cost of the remaining treatment works keeps growing and the completion date fading into the distance. The sewage grant plan has had its flaws, aside from the inevitable government red tape. Many cities, with 75 percent federal aid available, built plants that were too large or not cost-effective. Some used the money to build new sewers to areas still being developed. Not all the waste was in the sewer lines.

Kansas City voters in 1960 approved $75 million in bonds to build the treatment plant at the Blue River mouth on the Missouri and related interceptor sewers and other facilities. But as William D. Ruckelshaus, environmental protection administrator, pointed out recently, some cities and states delayed construction, waiting on federal funds rather than spend their own.

As the program stalled, congressional budget cutters in 1981 halved the spending to $2.4 billion a year and lowered the federal cost share to 55 percent. This reduced the government cost of the remaining projects from $118 billion to $35 billion. Now the EPA is studying whether to end the program in its present form, possibly substituting a federal construction loan plan. The sewage treatment grants program, a slow and ponderous thing, has had its successes and shortcomings. Now some kind of new approach is indicated to re-energize the water pollution abatement campaign.

The Chattanooga Times

Chattanooga, Tenn., March 13, 1984

The public is beginning to reap the benefits of the enormous investment the city of Chattanooga has made in expanding and upgrading the Moccasin Bend Sewage Treatment Plant. For the first time Chattanooga is meeting the water quality standards for municipal treatment plants which were set out in the Clean Water Act of 1972. And city officials, current and past, who gathered at the treatment plant last week to celebrate this benchmark of progress deserved to crow a little about their accomplishment.

It means that Chattanooga will not be subject to stringent federal sanctions for non-compliance with the Clean Water Act, and it means that Marion County will no longer suffer the effects of inadequate sewage treatment at Moccasin Bend. The quality of treatment there had been causing problems downstream for many years. Those problems were made worse, and caused an understandable uproar when construction on the new plant forced the city to discontinue temporarily the partial secondary treatment it had been providing. It was one of those bad situations that had to get worse before getting better. And better it is.

Capacity is to be added with activation of a number of duplicative facilities at various stages of the treatment process, and computer controls have yet to be hooked up. But all aspects of the upgraded treatment process are on line, working at a capacity sufficient to handle normal flow into the plant. The head of the state's division of water quality control office here predicted it won't be long before signs posted to warn the public against swimming or fishing in the river will be removed. The cries of outrage in Marion County over the problems at Moccasin Bend should be replaced with expressions of relief — and in fairness, appreciation for the accomplishments there.

It's been a long haul, and terribly expensive. The city of Chattanooga, along with the federal government, has poured almost $100 million into the Moccasin Bend plant. The tremendous cost makes us doubt the project would have been undertaken in the absence of a federal requirement and 75 percent federal funding. That does not diminish the hard work and dedication of the city officials who saw the project through in the face of many obstacles. It simply points out the important role the federal government has played in setting standards and assisting local governments in meeting their environmental responsibilities.

Unfortunately the federal role in funding assistance is diminishing, which shifts greater responsibility to local and state governments. Because Chattanooga is so big, its sewer program and problems have received the greatest public attention, but other cities, including Signal Mountain, Lookout Mountain, East Ridge and Red Bank have long been seeking federal assistance to correct problems with their sewer systems.

Some of these towns have commendably proceeded on their own, and since it appears further waiting for Uncle Sam will be in vain, more must be done to tackle these problems with state and local revenues. The General Assembly should approve the governor's program to help local governments in this regard, and, despite the unpopularity of sewer charges and property taxes, officials and citizens of local governments should accept their responsibility to pay for the services necessary to protect their health and their environment.

The Courier-Journal

Louisville, Ky., March 13, 1984

THE LOUISVILLE and Jefferson County Metropolitan Sewer District has caught a lot of deserved flak in recent years, and not just for its snafu in permitting serious flooding in southwestern Louisville last May. But the proposed $32 million West County sewage treatment plant is an altogether different matter. Blame for this growing debacle should rest squarely on the doorstep of the U.S. Environmental Protection Agency.

During the 1970s, the EPA pushed for construction of the West County plant and sewer lines to serve approximately 25,000 homes in southern Jefferson County. But the project was derailed after EPA yielded to pressure from the roughly 17,000 families in the area west of Dixie Highway.

Those residents said their septic tanks were adequate, and EPA — after a five-year study and after installation of more than eight miles of large sewer pipe that has never been used — agreed. That agreement was in 1982. Now, after all of that political capitulation and footdragging, EPA suddenly has given MSD less than a month to approve the treatment plant.

To tighten the screws even more, an EPA official has threatened that long-needed repairs at the Morris Forman plant, the largest sewage treatment facility in the county, will be held hostage. The official, who heads the construction grants section of EPA's Atlanta office, says a $12 million grant for Morris Forman might be withheld unless local officials approve the West County project.

That's an unseemly show of arm-twisting by a federal agency that is directly responsible for the problem it's now trying to solve.

Miles of dry lines

Meantime, under the scaled-down plan being discussed now, the West County plant would process up to 15 million gallons of sewage a day, one-third the originally projected capacity. It would serve about 16,000 customers in the Pond Creek area of southern Jefferson County. Their sewage is now treated by 34 small treatment plants, which would be acquired by MSD.

Even worse than the indecision and delays that accompanied the project is the roughly $37 million provided by EPA, and some $13 million in local funds, spent for equipment that has mostly lain idle. The sewer lines just sit there, dry. And a pumping station at the site of the projected new treatment plant, at Orell and Lower River roads, has had only limited use. Without EPA having the will to see the project through, that expenditure is fiscally irresponsible. Now, the agency is hell-bent to correct the mistake.

Even if officials can overcome past mistakes, planning and building the treatment plant won't be easy. It obviously can't be assembled in just a few weeks. There is uncertainty whether the plant's operating costs would be greater than revenues. No one knows who would pay for the small treatment plants with their millions of dollars in assets and indebtedness.

In addition, although EPA is pushing full steam ahead, final approval must come from Frankfort. The state holds the trump card on the funding issue because it decides how federal money for such projects is spent in Kentucky. The issue is whether the federal money will cover 75 percent or 55 percent of the project's cost because after September 30, federal funding will drop from 75 percent to 55 percent.

Although the change hasn't gone into effect, the state is reluctant to commit 75 percent to the West County project when it could provide 55 percent and use the savings for other water quality projects around Kentucky.

In all this, EPA has shown its true colors — as poor planners and arrogant bureaucrats. The threat to cancel funding of repairs at the Morris Forman plant is just the final straw.

MSD, meantime, although it has been thrust into a difficult situation, must exert some leadership. If the problem is to be solved, the agency must find a way to unite the state, EPA, city and county officials and the small treatment plant operators. The community needs the West County plant. But the plan for it must be financially sound and the treatment plant operators should be compensated for their investments.

For their part, the state, EPA, city and county governments must get serious about water quality. Failure to act can yield but one result: continued pollution of this community's waterways.

The Houston Post
Houston, Texas, March 25, 1984

Once again, the levels of fecal coliform bacteria in Lake Houston have risen far above what is acceptable. This time, though, there exists more hope than ever that something may be done about it.

It has been clear for some time that the chief culprits in this mischief are small sewage plants operating in the lake's watershed. But historically the city of Houston has taken such a hands-off attitude toward the problem that last summer an assistant state attorney general accused officials here of having "no guts." The city especially shunned lawsuits against violators in its extraterritorial jurisdiction, citing qualms over whether it had legal authority.

A few days ago, however, the Houston City Council's committee on Lake Houston urged suing polluters in the extraterritorial jurisdiction. This recommendation followed a review of Environmental Protection Agency regulations by the city's legal department, which concluded that city health inspectors could file suits as private citizens. That's a time-consuming process, involving delays of up to 60 days while state and federal agencies decide whether they themselves wish to sue. But time-consuming or not, it's a lot better than the action we've been seeing — almost none.

Sometimes fecal material gets into the lake because plants are poorly run or mechanically faulty. They spew raw sewage into the watershed, or a heavy rain washes sludge (the sediment left after sewage treatment) out of its settling tanks. Sometimes sludge is dumped illegally by tank truck drivers who are hauling sludge from the sewage plants or from septic tanks for disposal. Although it settles to the bottom of the lake, it can be churned up by storms. But all this is beside the point. Any sludge that gets into Lake Houston is there illegally. It's supposed to be disposed of in landfills.

As it happens, fecal coliform bacteria themselves are not horrifically dangerous, but their presence is an easy-to-check indicator of the presence of other, more hazardous bacteria and viruses. The water as it now exists is treatable for drinking purposes — a good thing, since the lake provides almost half of Houston's drinking water. But in its untreated form, it represents a continuing threat to recreational users of the lake.

In the long term, the answer to this problem is the development of larger regional treatment plants which presumably will be more responsibly run. In the meantime, the EPA regulations give the city a tool, however cumbersome, to clean up this unconscionable mess. The city should throw itself fully into pursuing such actions. That's the taxpayers' water that's being fouled, and it must stop.

THE BILLINGS GAZETTE
Billings, Mont., April 25, 1984

Some proposals are brilliant in their simplicity.

The latest proposal for a Heights sewer connection is one of those.

The plan, the result of some 20 years of study, consideration and wrangling, is simply a winner, nothing less.

First, 75 percent of the initial construction costs will come from federal coffers via an Environmental Protection Agency grant.

That represents a considerable savings — about $5 million — for Heights residents.

Second, the proposal won't require the selling of bonds before the project begins.

The city of Billings will initially pick up the remaining $1.7 million of the $6.7 million total.

That money will be reimbursed to the city as Heights residents hook up to the system.

Third, the plan couldn't come at a better time.

The spread of septic tanks in the Heights, particularly in specific areas, constitute a real potential health hazard for city and Heights residents, alike.

The situation promises to grow worse, unless something is done soon.

There isn't time to stand around with our collective hands in our pockets.

Fourth, duplication of services is avoided.

There is no need to build a second wastewater treatment plant.

Instead, the city plant, with excess capacity expected into the 21st century, will be utilized.

And there is room to expand the plant when the need arises.

Fifth, it takes the edge off the controversy over annexation.

Annexation would take place only as neighborhoods request sewer service.

And the city is not faced with wholesale annexation of areas prior to sewer hookup.

The city is stretched to provide services for its residents, now. It will not, officials say, annex areas until it can afford to provide full services there, too.

Neighborhoods that fall into that category need only to sign waivers of protest to annexation to receive service.

The city may then choose to annex those areas at a later date, when it can provide full city services.

The city, county, state and EPA deserve our thanks for taking the project this far.

Success hinges now on EPA approval next fall, approval that seems likely.

The system is working.

The Burlington Free Press
Burlington, Vt., April 25, 1984

Excess Burlington sewage should be piped directly into Lake Champlain.

That's what city officials have been advised to do by the state's Environmental Protection Agency.

No, there's no mistake. And, yes, the suggestion comes from the same agency that is dedicated to protecting the state's environment and insuring that Vermonters are not exposed to toxic wastes or pollution — or so its rules say.

The lake is perhaps one of the most precious natural resources in the state not only for its scenic qualities but also for its recreational uses. Protecting the lake from pollution has been the goal of several organizations in recent years. They have advocated stringent measures to guarantee the purity of lake waters. And they have had some success in carrying out their mission.

But now that work could be canceled if the city is allowed to install equipment which will bypass the sewage treatment plant and pump more untreated sewage into the lake, as suggested by the Environmental Protection Agency. Lack of separate sewer and storm drainage systems now cause massive overflows of sewage that the plant is unable to treat. Instead of overtaxing the plant, the agency officials say they would prefer that the excess sewage be dumped into the lake without making any effort to treat it, says Thomas Moreau, manager of the Wastewater Division of the city's Water Resources Department. He said the city is pumping 240 million gallons of raw sewage into the lake each year. That is a surprising admission in light of federal and state emphasis on water pollution control.

Because the city draws its water supply from the lake, questions must be raised about the condition of drinking water, even though it passes through water treatment facilities before it is piped to businesses and residences. Of equal concern is the health of people who swim in the lake. Contact with raw sewage doubtlessly could cause a number of illnesses among swimmers.

What is puzzling about the Environmental Protection Agency's suggestion is that it appears to collide with water pollution control statutes that allow the state Water Resources Board to order municipalities that are discharging raw sewage into lakes and streams to "correct or abate the condition." Under such an order, communities are required to furnish "reasonable" time schedules for remedying the problems. As a signatory with New York of the New England Interstate Water Pollution Control Compact, Vermont is obligated to work toward the abatement of existing pollution in Lake Champlain and bend its efforts to control future pollution, so that the waters of the lake are maintained in a "satisfactory condition."

In explaining the situation to the aldermen at Monday's meeting, Moreau said the federal Environmental Protection Agency is cracking down on sewage capacity problems. Federal funds for such work are "drying up," he said. "Tremendous costs" are involved in solving the problem, said Alan Charron, chairman of the Water Resources Commission.

But when the costs are balanced against the possible health hazards, there should be no question in the minds of city officials that the work should be done as soon as it is feasible.

The state Environmental Protection Agency should have advised the city to do just that instead of suggesting that more raw sewage be dumped into the lake.

Are the bureaucrats in the agency merely indifferent to the need for strict pollution controls or have they indeed forgotten that their duty is to protect the environment?

TULSA WORLD
Tulsa, Okla., May 12, 1984

OKLAHOMA must sooner or later face the growing problem of sewage disposal in semi-urban areas, as the growing controversy over stopgap disposal methods like sewage lagoons and septic tanks shows.

Truly rural areas have few problems in this regard. In those areas, housing is so sparse that septic tanks can be made to function with a minimum of problems.

The rub comes when housing demands are great enough to encourage subdivision-style development on land which cannot accommodate septic systems. A lagoon often is proposed. When properly installed and maintained, it functions quite well.

There is a public aversion to lagoons, however. Often they are not well designed or maintained. In fact, there is no substitute for true sewage disposal systems.

In many, if not most, areas of increasing development, a bit of foresight (and yes, money) could head off future problems.

A state program to encourage and help finance sewage disposal systems would go a long way toward solving the problem.

For the expense involved in installing great numbers of individual septic systems, many areas could have a functioning sewage disposal plant, if, and here's the big rub, there were legal mechanisms to require the formation of sewage disposal system districts.

Septic tanks and lagoons are expensive stopgap answers. A progressive state should devise a framework for providing lasting sewage disposal systems, a framework within which counties and private enterprise can spend money already sunk in temporary systems on permanent solutions to the problem.

AKRON BEACON JOURNAL
Akron, Ohio, May 31, 1984

WITH A DEADLINE approaching from the Environmental Protection Agency, Akron and its neighboring communities can't afford *not* to agree on sewer-use contracts. At this point there is no choice; the cost of not complying with EPA standards is too high to risk.

The required agreements stem from EPA's need to close off old sewer and related grants that helped cities like Akron meet tougher water quality standards from the 1970s. Akron has received, and spent, some $40 million for improvements at the Botzum treatment plant. More recently, the city has received another $25 million for a proposed sludge-composting plant. Area communities also benefit from these improvements.

Local users of sewer systems built with federal grant money must now enter EPA-approved contracts that require Akron-area customers to pay adequate rates to cover maintenance, operation and replacement of the major sewage-treatment plant. Each community must pay a fair share, depending on use.

That seems simple enough. Yet at least four years of negotiations have resulted in only one agreement — between Akron and Tallmadge. Akron still must come to terms with Cuyahoga Falls, Fairlawn, Lakemore, Mogadore and Summit County, which represents the Mud Brook area, including parts of Cuyahoga Falls, Silver Lake, Northampton Township, Munroe Falls, Hudson and Stow.

And the EPA deadline of June 30 is serious. Already Akron has received one extension — from April 30 — through lobbying by Mayor Tom Sawyer and others. But it is clear that Akron — the largest city without contracts in the six-state EPA Midwest region — is on borrowed time.

The stakes are high: Akron could be forced to repay the $40 million that went for sewage-treatment improvements. The city could also lose the $25 million grant that will go for sludge composting. And if no acceptable alternative is found for sludge disposal, the EPA could impose fines of $25,000 a day and jail terms for city officials.

Even if the situation never comes to that worst case, there is no excuse for delaying the contracts any longer. There have been charges that past Akron city administrations did not take negotiations seriously enough. But that should only increase the urgency now to do something about it. And Akron deputy service director Linda Sowa certainly seems serious about the negotiations between the communities and capable of working them out if all cooperate.

It is understandable that the smaller communities would fear entering agreements that would mean higher rates. But even if slightly higher rates occur from the contracts that scenario is vastly overshadowed by the possibility of users paying back $40 million while maintaining the system and financing a sludge-disposal facility.

It's obvious that this problem does not belong to just one community, but is shared by many, just as the benefits have been shared from the treatment facilities. It is time for a shared solution. Coming to terms on the user contracts is the only solution.

The Star-Ledger
Newark, N.J., April 9, 1984

The investigation of New York City's massive discharges of raw sewage into metropolitan waterways has given an overdue needed impetus to begin resolving what has become the nation's worst sewage and sludge problem. This is a refreshing change of official attitude, further fortified by recent pledges by the U.S. Environmental Protection Agency (EPA) for "strict enforcement" of federal clean water regulations.

A ranking EPA regional official, Dr. Richard Dewling, said the federal agency is giving a priority status to funding for sewage treatment plant upgrading in crucial urban areas in New Jersey and New York. This belated governmental action is a direct response to an ongoing congressional inquiry.

Two New Jersey congressional Democrats—Sen. Frank Lautenberg, a member of the Senate Environment and Public Works Committee, and Rep. James J. Howard, chairman of the House Public Works and Transportation Committee—are scheduling hearings on New York City's blatant violations of the 1972 Clean Water Act.

The congressmen want to know why New York City has been permitted to discharge more than 220 million gallons a day of raw, untreated sewage directly into the Hudson and East Rivers since 1972 when the Clean Water Act became law.

The regional EPA director, Jacqueline Schafer, denied reports that the agency has given New York City preferential treatment by permitting the discharge. She said the EPA is "going to enforce and implement the Clean Water Act on both sides of the Hudson." That's the most encouraging development to date for New Jersey in this ongoing pollution controversy.

If that commitment is vigorously enforced, it would constitute a responsible administrative action in eliminating what Sen. Lautenberg and Rep. Howard have criticized as a "double standard" that gives New York the right to build and pollute, while New Jersey must control growth and check pollution with expensive sewage treatment plants.

There are no operating treatment plants in Manhattan, yet New Jersey's largest sewer plants are operating on a secondary treatment, the highest treatment for urban wastewater. New York City has erected billions of dollars worth of high-rises during the past dozen years, resulting in a ceaseless flow of filth into the rivers and coastal waters along the Jersey Shore and Long Island.

The congressional probe underscores a compelling sense of urgency for a broad coalition of federal, state and local governments to begin addressing this enormous pollution problem by finally completing pollution control programs that began two decades ago.

As a potent initial phase, the EPA should mandate New York, the principal source of widespread contamination of regional waterways, to complete its pollution control facilities before further significant development proceeds.

The Forum
Fargo, N.D., June 6, 1984

Wisconsin's lawsuit against Minnesota aimed at ending the discharge of raw sewage at the Twin Cities into the Mississippi River makes one wonder how this river pollution has gone on so long without being remedied.

The complaint is that as much as 13 million gallons of raw sewage per day has been flowing into the river because of a lack of separation of storm and sanitary sewers.

Fargo once had the same problem with the Red River on a much smaller scale, but it has virtually been corrected by sewer separation programs at the insistance of the U.S. Environmental Protection Agency. While federal funds have helped with this project, it also has been very expensive for local property owners.

Minneapolis and St. Paul also have sewer separation projects, but they are too slow for the EPA and Wisconsin.

Wisconsin's suit challenges permits issued by the Minnesota Pollution Control Agency for the sewage operations in Minneapolis-St. Paul. The permits, issued in April, already have been rejected by the EPA because they do not set a deadline for a plan to halt the discharge, estimated at 4.6 billion gallons of raw sewage into the river annually. The discharges come especially after heavy runoff from rain or snow, but also occur in dry weather.

Minnesota Gov. Rudy Perpich said he was surprised by the suit, but he also said Wisconsin was right in filing it. He said he hopes the court gets tough and makes Minnesota have a timetable for ending the discharges.

It is difficult to comprehend that the Minnesota PCA would tolerate this kind of discharge for long, or that environmentalists in Minnesota would not have forced something to be done. It is from Minnesota environmentalists that much of the criticism of coal-burning generating plants in North Dakota, and protests of North Dakota's Garrison Diversion project come. The PCA is concerned that acid rain in Minnesota may be caused by emissions from the North Dakota plants.

Minneapolis Mayor Don Fraser, when he was in Congress, was a consistent opponent of aspects of Garrison Diversion. Garrison opponents raise a fuss because they suspect that it will introduce foreign fish and biota (plant life) from the Missouri River into other water systems, including the Red River. Garrison backers in North Dakota are cooperating with the federal government to research this aspect and the project has been diminished at least until environmental concerns are resolved. Canada is concerned that its fishing industry might possibly be harmed by Garrison Diversion.

As for dumping raw sewage into rivers, Fargo's problem with the Red River is all but solved. Sewer separation is in its final stages after about 10 years of reconstruction. The residential neighborhoods are done and the downtown project (Phase 7) is in the works. Fargo's problem, incidentally, came only in high water runoff periods.

In fairness to Minnesota congressmen, it should be pointed out that 7th District congressmen have consistently backed the Garrison Diversion project because they are knowledgeable as to the good effect it would have for the portion of Minnesota that shares the Red River. Better water levels and water freshness are the prospects for Red River Valley communities if Garrison Diversion is completed as originally planned.

Detroit Free Press
Detroit, Mich., June 1, 1984

IT HAS BEEN a long five years since the federal district court took Detroit's sewage plant under receivership. In the late 1970s, the plant was fouling the air from its incinerator stacks and fouling the river with chocolate-colored sewage. Maintenance was abysmal. Routine repairs took months; it took 40 weeks to get a ball bearing for a broken conveyer. Smelly sludge stacked up for lack of adequate disposal facilities. Poorly trained and demoralized employes occasionally vandalized the machinery; once an employe even fell into the sewage tanks. The plant would have been a joke, if it had not been such an embarrassment.

The last report from the court-appointed monitor told a remarkably different story. Wastewater pouring out of the plant now meets federal water quality standards. The incinerators are operating in compliance with the Clean Air Act, although some work remains to be done on them. A hiring and training program has improved the staff to the point where the problem now is that key employes are being lured away to work in other cities.

Much of the progress is the result of the continued prodding from Chief U.S. District Judge John Feikens, who presided over the lawsuit brought against the city by state and federal regulators. Last week Judge Feikens announced that his oversight of the plant will end as soon as Mayor Young submits his final report. The mayor was appointed receiver of the plant by Judge Feikens in 1979; that gave the mayor power to bypass the lethargic city purchasing and personnel departments and let contracts himself. Unfortunately, it also led to the Vista affair, which has far overshadowed the successful cleanup at the plant.

The monitor has warned that continued diligence is necessary if the gains in clean air and clean water are not to be lost. But the direct intervention of the court in running a wastewater treatment plant is an extraordinary measure to be employed only in extraordinary and limited circumstances. Judge Feikens is correct in declining to oversee the plant in perpetuity. The running of the Detroit Water and Sewerage Department is the city's responsibility, and the city should be prepared to assume it. The Vista business will have to be lived down. The remarkable environmental cleanup engineered by the court remains something to be lived up to.

THE MILWAUKEE JOURNAL
Milwaukee, Wisc., May 12, 1984

A well-dressed guy stands on Wisconsin Ave., bellowing Bible verses into a bullhorn. Frustrated by his failure to win converts to his brand of religion, the preacher threatens to become an atheist if passersby do not immediately join him in prayer.

That, metaphorically, is Milwaukee's Mayor Maier, preaching on the topic of dirty runoff into waterways (nonpoint pollution). People upstream should stop sinning, Maier says, but if they don't, Milwaukee should continue its own pollution. He's right on the former, wrong on the latter.

Last year, Maier threatened to file lawsuits aimed at stopping the Metropolitan Sewerage District's massive anti-pollution program until the state did something about upstream runoff. This month, Maier suggested that citizens consider "civil disobedience," possibly by refusing to pay sewer taxes, if the upstream polluters were not stopped.

It seems to us that the mayor's threats are attention-getting fireworks rather than serious proposals. They also are unfortunate. Such comments could undermine the credibility of the community's commitment to clean up the river, and they give comfort to those few remaining troglodytes who think the huge cleanup here can somehow be slowed down or halted altogether. That's simply not allowed by law, and most of Maier's appointees to the sewerage commission have long since accepted the fact.

Further, it is misleading for the mayor to imply that the sewerage cleanup here will be a total waste of money unless something is done about runoff. Harold Cahill, former top administrator of the sewerage district, responded to that claim a year ago in words that still ring true: "You cannot tell me that when I take the human feces and debris out of the river, that the river is not better off.

Let's stop throwing stones at everybody else."

It is also misleading for the mayor to imply that the runoff comes entirely from rural areas north of Milwaukee County. A goodly share of the pollution comes from within Milwaukee County. Although that particular kind of runoff is often more difficult to control than rural runoff, it is nonetheless pollution that needs to be curbed.

Having said all that, it must be acknowledged that the mayor's bellowing has been of great service in making the public aware of facts that had long been known only to river researchers and a few concerned citizens. The mayor has created renewed public pressure for a cleanup of the entire length of the river, and for that he should be thanked.

The next step, we think, should be for the mayor to turn off his bullhorn and give existing state law — admittedly quite weak — a fair try. (The governor just signed a bill providing at least $200,000 to begin dealing with runoff in the Milwaukee River.)

It is just possible that with a little good will and patience on all sides, the state can get everyone in the watershed to sing from the same hymnbook.

THE DAILY OKLAHOMAN
Oklahoma City, Okla., June 1, 1984

OKLAHOMA should move now to forestall a long-standing threat to one of our state's most valuable natural resources — the Illinois River and Lake Tenkiller.

Nobody should relax simply because the Environmental Protection Agency has refused to fund a small part of a proposed sewage treatment plant for Fayetteville, Ark. The EPA action, as Oklahoma Sen. Don Nickles has correctly noted, is much less than it appears.

Only that portion of the $27-million plant designed to discharge effluent into a tributary of the Illinois — about $2 million worth — is involved in the EPA withholding decision. As matters stand, there is nothing to prevent Fayetteville from proceeding with the plant if it wants to make up that $2 million from other sources.

Rep. Mike Synar, the Muskogee Democrat whose district includes the Oklahoma portion of the Illinois and Tenkiller, says it may take a federal lawsuit against Fayetteville to eliminate the pollution threat once and for all. That may well be, and Attorney General Mike Turpen should get his staff started on researching that possibility without delay.

Partially or inadequately treated effluent flowing into the Illinois would cause irreparable damage to the quality of Illinois River water that forms one of our most beautiful lakes behind Tenkiller dam. Oklahoma simply must not let this happen.

ST. LOUIS POST-DISPATCH
St. Louis, Mo., April 9, 1984

In an unusual rebuke, the state Department of Natural Resources has ordered the University of Missouri's College of Agriculture to stop violating the state's clean water laws by discharging animal wastes from its beef, swine and dairy operations into nearby creeks and sinkholes. The College of Agriculture has 30 days to appeal the citation or submit to DNR a plan for managing animal wastes in an environmentally sound manner.

Polluting the state's ground and surface waters is a serious matter. Even more distressing, however, is the indifference shown by college officials in allowing this bad example to be set for their students. Are the future farmers of Missouri to believe that it's OK to flout state laws, or simply that it's one of the costs of doing business in agriculture? We think not.

Agriculture remains the leading industry in Missouri and, if statistics can be trusted, will remain so for the foreseeable future. Students of farming and animal husbandry need to realize that clean water is a precious and finite commodity without which neither they nor the rest of the world can survive or prosper. Let the students involved in the beef, swine and dairy operations now take the lead in devising a remedy to provide them, if not their teachers, with some practical experience in dealing with regulatory agencies. That, too, is one of the costs of doing business.

Rockford Register Star
Rockford, Ill., June 10, 1984

Rockford has an opportunity to become virtually unbeatable as an industrial site.

It's a rare opportunity as the community fights for its economic future against raiders from other parts of the nation. It's an opportunity that must not be missed.

Rockford's ultimate weapon? Waste disposal.

It may not be a very glamorous weapon, but it can be a very effective one.

We've already got a leg up on every other industrial city in the nation.

That came when the local Sanitary District became the nation's first to win federal pollution-exemption "credits".

The credits derive from our district's outstanding efficiency in sewage treatment and pollutant removal.

They mean two things, both very important:

● Rockford has a sewage disposal system that works unusually well.

● Industries located here can escape from some of the costly anti-pollution systems they would need if operating in another community with a less effective sewage disposal system.

That can be a crucial advantage in the ongoing fight to keep our job-producing industries or to lure others here.

But it also opens another great opportunity.

Rockford now is attempting to develop something that doesn't now exist anywhere else — a complete solid waste disposal system that is workable, cost-effective and pollution-safe.

Since this is an industrial city and our best hope for the future lies with our industries, that disposal system cannot be complete until it includes facilities for industrial and toxic waste. It simply must solve the problem of disposing industrial wastes.

If this can be done, we believe Rockford can become unbeatable as an industrial site and the community's economic future will be assured.

If we can offer manufacturers both the nation's best sanitary sewer system and the best industrial waste disposal system, we'll have a couple of the best possible economic weapons in the fight to keep the well-paying jobs only industry can provide.

The challenge is clear. We can become the best place in the nation for industries with waste disposal problems — a category that includes almost all of the newer, growing industries.

Let's meet that challenge. Let's win that distinction for Rockford.

Then let's tell the world about it in an aggressive campaign to bring those industries and their jobs to Rockford.

Oceans:
Sludge Dumping, Toxic Waste Burning

The 1972 Ocean Dumping Act (part of the Marine Protection, Research and Sanctuaries Act) authorized the Environmental Protection Agency to issue permits for the dumping of waste materials in the ocean, provided that the dumping would not harm the ocean environment or endanger public health. In 1977, the EPA designated 140 dump sites for interim use while it continued to conduct studies on the effects of such dumping. Opinion is still divided over the impact such practices have; although the most commonly approved waste for ocean-dumping is sewage sludge, other forms of waste have also been permitted. In 1983, the House passed legislation to tighten the standards of the Ocean Dumping Act, but no action was taken on the bill in the Senate. The House bill would speed up the study of sites by the EPA, tighten the standards for choosing and monitoring dump sites, allow the EPA to charge permit processing fees, improve the scientific basis for granting permits and impose criminal penalties for violating permits.

Another key issue regarding ocean pollution has grown out of the mounting problem the United States faces in disposing of its hazardous chemical wastes. One avenue of research that has attracted much attention concerns the possibility of burning wastes aboard specially constructed ships at sea. This method was resorted to in 1977 to get rid of over 10,000 metric tons of defoliant Agent Orange, which contains the herbicides 2,4-D and 2,4,5-T, and a dioxin contaminant known as TCDD. (See pp. 208–217.) The Agent Orange was burned on the *Vulcanus,* a ship converted to a waste incinerator. There is much support for using the *Vulcanus* or a similar ship to dispose of other long-lived toxins, such as PCBs. One problem to be overcome is that the burning of PCBs creates TCDD; other toxins may also release contaminants when burned.

The Dispatch
Columbus, Ohio, October 3, 1981

THE U.S. ENVIRONMENTAL Protection Agency (EPA) has given a tentative go-ahead to Allied Corp. to dump 177,000 tons of hydrochloric acid in the Atlantic Ocean.

Showing an apparent lack of ecological sensitivity, one lower-level EPA official said it has been determined that the potent waste won't damage marine life because the acid will be neutralized by the salt water.

We don't buy it. Ocean ecology is one of the most fragile in our world. Even a grade-school child would realize that 177,000 tons of anything — let along hydrochloric acid — would adversely affect marine life, even in a healthy region.

The area where the acid dump is scheduled to take place is far from a healthy one. A recent National Oceanic and Atmospheric Administration study says the region, off the New Jersey coast, already contains substantial amounts of man-made chemicals and that oxygen at the ocean floor has been reduced to subnormal levels and that pollution has reduced the diversity and numbers of marine creatures there. "The water off the Northeast coast is living on the edge of disaster," Derry Bennett, a noted environmentalist, says.

The very fact that Allied had to get EPA permission to dump the acid testifies to the seriousness of the situation.

Starting late this year, a dumping ban for the region takes effect because more sensible officials at EPA realized earlier the damage that was being done to the ocean's ecology. Allied's request was for an exemption to the ban since its dumping will take place next year.

The EPA should rescind the dumping approval. Allied should find some other way to dispose of the acid.

THE DAILY HERALD
Biloxi, Miss., September 30, 1981

The U. S. Environmental Protection Agency made a mistake the other day when it issued a special permit to allow a company to dump tons of hydrochloric acid in the Atlantic Ocean.

If the mission of EPA is reflected in its name, the agency ought to operate under guidelines that are designed to protect the environment, not harm it.

EPA granted the permit to Allied Corp., which plans to pour 177,000 tons of hydrochloric acid about 15 miles off the New Jersey shore. One justification for the permit was that the dumping process will cost 25 to 50 times less than other disposal methods. EPA might not be looking out for the environment, but it certainly is taking care of Allied's bottom line.

Peter W. Anderson's claim that the acid dumping does not degrade the environment and that the hydrocholoric acid would be neutralized by the ocean's salt water are questionable. Anderson is chief of the Marine and Wetlands Protection branch for EPA's New York-New Jersey region.

The ocean in that region has been so polluted for so long that past permissiveness seems to have contributed to a lessening concern for the quality of the ocean environment.

The National Oceanic and Atmospheric Administration recently did a study of Atlantic waters between the eastern tip of Long Island and the southern tip of New Jersey. It found the waters contained substantial levels of man-made chemicals. The oxygen at the ocean floor had been reduced to subnormal levels and pollution had reduced the diversity and num-

bers of marine life.

Ocean dumping of various pollutants has been going on for too many years in rivers and off the coast of the heavily populated northeast region. It hasn't been healthy for marine life; some of the areas that have been long-time dumps have waters described as "dead," incapable of supporting marine life. Adding 177,000 tons of hydrochloric acid will do nothing to improve the situation.

The country has one day got to realize that it cannot continue to allow continued disposal of harmful and toxic substances on the old theory that out of sight is okay. Dump it in the river, out of sight. Dump it in the ocean, out of sight. Dump it in a landfill, out of sight. Keep dumping on that basis and rivers and oceans become transformed into cesspools, landfills become Love Canal nightmares.

The Philadelphia Inquirer

Philadelphia, Pa., January 23, 1981

America is awash with hazardous chemical wastes. Each year, 96 billion pounds of toxic chemicals are produced, with little attention paid to one crucial question: How to dispose of them? The enormity of the disposal problem is only gradually becoming apparent. The urgency of finding an answer to the disposal question cannot be overstated. It truly is a matter of life or death.

The U.S. Environmental Protection Agency, charged with regulating toxic materials from "cradle to grave," currently has under consideration, and open to public comment, a plan to permit the incineration of three types of extremely toxic materials (DDT, the banned pesticide Sylvex and polychlorinated biphenyls) at sea in a specially built ship.

If the plan receives EPA approval, many chemical producers hope it will lead to a broad program of at-sea incineration of other types of chemicals. A number of companies on the East Coast — including at least one in Philadelphia — hope to construct facilities which would enable an incinerator ship to dock and take on cargoes of the chemical wastes for incineration 130 miles east-southeast of Atlantic City.

Incineration of hazardous wastes, particularly those consisting of chlorinated hydrocarbons, is generally regarded as the safest means of disposal. These chemicals can be destroyed at extremely high temperatures; however, the combustion process produces quantities of toxic ash and gases. The air-pollution problem, as well as public opposition to the presence of hazardous materials in their communities, has severely hampered efforts to construct land-based incinerators.

By moving the incineration process out to sea, EPA believes, public opposition as well as the potential for human exposure is reduced. At-sea incineration was used in the 1970s in closely monitored pilot projects to destroy quantities of the defoliant Agent Orange, as well as several thousand tons of industrial wastes. In both instances, well over 99 percent of the substances were destroyed, with the remainder entering the ocean environment in the form of fall-out.

The risks of at-sea incineration are obvious. Weather conditions, the chance of accidents and contamination of water supplies during loading immediately come to mind. But there is one other major problem with the proposal: "Out of sight, out of mind." The potential for abuse and mishandling of the incineration process increases as the opportunity for careful scrutiny is diminished.

Man has used the ocean as the ultimate absorber of his wastes for centuries, and as long as the ocean was available, no one bothered to come up with other disposal methods. Only recently has the price of that abuse become known. Many of the chemicals suitable for at-sea incineration, such as DDT and PCBs, pose extreme genetic and environmental risks.

At-sea incineration poses a tempting solution to the desperate need for disposal of large volumes of toxic chemicals, particularly for the northeastern industrial corridor. In no way can it be viewed as a final solution. There will be tremendous pressure, however, for precisely that to occur. If EPA agrees to authorize at-sea incineration on anything but a case-by-case basis, with the most stringent of monitoring procedures and full opportunity for public participation in the decision-making process, it would guarantee that what little research is now underway on other methods of disposal will be abandoned.

It also would guarantee that not only will America be awash with toxic materials, but so will its, and the rest of the world's, oceans.

Philadelphia, Pa., September 5, 1981

Under orders of the Congress, ocean disposal of sewage sludge must end by Dec. 31, 11 years after passage of the federal Marine Protection, Research and Sanctuaries Act. The deadline was set after scientific studies proved that years of dumping sludge had contaminated surrounding ocean water, seabed and aquatic life with toxic pollutants such as oil, metals, PCBs and other substances. The long-range effects of that are not entirely known, but are perilous.

Philadelphia, New York, Camden and several northern New Jersey municipalities that barged their sewage sludge to sea fought the deadlines imposed by the U. S. Environmental Protection Agency. After granting the cities additional time to find suitable land-based sludge disposal methods, the EPA adopted a get-tough policy to force the cities to halt ocean-dumping. In April 1977, the EPA fined Philadelphia $225,000 and set a strict timetable for alternative sludge disposal.

Despite Philadelphia's opposition to the EPA's actions, the city ended ocean-dumping a year before the 1981 deadline and, in fact, has put the sludge to beneficial uses on land, including reclamation of barren strip-mined land, and soil supplements in the city and on home lawns and gardens. Camden halted its ocean-dumping in 1978 and now uses its sludge for composting material.

The city of New York and the northern New Jersey municipalities still are dumping their sludge in the ocean. The EPA currently is reviewing its deadline for New York to halt that disposal practice, following a lawsuit brought by the city which still is pending in the courts. That leaves the six municipalities, which have asked the federal court in Newark to invalidate portions of the EPA's ocean-dumping rules in order that they can continue ocean disposal.

One of the most noteworthy aspects of the lawsuit is the apparent shift in thinking on the part of the EPA under the Reagan administration. EPA officials have expressed a willingness to allow New York and the New Jersey communities to continue dumping if they barge the sludge further offshore. The waste materials are now dumped 12 miles out in the New York Bight, a shallow body of water off northern New Jersey that is considered to be the most degraded coastal area in the country.

The National Wildlife Federation and its affiliate, the New Jersey State Federation of Sportsmen's Clubs, have gone to court to defend the federal ocean-dumping regulations as applied to the New Jersey communities. According to a lawyer for the wildlife organization, "if we hadn't gotten involved in this lawsuit, there would be no one arguing on behalf of the oceans."

The New Jersey municipalities claim the alternatives to ocean disposal of sludge are unsatisfactory — an argument raised in the past by Philadelphia and Camden. That argument proved fallacious in the 1970s, even when technology was not as far along as it is today. The experience of Philadelphia and Camden, plus the technical advances in the intervening years, render it even more meaningless.

EPA's lack of enthusiasm for enforcing the Marine Protection, Research and Sanctuaries Act is a cause of alarm. Scientific research in the past decade has indicated that perhaps the threat of ocean dumping of toxic materials, including sewage sludge, may not pose as serious an immediate threat to the world's environment as was feared when the act was passed. But even the most optimistic scientist cannot predict what the long-term effects will be. There are proven methods of disposing of the materials on land. Ocean dumping of sludge must be halted.

The Courier-Journal

Louisville, Ky., November 9, 1982

THE GOOD NEWS is that the oceans — if you go far enough from shore to escape large pockets of contamination that infest many bays and shorelines — are still relatively unpolluted. In fact, the high seas are cleaner than they were 10 years ago, according to an impressive United Nations study conducted by nearly 100 scientists from three dozen countries.

The study, however, leaves no room for complacency. It's something like a patient being told he doesn't have the bubonic plague, though the organisms that cause it are all around.

If the time does arrive that the oceans become as contaminated as are many of our fresh-water lakes and streams, this planet may well be uninhabitable. The tiny plankton that are the basis for all marine life not only produce, indirectly, much of the world's protein, but regenerate much of its oxygen. A decade ago, many marine scientists feared that a crisis was possible by the turn of the century.

But changes were being made. In 1972, Congress passed the Marine Protection, Research and Sanctuaries Act, more commonly known as the Ocean Dumping Act. At that time, significant concentrations of such dangerous chemicals as PCBs could be found in the open ocean. Now, says one scientists who took part in the study, PCB concentrations are barely measurable.

The Ocean Dumping Act, of course, is due only part of the credit for the improvement. Most of the developed nations have banned DDT and other such long-lasting destructive pesticides, and restricted the disposal of the more dangerous toxic chemicals. Marine scientists now concentrate much of their worry on the Southern Hemisphere, where underdeveloped nations have few restrictions on such chemicals, and are even increasing their use.

But most oceanic pollution has come from the Northern Hemisphere's industrialized nations. So it is a relief to find that efforts to clean up pollution are bearing fruit. It's also somewhat reassuring that scientists are more confident now of the ocean's ability to absorb and, in many cases, render pollutants harmless.

Amount of lead declining

"Somewhat" is the key word, because scientists aren't unanimous on the eventual result of the buildup of pollutants in the ocean. Though much knowledge has been acquired, the marine sciences are still in their infancy.

What isn't controversial, however, is the view that mankind is safer because of environmental protections put in place within the past two or three decades. The U. S. Clean Water Act, along with similar legislation in other nations, has played an important role.

Even laws passed without the ocean in mind, such as America's Clean Air Act, have helped. The upper quarter of the seas, according to researchers at the California Institute of Technology, now contains 10 times more lead than before man began releasing lead-containing fumes into the atmosphere. But the total is decreasing because of curbs on leaded gasoline. Presumably, it will decrease much more as leaded gasoline gradually goes out of use.

The danger now is that optimistic reports like the U. N. study will feed complacency, making things easier for those willing to turn back the environmental clock. Those forces already are hard at work, especially in the U. S. The Reagan administration's push to weaken the clean air and clean water laws — and to relax curbs on dumping in the ocean — could reverse this healthy trend in its infancy. Of all the foolish risks for mankind to take, that would be among the least defensible.

SYRACUSE HERALD-JOURNAL

Syracuse, N.Y., October 21, 1983

The Environmental Protection Agency (EPA) has agreed — at least tentatively — to a questionable plan for dealing with hazardous wastes. It may catch on in popularity because it would allow officials to sidestep the politically touchy problem of how to dispose of toxic waste in their home regions.

EPA has tentatively agreed to let Chemical Waste Management Co. burn hazardous wastes in the incinerators of two ships while they are at sea in the Gulf of Mexico.

The system would be used to burn 80 million gallons of liquid wastes, including PCBs, over the next three years. One ship has the capacity for incinerating about 837,000 gallons each voyage; the other, 700,000 gallons.

▽   ▽

What happens if one of the ships sinks? Since these chemicals are the most dangerous known to man, the devastation to marine life in the fertile Gulf would be incredible. A major seafood region would have to be put off limits, undercutting the livelihood of hundreds of thousands of people in the fishing industry.

The risk of a spill is admittedly small but it exists. Can we aford to take it?

▽   ▽

Even if there is no spill, this plan could still endanger marine life. Ash from the burned 80 million tons would be dumped. An EPA official said the tentative permit specifies that 99.99 percent of the wastes must be destroyed in any burn — a standard described as 10 times more stringent than what EPA required under a 1981-82 research permit for the burning of some Agent Orange. However, the 99.99 percent standard would still allow 30 tons of hazardous wastes to be dumped into the ocean.

It may be this plan makes the best of a bad problem but the EPA is proceeding too rapidly. This is not, after all, a garbage burn in the back yard. It's taking a risk with the life of a great body of water.

That merits a great deal more thought.

The Morning News
Wilmington, Del., April 30, 1984

THE EFFECT of the Environmental Protection Agency's action Friday depends on where you look at it from. New York headlines called it a move to *end* ocean dumping at a site 12-miles off northern New Jersey. Delaware and southern Jersey tend to look at it as a *beginning* of dumping 106 miles at sea, but at a lower latitude, closer to their shores.

Delaware's congressional delegation stands stoutly at the barricades in opposition to the change, with Rep. Thomas Carper warning that ocean currents would make the new site more of a pollution threat to the First State's coast and questioning why ocean dumping should be permitted at all.

New York officials bewail the additional cost of taking the treated sludge 90 miles or so farther to sea. The cost in money will be considerable if the EPA puts the shift into effect in October, as it hopes to. The EPA figures the increased cost would amount to just about $2 a year for each and every one of the 12 million persons in the communities, including New York City, that do the dumping. That's a lot of extra money, but these communities, as a result of their various activities, generate a lot of this waste, in fact 7 million tons a year.

The EPA argues that the new site, with water depths up to 8,000 feet, will be "much more dispersive" than what it calls the deeply degraded 12-mile site used for this dumping for 60 years. And, says EPA, "dispersion seems a better answer for sewage sludge."

The answers are not all in on effects of sludge at sea. Some folks, including New York officials, deny it harms marine life. New York City's environmental director last month denied any substantial metal contamination at the dump and insisted there was no fish disease. "There are no dying fish out there," said Joseph T. McGough Jr. (Could he have meant *any more*?)

Other environmental officials charge that 60 years have left the 12-mile dump site completely devoid of life except for "sludgeworms." (These, *tubifex tubifex*, apparently are not an especially attractive form of wildlife for neighbors but can live where there is very little oxygen, as in sludge.)

EPA does not, however, propose to cut any answer in stone.

It says it would seek to halt the 106-mile operation after five years. "We are not looking for ways to continue ocean dumping," said a spokesman. "The idea is to stop that."

Maybe having to spend an extra $25 million a year (EPA estimate) to carry the waste farther to sea will accelerate New York's efforts, which the communities claim are already advanced, to initiate an alternative and better form of disposal.

The State
Columbia, S.C., December 19, 1983

AS THINGS stand now, the Environmental Protection Agency is going to permit highly toxic carcinogenic wastes to be burned in the Gulf of Mexico on a specially built ship and then dumped into the water.

The EPA has not developed any regulations for this process, yet is prepared to issue a license for three years. EPA says, however, that the technology of incineration at sea is well understood and well developed.

A man who knows more about the oceans than any other living person, explorer Jacques Cousteau, declares, "I can think of no worse place for such dangerous chemicals to be transported or destroyed than the sea."

It may turn out that such disposal of these wastes is reasonable, but we fail to see why it should be permitted without regulations based on thorough research and with regard to the possible consequences.

"In hindsight, I guess we should have gone the one additional step in developing regulations specifically for incineration," an EPA spokesman admitted to a congressional committee last week.

It isn't too late — yet.

THE DAILY HERALD
Biloxi, Miss., December 1, 1983

When it comes to environmental issues, we should be immune to surprise by now at how unprepared the federal government is in dealing with potentially dangerous issues.

Case in point: A top official of the Environmental Protection Agency, testifying last week before a Congressional subcommittee, acknowledged regulations have not been developed to govern incineration of toxic wastes aboard ships at sea.

The EPA had tentatively decided in October to issue permits to two firms to begin incineration of toxic wastes on two ships at a site in the Gulf of Mexico about 200 miles off the Texas coast.

But as testimony developed before the Merchant Marine and Fisheries wildlife conservation and environment subcommittee, the EPA official backed off with the remark, "What I hear the Congress saying is they would like to see those regulations in place," before permits issued.

Question: Why would it take a Congressional inquiry for an agency charged with protecting the nation's environment to determine sufficient regulations should be in place before any substance is disposed of which might pose a threat to the environment and the well-being of humans and wild life?

We're not saying the proposed incineration at sea of toxic wastes such as PCBs and dioxin is a proven hazard to marine life and marine industries. We understand the distress of the waste management industries which applied for permits over two years ago. And we recall that several years ago leaking drums of Agent Orange, stored since the Vietnam war, were hauled away from the Seabee Base at Gulfport to be incinerated aboard ship in the Pacific Ocean as the best option to dispose of the potent chemical.

But we are saying, emphatically, that EPA must know what potentials it is dealing with before implementing permits are issued.

The promise of nuclear energy became a nightmare when plant operations were permitted before proven methods of disposing of long-term radioactive wastes were developed or sites chosen. We're paying for that error today and quiver at the thought that a salt dome at Richton, Miss., is being considered as a depository.

The seafood industry which is so important to Gulf Coast states relies on a delicate balance of marine environments.

Noted oceanographer and conservationist Jacques Cousteau warned at the Congressional hearing that waste disposal, without knowledge of environmental impact, threatens life chains on which we all depend.

We trust the EPA will develop conclusive information on which to base issuance of incineration permits. And have the intelligence to deny permits if the evidence weighs against such a method.

San Francisco Chronicle
San Francisco, Calif., April 3, 1984

AS DISPOSAL of toxic wastes in the most efficient — and least risky — manner possible becomes one of the vital goals of this decade, the plan being considered by the Environmental Protection Agency to allow the burning of thousands of tons of the stuff off the California coast has its fascination. But it is also one not to be embarked upon lightly.

The proposal would license oceangoing "incineration ships" to burn a soup of toxic liquids — including PCB, DDT and dioxin — at an offshore site halfway between San Francisco and Los Angeles. A report from a Texas-based consulting firm strongly recommended the burning site 200 miles off Point Arguello in Santa Barbara County. It said the potential site is one "at which ocean incineration can be carried out without fear of impinging upon other uses of the sea, of detracting from the value of human amenities, or of jeopardizing human health."

That sounds fine. The site is far from fish-breeding areas, shipping lanes and bird and mammal migratory routes. But how certain are these assurances? The recent lesson of the Gulf of Mexico should be taken into account before further steps are made toward achievement of this sea-borne incinerator.

THE EPA CONDUCTED test burns in that Gulf, and these came under strong criticism from Texas Governor Mark White.

"EPA has not set standards for burning temperature, the amount of time the waste will be burned, or what happens to the residue," said Sue Stendebach, natural resources analyst for the governor. "We believe EPA needs to promulgate specific ocean incineration regulations before they can issue permits."

Texas has called upon the EPA to do more safety studies, contingency planning in the event of an emergency and to ensure adequate liability for a major catastrophe.

That seems like just basic common sense, for the risk in this kind of an operation is high. The biggest danger stems, of course, from some unforeseen danger like a spill. It is also one thing to say that a burn ship could be 99.99 percent efficient. But some hard-and-fast set of regulations about allowable emissions will also have to be in place before this kind of operation can be allowed.

THE DAILY HERALD
Biloxi, Miss., May 25, 1984

The Environmental Protection Agency's ban — at least temporarily — on the burning of millions of gallons of hazardous wastes in the Gulf of Mexico is a commendable display of governmental caution. The agency is being conscientious, rightly so, in meeting its responsibility to safeguard the lives and property of American citizens.

Chemical Waste Management Inc. wants to incinerate 79.7 million gallons of banned toxic materials in 100 burns aboard two ships, the *Vulcanus I* and *Vulcanus II*. The plan has drawn objections in two states.

Citizens in Alabama, where the poisonous materials would be loaded onto the ships, are unhappy about the idea. And in Texas, a lot of people are worried about accidental spills or discharges off their Coastline, where the burns are scheduled to take place.

Mississippi has adopted a neutral position on the grounds that the hazardous materials won't go through either of its state ports and the planned burn area is far enough away. The state would have some exposure should one of the ships have an accident en route from Mobile to the burn area.

EPA's decision Wednesday not only withheld the permit for the massive incineration, it also rejected the recommendation of one of its own hearing officers, who wanted to allow Chemical Waste to do a test burn of 3.3 million gallons.

Instead, Jack Ravan, EPA assistant administrator for water programs, ordered a comprehensive research project to consider other ports and incineration sites in the Atlantic and Pacific Oceans in addition to those in the Gulf of Mexico.

EPA's cautious attitude and concern for public safety ought to be imitated by the Department of Energy, whose officials seem oblivious to the hazard that locating a nuclear waste repository in a Perry County salt dome poses for the growing population of South Mississippi.

FORT WORTH STAR-TELEGRAM
Fort Worth, Texas, May 29, 1984

It is unfortunate that the world is not flat, as it was thought to be for centuries.

If it were, the problem of hazardous waste disposal would be easily solved. It would only be necessary to package up the PCBs and other unhealthy byproducts of industrialized human society, take them to one of the edges of the earth and give them the old heave ho.

The best way to dispose of toxic wastes in an elliptical world, some say, is to ship them out into a large body of water and burn them. That, however, is not a universally shared view. While some countries burn toxic wastes in the Atlantic and Pacific Oceans, incineration of hazardous wastes has run into vehement opposition in this country.

The Environmental Protection Agency inclined a sympathetic ear toward that concern last week by denying a permit to Chemical Waste Management of Oak Brook, Ill., to burn 3.3 million tons of hazardous wastes aboard the incinerator ships Vulcanus and Vulcanus II at a location 195 miles off the Texas coast. The incineration, which was opposed by state agencies, environmental groups and some members of Congress, was to be done under the aegis of research. But critics contended that more research is needed in-to the effects of burning toxic wastes at sea before research projects of that scale are allowed.

EPA assistant administrator Jack Ravan made the decision to block the planned PCB and DDT burning against the advice of his staff. That was a prudent and appropriate exercise of Ravan's authority. The simple truth is that not enough is known about what residues are left when toxic wastes are burned and what transformations they may generate in the air and water. Even less is known about the ultimate effects upon marine life when such materials are burned at sea.

The effect of Ravan's decision will be to postpone large-scale commercial toxic waste incineration by U.S. companies for at least two more years, while the necessary studies are being carried out. Eventually, however, ocean-incineration appears to be destined to be the wave of the future for large-scale toxic waste disposal.

It is vital, therefore, that ocean-incineration regulations, based upon thorough research, be developed to reduce ecosystem damage and human health risks to the minimum. Considerable research also must be devoted to finding the best and most distant sites from Texas for those burnings.

Newsday

Long Island, N.Y., June 29, 1984

There are two things wrong with a pending federal Environmental Protection Agency decision on sludge dumping: It's overdue and it's still tentative. The dump site, which is now 12 miles off New York Harbor, should have been moved long ago to a new location farther offshore.

In 1977, Congress grew tired of the EPA's sluggish response to the problems associated with dumping sewage sludge in the ocean. So it amended the Marine Protection Act to prohibit such dumping as of Dec. 31, 1981. During debate on that amendment, Rep. Robert Leggett (D-Calif.) warned: "We cannot continue to expect the oceans to act as a food resource, recreation center and cesspool."

But a federal judge, after hearing a suit brought by New York City, ruled that the dumping of sludge at sea could continue pending studies of the long-term effects of other methods of sludge disposal.

Now — with Congress threatening to intervene once again — the EPA has decided *tentatively* to order sludge dumping in this region shifted to a site 106 miles off the New Jersey coast, thereby increasing New York City's disposal costs. At recent hearings, Long Island fishermen vigorously endorsed the site change. But New York City's Environmental Protection Commissioner, Joseph McGough Jr., said the city would go to court again if the change were made. Since a suit would probably cause additional delay in ending use of the present dump site, the sooner the EPA makes a final decision to move it, the sooner the entire controversy can be resolved.

The Oregonian

Portland, Ore., January 1, 1984

Reckless ocean dumping of nuclear or other toxic wastes cannot be condoned, but this should not rule out efforts to develop a safe technology for the careful burial of dangerous wastes in ocean-bed geological formations that prove stable.

Many nations of the world do not have a land option for their toxic wastes. Denmark, for instance, has been examining an ocean option for high-level nuclear wastes because it has no good land prospects.

Efforts to find safe ocean burial grounds must not become excuses to derail or slow down efforts under way in the United States to develop land sites for storing high-level nuclear materials.

These sites are needed before the next century and long before suitable oceanic sites can be located, tested and developed. Further, land sites are still the best where they are available. They do not involve international agreements against ocean dumping, and they benefit from a better understanding of their geology than that under the world's oceans and seas.

The Florida Institute of Technology, which is one of the centers of research on oceanic nuclear waste disposal, is the headquarters for a global effort that is supported in part by the National Oceanic and Atmospheric Administration.

These efforts need encouragement because more knowledge is needed about ocean beds, their underlying materials, the effects of earthquakes, volcanism and water movements. While the ocean has disadvantages compared to nuclear storage sites on land, it is closer to many nations that need storage help. If natural caves and man-made tunnels can be utilized, suitable nuclear burial sites may be found that will not degrade ocean life or the human food chain.

This kind of nuclear burial ought not to be confused with the dumping of low-level wastes in corrodible cannisters that was all too common in the early years of the nuclear age. Even the Pacific Ocean off the mouth of the Columbia River, scene of a major fishery, became a nuclear dumping ground until dumping was halted for fear of poisoning fish. No notion of safe waste disposal encompasses that kind of dumping.

The Star-Ledger

Newark, N.J., February 26, 1984

The United States Environmental Protection Agency is blithely playing a dangerous pollution game with New York City, one that is harming the Hudson and East Rivers, the New York Harbor, as well as the recreation waters along the Jersey Shore and Long Island.

The game has been going on since soon after EPA came into existence in 1970. In this hypocritical game, EPA allows New York City to do what it wants to the regional environment, while New Jersey and the rest of America are forced to obey tough anti-pollution laws. The game is called Pollution Control.

Congress passed the Clean Water Act in 1972, requiring all municipalities to build sewerage treatment plants to protect the nation's public waterways. New Jersey launched a $3 billion construction program, while New York City initiated a $2 billion effort.

But in 1973-74, EPA came down hard on New Jersey by imposing restrictions on population growth and development in Ocean County, a wide-open area in the process of building three large sewerage treatment plants. The federal agency cracked down on human activities in New Jersey's heartland by reducing the size of the biggest treatment plant and effectively limiting growth.

At the same time, EPA gave the Big Apple the green light to build as many skyscrapers and "mini-city" projects as it wanted, regardless of the water pollution impact on the metropolitan region.

In 1972, the then world's largest edifices went up along the Hudson River—the Trade Center twin towers whose 7,000 toilets flushed directly into the New York Bay. It took New York five years before the raw sewage was routed to a treatment plant in Brooklyn, a facility which to this day sends poorly treated effluent into the East River.

EPA then allowed New York State and City to fill in the lower Hudson River and create the $2 billion Battery Park City whose sewage would also go to the dilapidated Brooklyn treatment plant in need of more than $100 million worth of repairs.

Throughout the 1970s, Manhattan Island went on a building binge, throwing up one skyscraper after another, the wastes going right into the Hudson and East Rivers.

By 1976, so much filth was pouring out of the Hudson River Valley that the Atlantic Ocean suffered an unprecedented pollution disaster. From Sandy Hook to Long Beach Island, millions of fin and shellfish were wiped out.

* * *

A public protest led by fishing and recreation interests put pressure on Congress to adopt the 1976 Ocean Dumping Prohibition Act. The federal lawmakers gave municipalities and industry until Dec. 31, 1981, to stop dumping sludge into a place known as the "Dead Sea" off the Jersey Shore—plus other wastes discharged out in the ocean.

New York City brazenly challenged the federal law in court, taking the position that sludge dumping was an economic necessity. Ocean disposal obviously is expedient and inexpensive.

EPA, for reasons never made known, went along with the New York federal court judge's decision to allow the city (and, later, other dumpers) to continue using the Dead Sea as the world's largest cesspool.

To somehow make New York City comply with the 1972 Clean Water Act, concerned citizens demanded that EPA take the city to court to speed up construction of its wastewater treatment plants. New York was placed on an extended timetable to complete primary treatment plants for the East and West Sides of Manhattan, where some 250 million gallons of raw sewage flow into the tributaries daily.

The first stage of Manhattan's sewer plants will be completed by 1986-87 and the second and final stage by 1992 . . . perhaps. New York does not want to go to the better secondary treatment system because of the extra costs.

New Jersey's biggest sewer plants already are at secondary, or 90 percent treatment.

The water pollution issue remains unresolved. New York City flagrantly disregards and challenges the laws of the land while continuing to build one superstructure after another, further spoiling the region's environment where some 20 million people must live, work and play.

And EPA continues to look the other way, an indifferent bureaucracy that has been rendered shamefully impotent by New York's short-sighted politicians.

Clean Water Act:
Reaching a Compromise on Pollution

The Clean Water Act of 1977 gave the Environmental Protection Agency the authority to regulate discharges of industrial waste into lakes and streams, and provided federal funds to aid cities and counties in building sewage systems. The act gave industry until 1984 to clean up toxic and conventional pollutants, and until 1987 to clean up other pollutants. Since its passage, the legislation has been criticized by environmentalists as too lax, and by industrial representatives as too severe. Authorization for government spending to enforce parts of the act lapsed in 1982, but Congress, unable to find a middle ground in attempts to revise the act, has provided funds through annual appropriations bills.

The long impasse over the revision of the act appears in 1984 to be easing somewhat; a House compromise bill renewing the act for four years was finally passed June 26 on a vote of 405–11. The overwhelming support for the measure reflected a compromise among contending forces of environmentalists and industry lobbyists reached several days before the vote. The environmentalists won sterner deadlines for cleanup by certain industries, including auto-assembly plants and chemical manufacturers. Industry won a longer extension of water-discharge permits, to 10 years from the current five years. Other provisions of the bill called for stricter monitoring of ground water contamination, raised the maximum civil penalties and criminal penalties for violations of the act, created a new funding program to help states control "non-point" pollution (runoff from broad areas rather than emissions from specific pipes), authorized state grants to restore the quality of acid lakes, and established cleanup programs for the Chesapeake Bay and Great Lakes. The House bill also provided $5 billion annually in grants and loans for the construction of municipal sewers and treatment plants. (See pp. 154–162.)

The future of the House bill is uncertain, as it faces opposition from the Reagan Administration and from fiscal conservatives in the Senate.

The Washington Star
and Daily News
Washington, D.C., December 4, 1980

Everyone, we guess, wants clean rivers. Everyone also wants heat, electric power and what is now called "energy independence" for the U.S. But reconciling these competing wants is not always easy — or so suggests an important Supreme Court decision this week.

The Court was asked to say whether or not Congress, in passing the 1972 Clean Water Act, contemplated pollution-control regulations so strict as, in some cases, to drive marginal coal-mining and rock-crushing companies out of business.

The Environmental Protection Agency, which administers the act, had issued regulations calling for the installation of "best practicable technology" by 1977; but it was argued that several dozen companies might be forced out of business by the additional cost of new cleanup technologies unless more lenient variances were allowed.

The Court's finding, which was unanimous, is that indeed Congress *did* accept, and even anticipate, that the water-pollution cleanup might ruin some mining companies. And while it provided for government loans for the purchase of pollution-control technologies, it had not meant to abate the act's standards in the face of economic hardship.

The consequences of this decision are not yet fully known, and probably won't be for some time. Some of the petitioners for leniency (who had found a kindlier ear in the Fourth Circuit Court of Appeals, now reversed) may have been crying "wolf!" and may find some hidden mattress money to stay within the law rather than close. Or the decision may mean the consolidation of some smaller independent mining operations into larger, more efficient ones, flush with capital. But there is always the possibility, too, that the draconian policy will mean that less coal is mined — which would, of course, have its effect on other competing goals of national energy policy.

A further possibility is that with the change of administrations a change of personnel or regulatory philosophy at EPA might relax clean water standards a bit. Whether that relaxation would be compatible with the law as now interpreted, or contemptuous of it, is not entirely clear. The Supreme Court's judgment is unanimous; but the Fourth Circuit judges, who are not slouches at the interpretation of statutes, read the Clean Air Act in a very different way.

In any event, the case of *EPA v. National Crushed Stone Association, et al.*, is another reminder — as if one were needed — that we can't always have absolutely everything we want — not, at least, in the quantities we are accustomed to. The price of water cleanup appears to be too high for quite a few small energy producers to bear, and that may mean the production of less energy — period.

There is, of course, a sort of thaumaturgical school of energy sufficiency which tends to blame all such problems on the interference of government — which, ended, would solve them forthwith. The president-elect has sometimes talked of energy-environmental issues as if he belonged to that school. But here the choice appears to be considerably more complex than between more government and less. It may even be between more energy and less, or at least between more clean water and less. Other such hard choices are probably pending. But not to worry. Mr. Reagan will learn for himself soon enough.

AKRON BEACON JOURNAL
Akron, Ohio, December 6, 1980

EVEN THOSE most indignant about what they regard as unreasonable excesses in protection for the environment have no grounds for crying "judicial tyranny" in response to this week's Supreme Court decision that environmental standards can be legally enforced even when they put companies out of business.

The culprit, if indeed there is a culprit in this, is not the court but the Congress. In his writing of the 8-0 opinion, Justice Byron White made this abundantly clear — so clear, in fact, that it becomes hard to understand how a lower federal court could have reached an opposite conclusion.

The 1972 law whose enforcement was being challenged in this case does indeed raise the economic feasibility issue, he pointed out — but in connection with the much toughened standards for 1987, not those that took force in 1977.

The Clean Water Act directed the EPA to draw up for enforcement starting in 1977 a set of standards based on the "best practicable control technology currently available." The more stringent standards it called for starting in 1987, in contrast, were to be based on the "best available technology economically achievable."

Any argument that the word "practicable" was intended to bring into consideration whether violators of the standards could afford to clean up their acts is belied by both the presence of the two differing phrases in the act and some other elements in the record.

The Congress, it seems clear, knew what it was doing and knew that its action was likely to kill some enterprises — those with the oldest and highest-polluting equipment.

In fact, Justice White pointed out that one report to the Congress before passage of the act estimated that enforcement of the 1977 standards would cause 200 to 300 plants to close because their owners could not afford to comply.

It should be obvious, on reflection, that enforcement of any environmental standards at all is certain to impose economic hardship on some. There is no cost-free way to do this, and yet it has been the judgment of the Congress and most Americans that it has to be done anyway.

The trick has been to get it done in such a way that the cost-benefit ratio involved be sensible — that the guide be the greatest benefit to the greatest number with the least unnecessary disruption of the economy.

But whatever your judgment as to how well this guideline has been followed, don't blame the court and don't blame the EPA for what has been and is being done.

The responsible agent is the Congress, acting on the basis of what was perceived as a broad American consensus that we simply cannot afford to risk choking ourselves out of existence with our own wastes.

Newsday

Long Island, N.Y., December 7, 1980

When Congress passed the Clean Water Act of 1972, the lawmakers realized that its pollution-control requirements might force some businesses to close because they couldn't afford to clean up. Even so, there's been a lot of stalling since then as a result of court decisions in favor of companies contending that costs were too high.

But now the Supreme Court has ruled unanimously that a particular operation's inability to meet the standards set by the act doesn't mean it's entitled to a variance from the rules.

That's good news for the public, which has every reason to be concerned over the quality of the nation's water.

The rules in question became effective in 1977, but a myriad of lawsuits have prevented their implementation in many areas. Now that the high court has spoken—convincingly and unequivocally—the worst polluters will have to clean up or shut down.

The outlook for the future is not quite so clear, however. A great deal will depend on the zeal with which the next administration in Washington goes after the polluters.

If President-elect Ronald Reagan's comments during the election campaign are any guide, zeal may be sorely lacking in this area. A new and more conservative Congress may also be tempted to amend the act to give more weight to the cost argument.

This is not the time for that, in our opinion.

There's nothing wrong with taking costs into account, at least not on an industry-wide basis. The trouble is that the costs of pollution controls are normally not very difficult to measure, while the environmental and medical benefits of clean water may be. Yet every week—or so it seems—brings new evidence of the harm past carelessness or ignorance has wrought.

Congress and the Reagan administration need to keep in mind that the cost of pollution controls has often been unfairly blamed for the decline of companies that were in poor competitive positions anyway. Plants that are old and costly to clean up are liable to be using outdated or inefficient equipment in other areas as well.

The 1977 pollution controls are supposed to be based on the "best practical control technology currently available," but starting in 1987 the rules will become even stricter; the standard then will be the "best available technology economically achievable."

By that time, or so we hope, some of the worst pollution problems ought to have been taken care of. And by then scientists should also have a better idea of the relationship between polluted water and public health. If so, that seems a more appropriate time to factor costs into the statutory equation.

THE DAILY HERALD

Biloxi, Miss., February 4, 1982

A five-year extension to the deadline for compliance with standards of the federal Clean Water Act is certainly welcomed by Mississippi Coast communities which have been struggling to comply with the previous 1983 target date.

However, city and county officials should not interpret the new deadline of 1988 as reason to relax their efforts toward achieving compliance.

Rather, we hope our officials will consider the extension as an opportunity to consider with greater care the steps to be taken in organizing wastewater districts along lines which will best serve their political entities at the most reasonable cost to users and taxpayers.

Over the past several years, the efforts to comply with the clean water mandate have been educational, at times confusing.

There has been no dissent with the act's objective — to eliminate pollution caused by inefficient or incomplete wastewater collection and treatment facilities.

There has been considerable controversy over how to go about achieving the objective.

The initial concept advanced by federal and state planners of a regional wastewater system embracing the three coastal counties ran afoul of both political and geographical obstacles.

Although Jackson County and its communities opted to join in the regional plan, both Harrison and Hancock counties, and their municipalities, had serious reservations, principally as to cost, efficiency, long-term commitment and control.

As a result, it now appears most likely that the Coast will wind up with wastewater districts on a county-by-county basis.

Legislation to create a Harrison County Wastewater Management District is in the mill at Jackson and, as we have commented previously, appears to constitute a more homogenous entity than was offered by the regional plan since it links the county only with its four cities of Biloxi, Gulfport, Long Beach and Pass Christian.

Similarly, Hancock County has voted to affiliate with Waveland in a wastewater district which the city of Bay St. Louis is being urged to join also.

Significantly, Charles L. Blalock, executive director of the Mississippi Department of Natural Resources, is on record as stating his office now "will seriously entertain the concept of individual county-wide authorities" for handling of wastewater management in lieu of the original tri-county regional concept.

This, we believe, is a bow to reality and offers the opportunity for the counties and their municipalities to come up with agreements on engineering and cost proposals which will be more advantageous than the original approach which sought to promote a forced regional marriage.

Recent changes in the Clean Water Act made by an economy-minded Congress will continue federal grants on the basis of year-by-year appropriations but, beginning Oct. 1, 1984, will reduce eligibility for federal funding from 75 percent to 55 percent of project cost.

Thus, local political entities will be inclined to exercise caution in commitments to systems which will require a greater participation of local funding.

Extension of the compliance deadline eases the sense of urgency but should not be taken as an occasion for foot-dragging. Rather, it is an opportunity to continue careful, detailed planning and analysis on the costs and efficiencies of proposed treatment facilities.

The Philadelphia Inquirer

Philadelphia, Pa., December 8, 1980

The people who have been fighting for the last several years to clean up the environment won a big one last Monday. The U.S. Supreme Court, in a unanimous ruling, declared that those who pollute the nation's water can no longer give the excuse that it's too expensive to clean up their acts.

Representatives of the coal, crushed-stone, sand and gravel industries had argued that the federal Environmental Protection Agency, in enforcing federal clean water laws, should be ordered to make exceptions to its regulations for plants that were "doing all that the maximum use of technology within (their) economic capacity will permit." The trouble with that language is that any company could choose to spend all its money on production equipment and staff, leaving no funds for pollution-control equipment. Then, when the EPA came around, all the company would have to do would be to show the excuse that it couldn't afford to meet the standard.

The court, looking into the congressional discussions that led to the writing of the clean-water laws, noted that the Congress had carefully weighed the economic consequences of its acts. Specifically, the framers of the laws had received a report estimating that as many as 300 plants could be forced to close when the first set of pollution standards went into effect. Partially to offset this, the Congress made low-cost loans available to hard-pressed small businesses.

But the Congress, in weighing the short-term effects of plant closings against the overall long-term benefits to be gained by removing sources of pollution from America's water, properly decided that, even if some plants wouldn't be able to afford to do the things necessary to meet the standards, the nation couldn't afford to continue to accept the harmful effects of the pollution they caused.

As it turned out, the number of plant closings forced by the regulations has been much smaller, and the Supreme Court's decision merely recognized the critical public benefit to be gained from that legislative imperative.

FORT WORTH STAR-TELEGRAM
Fort Worth, Texas, September 28, 1982

There has always been an uneasy relationship between the business community and the Environmental Protection Agency. Industry, ever conscious of costs and profits, has frequently resisted EPA efforts to clean up the air and the water and has been criticized for its reluctance.

Now the EPA itself is under attack, not for seeking too much but for reneging on some rules, and the issue is one which could make it even more difficult in the future for the agency to convince industries of the need for prompt compliance with its regulations.

Relaxations of some environmental rules have angered those companies that had gone ahead and spent millions to change their operations in order to meet EPA standards. In the refining industry, Exxon, Texaco and others recently criticized an EPA proposal to allow small refiners to use more lead in gasoline, saying it

penalized companies that had already invested in modern facilities in order to comply with the standards.

Now the Bacardi Corp., distilling rum in Puerto Rico, says it is at a similar competitive disadvantage because of the EPA. Starting in 1979, after several years of negotiation and foot-dragging, Bacardi spent $10 million to clean up water pollution, installing highly sophisticated equipment in two of its distilleries to meet standards the EPA set for the entire industry.

Other U.S. Caribbean distillers had been less forthcoming about building the expensive treatment facilities. Recently, the EPA responded to pleas from Bacardi's competitors by relaxing the standards and allowing greater dumping of molasses residue into the waters around Puerto Rico and the Virgin Islands.

And Bacardi is complaining.

Clean air and water regulations are a good idea. It is in the public interest to protect the environment with workable standards. But those companies have a point. The EPA needs to be even-handed about its enforcement and consistent in the standards it sets. Otherwise industry, and the public, will lose what confidence they have in such environmental projects.

That's the danger posed when Bacardi gets a rum deal.

The Kansas City Times
Kansas City, Mo., November 10, 1982

When the Environmental Protection Agency said it wants to change the water quality rules to give the states greater flexibility in setting their own standards, the reaction of environmental organizations was predictable. In the Reagan administration effort to ease what it considers burdensome federal regulation, "flexibility" has become a code word to the environmentalists for degrading the safeguards against pollution.

The EPA explains that the proposed changes would concentrate on local conditions rather than uniform national requirements, and give state governments more discretion in determining the use of a particular body of water. Certainly there are vast differences between American streams and lakes and the amount of their exposure to human activities, from a mountain creek in a sparsely populated area to a river lined by industrial plants in a teeming Eastern city.

Under the 1972 Clean Water Act passed by Congress just a decade ago over President Nixon's veto, the current official goal is "fishable, swimmable water" all across the nation by 1983, an idealistic goal certainly not attainable by next year if ever. But in the brief history of pollution abatement in this country, the fact that our legislative reach has invariably exceeded our grasp is not necessarily bad; it just requires some midcourse adjustments in goals.

Because of differing geography, geology and weather, lakes and streams differ in their capacity to cleanse themselves naturally. In some instances the best attainable use of a heavily polluted urban stream may be as a disposal point for treated wastes, with swimming beaches out of the question. Trying to squeeze a few more percentage points of quality out of an already greatly improved stream can involve costs totally disproportionate to the achievable benefits.

The concept of each state deciding, in the light of its own specific needs, just what clean water standards to set for each lake or stream, has some appeal. After all state governments too are subject to the will of an electorate alarmed by pollution. But if this led to industries jumping to states willing to downgrade water protection to unacceptable levels, that would be a setback. Detailed publication of the proposed EPA changes, and the ensuing period of public comment, will provide a fuller reading of what their effect might be.

The TENNESSEAN
Nashville, Tenn., October 27, 1982

THE Reagan administration is now proposing broad changes in water quality rules that would halt and reverse a decade of progress toward making the nation's waters safer and healthier.

The Environmental Protection Agency says the changes are designed to give the states more flexibility in meeting the goals of the Clean Water Act. But what they would really do is effectively undermine the act itself. The changes would weaken efforts to clean up toxic wastes such as lead, mercury, and DDT.

This would go back to the approaches of the states before passage of the Clean Water Act when the individual states set discharge limits for polluters based on the amount a body of water could tolerate. It was not a good approach, because there is difficulty in assessing the cumulative effects of various pollutants.

When Congress passed the Clean Water Act, it chose to have a "discharge control" approach that required the EPA to formulate nationwide uniform control regulations for industries that discharge pollutants. States were to supplement these controls with their own.

But the EPA, in looking toward the "water quality" approach of old, is also preparing to undercut the very state water quality standards on which the approach is based.

The new rules would alter the standards by which a state may change the designation of a stream classified for uses such as fishing, agriculture or industry. It would delete the expressions "upgrading or downgrading" on the grounds the words are misleading. Instead, the rules would permit actions such as "modifying" the use of a body of water.

Environmentalists say that is simply a euphemism for allowing states to downgrade more easily the designation of a particular body of water. The changes would permit states to compete for new "industry" by offering to weaken pollution standards or by changing the designation of a specific body of water.

In addition to changes in water quality standards, the EPA has already weakened the toxics standard for the textile industry, requiring, for example, the removal of only one chemical from wastewater. That would allow an estimated 450,000 pounds of toxic pollutants such as zinc and cyanide to flow into the nation's waters each year. Any further toxic cleanups would have to depend on the states, many of which are already having budget difficulties.

And, speaking of budgets, the cuts in the EPA's have reduced funding for municipal sewage plants by 28%, and cut 40% of the EPA's water program funds needed for monitoring, compliance, enforcement and research.

The indifference of this administration to environmental protection can be charted by a number of moves to weaken that protection, but none of them quite match what is to be a comprehensive assault on clean water. The administration tried to do some of that legislatively and that failed, so it is now determined to achieve its objectives through regulatory changes and budget cuts.

The nation has made substantial progress in cleaning up its waters in the decade after the Clean Water Act was passed. There is more to be done, rather than less, but the administration now is willing to undercut what has been done, no matter what the cost in human and environmental health.

The Times-Picayune
The States-Item

New Orleans., La., October 26, 1982

There are serious problems with the U.S. Environmental Protection Agency's proposed changes in federal water quality regulations. The first problem is that the changes might circumvent the will of Congress by negating the principal objective of the 1972 Clean Water Act. That objective was to be a federal-led effort to restore the nation's waters to a "fishable and swimmable" condition.

The EPA's proposal would allow state governments to determine the use and pollution standards for particular water bodies. The proposal undoubtedly faces court tests.

The EPA proposal is based on the appealing idea that states are in a better position than the federal government to decide if the best use of a water body should be for fishing or industrial uses and, therefore, should decide how polluted the water can be allowed to be.

Other than the obvious fact that Congress, in passing the federal Clean Water Act, decided that the responsibility for setting water quality standards should belong to the federal government, there are at least three major problems with shifting the responsibility primarily to the states.

The first is that many water bodies, such as the Mississippi River, are used by various state, municipal, industrial, commercial and agricultural interests simultaneously. Which state and/or interest will set the standards for a particular stream or other water body? Further, many water bodies converge. Would standards set for a section of the Ohio River, a tributary of the Mississippi, be sufficiently high to safeguard the drinking water requirements of New Orleans and other Louisiana communities?

The second main problem is that many water bodies serve a variety of uses simultaneously, so who is to determine whether pollution standards should accommodate a single use to the exclusion of others? Would, for instance, the industrial arm of the Louisiana Department of Natural Resources take precedence over the environmental branch in setting pollution standards for Louisiana streams?

The third and perhaps most serious problem has to do with foreclosing in perpetuity the potential for restoring any water body to a fishable or swimmable state. Some streams and water bodies, once given up as industrial sewers, already have been restored. The notion that any water body should be abandoned to industrial purposes exclusively would set a dangerous precedent that could lead to steady erosion of water quality nationwide.

With its proposed changes, the EPA could be getting into a regulatory thicket that, in practice, might prove unworkable and contrary to the best interests of the states and the nation.

The Providence Journal

Providence, R.I., October 21, 1982

The Environmental Protection Agency is saying it wants to give states more flexibility in pursuing goals of the Clean Water Act. To this end, it has proposed major changes in its rules. The changes may look good on the surface, but unwanted side effects could float up later.

Item: No rational person believes that EPA wants to give water polluters a free hand, but this agency's performance during the Reagan administration invites skepticism, not confidence.

Item: In environmental regulation, although not in all matters, to let individual states write their own

The water cleanup program will work best under some national direction

tickets is to prescribe a hodge-podge of 50 sets of water standards.

Many of the nation's streams cross one or more state lines. If Massachusetts were to let the Blackstone River deteriorate to a sheer Class E or "nuisance" category (nothing of the sort has been suggested), Rhode Island could never make its downstream part of the river "fishable" or "swimmable."

Nationally established controls, based on a consensus of how successfully the United States can attain water quality goals, must prevail.

Nevertheless, in a rules draft coming out soon for public comment, EPA will offer states much more flexibility in deciding the most technically attainable and economically satisfying use of rivers. If they wanted to, states could swap "an existing stream use for economic growth and development." Trout fishermen, goodbye; oil refiners, hello.

Arguments for state-by-state flexibility do have some validity. States could focus on highest-priority streams, regulating where local officials see a real urgency, rather than making every trickle of water fit a nationally decreed regulatory format. Ecological needs vary: what may be required by predominant fish species in Rhode Island doubtless differs from conditions that favor North Dakota's fish.

Indeed, a national association of state water pollution control administrators has welcomed EPA's proposed rule changes. The uncharitable thought arises that anything giving more jurisdiction and empire-building potential to state bureaucrats would surely be endorsed.

But overriding all concerns is the fear that gives environmentalists restless nights. They contend that because the Reagan administration has not yet persuaded Congress to amend the Clean Water Act, EPA is relaxing regulations and forcing reinterpretation of what the existing water act orders, not only in goal-setting, but in the degree of treatment required of municipalities and industries.

It is true that undiscriminating and uniform cleanup of every body of water is prohibitively expensive and scientifically unwarranted, but the states cannot be allowed an exclusive, case-by-case judgment. The gnawing feeling is that states might even compete in *lowering* water quality in order to attract industry.

If EPA is allowed by the Reagan administration to summon up its resolve and reassume some of its once-impressive enforcement powers, the water cleanup program will work best with a nationally applied orientation.

The Miami Herald

Miami, Fla., September 18, 1982

HOW much is clean water worth? The Reagan Administration has asked that question and offered the unacceptable answer that it's not worth as much as the American public has been paying for 10 years. The Federal Environmental Protection Agency therefore has "reinterpreted" the Clean Water Act of 1972 to allow cities and counties to increase the water pollution from sewage-treatment plants.

Congress should forbid this executive-branch usurpation of its environmental legislation. For 10 years, the law's standard was that municipalities should remove at least 85 per cent of the organic material from sewage effluent before discharging it into waterways. The Administration's new, lax standard of only 25 per cent should not be permitted.

There's only one motive for this assault on the Clean Water Act. That is the false economy of cheaper sewage-treat-ment plants. The General Accounting Office estimates savings of $4 billion to $10 billion if the EPA grants some 800 applications for the looser standards.

But what will it cost the American taxpayer in the 1990s to clean up rivers that have become open sewers? To restock fish and marine vegetation? To replace recreation areas whose waters have been fouled by a decade of gradually increasing pollution levels?

If this nation has learned anything in the post-World War II period of rapid growth and development, it has learned that the cost of repairing environmental damage is greater than the cost of preventing it. People who believe in the future and who care about the continent that their children will inherit do not plan systematically to dump feces in the water.

The Reagan Administration is wrong to propose allowing increased pollution. Congress should not stand for it.

Arkansas Gazette.

Little Rock, Ark., November 7, 1983

A little over a year ago, Mrs. Anne Burford, who was then administrator of the Environmental Protection Agency, proposed a relaxation of water quality standards that would have endangered many of the nation's lakes and streams. A new set of regulations issued this week by the new EPA administrator, William Ruckelshaus, will remove the threat posed by the Burford regulations and certainly will improve the nation's water quality standards.

In essence, Mrs. Burford's regulatory change would have greatly weakened long-standing rules on how states establish pollution standards under the Clean Water Act, a change that would have allowed industrial contamination of some lakes and streams.

Mr. Ruckelshaus' regulations will restrict the power of the states to weaken water quality standards, and, in addition, require states to set new limits for a wider range of toxic substances. In short, Ruckelshaus would make cleaner those waters that Mrs. Burford would have made dirtier.

It is a valuable and welcome step that Mr. Ruckelshaus has taken, especially at a time when Congress is considering reauthorization of the Clean Water Act. The EPA's hand would be strengthened if the Ruckelshaus regulations, or at least the principles of the rules, were incorporated into the act.

Mr. Ruckelshaus' progressive approach to water quality will stand on its own merits, but it also highlights the new administration's lack of action on protecting air quality, specifically in proposals for dealing with acid rain. Would that Mr. Ruckelshaus, without delay, gives as high a priority to clean air as he does to clean water.

Houston Chronicle

Houston, Texas, July 11, 1983

A new Clean Water Act is working its way through Congress, and there is the usual jousting between industrial groups and environmentalists about the wording and timing in various clauses. But basically the issue has been decided for years: The public wants strong legislation on clean water.

Those toxic dump sites dotted around the nation must be cleaned up. Cities must comply with strict regulations on sewage. Our streams should be able to support fish. There has been a great deal of progress in this direction, and much more will be done.

One aspect of this proposed legislation, however, may come as a surprise to the general public. For years, the polluter has been identified in the public mind as an industry with its pipes pouring pollution into a river. If it isn't an industry, it is some city that hasn't voted the bonds and built new sewage systems.

Less well known is the fact that these industrial and urban sources account for only about half of the pollution that enters our streams. The other half is called non-point pollution — and the source of that pollution is not some industrialist or bureaucrat but the public itself.

This includes someone who changes the oil in his car and then dumps the old oil into a ditch. Someone who tosses his trash out of a car. Someone who doesn't keep a construction site clean. Someone who uses chemicals to grow things, whether he is a farmer or a rose gardener. In fact, anyone who does anything that would lead to a pollutant finding its way eventually into a stream.

Congress, of course, will not pass a law directed in detail at such sources of pollution. Improved "management practices" at the state and local level probably will be sought. Farmers, contractors, loggers and the like will be affected the most.

The point remains, however, that as pollution from the "big guys" is brought under control, the actions of the "little guys" will be the key to further improving the quality of our water. Instead of pointing the finger indignantly at someone else, we are going to have to change our own ways if we really want clean water. Undoubtedly, that will prove more difficult than passing a new law.

Detroit Free Press

Detroit, Mich., November 9, 1983

BACK IN 1982, when the Environmental Protection Agency under Anne Burford was suffering through its Dark Ages, the EPA recommended changes in its regulations under the 1977 Clean Water Act. Like all such regulations, good and bad, they were couched in a murky bureaucratese, but when deciphered they appeared to be based on a few simple principles:

1) States should be permitted to weaken their water quality standards. If this resulted in a competition to lower the standards among states desperate to attract industry, well, isn't competition what made America great? 2) If cleaning up a stream or river would cost somebody money, maybe it shouldn't be cleaned up. 3) If nobody swims in a river or stream because it is already polluted, it's okay to dump more pollutants in it. 4) On the other hand, if you've got a really pristine waterway, it's okay to pollute that one just a little bit, too.

The reaction from state water quality agencies and the public was overwhelmingly negative. That in itself might not have deterred Ms. Burford's EPA from plunging ahead with the revisions, but earlier this year Ms. Burford departed under pressure, to be replaced by William Ruckelshaus.

Last week the new EPA director issued a new and very different version of the disputed regulations. Mr. Ruckelshaus' rules limit the states' ability to reduce water quality standards. They require states to establish limits on toxic discharges, to err on the side of safety when setting water quality standards, and to set out compelling economic reasons for allowing activities that will lower the quality of presently pristine waters.

In short, they reflect the original spirit behind the Clean Water Act: If it's dirty, clean it up, and if it's clean, keep it that way. Fortunately, unlike his predecessor, Mr. Ruckelshaus doesn't find that a radical idea.

Minneapolis Star and Tribune

Minneapolis, Minn., September 30, 1983

A river bursting into flames or thousands of poisoned fish washing ashore are dramatic events that incite public outcry. But fewer people notice when a river dies gradually, losing fish and plant life because of slowly rising levels of toxic substances. Since the passage of the 1972 Clean Water Act, dramatic environmental catastrophes have occurred rarely. Yet the quieter tragedies keep happening every day. To end them, Congress must plug leaks in the law.

That is what Minnesota Sen. Dave Durenberger, a member of the Senate committee charged with renewing the Clean Water Act, has been pushing Congress to do. In many areas, he and his environmentalist colleagues have succeeded: The reauthorization bill now awaiting action on the Senate floor includes strong water quality standards, new rules for controlling toxic discharges and firm deadlines for industry compliance. But Durenberger has been less successful in persuading the committee to accept his most valuable contribution: a proposal designed to curb the great volume of water pollution that comes from diffused sources.

Such "non-point source" pollution — rainwater runoff from farms, city streets, construction sites and waste dumps — is responsible for at least half of all water pollution. The current Clean Water Act does almost nothing to combat the problem.

To fix that deficiency, Durenberger argued for provisions to require better soil conservation, street cleaning and construction-site management practices. His proposal included a grant program to help states develop sound pollution-control plans. It also contained a three-year compliance deadline and enforcement provisions to ensure that the plans are quickly instituted and followed.

The Senate committee last week accepted many of Durenberger's ideas for controlling non-point source pollution, but rejected his deadline and enforcement proposals. But, as Durenberger pointed out, such voluntary programs are already in place to control soil erosion — and have failed miserably. The plan the committee sent to the floor is not likely to succeed where an identical strategy has failed. If the Senate wants to save the nation's slowly dying waters, it will have to do more. It can do more by including Durenberger's enforcement provisions in the new Clean Water Act.

THE MILWAUKEE JOURNAL
Milwaukee, Wisc., November 7, 1983

William Ruckelshaus, administrator of the federal Environmental Protection Agency, has taken stern action to protect the nation's waters from pollution. The move deserves particular attention, for it suggests that he is serious about undoing the negative impact that his predecessor, Anne Burford, had on environmental policy in 1981 and '82.

Burford sought to give the states a much freer hand in setting pollution cleanup standards for lakes and rivers. Her proposal would have permitted states to decide that certain streams formerly suitable for fishing and swimming should be converted into industrial water supplies. She also would have allowed degradation of certain high-quality waters, including some in parks and wildlife refuges. And she would have given undue weight to crude cost-benefit analysis in deciding what level of protection to impose on other waters.

Ruckelshaus has replaced the Burford plan with new rules that would require states to (1) place limits on toxic materials allowed in their waters, (2) maintain high-quality waters unless the states can demonstrate that degradation is essential to "important economic or social development," and (3) observe adequate margins of safety to assure that waters are kept clean enough to fully protect existing uses.

Ruckelshaus' policy has won praise from the major environmental groups. A lawyer for the Natural Resources Defense Council called it "the most important water quality regulation issued by the Reagan administration."

The change of policy doesn't necessarily mean that Ruckelshaus will be able to help the environmental cause across the board. So far, others in the Reagan administration have thwarted him in his efforts to impose stricter smokestack emission standards to combat the acid-rain problem. But his water policy does indicate that Ruckelshaus — who served as EPA's first director in the Nixon administration — is still a valuable friend of the environment.

THE CHRISTIAN SCIENCE MONITOR
Boston, Mass., November 2, 1983

The Reagan administration has just taken a good step forward in preserving the quality of the nation's fresh water. In a political year it is a useful move, too, in combatting the frequent perception that this administration is anti-environment. The challenge is for it to be similarly alert to other legitimate needs of present and future Americans for an environment free from potentially harmful pollution.

In recent years the United States has made significant progress in cleaning up fresh water. Now it is about to embark on a new phase. Subject to approval by the federal Environmental Protection Agency, individual states will establish minimum quality standards for their rivers and lakes, based on the use each will have: swimming, fishing, industrial. Then the states will set the maximum level of pollutants allowable under these standards.

Each state must justify its decisions to the EPA, which is where the new administration action comes in. In a reversal of his predecessor's decision, EPA administrator William Ruckelshaus has decided to insist that the states adhere to high standards.

Environmentalists believe this is a particularly important decision, since the states soon will begin to decide what the safe levels are for toxic chemicals fed into waterways. Setting standards for toxics is "the next generation of [water] pollution control," one specialist says.

Beyond this EPA decision, loom others. One expected soon deals with ways to clean up the air pollution that causes acid rain. Beyond that is the question of whether William Clark will bring significant changes to the environmental policies of the Department of Interior.

BUFFALO EVENING NEWS
Buffalo, N.Y., November 8, 1983

Environmental leaders have rightly hailed the action by William D. Ruckelshaus, administrator of the Environmental Protection Agency, in reversing the decision by his predecessor that would have weakened the water-quality goals of the Clean Waters Act.

Under changes proposed last year by former EPA administrator Anne Gorsuch Burford, the agency would have given state governments much more discretion in setting pollution cleanup standards for rivers and lakes. One effect of the rules would have been to grant the states authority to downgrade the use of a stream from fishing and swimming, for example, to industrial water supply.

In abandoning those controversial provisions, Mr. Ruckelshaus responded, as he noted, to protests by environmental groups and to threatened moves in Congress to impose stricter water-quality standards as a matter of law rather than EPA regulation.

The new rules make it more difficult for states to relax standards. They drop proposed provisions that would have required weighing the costs of regulations against the benefits in deciding the level of protection. The new rules require that proof be established that downgrading a stream quality level is needed for "important economic or social development." That would still allow downgrading for dams and other projects or where a variance was requested because of hardship.

Last year's proposed regulations required that only a few toxic pollutants be listed for control. The new rules establish limits on a much wider range of toxic substances, and require each state to pinpoint streams with high levels of toxic pollutants and to set standards to control them.

These welcome changes have drawn praise from members of the Senate and from spokesmen for environmental groups who see the Ruckelshaus decision as an encouraging sign of a new beginning in an EPA tarnished by the inept leadership of his predecessor.

Without firm federal restrictions against unjustified downgrading in the uses of streams, the intent of the Clean Waters Act can all too easily be eroded by competition among states to attract plant locations by industries shopping for lax pollution standards.

Preserving the quality of America's waters is critical to our quality of life and health, and environmental concerns reach beyond those of individual states in the many cases of waterways that affect neighboring states and areas beyond U.S. borders.

In a report last month, U.S. and Canadian officials warned that there has been little progress in reversing "serious degradation of water quality" in the most seriously polluted areas of the Great Lakes despite the massive cleanup effort of the past decade. According to the Great Lakes Water Quality Board, the Buffalo River, the Niagara River and 16 other areas in the Great Lakes remain grossly polluted despite remedial efforts.

By his enlightened step in dropping the proposed easing of water-quality rules, Mr. Ruckelshaus has flashed the right signal of his determination to reinforce federal responsibility for protecting our vital waterways against harmful waste discharges.

ST. LOUIS POST-DISPATCH
St. Louis, Mo., July 12, 1983

If the clean water bill that the Senate Environment Committee sent to the floor the other day had been reported out five years ago, it would have been regarded as a setback for environmentalists. But this is 1983 and the administration of Ronald Reagan, so the measure is being hailed as a victory. It is so perceived because it gives the polluters little that they sought.

That is not to say that environmentalists plan to settle for a standoff. They intend to try again later this month for an amendment dealing with so-called nonpoint pollution, that is, pollution that does not come out of the end of a pipe but is carried into waterways in the runoff from farms, city streets and construction sites. Federal authority to deal with groundwater contamination needs strengthening, and better safeguards against filling wetlands will be sought, too.

But at least the existing wetlands protections were not tampered with, though industry lobbyists and the White House had sought language making it easier to dredge or build in wetlands areas. And they failed also to remove the requirement that industries treat certain of their wastes before sending them on to municipal sewage treatment plants. Their major victory was the committee's agreement to extend from five years to 10 the maximum life of industrial discharge permits.

In a stinging rebuke to the Reagan administration, the committee voted to incorporate into the law water quality regulations that the Environmental Protection Agency under Anne Burford had planned to relax. The new EPA administrator, William Ruckelshaus, has promised to restore the old rules, but the committee, mindful of the Reagan environmental record, wisely decided to play it safe.

The clean water bill introduced in the House by Rep. Howard of New Jersey is much stronger than the Environment Committee's bill, and environmentalists would no doubt prefer to see it enacted, but they will gladly settle for the Senate version; for such are the times that holding one's ground is a form of forward movement.

The Oregonian
Portland, Ore., May 20, 1984

The House is about to consider a new Clean Water Act. The legislation would provide more funds for waste treatment plants along with other worthy programs. But it is clouded by concerns that wetlands protections are weakened, that pollution practices are extended and that it is generally weaker than the bill in the Senate or the version introduced in the House Public Works Committee that sent the modified act to the floor.

Concerns have been raised that provisions in the bill to curb industrial and agricultural pollution are too weak to enforce and rely, instead, on voluntary compliance.

In some ways, the current law, which has been kept in effect through continuing resolutions, offers more environmental protections than this latest House effort. But if the House passes the bill, the battle will have to shift to the Senate, where a stronger bill may emerge in conference committee.

The Record
Hackensack, N.J., June 27, 1984

A workable water-pollution bill appears to have survived its perilous trip through special-interest shoals in Congress. The House yesterday reauthorized the Clean Water Act, the federal law that aims to clean up America's rivers and water supplies. But it was a squeaker. Only vigorous action by environmental lobbies saved the measure from a bad mangling.

As recently as June 21, the House Committee on Public Works was ready to allow more leeway to some of the most serious polluters, including the electroplating industry. A committee draft bill exempted from compliance a couple of huge paper mills in Alaska. Old mines were to be allowed to reopen without correcting water-pollution violations. Compliance deadlines were to be extended across the board.

Interceptor sewers, which join neighborhood sewer lines to large trunk lines, have never been eligible for federal funding. There's a good reason. Such subsidies encourage sprawl — the development of housing on raw farmland and open space, remote from centers of employment. But land speculators had persuaded the Public Works Committee to fund interceptors in the new Clean Water Act.

All of these bad features were eliminated or modified in intensive negotiation late last week with Representatives James Howard and Robert Roe of New Jersey, both influential members of the committee. The resultant bill, though vastly improved, still isn't as strong as the Senate version. For instance, the Senate would require industrial polluters to "pretreat" their wastes — that is, to remove some pollutants before discharging them into streams or sewers. The House bill requires pretreatment only by producers of toxic materials. Let's hope some of the Senate's superior commitment to clean rivers and sparkling aquifers will rub off when the two versions are reconciled in conference later this summer.

The Wichita
Eagle-Beacon
Wichita, Kans., June 25, 1984

A month ago it appeared the attempt to rewrite the nation's principal water control law would surely fail. The Clean Water Act reauthorization bill emerging from the House Public Works Committee was inferior to the present act — which expired in 1982 and since has been kept alive by resolutions. Environmental groups properly swore to stop it if they couldn't amend it.

Now, happily, environmentalists, industrial lobbyists and Public Works Committee members have agreed on a series of compromise amendments that all believe would help improve water quality nationally while relieving some of the regulatory burden now imposed on industry. The House, which takes up the matter Tuesday, should adopt the compromise package — with one change.

The compromise package unwisely would allow industry to get permits to dump sewage into waterways for 10 years. Environmentalists acceded to this provision because permit holders would be required to cease dumping if toxic pollution was found. The present strategy, giving permit holders five years to review and clean up their effluents, makes more sense, and should be retained.

The package likely would accomplish what the present act hasn't — actual reduction of water pollution nationwide. Waters highly polluted with toxic wastes that have resisted routine pollution control procedures, and runoff from agricultural and urban lands would be controlled for the first time. Moreover, federal funds would become available for cleaning up such runoff, which may be the most serious of all classes of water pollution. It's good, too, that the compromise package retains the wetland-protection section of the act, which the Public Works Committee had considered excising. That's important because clean wetlands are vital to wildlife and public water supplies.

But the package won't easily become law. If the amendments, and the bill itself, clear the House intact, they still face a skeptical Senate. And the renewed act is viewed with disfavor by President Reagan — who is said to object to the costs it would entail. But as the bill moves through the legislative process, all concerned should remember that water regulation and cleanup isn't some sort of wasteful social program: It's an investment in good public health.

The Courier-Journal
Louisville, Ky., July 1, 1984

AN OPTIMIST has observed that most of the bad things people worry about never happen. But the converse of that remark may be a better maxim. Many bad things *would* happen if nobody worried.

Last week's House vote to renew the Clean Water Act was a case in point.

Worriers had a field day after the pork-barrelling House Public Works Committee approved a version of the act that, among other outlandish things, would have exempted reopened coal mines from water pollution rules that other mine operators have to obey. The result of the uproar was that the full House scrapped the bill and passed a substitute version agreed on at a meeting between committee leaders and environmental groups.

There was a bit of give and take. The mine provisions are much improved — but still contain a break for reopened mines. Exemptions from ordinary water discharge standards could be granted by the EPA on a case-by-case basis. That could be troublesome, if, as many think likely, the EPA sinks back into the morass that prevailed before William Ruckelshaus was summoned to clean up the scandals of the early Reagan years. But environmental groups say safeguards against abuse are written into the bill.

The bill also still contains a provision sought by industry that would allow 10-year permits for plants to discharge wastes into regulated waters. Five-year permits are now the rule. But the new version of the bill provides that the permits may be revised if there are changes in the rules governing water quality or discharge of toxic wastes. That would allow improvements in pollution technology to be adopted if they were found to be needed during the 10-year span.

The Clean Water Act's troubles aren't over. The Senate Environment and Public Works Committee has reported a bill that seems to be in trouble on the Senate floor because of the reduced money it would provide for sewer facilities. And administration officials have said they find the House bill unacceptable.

A culmination of this effort is long past due. The 1972 Clean Water Act actually expired in 1982. Its provisions, like those of the Clean Air Act and other anti-pollution legislation, have been kept in effect under continuing budget resolutions. The House's overwhelming vote in favor of its version of the bill offers encouragement that the legislation soon will be passed in an acceptable form.

Part IV: Land

In August, 1983, the Environmental Protection Agency estimated that at least 150 million metric tons of hazardous wastes subject to federal regulation were being produced each year. The new estimate, up sharply from the agency's earlier estimate of 40 million tons, represents about 1,500 pounds of hazardous chemical wastes produced within the United States each year for every person in the country. (The estimate did not include toxic wastes produced by facilities not subject to existing federal regulation, or any wastes disposed of by illegal means.) The necessity of disposing of these wastes is one of the most serious environmental problems facing this country. Many of the substances now known or feared to be toxic were by-products of the chemical revolution in the years following World War II; the chemical industry today continues to introduce hundreds of new chemical compounds each year. Used in the manufacture of plastics, paints, fuels, synthetic fibers, building materials, food additives, detergents and countless other products, these chemicals have made profound changes in the daily lives of Americans. Environmentalists, health officials and others fear, however, that the industry's ability to create new chemical substances has outstripped the ability of scientists to determine their toxicity and of industries to dispose of them.

The discovery of growing numbers of sites where such potentially hazardous substances had simply been buried or dumped in the past has brought the issue to national attention. Dumps such as the Stringfellow Acid Pits in California, or New Jersey's Lipari Landfill, threaten the health of all who live near them. When ground water penetrates the buried wastes, it spreads dangerous materials, carrying them to the surface and into lakes and rivers, where they may eventually enter the food chain. Current methods of hazardous waste disposal include storage in secured landfills, injection into deep wells and the chemical combination of selected wastes into stable compounds. Although the safest means of disposal are expensive, the cleanup of improperly dumped or buried wastes costs far more in the long run.

Hazardous Waste: Disposing of our Chemical Trash

In the decades after World War II, containers of toxic chemicals were simply buried in conventional landfills, dumped in open pits or left in vacant lots. The poisonous legacy of such inadequate disposal methods is still being uncovered years later at thousands of sites throughout the country, as toxic chemicals seep into basements and yards and water supplies become contaminated. Government action to prevent the continued dumping of toxic materials was taken in 1976, when the Resource Conservation and Recovery Act (RCRA), the first legislation to deal with land pollution, was passed. But it was 1982 before the Environmental Protection Agency finished writing the regulations of the law, and the end of 1980 before it began implementing "interim" regulations. Since the Reagan Administration took office, there has been a great deal of controversy over the enforcement of RCRA, which was designed to provide "cradle-to-grave" control of the generation, transportation, treatment, storage and disposal of hazardous wastes. (Waste producers were required to keep a written record of all hazardous substances from their origin to their disposal, providing the EPA with the means to investigate companies' practices.) In 1982, for example, the EPA temporarily lifted a RCRA ban on the disposal of barrels of liquid wastes in landfills not equipped with impenetrable liners, saying the requirement was too extreme; the liners were to prevent chemicals from leaching into ground waters, as they had done at the Love Canal near New York's Niagara Falls. (See pp. 198–207.) The public outcry was so great, however, that the ban was reinstated a month later, instead of 90 days later as planned. Similar doubts about the management of EPA's "Superfund" for existing toxic dumps resulted eventually in the replacement of Administrator Anne Gorsuch in 1983. (See pp. 218–228.)

It was primarily in reaction to the controversy at the EPA that the House in November, 1983 passed a far more stringent version of the RCRA. The Senate passed its version of the bill, by a vote of 93-0, in July, 1984. It was the first approval of a major environmental program by the Senate since the Reagan Administration took office. The Senate bill, although it would not tighten the requirements of the RCRA as much as the House version, would make several significant changes. Among other things, it would extend federal regulation to small businesses and medium-sized businesses, which together accounted for about half of the hazardous wastes currently being generated without proper monitoring. (Current regulations exempted businesses that generated less than 2,200 pounds of hazardous waste each month.) The new bill would also set new controls on wastes burned as fuel and would require identification of dioxin-contaminated wastes, which were to be treated before disposal.

Hazardous wastes are defined by the EPA as those that are toxic, corrosive, flammable or explosive. Over half of hazardous wastes are organic, composed of the carbon compounds that make up living things. The most potent organic chemicals are pesticides and defoliants. (See pp. 1–35.) Many of these chemicals, which can react with human tissues, are suspected of causing cancer, defects in fetuses and mutations in genes. The effects of exposure can be cumulative, and often are not felt for years. Other toxic substances include PCBs (a group of synthetic organic chemicals) and dioxins. (See pp. 36–41, 208–217.) Hazardous wastes are nondegradiable and can gain entry into the body through the mouth, nose or skin. Although there is much research yet to be done on the toxic effects of these substances, they have been implicated in the development of many degenerative diseases and metabolic disorders.

The problem of hazardous waste disposal is compounded by "midnight movers," or those who illegally dump toxic chemicals, usually by night, in order to avoid the costs of proper disposal. Truckloads of such clandestinely discarded chemicals have been found in sewers, forests, fields, streams and rivers and along roadsides. Even when hazardous wastes are disposed of legally, it is not certain that they pose no threat to the public. A spot-check of 65 hazardous waste dumps in 1983 revealed that 51 of them were in violation of a federal requirement to monitor nearby underground water for evidence of toxic leakage into drinking water supplies.

ST. LOUIS POST-DISPATCH
St. Louis, Mo., February 28, 1980

The first significant steps toward solving the monumental national problem of safely disposing of hazardous wastes have finally been taken by the Environmental Protection Agency. In compliance with a 1976 law, the EPA has issued rules providing for a record-keeping system to keep track of the handling and disposal of such wastes. They are the first in a projected series of regulations. Further, more sweeping rules — to define hazardous wastes and regulate their disposal — are due in April. Both sets of rules would take effect next October.

Under the newly issued rules, multi-copy manifests would be required so that waste, as it moves from producer to transporter to final disposal site, could be kept track of by handlers and the EPA. Since each handler would have to account with a manifest for its proper handling of the waste, the system would presumably stop the midnight dumper who surreptitiously disposes of waste in sewers, in fields or in woods. The manifest system is expected to cost industry about $7 million at first and between $16 million and $24 million a year thereafter. The April rules are expected to cost billions of dollars annually — between 2 and 3 percent of sales of the affected firms, mainly in the chemical industry.

Still, the rules are being supported by the Chemical Manufacturing Association. Industry support should not be too surprising. The cost of not properly handling hazardous wastes might be much higher. An indication of the cost of inaction or improper action is already evident in cases such as that at Love Canal near Niagara Falls, N.Y., where many families had to be evacuated and millions of dollars are being spent to try to clean up and compensate for a hazardous chemical dump.

With 57 million tons of hazardous wastes being discarded yearly by 750,000 waste producers, no one really knows how much damage has already been done. EPA Administrator Douglas Costle says that 90 percent of present waste disposal employs "environmentally unsound methods." Clearly, then, the rules defining hazardous wastes and regulating their disposal cannot go into effect too soon.

DESERET NEWS
Salt Lake City, Utah, March 31, 1980

Each year American industry produces 126 billion pounds of toxic wastes — enough to fill the mammoth New Orleans Superdome from floor to ceiling once every day.

But only 10% of this dangerous garbage, the Environmental Protection Agency estimates, is disposed of safely.

The EPA reports there are 500 to 800 abandoned toxic chemical dumps around the country. If they all started leaking their deadly poisons, about half the nation's drinking water would be threatened.

When the owners of these dumps can't be found or won't clean them up, the EPA is supposed to do the job. But it would take a staggering $22 billion to $44 billion to complete the work.

To make at least a start on cleaning up the mess, President Carter is proposing a $1.6 billion "superfund" to deal with the very worst sites. But Congress is considering its own bills — and even these are moving slowly.

Even if Congress opened its purse all the way, money alone isn't enough to deal with this challenge. New laws also are in order. For example:

— Present law does not allow the EPA to clean out chemicals illegally dumped into an abandoned coal mine in Pittston, Pa. even though the chemicals have formed hydrogen cyanide — the gas used in execution chambers.

— EPA cannot legally prevent 100,000 rusted barrels of highly toxic chemicals in Bullitt County, Kentucky from leaking into the air or ground.

— The agency's hands also were legally tied in trying to deal with the infamous Love Canal near Niagara Falls even though President Carter declared the area an environmental disaster.

Eighty-two chemicals seeped from an abandoned underground dump into basements and yards. After residents complained of illnesses, miscarriages, and birth defects, more than 200 families were evacuated.

— Though the EPA can respond to toxic chemical spills into navigable waterways, the agency cannot move to control poison wastes that threaten to pollute the air or ground water.

If Congress appropriates money to clean up chemical dumps, the taxpayers will feel it in the pocketbook. Likewise, if the chemical industry is forced to do the job, it will incur steep costs that ultimately will be passed along to consumers.

But there's no other alternative if deadly poisons are to be kept from seeping into this nation's drinking water. Americans can't always assume that disasters like the one at Love Canal happen only to the other fellow.

The States-Item
New Orleans, La., February 23, 1980

The U.S. House of Representatives has passed a bill that gives the U.S. Environmental Protection Agency and state agencies badly needed capabilities with which to control hazardous wastes.

The House approved the measure overwhelmingly this week before sending it to the Senate, which passed a similar proposal earlier. Conferees should iron out the differences and speed the bill on to the president for his signature.

The bill authorizes $156.5 million for hazardous waste control programs. But new enforcement powers and directives aimed at federal-state cooperation to clean up hazardous waste dumps are more important.

The states are directed to begin inventories of hazardous waste sites, and the bill establishes a $20 million federal grant program to help pay for the job.

The EPA can require owners or former owners of any chemical or toxic waste dump site to pay for monitoring, testing, analysis and reporting of its contents.

EPA inspectors are guaranteed access to sites that have handled hazardous wastes.

The destruction, alteration or concealment of records on hazardous wastes are prohibited. And the reckless dumping of toxic wastes becomes a felony.

Protecting the public and the environment from improper disposal of hazardous byproducts of our industrial society has become the nation's most pressing environmental problem. It is good to see that Congress belatedly is responding to the threat in a meaningful way.

The Dispatch
Columbus, Ohio, March 5, 1980

LITTLE BY LITTLE, government officials are piecing together the jigsaw puzzle of hazardous waste disposal.

Latest proposal comes from the federal Environmental Protection Agency which is seeking new and safe methods for handling 57 million tons of toxic wastes America's industries generate each year.

The dollar cost will run into the billions but there is a consensus the nation will pay even more in terms of human suffering if it neglects this important area.

Memories of Love Canal near Niagara Falls, with its instances of miscarriages and birth defects, and Kentucky's Valley of the Drums are too vivid to ignore.

The EPA plans to require the 750,000 toxic waste producers, trucking companies hauling it and dump site operators to jointly file a manifest stating disposition.

Linked to this action is a proposal by President Carter to clean up known existing dump sites, especially those in the "orphan" category, those long since abandoned and to which no responsibility can be attached.

Remaining is an unsolved problem — locating an adequate number of sites for dumping hazardous wastes. Many a community can be expected to object to having such dumps in their vicinity no matter how well-supervised the process may be. Nevertheless, it is a vital piece of that jigsaw puzzle.

Los Angeles Times
Los Angeles, Calif., February 19, 1980

Like a small cloud on the horizon that warns of a gathering storm, the discovery of toxic substances in many wells throughout California is yet another distress signal coming from the overburdened environment. More ominous is the failure so far to trace the poisons to specific sources.

Not many years ago, when the population pressure was less, it appeared that the environment had an unlimited capacity to absorb industrial and other wastes. But our multiplying numbers and the rapid growth of our vast industrial machine have radically altered the equation between man and the environment; it is crumbling under constant assault.

Everyone, from the individual to the giant corporation, must be concerned, for no one will escape the consequences if we fail to halt pollution.

A series of legislative and regulatory reforms proposed by Gov. Jerry Brown to meet the increasing threat of chemical pollution look promising. By executive order, he has created a council to coordinate the actions of state agencies involved in efforts to control toxic wastes.

His legislative measures include a number of significant elements: tighter controls on waste disposal, closer monitoring for chemical contaminants in drinking water, increased fines for illegal disposal of wastes, and a provision also for civil penalties. A new unit would be added to the Department of Health Services to identify the source and scope of toxic wastes as they are discovered. An analysis of toxic pollution prepared by the governor's staff emphasized, "Current enforcement is limited by inadequate civil and criminal penalties and our minimal inspection capability."

The objectives outlined in the Brown plan include more effective controls on disposal sites, the identification of abandoned dumps and an effort by the state to provide technological alternatives to hazardous landfill disposal of wastes.

The analysis done by Brown's aides conceded, "Years of inadequate disposal practices will not be remedied without extraordinary effort commenced immediately to protect the public against chemical hazards."

Only last week, noted ocean explorer Jacques Cousteau told public officials in Sacramento: "Oceans, rivers, lakes and rain are being poisoned. Forests, prairies and seas are being systematically destroyed. Wetlands are drying out and being covered with concrete structures." The remedies, he said, for worldwide control of pollution "involve 'real education' that restores moral values."

Survival is at stake, and that, we believe, is a nonpartisan issue that should have broad appeal.

THE CHRISTIAN SCIENCE MONITOR

Boston, Mass., April 8, 1980

How to dispose of hazardous waste is bursting forth as a key environmental issue of the '80s. You can do something about it, as suggested by the cover story in this week's regional pullout sections of the Monitor.

With air and surface water being cleansed of foul discharges, the ground has become the last sink for disposing of pollutants. After a century of absorbing wastes that industry, research institutions, hospitals, and government have dumped on and in it, the soil, like a soaking sponge, is signaling that it's had enough. Drinking water in many parts of the nation is threatened with contamination. Hardly a day goes by, government officials report, that another 1,000 drums of obnoxious materials are not found close to some water supply. No one knows how many more Love Canals are out there waiting to happen, endangering life, health, and the environment.

But now a new era seems about to begin. The section of RCRA (Resource Conservation and Recovery Act) which Congress passed in 1976 to deal with this long ignored and rapidly mushrooming problem is going into operation. Some regulations already have been issued. Others, effective in October, are due out this month.

But what good will it do to track every ton of hazardous waste from point of generation to final disposal, to set standards for designing and operating environmentally sound disposal facilities and landfills, and to arrange to issue permits for them, if citizens block efforts to locate such facilities in their vicinity? Environmental Protection Agency (EPA) officials see such resistance as a major hurdle to the success of the hazardous waste program. Two communities targeted as possible sites have already set a precedent in Massachusetts by obtaining legislative exclusion from having waste facilities located within their borders.

RCRA casts the federal government in the role of regulator, assigning implementation of the law to the states. It gives EPA no power to site hazardous waste facilities. A major issue now under discussion in some states is whether a state siting board should be set up and given power to override local opposition to locating such faciliities. There are very few licensed facilities in the nation. "We need new facilities," says one EPA spokesman. "Developing them is going to be a key to making sense out of this whole program."

The alternative to legal disposal is illegal disposal. "Midnight dumpers" are the villains of this piece, underpricing legitimate transporters of waste, then pouring their toxic cargoes down sewers, dumping them in forests, on farm land, open fields, and roadsides with no regard for the general welfare.

At this early stage, there is little public understanding of possible solutions to this complex issue. When a hazardous waste facility is proposed, public perception of it generally is not of a well-designed and monitored site, as RCRA requires, but of the kind of dangerous, oozing dumps that have been making headlines. The very mention of a hazardous waste facility conjures up visions of Three Mile Island incidents (though, of course, hazardous waste does not include nuclear waste).

But lumped under hazardous wastes are many that are no more dangerous than table salt, and not as dangerous as gasoline, a most volatile fluid which people manage to live with daily. EPA defines a hazardous waste as one that may contribute to illness or death "when *improperly* managed."

How safe are environmentally acceptable hazardous waste disposal faciliities? Though not totally risk free, there are numerous technologies for disposing of the wastes with a high degree of safety: incineration on land or sea, chemical neutralization, tightly contained storage, waste exchange (in which one industry's waste becomes another's raw material).

If industrial firms are required to stop improper dumping of hazardous wastes, most of which is done on their own property, it stands to reason that proper disposal facilities will have to be made available. Well-managed and monitored facilities present far less danger to communities than the current virtually uncontrolled dumping. And, although the new facilities will increase the cost of waste disposal, the price tag will be peanuts compared to the staggering cost of cleaning up unmanaged dumps of the past, a large chunk of which the taxpayer has to pay.

Hazardous waste will not go away. It must be intelligently managed. We the consumers are we the polluters. Our demand for cars, fuel, pesticides, leather goods, etc. produces these offensive leftovers.

The previously mentioned article in this week's regional pullout sections tells how citizens in one community have taken action to protect their town from chemical contamination. That's one step. As states set up their hazardous waste programs, which are required to be "as stringent as" EPA regulations, citizens have an opportunity to participate in the process. We would hope they will work toward a solution to the problem and not become part of it by blocking efforts to locate proper disposal facilities where they can be appropriately and harmlessly sited.

Chicago Tribune

Chicago, Ill., April 11, 1980

One of the subjects before Congress is creation of a $1.6 billion "superfund" to clean up abandoned chemical dumps that pose a health problem. President Carter proposed such a fund almost a year ago. But the chemical industry is opposing it vigorously and not many people are working for it. It is a poor year for new funds, and very likely nothing much will happen soon.

Yet the subject of toxic waste is a big one that will not go away. The Environmental Protection Agency [EPA] estimates that there are between 30,000 and 50,000 dump sites in the country. A survey of the 53 biggest chemical companies has yielded a list of 3,383 sites where these companies had disposed of hazardous wastes during the past 30 years. More than a thousand of these are no longer in use, which means that the companies have no strong incentive to keep track of what is happening there.

Douglas Costle, administrator of the EPA, says that "the size of the problem is staggering." He estimates that 740,000 waste producers discard annually 57 million tons of hazardous wastes, 90 per cent of it by "environmentally unsound methods."

There ought to be a law, and there has been, since 1976. Regulations implementing it were issued in February of 1980, calling for a national inventory of disposal sites and for giving notice, not later than next July, by those engaged in the production, transport, or disposal of hazardous chemical wastes. But detailed definitions and standards are still to come. And when the system is more or less complete, it will deal with current production of toxic wastes, not with the problems already bequeathed us by the past.

Nobody denies that toxic wastes now in the ground constitute a grave problem. The president of the Chemical Manufacturers Association, Robert Roland, objects to President Carter's proposal that the waste producers provide 80 per cent of the proposed cleanup "superfund." The toxic waste problem, Mr. Roland says, is one "which everyone has helped create and for whose solution everyone should help pay."

Everyone will, assuming that a solution is achieved. Everyone will pay through both the chemical and oil companies' passing through new costs in higher prices and through taxes financing whatever the government's share will be. Some will pay more heavily than others—in places like the much publicized Love Canal site in Niagara Falls, N.Y., and the little town of Triana, Ala., as reported by The Tribune's Barbara Reynolds. Who knows where poisonous stuff will surface next with disastrous consequences?

In the Triana episode, the state successfully sued the federal government, because the pollution of the river was traced back to Army property that had been leased to a chemical firm. That is a salutary reminder that the states, too, have a constructive role to play, sometimes as an adversary of some aspect of the federal giant. No doubt the federal EPA will be the principal in future efforts to cope with the toxic waste problem. But there is lots of room for many agencies and individuals to help with both the preventive and cleanup dimensions of the toxic waste problem.

Edmonton Journal

Edmonton, Alta., June 19, 1980

Alberta must act with urgency, equity and efficiency to manage her hazardous wastes.

We produce tens of thousands of tonnes of such wastes every year. Yet nobody knows who produces how much of what, where it goes, how it moves or what becomes of it.

We do know its potential effect — as they do at Love Canal in New York, where 720 families were evacuated when President Carter declared a state of emergency over health problems associated with toxic landfill.

In Alberta, proposed waste treatment plants have aroused as much concern as the waste itself. Such plants are seen in some quarters as increasing rather than alleviating the hazard to health. The concern must be answered. Equity demands that the risk from hazardous wastes be reduced for all communities, not just concentrated in some. Modern technology now makes that possible; government and industry must make it happen.

Alberta Environment's Hazardous Waste Management Committee has taken a first step and recommended a high-technology integrated waste management system like one already operating in Bavaria. A second step is the public forum of the Environment Council of Alberta, now receiving briefs from professionals, public and industry.

Meanwhile, government should be gathering information on the extent of hazardous wastes in Alberta and regulating obvious insanities such as careless transportation and dumping.

For Alberta's future, we must leave a heritage of health — not of hazard.

ALBUQUERQUE JOURNAL

Albuquerque, N.M., July 25, 1980

Hazardous wastes, the unpleasant byproduct of our technological age, is not a newcomer to Albuquerque. But its profile as a subject for public policy and concern is on the rise.

A Chicago-based firm wishes to establish a "transfer station" for collection of hazardous waste in Albuquerque, for collection and shipment to disposal sites outside New Mexico.

The waste to be handled is already here, from electronics manufacturing plants to photofinishing labs and pest exterminators. And up to now, nobody has consistently monitored where it has been disposed of.

A Texas firm wishes to establish a hazardous waste dump in the vicinity of Hatch, trucking waste from out of state for disposal there.

Both with regard to hazardous chemicals from outside New Mexico and our own home-generated variety, it's time more care were given to their collection, transportation and disposal.

The horror stories from the East of yesterday's hazardous waste rising to contaminate today's citizens shows the potential for the future if nothing is done.

And if a transfer station is established in Albuquerque — possibly a viable approach to the problem — we would hope the location chosen is far removed from concentrations of people.

One thing is certain, the decisions on what is done and where should be made primarily by responsive public officials studying all the facts, rather than by private, for-profit firms which by their nature must consider other objectives.

THE ATLANTA CONSTITUTION

Atlanta, Ga., November 30, 1980

THE PROBLEM of toxic wastes — that which was dumped in unsafe sites in the past, that which must be disposed of properly in the future, and that which will be spilled in inevitable accidents — is a mammoth one for this country, one which requires a serious, no-nonsense commitment by the government and by the industry which produces it.

It will not get that commitment, however. Evidence is the passage by the Senate the other day of its version of the so-called "superfund" legislation, a bill that was supposed to set up a $4.2 billion reserve over six years to pay for cleaning up old hazardous dumpsites and new spills and to compensate people victimized by toxic waste contamination.

The Senate, urged on by lobbyists for the chemical industry, pared the bill down to the barest minimum — just $1.6 billion, only enough to pay for cleanup of old dumps and future spills of some wastes. Eliminated were oil spills, which in the past have been among the most economically devastating, and the compensation for those who suffer from the effects of the poisons. One senator, George Mitchell (D-Maine), was so upset by the latter omission that he characterized the action as a "judgment that property is more significant than human beings."

That may be the practical consequence of the Senate's bill, but what the action really means is that industry profits are more important to congressmen than the well-being of ordinary Americans. The chemical lobbyists fought the liability provision because they would have had to pay more into the fund to cover it. Now the only way victims will be able to win compensation for their suffering is to fight the responsible company in court, a long and uncertain process.

Still, we may have to be grateful that even this lesser version — sort of a "punyfund" — has passed the Senate, and hope that the House, which had approved a slightly better version, will go along. That's because failure to agree now would throw the issue into the next Congress, where there will be even less chance of passing a meaningful bill. It's a case of the people having to take what crumbs are left for them after the industry lobbyists have their say in the halls of the Capitol.

We'll be glad to get whatever version of the fund we can get. But we aren't exactly glowing with pride and admiration at the performance of our elected representatives in this attempt to protect us from the poisons that plague our society.

The Wichita
Eagle-Beacon
Wichita, Kans., September 13, 1980

The U.S. surgeon general's office has put into words the vague, fearful conclusion that has haunted Americans for the last few years: ". . . toxic chemicals are adding to the disease burden of the United States in a significant, although as yet not precisely defined way."

Common sense would seem to dictate such a conclusion, coming, as it does, after revelations that dangerous chemical materials for years routinely have been disposed of in haphazard, casual ways, with no concern for the safety of others, either in this generation or those to come. To believe there would be no price to pay, in terms of human health, for so carelessly salting the environment with deadly poisons is incredibly naive.

So it comes as no surprise that the nation's top health officials have found the country now is paying that price, even though they cannot speculate how dear a wage is being collected for past transgressions.

There has not, however, been a public outcry to correct the problem, although an ABC/Harris poll has shown 86 percent of the populace believes the dangers of inadequate chemical waste dumps and toxic spills deserve high priority status at the federal level. For most, though, such worries remain faraway considerations until the problem happens to them, personally.

While there has been no well-organized citizens' campaign to force national legislation that would address the problem, there has been a well-oiled, heavily financed lobbying effort by the chemical industry against three such corrective measures in Congress.

The Senate bill, S. 1480, which would create a $4 billion contingency fund to finance both clean-up operations and compensation of victims with legitimate damage claims, has come in for the most intense criticism. The chemical lobbyists claim it would amass far more money than is needed to handle the problem, and would encourage people to tap it with spurious claims.

The House bills call for much smaller funds and don't stress the need for a victims' reparation process, so are not as objectionable, from the industry's viewpoint.

But the fact is, as the surgeon general now has pointed out, that no one knows just how serious the resulting health problem is. In fact, no one knows just how many dangerous dump sites are located around the country, or how many millions of tons of toxic chemicals may be thrust back into contact with humans in the months and years ahead.

Thus, to claim the Senate bill would set aside far more money than necessary to cope with the problem is to accept the chemical industry's premise that the problem indeed is a relatively small one that can be handled expeditiously. There is no convincing evidence to that effect, however, and plenty to indicate that only the surface of the hazardous waste dilemma has been skimmed so far.

It is encouraging to note, therefore, that fully a fourth of the Senate membership now has joined in cosponsoring S. 1480, making its passage look more promising. The full Senate should approve it, and the House should add its endorsement to this needed legislation before Congress adjourns early next month.

It then will be incumbent upon President Carter to sign such a comprehensive measure into law, to aid today's victims of chemical pollution, and to prevent a worsening of the health problem that only now is being recognized.

St. Petersburg Times
St. Petersburg, Fla., October 6, 1980

Americans are so accustomed to simply turning on the faucet for a drink of water that it's hard to imagine the prospect of chemical contaminants also spewing out.

But it's time to face reality.

Toxic chemicals haphazardly dumped years ago are now threatening the nation's drinking water. A House subcommittee last week identified 250 sites across the nation where hazardous chemicals could be seeping into ground water, the vast underground reservoirs of water tapped by wells and used by about 100-million Americans.

Of those sites, Florida accounted for 54 — more than any other state.

Sounds bad, doesn't it.

Actually, it may be even worse.

IF THE subcommittee report is to be heeded, it must be put in the proper perspective. U.S. Rep. Anthony "Toby" Moffett, D-Conn., released it for political reasons, primarily to counteract a well-orchestrated chemical-industry effort to thwart congressional approval of a hazardous-chemical cleanup fund.

Although differing in details, bills in the House and Senate would set up a fund of about $1-billion to clean up chemical dumps that threaten public health. Industry would pay most of the cost, hence the frantic lobbying. A study released by Ralph Nader's Congress Watch organization reported that the chemical industry has wooed members of Congress with more than $2-million in election campaign contributions in the last three years.

Moffett's support of the crucial superfund legislation is laudable, but his tactics are not.

If anything, releasing a list of only 250 potentially dangerous chemical dump sites minimizes the problem. The figures cited also have confused and angered some state officials who have different data.

For instance, Dr. Rodney de Han, director of the Florida Department of Environmental Regulation's (DER) drinking water division, says the list may have grossly *under*estimated the potential threat to Florida's water supply. He said his division is conducting a study of *150* possibly hazardous sites to determine if they do, indeed, pose a health danger.

The truth is, no one really knows how many hazardous-waste dump sites exist or how many pose an imminent peril of contaminating drinking water supplies. The subcommittee will not guarantee the list's accuracy.

That means the report may have unnecessarily panicked people who live near the sites listed, while unrealistically reassuring residents of communities not mentioned.

One flaw in the report relating to Florida is the omission of dump sites associated with the phosphate industry. In June, the state DER concluded that the ground water in Central and Southwest Florida is being endangered by the proliferation of phosphate waste pits and gypsum stacks. Ground water provides between 85 and 90 percent of the state's drinking water.

A SUBCOMMITTEE aide warned that the list, which was based on information compiled by states, was incomplete. Florida may have led the list because it has done a better job of reporting potentially dangerous dump sites than other states. It is unlikely, for instance, that New Jersey, which is reported to have significant problems of hazardous-waste disposal, was listed as having only a third as many dangerous sites as Connecticut.

In spite of the flaws, the report made a good recommendation when it called on the U.S. Environmental Protection Agency (EPA) to draft a national ground-water policy "to locate threats to ground water and eliminate them before precious ground-water supplies are irreversibly damaged."

But there's even a caveat here. The EPA's record of dealing with the hazardous-waste problem is miserable. So it would be foolish for state environmental officials to abandon their own studies and plans for dealing with the hazardous-waste crisis. The Florida Legislature would be wise to follow the suggestion of House Natural Resources Committee Chairman John Lewis to set aside an additional $1-million to clean up hazardous-waste spills and dump sites (a $600,000 cleanup fund already has been established).

The danger of releasing a flawed report about an extremely serious problem is that people might ignore the issue the next time it is raised. And if this issue is ignored, it may be too late to reverse the damage.

Oregon Journal

Portland, Ore., December 4, 1980

A compromise "superfund" bill to clean up pollution in the environment may sound super with a bankroll of $1.6 billion, but even that sum will hardly cover all the problems pervading land, air and water.

It is a start, and perhaps more than might have been expected from the "lame-duck" Congress. The greater part of the money for the trust fund will come from the chemical industry itself. General revenue will supply from 14 to 25 percent during the five-year period covered in the bill.

Probably the most valuable segment of the bill requires keeping track of where and how hazardous wastes are stored. Some of the most dangerous situations have occurred when a chemical company turned over its wastes to an inefficient or unscrupulous handler for disposal. A poisonous chemical fire took place last summer in New Jersey at a place of storage not controlled by any one chemical producer.

Accurate records must be kept, with a penalty of imprisonment or fine for falsification. Immediate notice must be given of a "release" of hazardous material, which could be leakage in a waste dump, a spill of such material in a waterway or road, or fumes going into the air.

Probably the most difficult problem to solve will be the presence of toxic or hazardous pollution in ground water: How to detect the source of pollution, trace it and eliminate it. The unmarked, forgotten or illegal dumps dotted around the country should be found, cleaned and secured.

Superfund is at least a start in a responsible direction, along a road we should have taken years ago. The price of producing useful but hazardous or poisonous substances should never have been defrayed by allowing them to be introduced into our air, water and land.

Richmond Times-Dispatch

Richmond, Va., December 1, 1980

The lame-duck Senate has joined the House in approving legislation for something called the environmental "superfund," a panicked response to exaggerated claims about the hazards of toxic waste dumps. The slightly varying plans of the two chambers, if they can be resolved before Congress adjourns, would take about $1.5 billion more from American taxpayers, mostly through direct taxes on chemical firms, while extending ominous new powers to the Environmental Protection Agency.

The superfund bill would give EPA authority to clean up any chemical dump simply on the basis of its own judgment of a hypothetical threat of "substantial endangerment" to the environment. If Americans are to be free from capricious government, no regulatory agency should have such a carte blanche, but especially not EPA, with its disgraceful record of demagogic distortion of scientific fact in order to promote its power-grubbing aims.

There probably would be no superfund bill on the verge of becoming law had it not been for the panic over the supposed health hazards from toxic waste at upstate New York's Love Canal. In the vicinity of Love Canal, residents became so alarmed by an EPA report alleging widespread chromosomal damage that they had to be evacuated on orders from President Carter. The Hooker Chemical and Plastics Company — which had sealed some chemical wastes away nearly 30 years ago only to see them unsealed by the waste site's new owner, the Niagara Falls Board of Education — unfairly was cast as the villain in a calamity supposedly caused by corporate greed inadequately checked by government regulators.

Just how gross a misrepresentation of facts had been committed by EPA is shown in the report of a scientific panel on Love Canal commissioned by New York Gov. Hugh Carey and headed by the distinguished physician Lewis Thomas, well known for his book, *Lives of a Cell.* The Thomas report accuses EPA of demagogic quackery in its widely publicized chromosome study. Wrote Dr. Thomas and his panel:

"With so much at stake for the residents involved, to have set up experiments that lead to public conclusions of such magnitude, without prior review of protocol by qualified, uninvolved peer scientists, and without any after-the-fact, independent review by competent scientists before release of the results, was a disservice to the citizens most intimately concerned and, as well, to the public at large. The damage done by this EPA effort is perhaps beyond mending; many of the Love Canal residents have by now become so distrustful of governmental agencies and their scientific reliability that they are unwilling to believe anything but the worst of news about themselves.

"It is a pity that this matter was so badly handled. There was no good reason why the responsible authorities in EPA could not have consulted beforehand with their counterparts in the New York State Department of Health, and enlisted the advice and close participation of outside consultants with international reputations in the field of cytogenics, and then mapped out a thorough, careful and scientifically valid approach to the question of chromosome injury."

During the height of the Love Canal fever, EPA was contending that the number of hazardous waste dumps could exceed 30,000 and possibly could run as high as 50,000. By the time the superfund reached a Senate floor vote the other day, its sponsor, Sen. Robert Stafford, R-Vt., had scaled down the superfund proponents' "guesstimate" to about 3,000 hazardous dumps. Reputable scientists concerned about the hazardous waste problem do not pretend yet to know its dimensions.

The superfund legislation offered by the lame ducks is clearly the result of fear and ignorance rather than dispassionate compilation and consideration of scientific data; it represents a sort of final fling for the liberal statists who are losing control of Congress. If some force yet can be mustered to prevent superfund's final passage so that it might not be signed by its proponent, President Carter, and so that the question may be considered by the incoming Congress, it would be a great service to reflective and responsible government.

Sentinel Star

Orlando, Fla., November 26, 1980

THE SENATE has at last given some hope to the prospect of cleaning up America's chemical wastes with the passage this week of a $1.6 billion superfund.

To be sure, the bill does not do as much as its supporters had wanted. In the first place, President Carter sought $4 billion for the fund.

In the second, it is an after-the-fact cure that does not address the more difficult problem of preventing careless or negligent dumping. There are grave doubts about how the federal and state governments — including Florida's — will enforce "cradle-to-grave" legislation designed to control waste from production to disposal.

But the cleanup bill is imperative, and the House should accept the weaker Senate version. If it doesn't, it is unlikely a conference committee can work out a compromise before this Congress expires. And to start all over with the new Congress certainly means months of delay, which could result in an even weaker measure.

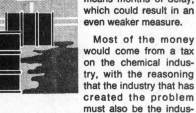

Most of the money would come from a tax on the chemical industry, with the reasoning that the industry that has created the problem must also be the industry that pays for its solution. Taxpayers, of course, know that it is the consumer who ultimately pays through higher prices for the products those manufacturers produce.

The Environmental Protection Agency estimates that there are 3,500 chemical spills each year, and there are more than 2,000 dump sites that are leaking so badly that they present a danger to public health. The environmental agency estimates it will cost $3.6 million to clean up each site.

Cleaning up the dumps that threaten water supplies and the health of entire communities cannot be done on a haphazard basis. The nation cannot move too rapidly to begin the cleanup.

WORCESTER TELEGRAM.
Worcester, Mass., December 8, 1980

The U.S. House of Representatives bowed to political reality when it approved the Senate bill to clean up chemical wastes across the nation. Some House members wanted a "tougher" law.

Although the new law avoids the excesses of earlier versions, it still runs the risk of becoming a multi-million-dollar boondoggle. It is aimed at resolving a problem the dimensions of which are not known, the responsibility for which has not been determined and the solutions for which have not been tested.

The law authorizes the government to amass $1.6 billion over the next five years, $1.38 billion of which will be extracted from the chemical industry. Chemical companies will be taxed without regard to whether they have been found guilty of dumping chemicals illegally or even whether they have caused any chemical pollution anywhere. The tax will be determined by a complicated formula based on how much of 45 listed products a company produces. The fees cannot exceed 2 percent of the market price, or 1 cent per pound on primary petrochemicals, or $10 per ton on inorganic raw materials or 3 cents per barrel on crude oil.

The fund will be used to pay for 90 percent of any remedial action taken at a chemical dump site (states will be liable for the other 10 percent); loss of natural resources and costs of restoring lost resources up to a limit of $50 million; and costs of health studies and diagnostic examinations for victims's health problems, but not other medical expenses. The bill does not cover the cost of cleaning up oil spills.

Money expended on hazardous waste sites will be recovered up to a limit of $50 million from those responsible for the pollution. The liability provision is similar to the one imposed by the Clear Water Act.

One of the most controversial subjects in floor debate was the lack of any special provision in the bill to permit individuals to sue directly for compensation without going through the usual court procedure. But the omission was wise; the near hysteria over Love Canal and other highly publicized chemical dumps shows how quickly psychosomatic symptoms develop under the pressure of mass anxiety. Although an authoritative study has found no proof that Love Canal caused any physical illness, the psychic and emotional damage has been enormous.

Something had to be done about chemical waste disposal. Although the problem most likely has been exaggerated, it is a problem and should be dealt with before it gets any worse. The new law should prove at least partially effective. But after a year or two of demonstration, Congress should bring it back in for a thorough checkup.

Arkansas Gazette.
Little Rock., Ark., December 14, 1980

Conservation and environmental protection legislation face extraordinary obstacles no matter what party may be in control at the White House or in one or both houses of Congress, but the going will certainly be rougher in 1981 than it has been in 1980. President-elect Ronald Reagan has made no secret of his insensitivity in this subject area and the Senate Republican majority will reflect pretty much the same attitude.

Because of this dismal prospect, there is considerable reason for rejoicing this lame-duck period in the enactment, first, of an Alaska lands bill and, now, of the "superfund" bill to clean up dangerous chemical spills and toxic waste dumps. In both cases the final legislation has represented a compromise that on balance has been attractive.

President Carter signed the "superfund" bill last week, calling it "landmark in its scope and in its impact on preserving the environmental quality of our country." He had lobbied for passage in recent days as the bill had come under heavy attack. It finally reached his desk after the "superfund" was reduced from $1.95 billion to $1.6 billion and a provision to include money for cleanup of oil spills had been deleted.

The fund will come mostly from excise taxes on this chemical industry. Properly, the legislation should be considered only a beginning of a broader effort to deal with harmful chemicals in the environment. The very least that those who are concerned about public health should try to accomplish in the coming years, even against the shifting political odds, is to restore provisions covering inactive dumps and chemical and oil spills. These environmental hazards are not going to disappear on their own. Only concerted federal action can do the job.

Newsday
Long Island, N.Y., December 6, 1980

If cleaning up the country's toxic waste dumps weren't such an urgent matter, we wouldn't be very enthusiastic about the Superfund bill the House and the Senate agreed to last week.

Spending $1.6 billion over the next five years will hardly scratch the surface of a cleanup task that the Environmental Protection Agency has estimated might cost between $22 billion and $44 billion. The bill provides no compensation for victims, nor does it cover oil spills.

But it will at least allow the government to start properly disposing of tons of ghastly chemicals that now lie scattered about the countryside. In that sense, it's a giant step forward.

Contributions from the chemical industry, which is supposed to pay 86 per cent of the cleanup costs, won't be coming in for some time yet. However, EPA can start drawing cleanup funds from the Treasury almost immediately. Considering the volatile state of some of these dumps, the sooner the better.

Regrettably, the final toxic-wastes bill was drastically scaled down from earlier versions. The most serious deletion was a provision allowing the fund to compensate victims for medical and other expenses.

That leads to a strange incongruity: Local and state governments may be reimbursed for the loss of trees, for instance, but human beings won't even be able to sue in federal court to recover damages for themselves.

Still, even a modest and imperfect cleanup fund—it can't really be called a Superfund anymore—is better than none at all, which is what seemed likely last month when the lame-duck congressional session began.

By 1985 we hope Congress will make another attempt at creating a true Superfund to remove illegally and improperly dumped poisons that may imperil millions of Americans. But the best solution is to dispose of them properly to begin with. Having to pay for the mess it has created over the years should give the chemical industry a financial incentive to do the job right in the first place.

Portland Press Herald
Portland, Maine, December 5, 1980

The record of the 96th Congress has been, until now, indifferent at best. In its last few days however, the lame duck session made a significant contribution to environmental safety by passage of superfund legislation to deal with chemical spills and toxic wastes.

While the New York Love Canal disaster focused national attention on the growing problem of chemical waste disposal, incidents closer to home sparked support for the measure. Empty chemical drums at the Saco dump, truckloads of poisonous industrial wastes parked at North Berwick, contaminated wells at Gray—all testified to the need for adequate controls over the handling and safe disposal of toxic wastes.

Money from the $1.6 billion fund, largely financed by the chemical industry, will be used to clean up literally thousands of abandoned chemical dump sites which may pose human health hazards. In many cases those responsible for the dumping cannot be identified, and often the cost of removal is beyond the ability of local governments. In other instances, the federal fund will allow the government to clean up chemical poisons quickly and then sue to recover costs from those who are believed responsible.

Best of all, from a parochial standpoint, the new law will not—as was once proposed—invalidate Maine's own oil spill fund and cleanup law. Our 11-year-old law is far superior to the oil spill cleanup provisions envisioned in the federal bill and Congress acted reasonably in striking them from the final version.

The new federal superfund law may not be ideal but it can be perfected over time. In any event it represents the realization that the public must be protected against the random and uncontrolled disposal of toxic chemical wastes in our land and waters.

THE COMMERCIAL APPEAL
Memphis, Tenn., October 29, 1981

MORE THAN a year after Congress authorized a $1.6-billion "superfund" to find and clean up the country's most hazardous waste-disposal sites, the Environmental Protection Agency has finally come up with its hit list.

Now that names have been named, the country can get to work at the Hollywood Dump and 113 targets elsewhere that EPA fingered. Right? Wrong.

THE AGENCY, it seems, still must come up with rules and procedures for the work, which shouldn't be surprising since it took EPA more than three years to write the regulations on what constitutes dangerous waste. Once an approach is agreed on, work will still be postponed many places because there is money enough only to treat a handful of sites during the next year.

If some of the delay is unavoidable, some of it is not. It should be noted, however, that the superfund wasn't part of the budget and personnel cutbacks for which EPA has drawn so much fire.

What happens in the meantime to those sites that must wait? For the Hollywood Dump and the other disposal sites in West Tennessee that weren't listed by EPA, waiting wouldn't be worry-free at all. A study completed last year, and financed by the federal Interior Department, found that although there was no "immediate threat" that our underground water supply would be contaminated by chemical dumping, a long-term danger does exist. If Memphis were to keep using water at its current rate of roughly 187 million gallons a day, study coordinator Dr. John Smith of Memphis State University said, traces of pesticides and other chemicals could start showing up in the Memphis aquifer within, at worst, 20 to 30 years.

These conclusions were not without controversy when the study was released. The Memphis Light, Gas & Water Division, for one, questioned the results.

STILL, THE QUESTION today is much larger than it was then. The Reagan administration has cut back on grants specifically aimed at sewerage and water treatment plants and is refusing to release additional money until Congress cleans up the Clean Water Act. Needless to say, states and localities at the end of the federal pipeline have complained bitterly about these shutoffs, as bitterly as they once complained about Washington's wheels grinding too slowly back when there were wheels to grind.

In the case of Memphis, these administration actions could be devastating if the Interior Department study is right. Where would our city and LG&W go if they had no choice but to start treating our water before it comes out of the tap? Then, the costs of the "new federalism" and chemical dumping both would come home with a vengeance.

Our city, county and state need to do what they can to avoid such an outcome. Whatever help is forthcoming on the Hollywood Dump, clean-water problems are their problems now. It's time they fess up to their role in the cleanup before the costs get too high.

The Knickerbocker News
Albany, N.Y., February 10 1981

Technology has been both boon companion and slavering, hungry wolf in our travels through the past hundred years.

More friendly than not, at least until recently. It's only now that we're beginning to fully understand the balance, the good-and-evil pendulum that the computerized, push-button, labor-saving life swings upon.

Hazardous wastes are just one aspect of the problem. Acid rain, ordinary pollution, radioactive wastes and many more of our modern evils are technology's stepchildren. But it's hazardous wastes, those toxic chemicals that turn an ordinary neighborhood into a Love Canal, that ride near the top of the mind.

Who can reliably say that his street, her back yard won't be the next poisonous cesspool to bubble to the surface? A good part of the problem, with Love Canal and other toxic dumpsites, both discovered and still hidden, is that no one really knows any more what's been dumped where, by whom, when.

Decades ago, no one thought much about the effluent left after a manufacturing process. They just jugged it and tossed it out, or let it run down the local sewer, or pooled it on a far corner of the property. Nobody kept records. Nobody knew they were laying down a deadly foundation on which their children and grandchildren would build their lives.

There's not much each of us, alone, can do. We can watch, warily, for potential toxic dumps or other hazardous waste problems. We can keep an eye on our lawmakers, to see that dumpsites are cleaned up. We can make sure those who wield the power know of our concerns and fears.

In a move that would help forestall future Love Canals, the state has proposed that private industry build and operate a high-technology hazardous waste treatment facility on state land. The state would monitor the plant for safety and effectiveness.

The idea is still in its preliminary stages, but the Department of Environmental Conservation has prepared a tentative environmental impact statement on the project and has scheduled hearings across the state. Locally, a hearing will be held at 1 p.m. Friday, Feb. 13, in Lecture Hall 19, on the SUNYA campus, 1400 Washington Avenue.

The statement lacks specifics — no proposed site locations, no detailed waste treatment plans. But it is a good overview of the impact of such a project on the state, on industry and on the locality where any project might be located.

It's clear some such treatment plant is needed; private ownership with intense state scrutiny seems good. Through the public hearing process, we should be able to consider pros, cons and alternatives. Everyone concerned with taming this particular wolf should either attend or keep themselves informed of the proposal's progress.

The Philadelphia Inquirer
Philadelphia, Pa., July 5, 1981

In August 1976, Congress passed the Resource Conservation and Recovery Act, which gave the federal government authority to enact regulations providing "cradle-to-grave" control of hazardous wastes. The program addresses a rapidly growing national problem of illegal dumping of toxic, and often deadly, chemical wastes. August 1981 may come and go without having those regulations in place.

The final phase of the long-delayed regulations dealing with standards for the design and construction of new disposal facilities is scheduled to go into effect on July 13. (Standards governing generation, storage and incineration of waste, also long overdue, went into effect last year.) Two deadlines for the regulations have been missed, and the Environmental Protection Agency is under a federal court order to meet this deadline.

EPA administrator Anne M. Gorsuch, however, is under pressure from the Office of Management and Budget to submit the design and construction standards to regulatory review, or, in other words, to a cost-benefit analysis. Mrs. Gorsuch has let it be known that she is seriously considering abiding by the OMB request, which could delay implementation of the program for as much as another year.

The delays in promulgating final design and construction standards for new disposal facilities have created a serious nationwide problem. New facilities have not been opened because no one was certain what standards the EPA would finally impose. Existing facilities are nearing capacity, or have stopped accepting hazardous wastes.

In Pennsylvania, between 20 million and 40 million tons of materials contaminated with four million tons of hazardous substances are produced each year. According to the state Department of Environmental Resources,

disposal facilities in compliance with the proposed new guidelines can manage only 20 percent of those materials. Forty percent is now transported to sites that soon will be full or will close when the new standards go into effect; 20 percent is shipped out of state; 10 to 15 percent is being temporarily stored for eventual disposal, and the remainder, presumably, is being disposed of illegally.

The pressure to find disposal sites for hazardous waste is mounting daily. "The clock is ticking," said Gary R. Galida of the DER's division of hazardous waste management. "The last thing we need now is any more delays."

The need for a national program of strict regulations is obvious. Illegal disposal of hazardous wastes is a lucrative and flourishing business. So-called midnight dumpers flocked to states where regulatory programs were weak, including Pennsylvania, prior to adoption of a new enforcement program modeled after the federal standards. Thus, any federal decision to back away from such a program would have major repercussions. As a result of the federal law, "all states have to make firm decisions on what levels of control are necessary," said Mr. Galida. "The federal program lent a measure of uniformity. Right now there is no place to run in the United States. There's no place to go where the requirements are very, very lax."

In terms of cost and benefits, there is no question that "cradle-to-grave" monitoring of hazardous wastes adds to the price of managing those materials. But the cost of *not* maintaining strict control over the wastes far exceeds that expense, in terms of actual dollars and the potential cost of serious, long-lasting harm to humans and the environment. That, quite simply, constitutes the benefit of the program. No further analysis is necessary.

THE SUN

Baltimore, Md., December 23, 1981

It's not surprising that scientists know so little about the effects on humans of toxic waste dumps, such as the notorious Love Canal in New York. The age of organic chemistry, characterized by man's ability to manipulate molecules of coal tar or petroleum, started only about the same time as the age of the automobile. Until recently, chemical companies and the public were so enthralled with the wonders of chemistry that they gave little attention to the toxic effects.

Because these chemicals are man-made, few bacteria exist to degrade them. So chemical dumps, such as Love Canal, remain hazardous for decades. When the chemicals are persistent *and* fat soluble—which is the case with DDT and PCBs, for instance—they can become part of the fatty tissues of humans (or other animals), where they can persist for years in some cases.

But as *The Sun*'s Mary Knudson pointed out in a recent series of articles, defining chemical hazards with precision is extraordinarily difficult, especially if people are exposed to several chemicals, as was the case at Love Canal. Humans obviously cannot be given measured exposures experimentally, and other ways of determining effects are limited. Environmental health research, says one physician interviewed by Ms. Knudson, "is in its infancy. I compare it to where infectious disease research was 50 years ago."

Epidemiologic studies—measurements of effects, including effects of small, long-term exposures, in large numbers of humans who are inadvertently exposed, as at Love Canal—is the most promising approach. But studies on the needed scale are terribly difficult and expensive. So far, epidemiologic evidence is very limited. It will accumulate more or less gradually, the rate depending largely on how much money Congress and the administration provide.

The obvious goal in the meantime should be to separate toxic wastes and humans wherever possible—while also developing new techniques for reducing wastes in manufacturing processes, and for recycling, instead of discarding, wastes that *are* generated.

There are some 9,300 hazardous waste sites in the United States. The federal Environmental Protection Agency is working to contain wastes at the 52 worst sites, but it has done little to clean up and destroy the chemicals. More money is needed for cleanup, for epidemiologic studies, and for developing techniques to reduce generation of wastes and for recycling. In the meantime, state authorities must see that future storage sites—such as one proposed for Hawkins Point in Baltimore—are equipped with all possible safeguards to prevent human exposure to the potentially dangerous chemicals.

TULSA WORLD

Tulsa, Okla., October 26, 1981

THE Environmental Protection Agency's listing of 114 hazardous waste dumps, along with plans for cleanup, suggests the agency is finally getting its priorities straight.

Armed with $1.6 billion under the Superfund program established by Congress last year, the EPA will set out to eliminate the worst dumps. Another 300 or so are targeted for later federal-state efforts.

Oklahoma's Tar Creek area in northeastern Oklahoma, where lead and zinc from old mining operations have infiltrated some water supplies, was listed among the 10 worst in the nation.

Too often, the EPA has worried over minor pollution of air and water where there is little or no threat to health while ignoring potentially deadly problems such as industrial waste dumps.

Congress and the EPA put the cart before the horse by decreeing that all the nation's streams be made swimmable and fishable before attacking obviously dangerous pollution. While the EPA was cracking down on cities to try to force horrendously expensive municipal waste treatment standards, it badly neglected the dumping of chemicals and other industrial byproducts.

The emphasis seems to have changed. With the creation of the Superfund, much of which comes from the industries that created the dumps, the EPA is headed in the right direction.

Certainly the goal of making the nation's air virtually pollution-free and its streams as near their natural state as possible should not be abandoned. But a great deal has been accomplished in these areas and it is likely that time and continued attention will solve the problems.

Immediate threats to life and safety, however, should command the first attention of EPA. Perhaps now they will.

RAPID CITY JOURNAL—

Rapid City, S.D., August 31, 1981

The control and disposal of hazardous industrial wastes may or may not be the environmental issue of the 1980s, but it will be in the near-term spotlight as Congress continues to consider legislation for cleaning up abandoned waste disposal sites.

The hazardous waste issue is a paradoxical result of American industrial advances which have led to a bounty of goods while producing a blight of wastes.

The hazardous waste issue is a natural continuation of the environmental movement which began in the 1960s. Congress acted first to clean up the air and water and now is moving on land pollution.

The problem is two-fold — preventing hazardous waste pollution in the future and cleaning up past disposal sites which pose a potential threat to health and the environment.

The complexity of preventing future pollution is evident from the problems the EPA has encountered in writing regulations to implement the Resource Conservation and Recovery Act of 1976. After four years of trying, the agency only recently issued its final regulations which still aren't satisfactory to industry and some environmentalists. One of the biggest problems was defining a "hazardous waste."

Still to be decided is the question of who should contain and clean up abandoned ("orphan") dump sites. The owners and operators of many closed sites cannot be found, have gone out of business or don't have the financial means to clean them up.

Congress is considering several proposals to create a "superfund" to finance the cleanup. Some are limited to "orphan" sites while others cover the cleanup of all chemical and oil spills and wastes — past, present and future.

It's generally agreed that because both industry and consumers have benefitted from cheap and unsafe practices in the past, both should share in paying for remedies. The difficulty comes in deciding how the financial responsibility should be shared. Some proposals would have industry pay 80 to 90 percent or more.

It is important that any fee on industry be properly placed because, like a tax, such a fee would influence industry's productivity and innovation. The fee plan should be designed to provide the greatest incentive to achieve the goal of reducing the volume of hazardous wastes.

The major bills before Congress would place a fee on raw materials. While such a fee might encourage producers to find ways of recycling some wastes as feedstocks, there would be little incentive to reduce the volume of wastes.

A more productive approach would be placing the fee on wastes themselves. That would provide a direct incentive not only to use old wastes as new raw materials but also to develop new technologies to reduce wastes and convert them into harmless substances. Under that approach, the dirtier the company the more it would pay.

The solutions to the hazardous waste problem should be future-and incentive-oriented rather than past-and punitive-oriented. If they are, disposal of hazardous wastes can be an environmental solution of the 1980s rather than a problem.

The Providence Journal

Providence, R.I., October 31, 1981

There sit the dumps — befouled repositories for the rusty, dusty, stinking and often deadly castoffs of man's industrial life. The stuff blights the land, sours the air and can seep unseen into who knows how much of our water supply. Of the 114 "worst" chemical dumps in the country, as determined by federal environmental officials, three have been found here in Rhode Island and four in Massachusetts. Nineteen Rhode Island firms have been cited for violating federal and state laws governing disposal of hazardous wastes. In all, the past record is a shabby one of carelessness, apathy and disregard for the public's health.

The menace of *existing* hazardous-waste dumps is being met, after a fashion. A $1.6-billion federal "super-fund," financed largely by taxes on the chemical industry, will be applied over the next five years to cleaning up the worst sites. This effort, coupled with responsible waste-handling by most chemical firms, should help overcome the legacy of the past.

But what happens to industry's noxious byproducts produced in decades to come? Cleaning up today's dumps does nothing to solve the problem tomorrow. And it is in failing to plan for future hazardous-waste disposal that most regions, including New England, are endangering their people.

The problem is one of inertia and politics, not one of technology. Safe disposal practices, ranging from recycling to burning to chemical detoxification, are known. Such techniques as secure solid-waste layering and leakproof leaching basins are in proven operation elsewhere.

What is needed, if the New England states are ever to confront this shared problem in more than hit-or-miss fashion, is public education and political agreement on a disposal site or sites acceptable to the public.

The six New England governors should familiarize themselves with the available technology. Hazardous wastes *can* be handled safely. The disaster of Love Canal in upstate New York was not the norm, and if the governors can convince citizens of this, they can help ease what one Massachusetts environmental official calls the "NIMBY" or "Not-in-my-Backyard" syndrome.

The governors also should study the economics of the problem, especially the question of whether a single, regional disposal facility may not make more sense than several smaller sites. Because of the region's size and the relatively manageable volume of hazardous substances produced, some chemical industry experts favor the regional approach. So do we. Once it is understood that such a facility, properly designed, can operate without environmental hazard, honest discussions can begin on where it might be placed.

But the problem won't go away. The top federal environmental official in the region is predicting economic disaster for New England unless it decides to meet its toxic-waste menace head-on — and *this* will take more political courage than the issue has yet drawn forth.

The Kansas City Times

Kansas City, Mo., February 16, 1982

The Environmental Protection Agency is going to try a new tack to get its priority list of the 115 worst hazardous waste dump sites cleaned up. Instead of dragging the companies responsible into court — which it hasn't done now since last June — the EPA will send letters to the 1,500 or more firms involved, inviting them to get started on a cleanup by a certain date. William A. Sullivan Jr., enforcement chief, describes it as a simple choice: "Look, do you want to clean it up, or do you want us to? If we do it, we'll collect the money from you later."

Mr. Sullivan isn't interested in cash settlements with the waste dumpers, nor does he intend to get into protracted negotiations. The letters are an order to "Do it," one way or another, and will carry a "drop-dead" date as a deadline. If the EPA can't trace those responsible for past dumping at the sites, or cannot get results from the firms solicited, it has up to $1.6 billion in the Superfund created by Congress to do the job itself. So far the agency has collected $200 million from chemical companies and spent $29 million on preliminary work at 30 sites.

The environmentalists, wary that this administration may be soft on polluting industries, don't think the letter approach will work. They want to see somebody hauled into court. But legal actions against reluctant defendants — and is there any other kind? — have a way of stretching out into long procedures before any results are obtained, if ever. Surely among those hundreds of companies are many which recognize that the government is going to wear them down eventually on this issue, and it might be cheaper and easier to go ahead and do the job themselves. At the least it's worth a try to locate any such and get something started on cleaning up some of those old dumps. The tough and recalcitrant polluters they can get around to later — and there will still be the courts as a last resort. All this potentially poisonous trash has been accumulating for decades, and getting it cleared away will not be quick or easy.

DAYTON DAILY NEWS

Dayton, Ohio, August 2, 1981

It is safer to bury the dead than to bury poisonous chemical wastes.

If we keep burying toxic wastes, we may be burying more dead soon.

Ohio is hostage to the old methods of landfilling poisons. This state is not yet equipped to deal with the threat to the underground water supply that is such a treasure to this region. Neither federal nor state governments is providing enough geologists and other technicians to keep tabs on the old landfills.

The new landfills are better. By law they have to be. But this country should not be using landfills as the primary method of disposing of poisonous chemicals. As writer Michael A. Verespej points out in his report in the *Dayton Daily News* business section today, there are better ways to get rid of the stuff.

Dow Chemical Co. is using a better way. Dow burns to smithereens 74 percent of its wastes and recycles the rest.

3M is using a better way. Over six years 3M has developed a program for *preventing* pollution, recycling toxic wastes and conserving raw materials, saving tens of millions of dollars in the process. Companies such as Monsanto have safely disposed of the notorious PCBs for years, but companies, for some odd reason, are not allowed to dispose of others' dangerous chemicals.

Nearly all of Western Europe is using a better way, and would look on the United States' heavy reliance on landfills as a ludicrous way to save money in the short run and risk public health and monstrous cleanup costs over the long run.

Not that all landfills are dangerous. Properly built, they can be reasonably safe. New laws require stricter standards for landfills. But they are not always monitored well. For example, an approved dumping site in Clermont County handles waste from many sates. Though neighbors are upset over the site, the dump cannot be checked constantly by the small staffs of the Environmental Protection Agency and state attorney general. The EPA acts only when there is a complain. Ohio's regulatory laws are weak.

This nation shouldn't have to run these risks and expenses of landfilling toxic wastes. Though the state and Miami Valley have to keep tabs on the old chemical time-bombs that have been planted over the water supply, landfills ought to be reserved for non-toxic wastes or those that can't be burned or made harmless without spending a fortune.

Landfilling is cheap in the short term — one-fourth the cost of treating or burning. But the government can change that here as have many governments of West Europe — through regulation. In the United States' case, that means the still-not-firmed-up rules for carrying out the Resource Conservation and Recovery Act of 1976 ought to discourage landfilling for certain poisons, especially since technology is available to make them harmless.

"We could catch up to Europe in a short period of time," said Philip Beltz, project manager for solid wastes at Battelle Memorial Institute's Columbus Laboratories, "if we made decisions based on both economics and fear of groundwater problems. But we are where we are because we want to be."

Sentinel Star

Orlando, Fla., January 5, 1982

FLORIDA has more poisonous chemical waste sites than any other state. According to the Environmental Protection Agency, no fewer than 200 known chemical waste storage sites pockmark the state and not one is approved as a chemical dump. There is some indication that officials may be making some headway in dealing with this problem, but they have a long way to go.

The amount of chemical waste strewn around us grows by a staggering 600 metric tons a year. In Florida, the hazard is compounded by a number of unfortunate circumstances, some man-made and some natural.

The most obvious is that most of the known sites are simple repositories for steel drums containing toxic devil's brews of solvents and heavy metals. Exposed to the extremes of Florida weather, the drums deteriorate rapidly, leaking their 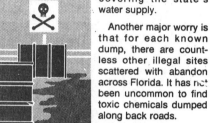 lethal loads directly into the thin layer of sand covering the state's water supply.

Another major worry is that for each known dump, there are countless other illegal sites scattered with abandon across Florida. It has not been uncommon to find toxic chemicals dumped along back roads.

One of the more promising developments came 10 days ago when the Governor's Hazardous Waste Policy Advisory Council in Tallahassee recommended that the state take command and assume the authority to force local governments to accept hazardous waste disposal sites.

As it is now, local governments routinely balk when asked by private firms for permission to put an industrial waste site in any area. The truth is that nobody is enthusiastic about having a chemical dump in his back yard. To help assuage public concerns, the council called for stringent state standards for the disposal sites and a monitoring program to cost $7 million to $10 million a year.

The council also recognized that Central and South Florida are unsuitable for waste disposal because of the high water table. North Florida's clay is acceptable.

If the Legislature follows through on the council's recommendations, Florida stands a strong chance of getting control of the flow of poisons threatening its future.

The second bit of promising news comes from the Environmental Protection Agency. It's thinking of turning an abandoned offshore oil rig in the Gulf of Mexico into a giant crucible designed to reduce chemical wastes to harmless ash. The experts believe that recent technological developments could turn the empty rig and others like it into safe and ultimately economical means of dealing with the national chemical waste problem.

That study is far from conclusive but it is solid evidence that technology can be used to deal with problems brought on by technology.

Whatever state and local governments do, they must not lose the sense of urgency in dealing with this threat. For years, Florida has depended on other states to carry the load of providing approved dumps for hazardous chemical wastes. Those days are all but over. Before Florida can proceed with its attempts to lure more manufacturing, it must be certain it can deal with the problems posed by those industries already here.

Post-Tribune

Gary, Ind., March 23, 1982

Well, Environmental Protection Agency Director Anne M. Gorsuch and her troops are at it again. In the name of "streamlining" EPA regulations, they are stripping this country of what protection it has against its polluters.

And this time the target is something that we dare not let get out of control — toxic waste.

It was national concern, spawned by such horror stories as New York's Love Canal, that sparked the 1976 Resource Conservation and Recovery Act to put those who generate and handle toxic waste under strict controls, administered by the EPA.

One rule required those industries to report regularly on what they're doing with their deadly products. This helped curtail "midnight dumping" which had become so common nationwide.

But the Reagan EPA sees this as too much of a burden on businesses; so the EPA has dropped its requirement for yearly reports on toxic waste disposal. Instead, the agency plans to send questionnaires to 1 percent of waste-makers, possibly expanding to 10 percent in later years, to keep up with what's going on. No one-the-spot inspections are planned.

That's like having an honor system for paying taxes.

Then there is the rule, which was to go into effect last July, requiring liability insurance for operators of dumps, in case of leakage, fire or explosions, plus guarantees that sites will be safely maintained after their commercial lives are exhausted (earlier such requirements could have avoided a Midco II in Gary and saved the government the money it's spending cleaning up several such dumps). Gorsuch and Co. put off the deadline to October, then extended it to April (with plans to gut the rules in the meantime).

EPA officials say the mandatory reporting system is flawed. As with most regulations, no doubt they can be improved, made more efficient. But that is what should be done; there is no reason to obliterate the intent to protect citizens from the hazards of toxic waste.

It's no wonder the jokesters are changing the "P" in EPA from "protection" to "pollution." The agency is fast forgetting its purpose and its duty to the citizens. Maybe the citizens should start reminding the EPA, President Reagan and Congress that people deserve to live without fear of being poisoned by toxic wastes seeping around uncontrolled.

Chicago Sun-Times

Chicago, Ill., March 18, 1982

The public doesn't need much convincing on the dangers of toxic waste. People know the horror stories about New York's Love Canal (though even that wretched dump pales in comparison with less notorious ones around the country). They've read Sun-Times reports about tragedies here: Illinois families poisoned when pollution controls were ignored.

National concern, in fact, sparked the 1976 Resource Conservation and Recovery Act to put those who generate and handle toxic waste under strict controls, administered by the U.S. Environmental Protection Agency.

Among the rules is one requiring those industries to report regularly on what they're doing with their deadly products. That can help check "midnight dumping," which has plagued so many communities nationwide.

But apparently that's too tough a burden on business, in the view of the Reagan administration's pathetic version of the EPA.

The agency has dropped its requirement for yearly reports on toxic waste disposal, proposing instead to use questionnaires to "sample" 1 percent of waste-makers (and perhaps 10 percent in later years). No visits will be made to plants getting questionnaires.

That's like the Internal Revenue Service asking people to send in tax checks on the honor system: Thanks for whatever you can spare, and talk to you again next year.

EPA aides say the current mandatory reporting system is flawed; that's reason to fix it, not to ax it. But that's not the style of EPA Administrator Anne M. Gorsuch.

Consider how her agency handled yet another toxic waste issue—that of requiring liability insurance for operators of dumps, in case of leakage, fire or explosions, plus guarantees that sites will be safely maintained after their commercial lives are exhausted.

The rules were intended to go into effect in early July of last year. Then Gorsuch aides put off the deadline to October. In September, they extended it again to April—and said the rules would be gutted in the meantime.

This is abandonment of duty, not "deregulation" or "streamlining" or any of the other euphemisms the dismantlers toss around. And the EPA can't be allowed to get away with it.

Wisconsin ⚓ State Journal
Madison, Wisc., March 31, 1982

The nation has too many federal regulations; it has had too much federal spending for too many years.

But the Reagan administration is cutting too deeply into essential programs of the Environmental Protection Agency.

One of the more questionable EPA cuts would adversely affect control of industrial toxic wastes.

EPA Administrator Anne Gorsuch has announced that in 90 days the agency will weaken rules that ban the dumping of liquid hazardous wastes in landfills and open pits, a practice which has, in the past, poisoned water supplies.

The new rules will allow up to 25 percent of the capacity of dump sites to be filled with barrels of toxic liquids.

Suspension outcry

While the new rule is being formulated, Ms. Gorsuch suspended the old rules for three months, allowing unregulated industry dumping of wastes. An outcry from environmentalists, Congressmen and companies that already had invested in new disposal procedures forced the EPA last week to reinstitute the dumping ban while the new rules are being formulated.

Ms. Gorsuch also has proposed cutting costs by eliminating quarterly checks on groundwater that flows under waste dumps and turning aside a proposal by the chemical industry that dump sites be lined with impermeable material to catch leaks.

While congressional requirements for the EPA to deal with toxic wastes are expected to double the agency's workload, the administration has called for a 42-percent reduction in the EPA budget, including a cut in staff from 11,400 to 6,000.

Retreat wrong

Congress can show in its budget actions that it wants the EPA's activities to remain strong. Nevertheless, the agency's retreat from toxic-waste control is disturbing, and wrong for two reasons.

First, the expense of cleaning up future Love Canals may well outweigh any government or industry savings that the regulatory and budget cuts will produce. Presently, the nation produces about 94 billion pounds of toxic wastes each year, and the average American's exposure to the chemicals is expected to double this decade.

Second, the cutbacks penalize those environmentally conscious businesses which, in anticipation of regulations, have set up alternative forms of disposal. They will find themselves at a competitive disadvantage with those businesses chosing not to go to the expense of finding better ways to manage their wastes.

Two firms which invested heavily in alternative waste-handling procedures — ENSCO and SCA Services Inc. — have joined the Environmental Defense Fund in a suit to reimpose some of the EPA regulations scheduled for abandonment.

Problem nationwide

Toxic wastes are among those national problems which cannot be cured without federal controls. Businesses are unlikely to spend the money necessary to properly manage wastes if they don't have to, and if their competitors do not.

Reliance on state rather than federal controls would foster competition among regions for minimum regulation. Reliance on states also ignores the fact that pollution does not respect political boundary lines. It is a national problem, requiring solutions at the national level.

The EPA, like other federal agencies, undoubtedly has waste and unnecessary regulations, but the EPA solution to this problem is too drastic.

The Reagan administration should back off plans to dismantle the federal role in toxic-waste management. Further, it should accept the need for its commitment to administer the laws that Congress has passed on toxic wastes since 1976, when the problem came into the national limelight.

The Boston Herald American
Boston, Mass., April 1, 1982

Fortunately, a storm of criticism has forced the Environmental Protection Agency to revoke its 90-day suspension of the regulation prohibiting landfill dumping of toxic liquid wastes. And there now seems no prospect of extending the suspension or reimposing it at a later date.

Had the suspension remained in effect for even 90 days, some experts estimate that up to 20 million gallons of toxic liquid wastes might have been buried in landfills across the country. Obviously, the potential for creating future environmental hazards comparable to those at Love Canal would have been multiplied manyfold.

The suspension was all the more inexplicable in light of the fact that there are practical, if somewhat more expensive, alternatives for disposing of toxic liquids. A budding new industry specializing in incineration and other relatively safe technologies for disposing of toxic liquids has sprung up in recent years.

Had the Environmental Protection Agency not been forced to backtrack, there would have been little incentive for chemical companies and other industries that generate such wastes to take advantage of improved disposal methods.

We are glad to note that the EPA has conceded the point, and belatedly moved to minimize prospects for new Love Canals.

The Houston Post
Houston, Texas, April 3, 1982

Under a barrage of protests, the Environmental Protection Agency has reversed its 90-day suspension of a ban on dumping chemical wastes in unlined landfills. But the retreat has not smoothed congressional feathers ruffled by this and other recent EPA policy decisions involving hazardous waste disposal. Among them: Suspension of a rule requiring chemical waste producers to submit annual reports to the EPA on how they disposed of their toxic wastes and a proposal by the agency to allow 25 percent of a landfill to be used for hazardous waste dumping.

Some lawmakers want the agency to spend more money from a superfund created by Congress in 1980 to clean up abandoned waste dumps. Most of the revenue for the $1.6 billion superfund comes from taxes on chemical manufacturers and is deposited in a special Treasury trust fund. Rep. James J. Florio, a New Jersey Democrat who heads a House Energy and Commerce subcommittee, estimates that the surplus in the fund will reach $352 million by the end of fiscal 1983. Florio charges that the EPA is not disbursing the money to the states fast enough, forcing some states to borrow money at high interest rates to finance the cleanup of abandoned dumps.

Chemical wastes leaking from an old landfill in the Love Canal neighborhood of Niagara Falls, N.Y., in 1977 focused national attention on the hazards posed by improper disposal of such toxic industrial residues. The EPA has had more than its share of problems developing standards for chemical waste disposal. But when it decided in February to temporarily lift the ban on dumping the wastes in unlined landfills, even some industries that had invested large sums in alternate waste disposal methods protested. The Hazardous Waste Treatment Council and the Environmental Defense Fund joined in asking the U.S. District Court in Washington, D.C., to make the EPA restore the ban. Unlined landfills permit chemicals to leak from deteriorating drums and contaminate water supplies.

Another sore point with some lawmakers is the EPA's contingency plan for cleaning up the 400 dumps in the nation that pose the greatest health and environmental threats. Critics complain that the plan, mandated by Congress, is inadequate and lacks strict standards necessary to keep chemical poisons from polluting the environment.

In response to charges that the agency's toxic waste regulations are too soft, EPA officials say the national contingency plan was made flexible so states would have sufficient leeway in their cleanup efforts. In defense of the growing surplus in the cleanup superfund, they contend that the money is being disbursed as fast as possible, consistent with sound planning, and that large sums cannot be spent until fiscal 1983-85. The agency says the filing of annual reports by firms generating hazardous wastes was dropped to reduce paperwork. Instead, 10 percent of the more than 50,000 producers of such wastes will be asked if they are complying with the law. But the absence of on-site inspections by the agency would leave the accuracy of this monitoring method open to question.

Ironically, Congress can do little now to force the EPA to follow congressional intent more closely in enforcing laws on toxic waste disposal. It can't threaten to withhold funds from the agency because the administration is already trying to cut the EPA's budget below what some lawmakers believe is an adequate level. But the public can certainly make the weight of its opinion felt on this grave national problem that directly affects its health and safety.

The Idaho STATESMAN

Boise, Idaho, July 25, 1982

The Environmental Protection Agency failed to protect the public adequately with its new regulations for hazardous-waste dumps. Even if industry complies with the rules, toxic chemicals still may leak from existing dumps into ground water.

The best way to understand the deficiency of rules for existing dumps is to contrast them with the much better rules for new disposal sites.

Where future disposal sites are concerned, the EPA takes a two-pronged approach. It requires builders of new landfills, settling ponds, waste lagoons and so forth to line them with a synthetic material to prevent leakage. It also requires these builders to construct ground-water monitoring wells to detect leakage if the liner fails. That way, corrective action can be taken.

The idea is that neither lining pits nor monitoring ground water is fool-proof. Liners can be punctured or may deteriorate with age. Monitoring may not pick up leakage or may detect it too late.

That's where the rules for existing dump sites fall short. They require only a monitoring system, no liners.

The EPA, which admits that monitoring is imperfect, claims the public must accept greater risk from existing dumps because it is not technologically feasible to line the more than 2,000 existing landfills, chemical waste ponds and other facilities governed by the rules.

That doesn't make sense, considering the damage that may be done if toxic chemicals get into ground water. It might be impossible to dig up and line some landfills, but many settling ponds and lagoons are another question. They often can be drained temporarily, lined and then refilled.

Rather than exempt the wide variety of existing facilities in one fell swoop, the EPA should allow case-by-case exemptions. A dump operator should be excused from lining his facility only if he can prove it is technologically unfeasible to line it.

Rocky Mountain News

Denver, Colo., May 9, 1982

IT'S called the EPA, and that used to stand for Environmental Protection Agency. Under the Reagan administration, some critics are charging, the acronym has come to stand for Environmental *Pollution* Agency.

Unfair? Probably.

It was nevertheless disquieting to learn through a recent Rocky Mountain News report about the generous manner in which the EPA office in Denver has been reducing fines for hazardous-waste violations.

During the past year, the EPA has reduced fines for seven Colorado companies in behind-the-scenes settlement conferences. The amounts — which originally ranged from $33,000 to $5,000 — dwindled to nothing higher than $2,000 and, in some cases, to zero.

The way the EPA tells it, a number of those reductions were justified. Where officials thought they saw violations, there were none. Perhaps EPA acted properly in erasing those fines. Nobody should want the feds to go tromping around on private businesses with the kind of regulatory overkill that has seemed to mark some previous administrations. If EPA goofed up all that much, however, the lesson would seem to be that the inspection operation is somehow inadequate.

Some of the fine reductions don't strike us as having been quite so excusable as others. In some cases, the companies fixed up what was wrong and EPA figured that was sufficient. But is that the way to make it clear that incorrect storage and disposal of waste are matters of large public consequence? And shouldn't the feds conduct the public's business in public? We don't much care for all those private conferences.

Hazardous waste, it should be understood, is a terrible danger. It can kill you as surely as a mugger on a street corner. Taking it easy on careless companies is no more desirable than coddling criminals. EPA should be reasonable, yes, but it should also send a clear message that the roof's going to fall on any company that threatens public health with hazardous wastes.

A step in that direction was the EPA announcement Friday that nine lawyers and three other employees are being transferred to the inspection division to beef up hazardous waste regulation. EPA is "serious about enforcement," said regional administrator Steve Durham. For everyone's sake, we hope so.

The Hartford Courant

Hartford, Conn., March 30, 1982

Anne M. Gorsuch, administrator of the federal Environmental Protection Agency, made a mistake a few weeks ago.

She proposed to allow the unregulated disposal of liquid hazardous wastes for an interim period of 90 days. The proposal shocked many people, including some in industries that might have reaped immediate benefits. The rule relaxation would have been an invitation to plant the seeds for future Love Canals. Members of Congress were outraged.

Mrs. Gorsuch had to beat a hasty, if temporary, retreat by altering the interim rules to prohibit the burial of any container in which toxic liquids are standing in observable quantities.

She ordinarily doesn't make such mistakes. Unlike Interior Secretary James G. Watt, Mrs. Gorsuch has not attracted many thunderbolts from critics with a series of highly visible, confrontational policy decisions. But in her own, less obtrusive way, she has been overseeing a destructive assault on the nation's environment.

Hazardous wastes are a good example of how that assault is being directed.

Mrs. Gorsuch has publicly stated that cleaning up hazardous wastes would be a top priority of her administration. Yet, in addition to rules changes that would exacerbate the problem or at least delay its solution, the EPA's proposed waste enforcement budget has been cut by about 50 percent.

In addition, the person who has been nominated by the president to head the EPA's program for cleaning up the nation's hazardous waste sites is an industry spokesman, Rita M. Lavelle, who has long been engaged in fighting environmental regulations.

The same strategy of undermining the environmental effort applies in other areas.

The shift is toward making anti-pollution regulation less stringent. The EPA has relaxed rules on lead emissions from automobiles, has supported a bill that would eviscerate the Clean Air Act, has pulled back from implementing a national groundwater strategy, and turned away from the effort to eliminate chemicals in the nation's waterways.

The Office of Enforcement has been dismantled, and enforcement efforts are down 48 percent for air pollution and 55 percent for water pollution.

The demoralized EPA staff is being led by a Fifth Column. Not a single environmentalist is included among the first score major EPA appointees. Nearly all, like Ms. Lavelle, were involved in industries that fought the agency.

Research is being truncated or eliminated. Money for studying the long-term impact of toxic substances is down; research on asbestos is being eliminated; research on cleaning up the Great Lakes is scheduled to go. These cuts could be particularly devastating. In some cases, they would end research programs that took years to get into gear and might take more years to revive if a more enlightened administration should take hold of the EPA in the future.

Congress can help. It can reject the massive budget cuts for the EPA that have been proposed by the administration for the 1983 fiscal year. Environmental groups can help by continuing to lobby and, where necessary, bring the agency to court if it fails to meet its responsibilities adequately.

But the administration must be convinced by public outcry that Americans will not tolerate a recanting of the nation's commitment to protect the environment by the very agency charged with its protection. The destruction wrought by Mrs. Gorsuch is not less serious, only more insidious, than that of the more flamboyant Mr. Watt.

BUFFALO EVENING NEWS
Buffalo, N.Y., March 16, 1982

The federal Environmental Protection Agency has made a serious mistake in dropping a requirement that manufacturers report annually to Washington on how they dispose of hazardous wastes.

An EPA official contended that 11,000 reports each year were an inefficient method of collecting data, that much of the information was inaccurate anyway and that good results could be obtained by requiring reports from only a small random sample of producers.

These sound to us like rationalizations for a retreat from adequate and responsible efforts in an area of legitimate federal regulation to protect public health and safety.

If the annual reports smother or confuse the problem, then the answer is to simplify and improve the reports rather than abolish them. Similarly, the solution to inaccurate data is not to drop the reporting requirement. Questionnaires mailed out to 10 percent of the manufacturers could scarcely compensate, in enforcing compliance with disposal standards, when the remaining 90 percent would wholly avoid any reporting obligation.

Sad experience, from Love Canal to darkened back roads of midnight dumping, should demonstrate beyond doubt the problems, temptations and dangers associated with the disposal of toxic wastes by our modern industrial society.

Yet in recent weeks the EPA has removed a requirement for hazardous waste dumps to carry liability insurance. It has temporarily suspended the ban on dumping containerized liquid wastes into landfills.

Reportedly, these ill-conceived decisions may soon be reversed. So should be the dropping of the requirement of annual reports from all manufacturers. And the EPA, instead of acting first and then considering objections later, ought to consider all aspects of the changes before carrying them out.

The safe disposal of these wastes, which pose grave health and safety problems along a ragged edge of our technological age, should be a major concern of federal regulation. In fulfilling that role, the EPA should err on the side of caution rather than of risk.

Roanoke Times & World-News
Roanoke, Va., March 19, 1982

IF THINGS continue the way they're going, the Environmental Protection Agency could wind up on a list of health hazards. EPA recently announced it was lifting, for 90 days, the ban on disposal of liquid industrial wastes in landfills. After it was deluged with protests that this would lead to an orgy of uncontrolled dumping, the agency reinstated the ban. But the episode offers little reassurance that EPA takes seriously the environmental dangers posed by disposal of more than 100 billion pounds of toxic waste annually.

It made a certain amount of sense last year for EPA to embark on a wholesale review of the 2,000 pages of hazardous-waste disposal regulations — putting many of them on hold meantime.

Under Ronald Reagan, though, the EPA seemingly prefers not to enforce the regulations at all. It had given advance word that it would lift the ban, so the chemical industry and others had let wastes accumulate. During the 90-day moratorium originally planned, there would have been no limit on the quantity that could be disposed of in landfills.

The health dangers from toxic wastes can't be dismissed by people who dislike regulation and its effect on profit-loss statements. About 130 billion pounds of these wastes will be generated in 1982. Some 48,000 chemicals are used in commerce; among those considered poisonous in varying degrees are 1,500 ingredients in pesticides, 4,000 in drugs, 5,500 in food additives, 1,200 in household products, and 2,500 chemicals and mixtures used by the plastics industry.

Jimmy Carter's EPA estimated that only 10 percent of all hazardous wastes were being disposed of in an environmentally sound way. If Ronald Reagan's EPA thinks things are better than that, it ought to explain why — not just to the waste generators but also to the people whose health may be jeopardized.

The Morning News
Wilmington, Del., March 20, 1982

Gradually, perhaps, in its rush to get government off the backs of the people, the Reagan administration is learning the difference between regulations that stifle and regulations that protect.

Perhaps a good indicator in deciding between the two kinds is the public outcry that arises when a regulation is either promulgated or withdrawn.

That certainly has been the case in the reversal by the Environmental Protection Agency of its suspension three weeks ago of the ban on putting hazardous liquid wastes into landfills.

EPA administrator Anne M. Gorsuch, who for months has been almost as much of a target for environmentalists as Interior Secretary James Watt, declared in revoking the 90-day suspension Wednesday that in all such cases it behooves her agency to "err on the side of caution."

In that light, Mrs. Gorsuch revoked what she insisted was not all-out permission to dump at will in the first place. She then instituted an alternative that many of the parties to the dispute seem to feel is workable. Under the new plan, containers found to contain toxic liquids will either have to be decanted of them or materials will have to be mixed in to absorb the liquids.

The suspension three weeks ago had been viewed as a concession to some chemical companies and operators of landfills that an all-out ban was not feasible.

In fact, said Delaware's U.S. Sen. William V. Roth Jr. in a letter of protest to Mrs. Gorsuch, suspension of the ban penalized those companies that already had developed, at some initial cost to themselves, alternatives to the prohibited mode of dumping. (Sen. Roth, it should be noted, has a reputation as a supply-sider and deregulator.)

These alternatives can take the form either of other methods of disposal or, as some companies seem to be discovering, the further alternative of not making waste in the first place. Since stronger federal regulations went into place under the Resource Conservation and Recovery Act of 1976, some firms have found that modest capital investments can pay off handsomely in savings realized through recycling what previously would have been wasted material.

The outcry at Mrs. Gorsuch's original suspension of the anti-dumping rule demonstrates that a large segment of the public refuses to bear the *direct* cost of such waste disposal any longer. Sentiment exists that an industry and its specific customers, not the public in general, should bear these costs.

Of course in the long run we all pay for whatever forms of disposal are employed, although under safer systems we are spared health and environmental damage. For this reason, we should all endorse the formula of not making waste in the first place. It is to be hoped that this would allow manufacturers to realize savings that could be passed on to the consumer.

Protests still are being sounded that the EPA suspension already has permitted three weeks of free-wheeling dumping that inevitably caused some damage.

When such a conservation-minded organization as the Environmental Defense Fund terms the new approach "reasonable," however, we can join in hoping that it also represents some kind of a "turning point by the Reagan administration" in exploitation policies that too often have justified the description of "rip and tear."

The Salt Lake Tribune
Salt Lake City, Utah,
December 24, 1982

Announcement by the Environmental Protection Agency of a "national priority list" of the 418 most hazardous waste sites in the nation comes just about two years after Congress mandated the so-called Superfund, an authorization of $1.6 billion to clean up dangerous dump sites. The list replaces an interim identification of 160 sites announced earlier.

The 418 sites are the worst case examples of some 12,000 similar dumps Anne M. Gorsuch, EPA administrator, describes as "ticking time bombs" scattered across the country.

Perennial critics of the EPA chief have been quick to denigrate the announcement. Typical was the knee jerk of U.S. Rep. James J. Florio, D-N.J., one of the principal authors of the Superfund legislation. While welcoming the list he complained that "it is a year and a half overdue."

For someone who ought to be intimately acquainted with the general snail's pace of the federal bureaucracy, Rep. Florio seems to have forgotten the legislative history of a measure he helped write. The Comprehensive Environmental Response, Liability and Assistance Act of 1980 wasn't passed until the waning days of the 96th Congress, on Dec. 3, 1980. Also it was one of the final laws approved by former President Jimmy Carter; he signed it (PL96-510) on Dec. 11.

So, Mrs. Gorsuch's announcement comes two years and 10 days after Superfund became law. It, thus, seems somewhat unrealistic for Rep. Florio to have expected the "national priority list" to have been completed some time in June of 1981, particularly so when the agency charged with administering the Superfund was without a permanent administrator, following the departure of David Costle when administrations changed in Washington, until May 5, 1981, when Mrs. Gorsuch was finally confirmed by the U.S. Senate.

When an agency has to select from a possible 12,000 "ticking time bombs" the 400 or so (the EPA was told by Congress it has to find at least 400) that are likely to create the most damage if they explode, it is probably prudent to be carefully selective.

Instead of carping unrealistically about how long it took to compile the list, people of Rep. Florio's ilk ought to be encouraging, even pressuring, the EPA to quickly, now that the targets have been identified, get on with the work of cleaning up and mitigating the hazardous sites before they do any further damage to the environment.

The Star-Ledger
Newark, N.J., January 3, 1983

What do Pitman and Mantua in Gloucester County, Pleasantville in Atlantic County, Old Bridge in Middlesex County, Gloucester Township in Camden County and Freehold Township in Monmouth County have in common? Nothing to be proud of. Each of these municipalities harbors within its borders a site for dumping hazardous waste that ranks among the worst 20 of these sites in the entire nation.

The U.S. Environmental Protection Agency (EPA) has identified the 418 worst hazardous waste dumps and, regrettably, 65 of them are in New Jersey. This far exceeds the number in any other state. The doleful catalogue is completed with the notation that the aforementioned six sites rank among the worst of the worst.

Just how New Jersey became saddled with so many dangerous public health problems is difficult to explain. No doubt it goes back to another era before the environmental movement brought such matters into full focus and there was a hue and cry to remove or remedy these conditions.

More recently, New Jersey has been a national leader in the drive for cleanup efforts. In fact, the federal list is in a way the result of strong New Jersey action, which brought suit against the EPA to compel the promulgation of the list. It is the view of Gov. Thomas Kean that the list "should not be construed in a negative fashion, but rather be viewed as a successful effort to deal with ... waste sites in the state."

In a sense, the Governor is right. The list may make dreary reading but it does bring the problem out in the open and outline what must be done to solve it. Still, the enormous number of hazardous facilities is also an indictment of state and federal regulation efforts. Despite the presence of strong state laws, there have been several examples documented in The Star-Ledger in which landfill cleanup attempts failed to proceed in an effective and orderly manner.

And the federal cleanup effort has been locked in partisan controversy, with Democrats claiming that EPA Administrator Anne Gorsuch just isn't doing her job. Rep. James Florio of New Jersey has spearheaded an attack within Congress on the controversial Mrs. Gorsuch. She has been cited for contempt for her refusal to release to Congress documents which her opponents claim would prove she hasn't been doing her job.

Officials, businessmen and the public will all have to be more vigilant if New Jersey is to make the trip from the top to the bottom of the pollution list. It may be a long—and oftimes a frustrating one —but it should be amply clear there is no other way to go.

THE SACRAMENTO BEE
Sacramento, Calif., January 5, 1983

In the hands of the Reagan administration, the hazardous waste Superfund has become a travesty. The Superfund was set up with great hopes by Congress in 1980. It was to clean up the most dangerous of the nation's abandoned toxic dumps, using $1.5 billion raised largely by taxing chemical manufacturers. By 1985, cleanup was supposed to be completed, and future tragedies were to be averted by a companion program of safety regulation of industry's current hazardous waste practices. But over the last two years these good hopes have been trammeled.

It was not until last month, a year and a half after the Superfund law required it, that the Environmental Protection Agency (EPA) first got around to releasing a list of the country's most pressing cleanup problems. And after all that time, the list did not even include specific cleanup plans. Two years into the program, the EPA claims to be cleaning up *five* sites — five, out of the 14,000 potentially dangerous sites Congress has been told of and the 418 in the EPA's own catalog. And those five cleanup efforts, literally, have not yet scratched the surface.

Meanwhile, the EPA has all but ceased development and enforcement of regulations concerning current hazardous waste disposal practices. And this grasshopper's approach to winter is certain to create new toxic contamination problems in the future — in addition to the existing problems the EPA obviously is never going to get around to. Nonetheless, EPA Administrator Anne Gorsuch recommends that the Superfund program end in 1985 as scheduled. Her lack of commitment to the program is palpable.

No wonder Congress is furious. The House voted to cite Gorsuch for contempt of Congress — ostensibly because she refused to let Congress see certain internal documents about her enforcement of the Superfund law, but more likely out of sheer frustration. Congress usually handles a recalcitrant administrative agency by threatening to cut its budget. But Gorsuch *wants* to spend less on environmental protection. Thus far, four-fifths of the Superfund money that's been collected for toxic dump cleanups remains unused.

One shouldn't mistake this for genuine concern about the federal budget. Gorsuch has had several opportunities to replenish the Superfund — the law directs her to find the companies responsible for abandoned dumps and to charge them up to three times the cleanup cost — but she hasn't done it. Her preference seems to be to get the companies to do a little surface cleanup in exchange for total immunity from further responsibility. She calls this a cooperative approach to government-industry relations.

What it really adds up to is a dereliction of duty. And after the first victims of abandoned toxic dumps did their part, spending years lobbying Congress and informing the public; after Congress did its part, enacting a program that could actually prevent future contamination tragedies; after the chemical industry did its part, paying hundreds of millions of dollars into the Superfund to pay for cleanups — it is worse than contemptuous that Gorsuch and her superiors are failing so completely to do theirs.

ARGUS-LEADER

Sioux Falls, S.D., August 7, 1983

There have been many reminders for Americans in recent weeks about pollution threats to the water supply and problems associated with waste disposal. The lessons should be taken to heart.

Consider the following:

☐ Purity of the nation's water is threatened by toxic waste. Unwanted industrial byproducts, household garbage, sanitary wastes and discarded chemicals and pesticides have been dumped for decades on land. A survey shows that half the nation's underground water supplies is in danger of being permanently contaminated from runoff.

☐ One metric ton of hazardous waste is added to the environment each year for every American, according to a congressional report. Water consumption has reached a record 450 billion gallons a day, double the 1950 rate. Disintegrating pipes and water mains in older cities will require an estimated $138.6

billion a year to repair during the next 20 years.

☐ An Environmental Protection Agency study shows that despite supposed heightened awareness about toxic dangers, 109 of 171 operated hazardous waste plants were not meeting minimum requirements to monitor possible groundwater contamination.

☐ Congress Office of Technology Assessment says it will take up to $40 billion to clean the 15,000 known toxic dumps, about 30 times the available money in the environmental agency's superfund.

☐ Some fishing waters throughout the nation are contaminated, by such chemicals as polychlorinated biphenyls (PCBs), dioxin, mercury, etc.

Many of the problems stem from the nation's modern way of life and dependence on chemical-industrial technologies.

Sometimes a problem isn't ap-

parent until years after the event. For example, in 1965, a train derailed in Perdido, Ala., spilling benzene and other chemicals. The spill was mopped up. Tests completed this year confirmed the benzene reached the water supply by 1981; the wells have been closed.

The American public applauded the national goal of clean rivers and waters set in the 1970s. Some progress has been made. But the present threat to water supplies and lack of progress in cleaning up toxic waste dumps is disturbing.

The situation places a larger burden than ever on the EPA and its new administrator, William Ruckelshaus, to carry out Congress' mandate to get on with the job of cleaning up the most dangerous toxic dump sites.

There's also a need for state and local governments, industry and agriculture to be more vigilant than ever to forestall new problems before they occur.

THE LOUISVILLE TIMES

Louisville, Ky., June 9, 1983

The nation's continuing struggle to gain control over hazardous waste offers yet another rebuttal to the Reagan administration's casual attitude that environmental problems lie in the past.

Laws to soothe public anger over contaminated landfills or the horrors of Love Canal were, in fact, only the beginning of governmental efforts to contain poisonous wastes. Because of the huge volume of dangerous materials produced by American industry, the lagging federal and state response and defects in corrective programs, the cleanup has barely started.

There has, of course, been progress. But uncontrolled chemical dumping, some of it legal, much of it illegal, continues, with especially devastating

results in the industrialized states of New York and New Jersey, where organized crime reportedly controls much of the waste disposal business. Political finagling at the Environmental Protection Agency has slowed the cleanup of infamous older dumps.

Despite conservative grumping about "overregulation," a big part of the problem is that federal rules have turned out to be too lenient, both in their restrictions and their penalties. A bill now before Congress addresses several major deficiencies.

Moreover, the emphasis in the principal federal waste disposal control law, the Resource Recovery and Conservation Act, is misplaced. A study by the congressional Office of Technology Assessment points out that the act at-

tempts primarily to manage waste after it has been produced instead of encouraging techniques to reduce the quantity of hazardous garbage, to recycle it or destroy it.

Consequently, land disposal remains the cheapest and most widely used method of getting rid of industrial poisons, even though it is clearly the most threatening to human health. Many experts point out that a high percentage of dangerous materials could be more safely disposed of if the right economic incentives were built into the regulatory system.

A company that buries barrels of toxic gunk, even in carefully designed landfills, is in effect managing to avoid some of the costs of doing business by passing them on to future generations and society at large.

The OTA estimates that alternatives to land disposal might cost 50 to 100 per cent more but that the long term expense of cleaning up a dump can be 10 to 100 times that additional cost. If the chemicals at Love Canal had been properly gotten rid of to begin with, the bill might have been $2 million rather than a predicted $100 million.

A measure that has cleared the House Energy Committee is by no means the complete answer, but it would strengthen the federal hand in protecting the public. For instance, an exemption for small waste generators, whose total contribution to the problem is large, would end. Dry cleaners, filling stations and others may thus come under the law. The practice of mixing waste with oil for use as fuel would finally be regulated.

And criminal penalties for illegal dumping or falsification of documents

THE ARIZONA REPUBLIC

Phoenix, Ariz., September 5, 1983

STAND by for this shocker — U.S. industry is generating four times more hazardous waste than previously believed.

These new figures are based on an Environmental Protection Agency survey of firms producing toxic and other hazardous waste, and included in a report to be released soon.

But EPA doesn't seem alarmed.

EPA's new figures suggest that U.S. companies generate 165 million tons of dangerous waste annually, instead of 44 million tons as earlier believed.

But EPA also found that the capacity and efficiency of storage and treatment facilities is greater.

Most of the waste is being properly disposed.

Since EPA has not changed its formulae for classifying and measuring wastes, this represents a monumental increase in waste production.

One example is a Du Pont Company facility in New Jersey that now produces as much so-called hazardous waste as did all U.S. industry two years ago — 44 million tons.

Is such a startling increase possible?

Or did EPA bungle the job in the past, and fail to measure the extent of dangerous wastes?

If EPA wants to enjoy the confidence of the American public, it now must provide reassurance that although its calculations about the extent of waste in the past may have been wrong, it is not lax in enforcing disposal standards that protect the public from chemical dangers.

The Burlington Free Press

Burlington, Vt., December 20, 1983

Improper disposal of toxic wastes has long-term effects on water supplies and poses dangers to residents who are living in the vicinity of the disposal sites.

In the 1970s, illegal dumping became a big business. Tank trucks picked up the chemical wastes and emptied them in out-of-the-way places with little concern for the damage that was being done to the environment. Occasionally, the illegal dumpers opened the spigots on the trucks and spilled the contents on highways as the trucks rolled along. Wherever the dumping was done, there was little doubt that the wastes ultimately drained into ground water supplies. Poisonous, combustible and carcinogenic chemicals leaked into aquifers at a tremendous rate. Even now, it is impossible to determine how much permanent damage was inflicted on some water supplies and whether they can ever be reclaimed for use as drinking water.

As accounts of indiscriminate dumping proliferated, federal and state officials expressed alarm about the situation and took steps in 1976 to curb the practice. Congress passed laws which made generators liable for their chemical wastes and required them to keep records on disposal from factory to dump. It now appears that there is sufficient control over the methods of handling hazardous wastes that illegal disposal has dropped off dramatically. While the crafty may still try to break the law, they face stiff punishment if they are caught. Violators can be fined $250,000 and sentenced to jail for two years for each offense. Civil penalties range from $10,000 to $25,000 for each violation.

In November 1980, Vermont in conjunction with five other New England states set up a load-by-load tracking system to follow each hazardous waste shipment from its origin to its destination. Each shipment must be accompanied by a uniform, seven-page manifest, telling what is being shipped, who is transporting it and where it is going. Special handling instructions are included for use in case of a spill. The shipment is compared with the manifest as it makes it way from the generator to the trucker to the disposal site. Once the cargo reaches its destination, copies of the manifest are sent to the state Environmental Conservation Agency for safekeeping. If the manifest is not returned to the agency within 35 days, steps are taken to locate the waste.

Even with the manifest system, state officials have said there is still a possibility of illegal dumping. "If the people want to 'midnight dump,' they can continue to do it," said Richard Valentinetti, head of the the state's Air and Solid Waste Division. "Let's not put anybody in a sense of false security that that piece of paper stops all midnight dumping."

But increased vigilance coupled with people's awareness of the problems that hazardous wastes pose for health and safety can encourage a more responsible attitude toward proper disposal by those who use toxic materials and insure that the state's environment is not poisoned by the wastes.

As a result of the stiffer regulations, industries and institutions in the years to come may well find it more feasible to reduce the amount of waste by processing it for re-use instead of paying the high costs of disposal.

AKRON BEACON JOURNAL
Akron, Ohio, November 16, 1983

FOR THOSE who live near the Deerfield toxic waste dump, the worst part is the uncertainty.

After an initial cleanup, health officials say the 11-acre dump in Portage County is still a source of dangerous pollution. But a study that might pinpoint the dangers is stalled for lack of funding.

That study and a total cleanup are caught in conflicts between the state and federal environmental protection agencies. The two EPAs are deadlocked over who should pay. And until that is resolved, a final cleanup has been put on hold.

The matter is certainly not an easy one. Deerfield, closed five years ago, was recognized as one of the nation's worst toxic-waste dumps. One year ago, a good-faith agreement between the state and some Ohio companies that had dumped wastes at the site led to a $2.4 million initial cleanup — the removal of some 15,000 drums of toxic chemicals. Another $5.5 million will be required for the study and final cleanup.

And since similar, though lesser, problem dumps exist across the nation, it is easy to see why the EPA might be stingy in doling out the barely adequate $1.6 billion superfund for toxic waste cleanup.

But tangled though the problem may be, it is unconscionable that government squabbles should be allowed to endanger public health. And that is what this dispute is about.

Given the porous soil around the dump site, health officials fear serious contamination of the area's underground water supply. Clearly, something needs to be done to determine the extent of that threat so it can be remedied.

The primary fault seems to be at the federal level. It is no secret by now that the Reagan administration has been unenthusiastic about pollution controls or toxic cleanup.

And of the two main points of conflict between the state and federal governments, Ohio seems to be on firmer ground: The EPA wants Ohio to take future responsibility for the site, but the Ohio Constitution forbids open-ended contracts, and the final tab for the cleanup is incalculable. Washington wants any cleanup money collected by the state from industry to go to Deerfield. But a 1980 law only requires 10 percent funding from the state; Ohio taxpayers have already spent $839,000 toward the $8 million cost.

People, however, are more important than jurisdiction. And the first job of government is to protect its citizens. If that means some funding compromise in order to identify the hazards from Deerfield, both state and federal officials should pursue it.

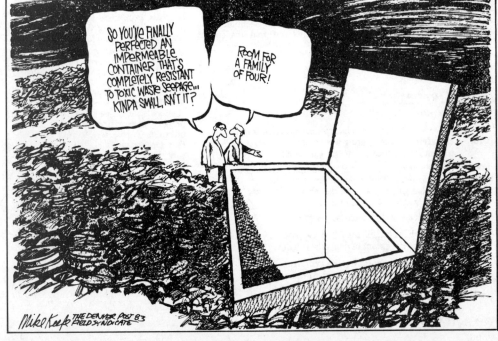

Chicago Defender
Chicago, Ill., November 1, 1983

A recent congressional report stated that 87 percent of the dump sites for toxic wastes in northern Illinois failed to meet the standards for identifying deadly chemicals in well water. These types of ground water monitoring systems, which detect leaks in waste dumps, were required by law in November, 1981.

Regarding one of the worst toxic waste dumps (the Stringfellow Acid Pits of California), Science World Magazine states that "The wastes are seeping into the ground, threatening the water supplies of nearby homes and farms."

Junior Scholastic Magazine notes that "there are 10,000 landfills and dumpsites in the U.S. which pose a threat to human health. Countless rivers and lakes are also polluted with toxic waste."

Dioxins are probably the most famous toxic wastes. They are poisons which are equally deadly when found in water or soil. Seventy-five different types of dioxins exist. Most of them are products of chemical manufacturing. They are the waste materials remaining after the maufacturing process has been completed. They are also created when garbage in city incinerators and fossil fuels at powerplants are burned below 750 degrees centigrade.

Scientists now know that some dioxins can exist in the soil in poisonous forms for more than ten years.

We urge the federal and state governments to create intense research projects for the purpose of converting toxic wastes into useful, non-poisonous substances. This would be a giant step towards creating and winning a war on toxic wastes.

Any money spent for this effort would be an investment in the healthful future of humankind. Currently we know that special incinerators, ultraviolent light, certain chemical processes and industrial recycling have proven successful in reducing the toxic quality of some waste products.

Experts say that a proper cleanup of toxic waste dumps across America would cost nearly 300 billion dollars. Some people say that's a lot. We say, so is the future of humanity.

Detroit Free Press
Detroit, Mich., September 3, 1983

MICHIGAN SHOULD be among the states most concerned about a new environmental survey that shows the production of hazardous wastes in the nation is nearly four times as much as was estimated previously. Michigan is part of a five-state region that has by far the greatest concentration of plants producing toxic chemicals, solvents, corrosive agents and other dangerous products.

The figures are now in from waste producers and handlers for 1981. Previously, they were not required to inform the Environmental Protection Agency of their production.

Before reporting was required, the EPA had believed hazardous waste creation amounted nationally to 40 million tons. The actual national production for 1981 was 150 million tons.

The EPA's director of the office of solid waste, John Skinner, is inclined to minimize the significance of the threat represented by the new estimate of the amount of dangerous gunk being produced. He must have a high threshold of alarm. That's nearly a ton of hazardous waste for every person in the United States.

Furthermore, 57 percent of the stuff is disposed of by deep burial, putting it dangerously close to water tables, and sometimes right into water tables. This represents a potential problem to the many communities served by wells, where it has not already done so.

Of the 14,100 regulated manufacturers who reported hazardous waste production, a total of 3,240 were in Michigan, Wisconsin, Illinois, Indiana and Ohio. The next highest regional concentration of waste-producing plants was in California, Nevada, Arizona and Hawaii, with 1,990.

This unexpectedly high volume of hazardous material indicates that federal laws on disposal of wastes need tightening and strengthens the case for rigid enforcement. The amount of waste being injected underground is startling and should be under more effective control. The problems of surface containment in double-lined ponds are complex, however, and pose expensive burdens for waste-producing plants, most of them manufacturing chemicals.

The EPA's new figures show that the hazardous waste problem is much, much larger than was known heretofore. They came as no surprise, however, to the hazardous waste division of Michigan's Department of Natural Resources and similar departments in other states. Their directors have held for years that the federal EPA was underestimating the production of hazardous waste.

The EPA could also follow the lead of DNR in requiring reporting of hazardous waste production if it reaches 100 kilograms monthly. At present the EPA exempts plants from reporting any production of 1,000 kilograms or less per month.

Michigan industries produced 2.7 million tons of dangerous waste in 1982. Most of it was waste water that was satisfactorily neutralized, the DNR believes, leaving about 450,000 tons that constituted a problem. Waste liquids are particularly difficult to deal with, requiring incineration. Michigan has no incinerators for that purpose, although there are three in nearby states.

The EPA was weakened by neglect early in the present administration. An attempt is being made to recreate the agency as a serious instrument to keep the nation clean. The new figures prove the need is great.

Fort Worth Star-Telegram
Fort Worth, Texas, October 17, 1983

The Environmental Protection Agency made a correct—if somewhat tardy—move last week when it announced a change in the manner in which the EPA proposes to deal with the cleanup of hazardous waste dumps.

The agency has proposed new guidelines requiring companies to pay for at least 80 percent of the cost of cleaning up hazardous waste dumps they have created. And, of perhaps more importance, those guidelines are going to be put in writing.

Under the proposed changes, the EPA says it will not even begin to negotiate the cleanup of a waste dump with responsible firms unless the firms are willing to assume at least 80 percent of the costs or perform at least 80 percent of the actual cleanup chores.

Also, the agency will retain the option of asking for more than 80 percent, settling for less than 100 percent from any one company only.if the EPA concluded it could collect the remaining 20 percent from other polluters who are not parties to the particular settlement.

The guidelines, if, indeed, they pass final muster and become standard EPA policy, should go a long way toward alleviating suspicion that the agency has been guilty of "sweetheart" deals with polluters. The EPA has been under increasing pressure from Congress to take its regulatory duties more seriously.

The value both to the environment and the American taxpayer of standing firm on hazardous waste cleanup becomes evident when one realizes that, just since the final days of the Carter administration, the government has identified 546 such waste sites as having created severe problems.

Without firm guidelines that clearly spell out how cleanup of such places is to be carried out—and how it is going to be paid for—taxpayers can never be certain that they are not shouldering more than their share of the burden.

Industries that create hazardous waste should be held responsible for disposing of it in a manner that protects the environment and the people who live nearby. When they fail in that responsibility, they should be held accountable.

The new, written guidelines provide the EPA with a workable tool with which to progress toward that end. They should receive congressional blessing and become part of official policy as soon as possible.

The TENNESSEAN
Nashville, Tenn., January 2, 1984

WHEN the Superfund legislation was passed in 1980, it promised to clean up America's environmental mess by applying strict fines on companies that dump their toxic wastes in the nation's waterways and landfills.

But Reagan administration officials who are responsible for enforcing that legislation are looking the other way — allowing oil and chemical firms to continue unloading their poisonous waste in the nation's streams, lakes, forests and fields.

The 1977 discovery of toxic dumping at Love Canal made this nation aware of the health and environmental hazards created by the disposal of chemical waste. Immediately after that much publicized incident, literally hundreds of other chemical dump sites were identified throughout the country. President Jimmy Carter called the presence of toxic chemicals in the environment "one of the grimmest discoveries of the industrial age."

In response to the environmental crisis, Congress enacted protective legislation to regulate the dumping of toxic substances.

One of those measures, the "Superfund" bill, gives the government the authority to sue companies for improper disposal of hazardous wastes. The money recovered from the companies is put in a "Superfund" to be used for cleaning up the dump sites that threaten public health.

But there is a loophole in Superfund — a three-year statue-of-limitations. And, according to Rep. Albert Gore Jr., the present officials of the EPA and the Department of Interior have missed the prosecution deadline on 400 polluted sites.

Such negligence on the part of environmental officials is inexcusable. They are not only flouting justice by refusing to apply the penalties set out in the legislation, but are also costing the nation billions of dollars in clean-up money that could be recovered from chemical and oil companies.

When Superfund was passed, it was a thoughtful, fair attempt to clean up America. But the Reagan administration's refusal to enforce Superfund is creating a windfall for some irresponsible industries, and a threat to the future health of the nation.

Rockford Register Star
Rockford, Ill., March 26, 1984

An entire community murdered by poisonous waste — it happened to Times Beach, Mo., just a little over a year ago. They are still mourning at the grave.

We must not let it happen to Rockford.

That means the toughest, most thorough and wide-ranging defensive action we can take — right now. The reeking killer is already in town — it's been here a long time, just waiting to do its job.

We've been attacked by toxic waste dumping at least three times since Feb. 2, the latest within the last two weeks, the Rockford Sanitary District claims.

Sanitary District Director Jon Olson said February incidents were caused by deliberate, illegal dumping of chemicals.

The Sanitary District is the first to feel the effects of toxic waste crimes. Illegally dumped chemicals first destroy the sewage treatment process, then are discharged into the Rock River, where they destroy animal and plant life.

That's the way it starts. Unless it's stopped, there's a dead community like Times Beach. It was murdered by the chemical dioxin, illegally dumped into road treatment oil.

An arrest was made in the latest local incident and State's Attorney Daniel Doyle has properly placed a high priority on dealing with the chemical dumping situation. He has called on assistance from the Illinois attorney general's office, citing that agency's familiarity with a new state criminal law on hazardous waste.

We totally agree with Doyle's appraisal of this threat to the community and we welcome all the professional help we can get.

Let no one underestimate this threat — it is against our lives and our homes.

The Seattle Times
Seattle, Wash., April 24, 1984

HEADLINES in The Times of last Saturday and Sunday told the story well. "Park was built on foul mound of chemical soup," and "It's a tremendous loss."

The first headline offered all the explanation necessary for Mayor Royer's action in closing the popular park at the north end of Lake Union until the Environmental Protection Agency finishes testing soil samples taken from the park in recent days.

The second headline expresses the feelings of the many thousands of local residents who have enjoyed the award-winning oasis since its opening in 1976.

Gas Works Park is nationally recognized as a unique urban recreational area. It is truly a Seattle treasure. But in view of the advice Royer was getting, he had no reasonable alternative to putting the park off limits until it is certified as safe by pollution-control experts.

A study as far back as 1971 showed that the site was heavily polluted by the gas-producing plant that operated there for a half-century until 1956. A benzene-manufacturing plant on the site and a nearby tar company may also have contributed to the pollution.

That information was passed on to city officials and the park's designer. They proceeded on the theory that the contaminants were far enough underground as not to constitute a hazard on the surface. We do not fault that decision. It was in line with the general thinking and practices of the time. Knowledge of cancer-inducing agents and other chemical poisons has greatly increased in recent years.

The EPA's findings are not expected before late May. We hope the city will then be prepared to assign highest Parks Department priority to a reopening of the Gas Works site under suitable safety standards.

The Oregonian
Portland. Ore., February 4, 1984

The federal Environmental Protection Agency is talking tough these days about cleaning up this nation's sadly neglected hazardous waste sites. However, the EPA's credibility test comes this year in Congress when the agency seeks appropriations for more money to finish the job that the Superfund started.

Will EPA Administrator William Ruckelshaus fight hard for expanding the $1.6 billion Superfund appropriation beyond 1985? Or will he bow to political conservatives, who, in the interest of deficit control, would like to see the Superfund a likely candidate for budget-cutting?

Lee M. Thomas, who replaced scandal-ridden Rita Lavelle at the helm of the Superfund program, believes that this Congress-initiated war on toxic waste dumps is just a good start. The previous appropriation, he says, is capable of cleaning up only 10 percent of the estimated 1,000 to 2,000 most dangerous waste sites.

The EPA has identified only 546 sites thus far. After three years under the Superfund program, only six high-priority sites have been cleaned up. Planning is under way at 147 others, and emergency removal is completed at 185 sites and in the works at 46 more. That still leaves 162 identified sites to deal with, plus another 500 to 1,500 sites to identify, categorize and sanitize.

Clearly, the EPA will be unable to accomplish this mission without at least another five-year extension and expansion of the Superfund program. It is doubtful that $1.6 billion would complete this job, although that level of appropriation should go much further than the previous one.

Under Thomas' direction, the EPA wisely is approaching regulation with more sticks than carrots. Under its new policy, the EPA is refusing to negotiate settlements unless polluters are willing to finance 80 percent or more of the cleanup cost.

These impressive agency changes in approach to hazardous waste cleanup will prove meaningless if the agency loses the Superfund appropriation.

Politically, it is too early to tell whether congressional conservatives will fight hard against a healthy reappropriation. The administration suffered much criticism through EPA's miniscandal. Reagan, as a result, has proposed a large budget increase for the EPA, including increases for the Superfund. He should stand firm on those commitments, not only because it defuses one of his areas of domestic program vulnerability, but also because it is the right thing for the chief caretaker of this nation's natural resources to do.

The State

Columbia, S.C., April 9, 1984

FOR SEVERAL years, the federal government has been nibbling away at the problem of cleaning up the hazardous waste dump sites in the country, including a couple here in Columbia.

This important work has been paid for out of a $1.6 billion "superfund" created by Congress in 1980. Most of the money has come from a tax on raw materials used to make industrial chemicals, the chief source of the toxic waste.

The work is going slower and the task is much greater than originally anticipated. So far, permanent cleanup has been completed at only six of the 546 sites originally identified by the Environmental Protection Agency.

Since there are potentially dangerous disposal sites in almost every congressional district, it is not hard to keep lawmakers' attention focused on the problem. The Reagan administration, however, does not have it high on its list of priorities.

Although the present financing doesn't expire until Sept. 30, 1985, moves are already under way in Congress to reauthorize the fund and increase it. The thinking is that Mr. Reagan will be forced to sign a bill presented him before the election while he might veto it afterwards.

The Administration did once oppose a renewal of the fund, but the President did an about-face in his State of the Union address and came out in favor of it. A veto would mean going back on his word.

Besides, the problem is too big to be ignored. The EPA now lists about 17,000 possible sites. It thinks 2,200 of them require urgent attention and estimates the cost will be between $8 billion and $16 billion. The present fund will cover only about 170 dumps.

Those staggering estimates have caused alarm among White House budget-makers. But two bills in the House, one sponsored by a Democrat and the other by a Republican, call for spending of $8 billion or more over five years. Both would impose an additional tax, this one on wastes as they are disposed of.

EPA Administrator William D. Ruckelshaus said his agency has not finished a study on how big the fund should be and does not plan to send its own bill to Capitol Hill until December.

When an issue is this important and the costs are so high, care should be taken to draft a bill that meets the most urgent needs and no more. It is not necessary to rush through a bill this year just to beat the election. This danger to public health cannot be ignored after the voting.

The Miami Herald

Miami, Fla., April 21, 1984

REMEMBER Rita Lavelle? Initially the Reagan Administration put her in charge of the Environmental Protection Agency's (EPA) "Superfund" program to clean up hazardous-waste dumps. She deliberately bungled the job, then lied about it to Congress. Her very name now symbolizes the Reagan Administration's breach of faith in protecting the environment.

With that ugly history in mind, environmentalists are pushing Congress to extend the Superfund program this year, even though its statutory life does not expire until the end of 1985. The Administration wants to wait until next year. Superfund defenders fear, however, that a re-elected Ronald Reagan very well might veto the Superfund entirely if its reauthorization is put off until next year. Their strategy therefore is to force the issue before November's election. That strategy makes sense. That way, either Mr. Reagan will be forced to support the program, or the voters can hold him accountable for vetoing it. The most ardent champions of the Superfund are not content with merely reauthorizing it, however. They want it not only extended, but strengthened. They would increase the program's funds, set mandatory clean-up timetables by statute, compensate victims of toxic-waste disasters from the fund, and create a new tax on waste disposal.

In principle these reforms are necessary and desirable, because since its inception in 1980 the Superfund has been proven inadequate to achieve its goals. The $1.6 billion current fund won't cover the costs, and the lax administrative mandate under the original legislation has proven too loose to force bureaucrats to act expeditiously.

Unfortunately, the effort to beef up the Superfund law predictably is drawing intense opposition. Opponents of the stronger law won a preliminary victory in a House Commerce subcommittee by passing instead a proposal that would extend the program for five years without change: Superfund advocates say this is a disguised effort to block a truly effective toxic-waste clean-up program.

In principle, the Superfund devotees are correct. They should fight to boost the Superfund's effectiveness as long as their fight seems winnable. Yet given the short legislative calendar in this election year, and given the extra hostility that a tougher clean-up bill inspires, at some point it may prove necessary to settle for mere reauthorization of the existing Superfund as the best attainable pre-election goal.

Better to reauthorize a weak Superfund before the elections than to have a stronger one killed by veto in 1985. Renew the basic program this year, then fight later to strengthen it. That's the best strategy for those determined to exorcise the ghost of Rita Lavelle.

Rita Lavelle

St. Paul Pioneer Press/Dispatch

St. Paul, Minn., June 23, 1984

The federal Superfund bill, which was passed in 1980 and expires in December 1985, has demonstrated two things: First, a national system to clean up the nation's toxic waste dump sites is an absolute necessity. Second, the $1.6 billion originally allocated for the first five years is woefully inadequate for the next five.

Facing election-year deadline problems, Congress is trying to speed up reauthorization legislation. Last Wednesday, the House Energy and Commerce Committee approved a bill boosting funding to $9 billion. Proponents claim the bill will be ready for a floor vote by August.

The need to rush continuation of the program is obvious. With at least 2,200 sites across the nation leaking substances, many of them toxic, into groundwater supplies, delays only exacerbate what already is a serious and expensive problem.

But backers of the current bill should not permit the necessary urgency to blind them to reality and fairness.

While acknowledging the need to increase Superfund funding, critics in the chemical industry claim that the Environmental Protection Agency cannot wisely spend that much additional money. They also believe that the funding base should be expanded to include more chemical-producing companies and firms involved in the disposal end of the chain. Their reasoning appears sound.

Cleaning up a chemical dump site involves more than attacking it with bulldozers. Analyzing the extent and seriousness of the problem and then developing a plan requires testing and analytical knowledge that only can be accumulated over time.

Industry experts claim that the EPA does not yet have sufficient staff or expertise to judiciously tackle the toxic waste dump problem at the level authorized in the proposed bill. If they are correct and millions of taxpayer dollars end up being wasted, a much-needed program would suffer serious harm.

Concerning fairness, under current legislation, one relatively small segment of the chemical industry — the petrochemicals — pays 87.5 percent of the $1.6-billion bill. In practice, 12 large firms pay 70 percent of the industry's share, even though two-thirds of the hazardous wastes that end up in landfills are produced by firms unrelated to chemical production. The remaining 12.5 percent of the program's funding comes from general tax revenues.

A fairer method, industry representatives claim, would be to spread the costs — they say $850 million a year would be closer to what the EPA can wisely spend — over more chemical firms, and impose a tax on companies that ultimately dispose of the wastes.

Although the chemical industry obviously has a large stake in the size and shape of the funding pie, it also has a legitimate interest in polishing an image that has become severely tarnished by Love Canal-style debacles. Congress should neither discount the industry's taxing recommendations nor overlook the fact that dollar bills alone will not make even a single toxic dump disappear.

Love Canal: "The First Detonation"

A state of emergency was declared in May, 1980 at the Love Canal in Niagara Falls, N.Y. President Carter ordered the federal government to evacuate and shelter 710 families whose homes had been built on the site—an unused, clay-lined canal that Hooker Chemicals and Plastics Corp. had used as a chemical dump site in the 1940's and 1950's. The families evacuated under Carter's order joined 239 others who had abandoned their homes in 1978, following accounts of a high incidence of miscarriages and birth defects in the area. The results of a study made public less than a week preceding Carter's declaration of an emergency had indicated that 30% of the residents had suffered chromosome damage, and that 11 of 36 persons tested possessed rare chromosomal aberrations that frequently were linked to cancer and genetic damage in offspring.

The question of liability for medical problems suffered by residents of the Love Canal is complicated. There is no absolute proof that the chemicals found in the trench are directly related to the reported health effects, although there were 10 potential carcinogens among the 80-odd compounds found in the dump. Hooker had not concealed the fact, when it deeded the land to the Niagara Falls Board of Education for a token $1 in 1953, that the site had been used as a toxic waste repository. In any case, during the years when Hooker was using the site there were no state or federal restrictions on hazardous waste disposal. It was well after an elementary school was built on the site and families had moved into the neighborhood that the chemicals began to surface on the school playground and the yards and basements of homes. Their appearance in 1976 followed several years of heavy snowfall and rain that caused large quantities of the canal's contents to overflow; chemicals that had already seeped into the soil appeared in nearby creeks. Witnesses at congressional hearings have testified, however, that Hooker was aware of the dangers posed by the seepage during these years and did not warn authorities because of the legal problems it might cause.

There is considerable controversy, also, over the extent of the damage that was caused by exposure to the dozen or so toxic chemicals that were found to have been part of the mix in the dump; these included PCBs, pesticides, benzene and chloroform. A federal study conducted in 1983 by the Centers for Disease Control in Atlanta and two national laboratories found "no specific relationship" between "exposure to chemical agents in the Love Canal area and increased frequency of chromosome damage." The study reported "no increase in abnormalities" among 46 residents or former residents of the area compared with a study of 50 residents of a control area about a mile away. It is difficult to evaluate such studies, however, since the exact relationship between chromosomal damage and exposure to chemicals is not yet known, and also because the degree and type of exposure experienced by the individuals at Love Canal is difficult to ascertain. The laboratories involved in the 1983 study themselves cautioned about the possibly limited validity of the findings.

The situation at Love Canal, because of the national media attention it received and the public outcry over the obvious threat to residents, became the springboard for a new movement to prevent the careless disposal of hazardous wastes. (Pent-up anxiety among Love Canal residents erupted into anger May 19, 1980 as home owners barricaded two officials from the Environmental Protection Agency in an office to demand federal evacuation of the entire area.) The end result of such strong public sentiment was the passage of legislation, the $1.6 billion "Superfund," intended to protect the environment from similar disasters. (See pp. 218-228.) Ironically, Love Canal has probably received attention out of proportion to its importance; as EPA's administrator for water and waste management, Eckardt Beck, expressed it at the time the site was first discovered: "Love Canal was merely the first detonation of a string of chemical time bombs literally strewn across the nation." In the Niagara Falls area alone, over a dozen large chemical dump sites have since been uncovered; several of them are much larger and potentially more dangerous than the Love Canal site that forced evacuation of residents from the area.

The Boston Globe

Boston, Mass., May 4, 1980

It's not suprising that residents of Love Canal trapped two federal officials in neighborhood association offices for several hours Monday to dramatize their anger and frustration. They were told over the weekend that results of a recent study indicated that 11 of 36 people tested had chromosome damage. Then they were told — again — that there was no cause for alarm. As one of the 11, a woman who has lived there for 26 years and already battled breast cancer, put it, "No matter what happens everyone still says the area is safe."

After the study was released, the state public health commissioner insisted that there was no link between the 20,000 tons of chemicals buried beneath their homes and the chromosome breakdown, that it could have been caused by hair dyes, occupational hazards or heavy smoking. Gov. Hugh Carey's office issued the usual statement, something about how the state would do everything in its power to save lives. The deputy administrator of the Environmental Protection Agency maintained that an immediate evacuation order would only intensify anxiety and fear among the 710 families who remain in the area. And the Hooker Chemical Company, which dumped the toxic substances into the canal more than 27 years ago, contended that the results of the study are "preliminary and uncorroborated."

The families who live in the Niagara Falls, New York community have been getting this same story for the past three years. After 235 families were evacuated from their homes in August of 1978, officials declared that the worst was over. Later, when remaining residents expressed concern about high rates of cancer, miscarriage and birth defects, officials said the incidence was no greater than anywhere else. Those who remained clung to these claims, to their modest homes and to hopes that the federal government would win its still pending $125 million suit against Hooker. They believed that the dumpsite could be cleaned, that, if it couldn't, the state and federal governments would chip in to buy them out so they could move on and put the nightmare behind them.

Just what does it take to get the people in charge to admit that there is genuine cause for alarm? Just how hard would it be to stand up and say that something has gone terribly wrong? At this point, the cautions and bromides don't serve any purpose. Love Canal residents are terrified and they have reason to be. Eleven out of 36 people have chromosome damage and even if chromosome damage can't be directly linked to the chemicals, it has been linked to cancer, miscarriage and birth defects. By wrapping themselves in protective layers of bureaucracy and science, by distancing themselves from the problem, public officials are destroying what little credibility they had and intensifying public mistrust.

Granted, it isn't going to be easy or cheap to help 710 families relocate. It isn't going to be easy or cheap to deal with their health problems and the potential health problems of their children's children. It isn't going to be easy or cheap to deal with the disastrous psychological aftereffects. It isn't going to be easy or cheap to deal with similar problems at other dumpsites around the country, places like Woburn, Mass., where high rates of leukemia have prompted investigation. But it isn't going to be any easier or cheaper, if public officials refuse to face up to the very real problems at Love Canal.

The Evening Bulletin

Philadelphia, Pa., May 21, 1980

Two years after "Love Canal" became a synonym for environmental disaster, it may surprise many of us that this upstate New York community is still home to hundreds of families. For most Love Canal residents, the health emergency declared by New York state officials over chemical contamination in August 1978 solved nothing, and even fueled their fears about their health.

Since then, much has been said and written about the need to control the dumping of poisonous chemicals. State and federal laws have been enacted to track the safe movement of millions of barrels of poisonous goop. A federal "superfund" for chemical cleanups has been proposed. These are important steps toward preventing future Love Canals. But while all this has been unfolding the people of the *first* Love Canal waited for some official verdict on what the chemicals in their environment were doing to their bodies.

Now they know the worst: The U.S. Environmental Protection Agency (EPA) last week reported that some Love Canal residents may have suffered genetic damage from the chemicals. Two extremely rare types of chromosomal damage turned up in 11 out of 36 persons tested by one laboratory. EPA calls this alarming and significant because this kind of genetic damage can signal future cancers, birth defects and miscarriages.

That evidence should be enough to break the logjam that has kept the Love Canal families living close to the site where 20,000 tons of highly toxic chemical wastes were dumped. Although 237 families were moved from their homes by New York state officials, the state purchase of more Love Canal homes has been held up for six months. Now that residents have been given the frightening results of this genetic study, the program ought to be moved ahead quickly.

EPA will decide this week whether to relocate as many as 710 families. If the findings of the chromosome study are confirmed by independent geneticists who will meet this week, we don't see how the Federal Government can do other than move these families.

In that kind of worst-case scenario, there's still a huge gap in public policy over the question of who will pay. It has to be addressed soon. Even the proposed federal superfund legislation, as reported out of a House committee last week, would not help because it is meant primarily for the cleanup and containment of chemical dumps themselves. Lawsuits against the Hooker Chemical Co., which dumped the Love Canal chemicals more than 27 years ago may cover the costs someday, but that's still in doubt.

The money for this emergency has to be provided from federal — and state — sources if need be, while the suit against Hooker is pressed.

The Providence Journal

Providence, R.I., May 21, 1980

A crisis that shook the people of Niagara Falls, N.Y., two years ago and launched a nationwide search for toxic chemical dumping grounds, has flared anew with potentially far-reaching implications. Blood tests given to 36 persons who reside near the infamous Love Canal area, where the government says nearly 200,000 tons of chemical wastes were dumped between 1942 and 1975, have shown that 11 have suffered chromosome damage.

Scientists say that chromosome breakage may be linked to cancer and spontaneous abortion, and could lead to genetic difficulties in offspring. The Environmental Protection Agency (EPA), which announced the findings last Saturday, indicated that 710 families may have to be evacuated from their homes. In 1978, 239 families who lived closer to the dump site left their homes when the New York State health commissioner declared a state of emergency.

The latest discovery raises a number of questions. That 30 percent of the 36 persons tested show signs of chromosome abnormalities does not, according to geneticists, necessarily mean these individuals risk future health problems. Indeed, the waste chemicals at Love Canal are not the only possible cause of chromosome damage. Still the findings are unsettling, particularly to the 710 families most intimately involved. Their fear and anger

spilled over on Monday when two officials of EPA were held hostage at the office of the Love Canal Homeowners Association in an effort to force a commitment from Washington that the area would be evacuated.

As sympathetic as one is inclined to be toward these innocent people, taking the law into their own hands in no way can be condoned. While their mood is one of desperation, their illicit actions are nonetheless inexcusable.

It is estimated that scattered throughout the United States are some 30,000 toxic chemical dumps. Will people living in these areas now be tested for chromosome breakage? Will many have to be relocated? And as more is known about this new crisis, where will the financial responsibilities come to rest? The U.S. Justice Department is suing the Hooker Chemicals and Plastics Corp., a subsidiary of Occidental Petroleum, for $124.5 million to clean up four dumps in Niagara Falls. New York State is suing Hooker and Occidental for $95 million in restitution for funds spent and $540 million in punitive damages.

Meanwhile, New York and Washington have been arguing over money and responsibility with no clearcut resolution. The new crisis makes it eminently clear that jurisdictional problems must be resolved

with dispatch. Public health must take priority. Barbara Blum, deputy administrator of the EPA, was emphatic on that point concerning the 710 families. "We certainly can't let money stand in the way of relocating the families if that should prove to be necessary," she said.

When these events in New York are multiplied many times over from coast to coast, the scope of health and fiscal problems looms large. All the more important is proposed legislation in the Congress to create a "superfund" to help pay for the cleanup of abandoned hazardous waste sites like the Love Canal. A bill authorizing a $600 million fund, half of which would be provided by the chemical industry, was approved 21-to-3 last week by the House Interstate and Foreign Commerce Committee. Another proposal is stalled in the Senate Environment Subcommittee on Environmental Pollution.

The latest findings of chromosome damage replace some of the earlier speculation on the dangers involved. They add an element of urgency that no longer can be brushed aside. From the nation's capital to health and environmental officials at the local level there must be a new commitment first and foremost to the welfare of people whose homes are located within the danger zone of toxic chemical dumps. Any doubts about that have been swept away by the EPA's latest report.

Newsday

Long Island, N.Y., June 2, 1980

A State Assembly report charging that the federal government dumped toxic waste around Love Canal years ago is one more reason Washington should stop stalling and help relocate Love Canal residents today.

Gov. Hugh Carey is asking for $20 million in federal money to help move about 700 families permanently. Most of the cash, roughly $18 million, would go to buy their present homes so they could make down payments on new ones. The state, which has spent a substantial amount to relocate other families, would put in $5 million this time.

Carey argues that his plan would be far cheaper than the $32.4 million he estimates Washington's temporary evacuation program may cost eventually. Yet the federal response has been typically sluggish so far. Although a state of emergency was finally declared after an inconclusive study frightened Love Canal residents by indicating abnormal chromosome damage among them, Washington still insists that buying their homes is solely the state's job.

We don't agree. By any rational standard, Love Canal should have qualified long ago for federal relief. Now the Assembly report—on which hearings are due later—suggests that Washington may share directly in the responsibility for creating this environmental abomination.

Genetic testing, further studies and litigation will take a long time yet. But there should be no more delay in enabling Love Canal families to move into new homes.

THE DENVER POST
Denver, Colo., June 3, 1980

RESIDENTS OF the Love Canal neighborhood of upper New York state so far have displayed more prudence than the Environmental Protection Agency in reacting to news that chromosome damage has shown up in a high percentage of their neighbors.

Only about 30 per cent of the eligible residents have applied for temporary relocation from the contaminated area, despite President Carter's dramatic declaration of a state of emergency.

The administration's reaction was in response to two new scientific studies. One showed chromosome damage and the other showed damage to nerves among persons living near the disposal site. But both were based on small samples. And both have since eroded into uncertainty.

Nonetheless, upon release of the reports, the EPA declared that "The last two studies have pushed us over the edge." The agency promptly announced that the government would pay for emergency temporary relocation while further tests and studies are completed.

The directors of both pilot studies recommended further follow-up investigations. Those will take six to nine months, after which the federal government will decide whether to relocate residents permanently.

But the EPA's reaction instilled an immediate fear among area residents that they and their children are doomed to early death. And the agency established a precedent of offering millions of dollars to get people out of their homes before a danger clearly has been established.

The chemical wastes dumped in Love Canal came to public attention about two years ago, when 239 families were permanently relocated from their homes. In the interim, there has been plenty of time for state and federal officials to design a reliable research program and procedures for removing other families from any proven danger. But the government agencies didn't plan well enough or far enough in advance.

There may well be good reason for fear in the Love Canal neighborhood. Certainly there is no excuse to permit, ever again, that kind of reckless disposal of compounds whose effects are unknown. But the way in which the Love Canal episode has been handled can only exacerbate those fears and must make the residents very unsure of whom they can trust.

The Honolulu Advertiser
Honolulu, Ha., June 1, 1980

Over the past two years, the Love Canal has become synonymous with the threat of chemical waste dumps and industrial pollution.

The Love Canal, of course, is a never-finished man-made canal in upstate New York that was filled with deadly chemical wastes by the Hooker Chemical and Plastics Corporation in the 1940s and 1950s. About 20,000 tons of wastes were dumped into the canal before the company covered it over and sold the land to the Niagra Falls Board of Education which built a school there.

LAST MONTH, following an Environmental Protection Agency report that found an abnormally high incidence of chromosome damage among Love Canal residents, the federal government announced it was temporarily relocating more than 700 families in the area.

In 1978, shortly following the first large-scale indications there might be problems, the state of New York spent $10 million to permanently relocate families who lived closest to the canal.

As might be imagined, the latest federal move has raised anew the fears of remaining Love Canal residents, and has brought to a head the complex legal and jurisdictional problems involved in determining who is responsible.

Should the federal government buy the homes of the more than 700 families to allow them to move, should that be a state responsibility, or should Hooker Chemical carry the burden?

As it is, there are a multitude of suits pending in various courts, and it may take a decade before the majority of the legal actions are settled.

WHAT IS ESPECIALLY disturbing about the Love Canal situation is that it is not the only such case in the country.

One federal study estimates there may be as many as 30,000 hazardous waste sites throughout America. Even if only a few of them involve as many people as the Love Canal, the potential cost to government or private industry would be prohibitive.

In the Love Canal case alone, nearly 1,300 residents have filed personal injury and property damage claims which total more than $15 billion.

Hooker insists it is not to blame. That will be up to the courts to decide.

But to date, the company's record of safely disposing of hazardous wastes is poor. Hooker had other dumps near the Love Canal site, and in other parts of the country the company has a similar record of using poor judgment regarding waste disposal.

It hardly is the kind of record that inspires confidence in the firm.

CHARLESTON EVENING POST
Charleston, S.C., June 12, 1980

Many disturbing questions have emerged from the disaster at the upstate New York neighborhood called Love Canal. One of the most compelling is that of responsibility. Who should shoulder the financial burden of cleaning up the toxic chemicals there and those at perhaps thousands of other hazardous chemical dumps scattered throughout the country?

The Chemical Manufacturers Association, pleading that improper disposal in the past has been largely due to ignorance not malice, says business is really not to blame. Environmental groups, the Carter administration, and many legislators disagree. And there sits Congress, as it does so often, smack dab in the middle.

A bill in the House to raise money for a "superfund" for clean-ups splits the economic responsibility 50-50, with $300 million to be raised from industrial fees and an equal amount to be taken from the taxpayers. A Senate bill would raise $700 million from industry, while taking only $100 million from the treasury.

Ignorance cannot completely erase responsibility, either individual or corporate, in matters such as this. To rationalize improper waste disposal as a means to lower costs to the consumer, as some in the industry have done, is reprehensible.

Neither proposal in Congress asks industry to pick up the whole tab for clean-up. Splitting costs half-way may shift too much responsibility to the taxpayer. The Senate bill, on the other hand, may shift too little. The important thing, however, is to get on with it. Some Americans are already paying a heavy price through increased incidences of cancer, genetic disease and miscarriage.

The Washington Star

Washington, D.C., May 24, 1980

Rep. Norman Lent, R-N.Y., criticized the Environmental Protection Agency's handling of the preliminary findings of genetic damage to Love Canal residents by charging that it "borders on the irresponsible." Closely, we would say. The Love Canal episode in upper New York state may be the most frightening instance of toxic-waste danger yet encountered.

There is no minimizing the potential threat from the toxic wastes that we have heedlessly strewn across the landscape. Rectifying past degradations and preventing future ones are proper governmental functions.

That having been said, EPA and its Deputy Administrator Barbara Blum were inept or insensitive in the way they feverishly announced, in a special press conference last Saturday, that preliminary results of a genetics study were "so alarming" that it might be necessary to evacuate more than 700 families from the Love Canal vicinity. The agency had received preliminary tests results — we stress preliminary, as did the Houston geneticist who did the study — last Friday. There was a White House meeting, at which Rep. John LaFalce, D.-N.Y., warned that the "perception of catastrophe" would require immediate evacuation.

However, EPA made the public announcement on Saturday after reports of the frightening findings leaked to local New York newspapers — *even though there had been no review of the study by a panel of geneticists.* That is an incredible way to handle raw scientific data, leaks or no leaks.

Ms. Blum hedged at the press conference: It might be necessary, she said, to order a temporary relocation of the 710 families, but the agency would wait until the review several days hence.

Why defer the decision, she was asked?

"It would be unfair to unduly alarm the residents of the area. Another two or three days is not going to make any difference," she said, noting that any health damage from the chemical wastes dumped in the canal by Hooker Chemical Co., between 1942 and 1953, would be a long-term effect.

But EPA, in its rush to the television cameras and reporters, had not notified Love Canal residents *before* the report of possible genetic damage was announced and went out on the air waves; they were being told as the press conference was going on. If that was not calculated to unduly alarm, indeed terrify, those residents who already have been through a nightmare of uncertainty, we can't imagine a better way to do so.

The sensational Blum revelations of Saturday were prominent on television and in the newspapers throughout the weekend, of course.

Why wouldn't the Love Canal families be pushed toward panic? Didn't the loud announcement of chromosonal abnormalities indicate that they were in even worse danger than during the past two years and an earlier evacuation of 200 families had suggested?

What a mess! President Carter stepped in Monday, declared it a disaster area and ordered immediate temporary relocation of the families — who had become so overwhelmed that they briefly took two EPA officials hostage.

When the scientific panel got together Tuesday to review the preliminary study — commissioned as part of the Justice Department's $124 million suit against Hooker Chemical — a proper note of caution was injected — belatedly.

The initial report that found chromosonal abornormalities in 11 of the 36 persons tested "provides inadequate basis for any scientific or medical inferences from the data (even of a preliminary nature) concerning exposure to mutagenic substances because of residence in the Love Canal area," the panel said.

How to explain EPA's extraordinary bureacratic horror show? Perhaps it was intended as a message to Congress. The administration's "superfund" proposal, and several other pending bills, to control hazardous wastes have been breaking no speed records on Capitol Hill — less because of the principle involved than because of disagreement over the scope and funding of such major legislation.

Admirable ends are not helped by callous means. EPA and Ms. Blum have dissipated their credibility in trying to advance toxic-waste control by the reckless handling of a very delicate matter.

The Des Moines Register

Des Moines, Iowa,
May 23, 1980

The Carter administration's decision to relocate temporarily 710 families living near the Love Canal in New York is welcome, but the action doesn't go far enough. The administration should have followed New York Gov. Hugh Carey's plea to help these families move out of the Love Canal area on a permanent basis.

Many Americans assumed that the Love Canal tragedy had been resolved when, in the summer of 1978, the government agreed to buy the homes of 239 families who lived closest to the Love Canal.

In the intervening months, there has been growing evidence that the damage caused by the chemicals buried in the Love Canal by Hooker Chemical and Plastics Corp. extended beyond the area bounded by the 239 homes. The 710 families living in the wider area reported a variety of ailments possibly caused by exposure to the toxic chemicals.

During this past week, the results of two independent scientific studies added to the fears of the families. One study found that 11 of the 36 Love Canal residents tested had damaged chromosomes. A second study suggested that many residents appear to have suffered nerve damage.

A top official of the National Institutes of Health has raised questions about the accuracy of the chromosome findings. But as the deputy administrator of the Environmental Protection Agency noted, there have been other studies that "in the aggregate . . . suggest very serious health risks."

The families moved out of the area will have to spend up to a year living in motel rooms and other temporary quarters while the government tries to decide how dangerous the neighborhood is. This will only prolong the agony. It would be much better to give the residents the option of moving out of the Love Canal area once and for all. They need a chance to begin a new life.

The Love Canal would be bad enough if it were an isolated incident. As the article on this page points out, the Love Canal would be only one dot on a map of the U.S. pinpointing hazardous waste dumps. Such a map would "look like an explosion of dots."

The aid provided to the Love Canal residents will help set a precedent for future incidents. The precedent being set is not good enough.

CORKY '80 STAR-BULLETIN

U.S. AIR FORCE

It's our neutron bomb-bottled water from Love Canal.

The Burlington Free Press

Burlington, Vt., June 2, 1980

In light of charges that the U.S. military disposed of hazardous chemical wastes at Love Canal and other sites near Niagara Falls, N.Y., in the early 1940s, it is proper that Congress investigate the situation.

Despite a 1978 Defense Department study which found no evidence of chemical dumping by the military in the Love Canal area, a preliminary report by a New York Assembly task force claims the military dumped radioactive wastes, nerve gas and other highly toxic chemicals at Love Canal.

A congressional probe would determine why the two studies are in conflict with one another and whether the federal government should provide restitution to those families who have been victimized as a result of the dumping of dangerous chemicals.

Until now it had been thought that dumping by Hooker Chemical & Plastics Corp. had been the sole cause of the crisis that affects nearly 1,000 Love Canal families.

Two recently completed studies show an unusually high incidence of chromosome damage among residents of the canal area. The chemicals are also believed responsible for stillbirths and malformations of surviving babies.

The federal government has been reluctant to become involved in the Love Canal controversy although it did agree to temporarily move 700 families from the area.

Now, however, New York Gov. Hugh Carey insists the federal government do more to assist the victims. "It appears that not only should the federal government be part of the solution, it may have been part of the problem," said Carey.

A federal probe would also determine whether the government knowingly transferred many "dangerously contaminated properties" to Hooker, the Niagara City School District, the Academy of the Sacred Heart convent and others, as is charged in the task force's preliminary report.

At the same time it is investigating the Love Canal situation, Congress should determine whether the military dumped hazardous wastes at other sites in New York or other states.

All Americans have a right to know whether they too may be victims of their own government's irresponsible disposal of hazardous wastes.

The Seattle Times

Seattle, Wash., June 25, 1980

THE American chemical industry "faces serious public-opinion problems," according to a "strictly confidential" study by the Chemical Manufacturers Association, a copy of which was obtained and released by Ralph Nader this week.

That's not exactly a surprise to most people, with the Love Canal case being the single most widely publicized incident involving a chemical company in recent years.

But recent developments indicate that the burden of guilt for Love Canal is not limited to the industry alone.

Consider the Environmental Protection Agency's much-publicized chromosome-damage report, released last month. It revealed "significant chromosomal abnormalities" in 11 of 36 Love Canal residents — abnormalities that normally occur in only one of 1,000 cases.

Perhaps predictably, area residents reacted with nearly hysterical demands that they be evacuated immediately, and President Carter within days declared a state of emergency at Love Canal.

Now, it turns out, the chromosome study has been discredited almost completely. A panel of genetic experts severely criticized the study for lacking a control group and drawing unjustified conclusions. The competence and objectivity of the study's author were questioned.

"In the view of several critics, the E.P.A. made an incredible blunder by releasing such a poorly conducted study," said Science magazine this month. "The psychological damage from this long-running disaster may ultimately rival and perhaps exceed the physical damage."

And consider another recent development: Hooker Chemical Corp., the company that used Love Canal as a chemical dumpsite in the 1940s and 1950s, has been widely accused not only of covering up its past actions but of failing to warn anyone of the potential dangers.

However, new evidence has come to light that Hooker officials did warn of the dangers as early as 1957. At a meeting of the Niagara Falls Board of Education, which bought the land from the company for $1, a Hooker representative warned that the area was not suitable for construction of basements, sewers, water lines or other underground facilities — and specifically cautioned that the property should not be subdivided for building homes. Nonetheless, the board sold the land to private developers and houses were built there.

The Love Canal case has a long way to go. Hooker Chemical, the E.P.A. and local government officials presumably will all have their days in court. We are no apologists for the chemical industry. But it must be remembered that in this tragic case, there is more than enough blame to go around.

WORCESTER TELEGRAM.

Worcester, Mass., June 24, 1980

There's a new scandal brewing over Love Canal in New York, but the villain this time is not Hooker Chemical Company.

The malefactor this time is the Environmental Protection Agency, which last month released a scary report that some Love Canal area residents may have chromosome damage resulting from the chemical wastes buried nearby.

According to the magazine Science, "the EPA has ended up by needlessly terrifying the Love Canal residents." The report — done by a consultant on an EPA contract — is said to be worthless, proving nothing.

The EPA's motives for commissioning the study apparently were not to get the facts, but to get Hooker. The study "was not even meant to be scientific, according to Stephen Gage, assistant administrator for research and development at EPA."

Instead it was "a small fishing expedition. The Justice Department asked us to undertake it in connection with our suit against Hooker," he added.

The director of the study, Dante Picciano, works for the Biogenics Corporation of Houston. When the Department of Health and Human Services (HHS) asked to look at his data, Picciano refused even to speak with the panel unless he could place a person of his own choosing on it. The HHS panel refused.

Blocked by Picciano from seeing his data, the panel studied his written report and concluded that it "provides inadequate basis for any scientific or medical inferences from the data (even of a tentative or preliminary nature) concerning exposure to mutagenic substances because of residence in the Love Canal area."

It gets worse. Picciano had not even used a control population to see how it compared with the Love Canal population data in regard to chromosome damage.

And his charge that Love Canal people showed "supernumerary acentric chromosomes" turned out to be bizarre. "The EPA panel . . . saw nothing that could by any stretch of the imagination be called supernumerary acentric chromosomes . . . the EPA panel concluded that there was no evidence that the Love Canal residents had excessive chromosome abnormalities and that supernumerary acentric chromosomes exist only in the mind of Picciano."

Baseless as its findings were, the Picciano study has done enormous damage. The already nervous residents of the Love Canal area are reporting all sorts of abnormalities and illnesses. The psychological effects of the study are devastating. So intense was the reaction that President Carter on May 21 declared a state of emergency, clearing the way for the relocation of about 2,500 Love Canal residents at a cost of millions of dollars to the federal government.

The whole thing is a rotten shame. Worse, it is almost a crime. In order to try to make a case against a chemical company, the government has panicked thousands of people, perhaps needlessly. And it has made it even more difficult to find out the actual dimensions and seriousness of the toxic waste problem, at Love Canal and elsewhere around the country. Instead of factual analysis, the country is being fed a steady diet of stories and articles that border on the hysterical.

But will the government do anything about its own internal problems? Will it seek to punish those who have needlessly wrought such psychological harm to the tormented people of Love Canal?

Don't bet on it. It's a lot more fun to prosecute chemical companies than bureaucrats.

THE LOUISVILLE TIMES
Louisville, Ky., May 27, 1980

Now that the unavoidable decision has been made to move 710 families from their homes near the Love Canal chemical dump, the suspicion that hazardous chemicals can cause genetic damage must be fully and rapidly explored.

A study announced last week found a high incidence of abnormal chromosomes among people who live near the infamous dump in Niagara Falls, N. Y. The abnormalities appeared in 11 of 36 persons tested, a shockingly high number when you consider that ordinarily no damage would be found in a group that size.

Even though its scientific validity has been questioned, the study introduces yet another horror into the chemical waste saga. Chromosomes are the tiny particles that transmit hereditary information from one generation to the next. Damage to them may cause birth defects or indicate susceptibility to disease.

For the Love Canal residents, the findings were understandably the final blow. Those who lived directly over the dump have been moved. Those left tend to feel, with some justification, that they are victims of corporate malevolence and bureaucratic indifference.

Even the decision to relocate them was a half-way measure. The government will provide lodgings in hotels or an old Army camp for only a year while studies of the genetic risk continue.

Most taxpayers would surely agree, however, that as a matter of common decency, federal and New York state officials should go one step farther and arrange for permanent relocation of the families at public expense. Every effort should be made to recover the costs from the Hooker Chemical Co., owner of the dump.

The Love Canal homeowners, after all, had no inkling that tons of toxic waste, including 11 substances suspected of being cancer-causing, were buried under or near their homes. They are clearly in a different category from people or corporations who ask to be bailed out after their own bungling has gotten them in trouble.

Now that the families have been rescued from a neighborhood that has for all practical purposes become uninhabitable, the Environmental Protection Agency must confirm or discount the genetic threat. Public waste management policy depends on the result. So does the peace of mind of Kentuckians, Hoosiers and others who have lived near hazardous dumps.

Several scientists have severely criticized the methods used in the original study. They point out that the persons tested at Love Canal were not compared to a "control" group. No one established whether the abnormalties might have some other cause, such as exposure to chemicals or radiation at work. And those who took part in the study were not chosen at random.

These obvious flaws must be eliminated from future studies. The implications are too serious for such inconclusiveness.

This latest episode in the Love Canal tragedy should also dispel any doubts Congress still has about the need for a "superfund" to pay for the cleanup of illegal or mismanaged dumps. And the EPA's cradle-to-the-grave waste disposal rules will seem unreasonable only to advocates of chemical chaos.

Such measures won't reverse the damage done by sloppy waste disposal. The goal now must be to fairly compensate those who have been injured and reduce the chances that similar disasters will occur in the future.

THE MILWAUKEE JOURNAL
Milwaukee, Wisc., May 22, 1980

The Love Canal disaster will go down in history as a classic illustration of the danger inherent in the call for government to simply step aside and let the free market flow.

It can safely be presumed that the profit motive, which generates so much that is good in our lives, played a large role in creating a toxic-chemical threat to life itself in the Love Canal area of Niagara Falls, N.Y. And pending the outcome of a $124.5 million federal lawsuit, we are willing to believe that the second major cause of the catastrophe was corporate ignorance, rather than negligence, on the part of Hooker Chemical and Plastics Corp. Apparently, Hooker believed that burial in Love Canal was the cheapest, easiest method of toxic waste disposal.

Easy, perhaps. Cheap, no. However, it was only after chemicals began oozing to the surface in recent years that the cost became obvious. Investigations showed the health hazard to be so great that the State of New York relocated 239 families living closest to the danger area and bought up their homes.

Now, after further federal study found chromosome damage to some of the remaining residents, President Carter has declared Love Canal a federal emergency. That gives the government authority to relocate about 700 more families, helping pay rent for up to a year and perhaps ultimately buying their homes.

Where was the government when Hooker committed its irresponsible act? Why have the taxpayers of the nation and of New York State been put in the position of facing millions of dollars of cost if the money can't be extracted from Hooker in a lawsuit?

The answer lies in the past — the inadequately regulated free-market past. Only in recent years has the nation recognized that certain emerging technologies pursued in the name of private profit can, if left unchecked, pose particularly horrendous danger and cost to society. Among such endeavors has been the introduction of chemicals into the environment without thorough study of potential consequences. And as a prime example, Love Canal will take its place in history alongside DDT, PCBs and, perhaps, the excessive use of fluorocarbons.

Yes, in its zeal to protect health and the environment, government can get too rigidly demanding and stifle necessary free enterprise. Regulation should be based on solid research evidence when available, and on careful judgment when a potential hazard looms before hard data (i.e. enough dead bodies) accumulate to document the danger. But if it must err, the truly prudent society errs on the side of public safety.

The Virginian-Pilot
Norfolk, Va., May 21, 1980

The terrifying extent of the human tragedy at Love Canal is now emerging: 11 of 36 persons given blood tests suffered chromosome damage of the type associated with birth defects, cancer, and spontaneous abortions.

If this sampling of individuals living near the Love Canal chemical-dumping ground in Niagara Falls, New York, is indicative of its toxic impact, it could mean that hundreds of persons—and possibly future offspring—will suffer crippling afflictions.

Already 239 families have been evacuated, and federal officials this week will decide whether to relocate another 710 families. Many were so angry over bureaucratic delays in declaring this a disaster area and providing an escape hatch from Love Canal that they locked two federal officials in a room Tuesday, demanding action.

The sooner residents are moved and compensated the better.

Until 1977, Hooker Chemical and Plastics Company used the swampy Love Canal site to dump 20,000 tons of pesticides, solvents, and compounds such as dioxin, one of the most toxic substances known to man.

The dump is now covered with clay, and drainage pipes carry liquid runoff to a special treatment plant. But folks who bought homes in the area or sent their children to a school innocently located in the middle of this toxic morass (but now closed) are fearful of the unseen killer that may lurk nearby.

Homes nearest the dump, from which the initial evacuation was ordered, have been boarded up. New York state will buy these dwellings so that their owners may at least not suffer financial loss on top of threats to their health.

Courts will determine financial responsibility. New York state is suing the company for restitution costs and $540 million in punitive damages. A federal suit seeks $124.5 million to clean up four chemical dumps in the area.

Hooker Chemical refuses to acknowledge any fault. "The report [of chromosome damage]," said Donald L. Baeder, the firm's president, "is inconclusive." But then Mr. Baeder and his family don't live near Love Canal.

Absolute proof of a causal relationship between the toxic dump and chromosome damage may be elusive. But circumstantial evidence is strong: Significant abnormalities in chromosomes occur on the average in only one of every 250 live births, which is less than one-half of one percent. But at Love Canal the incidence is a terrifying 30 percent.

Love Canal will serve as a grim reminder of the potential hazard to human health of toxic substances and of the necessity of strictly-enforced safe disposal methods that keep innocent citizens from harm's way.

Anchorage Times
Anchorage, Alas.,
July 19, 1981

WHILE EFFORTS are being made this weekend to stir up noisy opposition to the petrochemical industry in Alaska, it is interesting to note that the truth about Love Canal's health hazards, now coming to light, makes it appear that they were nil.

The New York Times carried an editorial June 20 recognizing that the incident was not what it was feared to have been.

"When all the returns are in, years from now, it may well turn out that the public suffered less from the chemicals there than from the hysteria generated by flimsy research irresponsibly handled."

THAT CONCLUSION might well be applied to the efforts of Dr. Barry Commoner and the Alaska Public Interest Research Group to create our own hysteria over what they claim are the hazards and perils of the petrochemical industry.

They have programmed an open-air rally for Monday evening to make a lot of noise and stir emotions against chemicals, with the big-name scientist as the main attraction. It is the same technique that has been used in many places in past years as a successful means of agitating for a cause.

There has been no incident here. The rally is apparently in protest of a study that is being made of the feasibility of a petrochemical industry in Alaska.

THE LOVE CANAL chemicals prompted a scientist to claim he found chromosome damage among homeowners in that New York city. His report roused fears of increased cancer and genetic defects. A researcher said she found a high rate of birth defects and pregnancy disorders. Another scientist reported clues of possible nerve damage. All three generated scary headlines nationwide. Hundreds of residents were evacuated.

But now a blue-ribbon committee named by the New York governor has discounted all those reports. It has found that even the state health department's early estimates, which led to the evacuation, were inconclusive. Science has rejected them as unsound. Error and bias were found in the scientific studies.

The New York Times noted, "But from what is now known, Love Canal, perhaps the nation's most prominent symbol of chemical assaults on the environment, has had no detectable effect on the incidence of cancer."

THE KANSAS CITY STAR
Kansas City, Mo., July 21, 1982

When headlines the other day proclaimed "EPA Says Love Canal Is Safe" it touched off an angry outburst of environmental protest. The full story is more complex and less drastic. What the Environmental Protection Agency said—in announcing a $7-million program to continue the cleanup of the contaminated neighborhood in Niagara Falls, N.Y.—is that 400 homes in the "outer ring" beyond the worst affected area are habitable. The 237 homes and a school in the immediate canal area still are considered unsafe and are being razed, except for two still occupied by families which refused to move.

The deadly long-range consequences of the Hooker Chemical Company's dumping of 21,800 tons of chemical wastes in the abandoned canal between 1942 and 1953 surfaced in 1978 when residents of homes built on the clay-capped canal began to sicken and chemical residue seeped into their basements. President Carter in 1980 declared a state of emergency. Scores of families moved out and the state bought many homes and began a cleanup.

Because the city and state wanted to revitalize as much of the area as safely possible, the EPA two years ago began extensive sampling tests comparing the outer ring homes with those elsewhere in the city and other control groups. They found the general level of contamination no higher, but said millions of dollars in work is needed to extend the canal cap, improve the drainage system and clean up sewers which originate in the affected area.

For those people who endured an ordeal of illness and mental stress, and even those who moved away in fear and uncertainty, even a limited official "all clear" is likely to be met with doubt. Some environmentalists cried "whitewash" at this seeming downgrading of Love Canal, which ranks in their pantheon of horror stories right up there with Three Mile Island and the Donora, Pa., air pollution disaster. But the EPA and the local governments are merely trying to salvage what can be, and limit the damage, from this alarming reminder of human vulnerability to careless disposal of hazardous wastes.

The Salt Lake Tribune
Salt Lake City, Utah, July 16, 1982

The neighborhood surrounding the Love Canal, an area that has become the national symbol of the risks of toxic-waste dumping, has been declared "habitable" once more. During the 1940s and 1950s some 21,000 tons of chemical wastes had been dumped into the canal, which in reality was the abandoned excavation of what was originally intended to be a barge canal near Niagara Falls, N.Y.

Later, because no preventive measures were taken originally, the chemical residues began migrating by ground water flows from the original dump sites into the basements of nearby homes, backyards and swimming pools. Many of these chemicals were proven carcinogens and the drainage waters contained highly toxic concentrations of them.

In a series of evacuations between 1978 and 1980 some 600 homes were purchased by government agencies and work was started to prevent a further spread of the contamination. The reports released by the Environmental Protection Agency and the U.S. Public Health Service found that efforts to reduce the spread of the contaminants, principally by construction of a clay cap and drainage systems around the contaminated area, were working.

This has led Dr. Clark W. Heath of the Public Health Service to announce, "We're saying that the area is habitable "

Such an announcement, however, could give an erroneous impression; that cleaning up and rehabilitating heavily contaminated areas is a relatively simple endeavor.

When one begins looking at the "fine print" in the U.S. Public Health Service announcement it becomes readily clear that only a portion of the area surrounding the Love Canal is really "habitable" once more. It is the area considerably removed from the canal, a part of the neighborhood away from the canal designated as "Ring Three." "Ring Three" extends from one and a half to six blocks from the original dump sites.

Studies done by the federal government and New York state have, to the contrary, found excessively high levels of dioxin remain in areas close to the canal, in "Rings One and Two."

The finding of "habitablity" will permit about 400 of the 600 homes acquired by government agencies during the evacuations of 1978-80 to be resold to private parties, thus, allowing governments to recoup some of the money spent in correcting a situation created nearly four decades ago out of ignorance and caprice. Yet there remain, and will remain for an indeterminant time, those 200 houses that can't yet be reoccupied; symbols of man's sometimes wanton disregard for the destructive potential of his own inventiveness and igenuity.

The Chattanooga Times

Chattanooga, Tenn., June 24, 1983

When the Environmental Protection Agency declared last year that New York's Love Canal was safe for human habitation, those who had followed the horror story of chemical contamination there which drove hundreds of families from their homes were shocked. Congressional representatives from the Niagra Falls area were suspicious of the findings and requested another scientific assessment. Not surprisingly, the follow-up report declared the EPA study critically flawed and cast its conclusions in serious doubt.

Love Canal was a partially dug canal used by the Hooker Chemical and Plastics Corp. for more than 25 years as a dumping ground for toxic chemical wastes. In 1953 it was filled and sold to the city for $1; a school was built; a residential area developed. Twenty years later the buried wastes began to resurrect. The residents whose basements were invaded by oozing toxics, whose families were stalked by serious disease, had considerable difficulty convincing local government and public health officials there was a problem. But federal attention, fueled by congressional hearings,

eventually resulted in a massive evacuation.

The EPA began its study in 1980. It was a difficult task at best, considering how little is known with certainty about the long-term health effects of chemical poisoning. Joel Hirshhorn of the congressional Office of Technology Assessment, the scientist who has now reviewed that effort, said EPA tried to do too much too fast and overlooked factors which could have radically altered its conclusion that the homes around Love Canal were safe. He said further study is necessary before a finding is justified one way or the other.

This stymies the plans of Niagra Falls officials who are eager for a clean bill of health to remove the stigma and allow redevelopment of the Love Canal area. But the vigilance of members of Congress who would not be easily satisfied and sought further investigation is to be commended. The EPA should quickly undertake the expanded study recommended by Mr. Hirshhorn, recognizing that the federal responsibility to close the knowledge gap on toxic waste contamination is important not only to the residents of the Love Canal area in New York, but to the entire nation.

THE PLAIN DEALER

Cleveland, Ohio, July 20, 1982

Love Canal is over now; everyone can stop worrying about it and go on to more immediate crises. At least, that's what the federal government would have Americans believe but we — and many others — have our doubts.

The all-clear signal was given by the federal Public Health Service after an Environmental Protection Agency study released last week declared most of the area around Love Canal "habitable." But even before the EPA study was made public, officials at both the National Bureau of Standards and the Center for Disease Control in Atlanta had questioned some of the methodology used and had asked the EPA to reconsider some of its conclusions, including the one involving habitability.

"Love Canal" has become synonymous with chemical horror, backyard threat, environmental sickness. In the 1940s and '50s, Hooker Chemicals & Plastics Corp. dumped more than 21,000 tons of assorted chemicals into the abandoned Love Canal near its plant in Niagara Falls, N.Y. Hooker capped the dump with clay and sold it to the local school system in the late '50s; the school system built a school there, and sold off remaining acreage for homesites.

It was 1978 before the danger of Love Canal became public knowledge. There was a witch's brew of stinking, foul chemicals under the ground; some had migrated from the dump site, and no one knew how far they had gone. Chemical sludge had been seeping and

bubbling from the ground for a while in the neighborhood, contaminating basements, killing lawns, leaving a layer of slime in swimming pools. Residents had been suffering from various unusual health problems, some severe. But residents of the area had to get angry, almost panic-stricken, and a few committed local experts had to wave the Love Canal reality into the national eye before government officialdom awoke to the danger.

State officials played the reluctant bridegroom to the hilt when Love Canal first came to light; it was only under the shotgun of public opinion that they reluctantly joined the fray. Now, we shouldn't be surprised that these same state officials, led by Gov. Hugh Carey, are the first to swallow the EPA report whole and announce that all's well that ends. The state would sincerely like to "revitalize" the Love Canal area, and coincidentally resell 406 homes not actually atop the canal but purchased from fleeing homeowners.

All is not well, however, and saying it is won't make it so. The EPA still has to answer the criticism of both the National Bureau of Standards and the Center for Disease Control. Then it has to face present and former homeowners in the Love Canal vicinity. And then it must explain exactly how it plans to see to cleaning up contaminated sediment in storm sewers and creeks, and how it will ensure that the area is "constantly safeguarded against future leakage," both necessary for safety, according to officials.

Sen. Daniel P. Moynihan of New York, among the first elected officials to land feet first on the side of residents in the Love Canal controversy, termed the EPA study "messy and inconclusive."

Love Canal is not over, and to expect people traumatized by years of chemical fear to accept an EPA study of such questionable worth is ridiculous.

NEWS ITEM: THE EPA ANNOUNCES LOVE CANAL IS ONCE AGAIN HABITABLE.

BUFFALO EVENING NEWS
Buffalo, N.Y., June 15, 1983

A new study of Love Canal — a hopeful one this time — has been made public, while another study is under way and a third one is being proposed. All these studies are symptomatic of the cloud of uncertainty that has tragically hovered over the former chemical waste dump for years.

The latest report was made by three prestigious federal research agencies — the Centers for Disease Control, the Brookhaven (N.Y.) National Laboratory and the Oak Ridge (Tenn.) National Laboratory. Its main finding was that people who once lived closest to Love Canal had not suffered any apparent genetic damage. The finding was in direct contradiction to a controversial 1980 report made under contract for the U.S. Environmental Protection Agency. That study, using some of the same residents, found some chromosome damage.

The new report stirs hope concerning any long-range health problems stemming from the Love Canal pollution, although some scientists have reservations or outright differences concerning the findings, and another recent study suggested that children who grew up near Love Canal are smaller and have more health problems.

However, the main immediate issue at Love Canal is whether to allow resettlement of the area surrounding — but some distance from — the former dumping grounds. The 16 acres of the central, toxic area have been permanently abandoned.

Like so many Love Canal issues, this question was supposed to have been resolved nearly a year ago, when the EPA reported that the outward flow of chemicals had been halted and that, after more remedial work, it would be safe to reoccupy the outer ring of homes. The remedial work includes the sealing off of old sewers from the core area, the building of a clay cap over the abandoned area and the building of a concrete wall around the perimeter.

But a cloud has arisen over the findings because of some questions regarding the testing procedures used. This cloud grew alarmingly in the past year as the entire EPA and its former head, Ann Burford, came under fire in Congress. The EPA findings have been called "politically tainted."

Now two state assemblymen, Joseph T. Pillittere, D-Lewiston, and Maurice D. Hinckey, D-Saugerties, have proposed another means of review — a new five-member state board composed of the commissioners of health and environmental conservation plus three independent members. This board would review all evidence to assure the area was safe for habitation and, where necessary, seek new evidence.

After all the studies that have been made, it is hard to understand what a new round of investigations is likely to accomplish. So many questions of credibility have been raised that no study could be expected to remove all fears about resettlement of any portion of the Love Canal area.

However, an independent study of the EPA's 1982 findings is now under way. The congressional Office of Technology Assessment and the General Accounting Office, the investigative arm of Congress, are checking on the validity of the EPA report.

A great deal of work was done for this study over a period of 2½ years, and if the findings stand up under congressional scrutiny, they could form the basis for Love Canal policy. When the congressional report is made, the next step in this long-playing drama can be taken with more certainty.

Buffalo, N.Y., September 29, 1983

It would be hard to find any more vivid example of the deficiencies of government in responding creditably to questions of public health and environmental safety than the record of faulty studies at Love Canal.

For five years, since the discovery of toxic waste perils at the canal focused national attention on the hazards of abandoned waste dumps, the habitability of the area surrounding the canal has been the subject of various state and federal studies. Throughout this period, fears about resettlement of any portion of the canal area have plagued the proposed revitalization program of the city of Niagara Falls.

Last June, the Office of Technology Assessment, an arm of Congress, delivered a major blow to any immediate resettlement plans by casting grave doubts on the validity of last year's study by the Environmental Protection Agency. Labeling the EPA's research as unsupportable scientifically, the OTA complained of insufficient sampling of test components, inadequate control-area data and a failure to consider the potential for combined health effects from the many toxic chemicals buried in the area.

In particular, the assessment office found that the sampling program could not justify a conclusion that homes in the outer areas could be made safe to live in after completion of current clean-up operations.

In the face of such disturbing criticism by the scientific arm of Congress, the new EPA team headed by Administrator William Ruckelshaus has in effect conceded that the 1982 report by its predecessors was inadequate and left "gaps in information" needed to determine if some Love Canal areas must be closed forever and if some are suitable for resettlement. It now plans an expanded habitability study to determine which, if any, areas in the rings surrounding the dump may be safe for eventual resettlement.

That we must have still more study after all this time is in itself a sorry commentary on the performance of officials whose sampling program, at a cost of $8 million, was found, so wanting. One must wonder whether one more study can overcome the credibility shortcomings in all previous studies, and indeed, whether there will ever be a final resolution of the Love Canal problem.

Even so, the new EPA leadership has taken the only proper course. Too much is at stake, both nationally in learning about remedial requirements at abandoned waste sites and locally in meeting community concerns, to leave the doubts about habitability unresolved. While Niagara Falls must put aside any thought of moving ahead at this time with any resettlement in the outer ring of Love Canal, only further and scientifically thorough study can remove the permanent uncertainty surrounding any resettlement.

The extended EPA study would review existing plans for cleaning up toxic chemicals that have leaked from the dump and for blocking further leakage. It would seek to correct the reported gaps of research information in the earlier study. And it would make habitability determinations "incrementally," working inward on a piece-by-piece basis with decisions on homes closest to the dump delayed the longest, up to 4½ years.

For the residents who have remained in the Love Canal area, it is understandably frustrating that they must now wait from two to four years for a new assessment on the safety of their homes. To ease their concerns, Rep. John J. LaFalce, D-Town of Tonawanda, has taken a constructive step in calling for an extension of the deadline by which people must decide if they want to sell their homes to the revitalization agency and move out.

The new EPA leadership appears properly determined to avoid the credibility pitfalls in the 1982 study. But if the EPA is to satisfy potential critics of its eventual new findings, it should make adequate assurance that these will be subject to independent review of testing and monitoring reliability.

The Pittsburgh
PRESS
Pittsburgh, Pa., June 9, 1983

The latest study has found no evidence that exposure to toxic chemicals in the polluted Love Canal area of Niagara Falls, N.Y., caused any chromosome damage to anyone.

The study, conducted for the U.S. Public Health Service, contradicts an earlier one commissioned by the U.S. Environmental Protection Agency. That one, done in 1980, detected abnormalities in the white blood cells of 11 of 36 people tested and led to mass relocation of more than 1,000 families.

But that study has always been in dispute because the chromosomes of Love Canal residents were not compared with those of non-residents to determine whether there were any significant differences.

The new study did that — and knocked the original findings for a loop.

Although the latest survey did not explore other possible effects of exposure to the chemicals, the Public Health Service says neither federal nor state health officials have been able to find any firm evidence of cancer or birth defects or other reproductive abnormalities among Love Canal residents.

Unfortunately, former residents have not been consoled by any of this. Many of them have expressed outrage at the report, with one labeling it a deliberate "smoke screen" to cover up efforts to revitalize the area.

★ ★ ★

This kind of distrust may be the most unfortunate legacy of Love Canal, as well as other toxic-waste incidents such as the one at Canonsburg.

People are afraid of what might be happening to their bodies, and their children's bodies, and they have little confidence in the government. Or in the companies responsible for the environmental pollution. Or even in the independent scientists and laboratories hired to investigate them.

The report on Love Canal, of course, is no argument for letting up on efforts to clean up old toxic-waste dumps and to prevent new ones. But it does suggest that the public's fears about the dangers can sometimes be worse than the actual threat to human well-being.

Dispelling such fears may prove to be as tough a job as the environmental clean-up effort itself.

The Washington Times
Washington, D.C., June 13, 1983

It in no way diminishes concern about the danger from toxic wastes dumped about the landscape to argue that there has been excessive howling and fluttering about the problem. Even in so delicate an issue, there must be proportion, conditionality. Inflamed passion and, its cousin, hysteria do not lend themselves to sound public policy.

Remember the extraordinary reaction when it was discovered that residents of New York's Love Canal area had been living atop an old noxious chemical waste heap, seepage from which was contaminating homes and, it was feared, the people who lived in them? The prospect of increased disease, birth defects and death scared the wits out of more than just people who lived near the canal.

That was not, obviously, an inconsequential episode. The great fear that the Love Canal disclosure caused was heightened by a 1980 quickie study, under EPA sponsorship, that purported to find indications of severe, perhaps lethal, contamination among residents. This was trumpeted loudly by members of Congress and environmentalists anxious to get money to clean up every toxic waste dump by yesterday, and the feds had to come up with money to buy the homes of more than 200 Love Canal area residents so they could relocate.

The 1980 report was an effective element in maintaining the furor over toxic wastes at a high political level — notwithstanding that within a few weeks, the methodology of the study was criticized by other scientists and its conclusions rendered suspect.

Now, the federal Centers for Disease Control and two major private laboratories have concluded a study. It shows that residents and former residents of the Love Canal area in Niagara are no more likely to have suffered chromosomal damage than residents elsewhere in the city. The study found "no increase in abnormalities" among residents or former residents around the canal compared to persons living in a control area a mile away. The latest study also said that even if chromosome damage had been found, it would be "impossible to know" whether such damage might presage "later occurrence of clinical illness."

There is, to be sure, a tentativeness in the study's conclusions: Careful scientific investigation avoids sweeping assertions. So what are we saying, you ask? That toxic-waste dumps don't threaten populations, that Love Canal residents were grabbed by mass hysteria and the politicians ran with it? Not at all.

What we are saying is that the sky-is-falling reactions that send political seismographs off the charts don't commonly do anyone much good and, indeed, are likely to complicate the attempts to deal with such nasty problems as toxic wastes. So we're saying the Love Canal affair, serious in its own right, was given a doomsday dimension that may have been grievously overstated.

THE RICHMOND NEWS LEADER
Richmond, Va., May 20, 1983

With his customary indiscretion, Ralph Nader proclaimed the alleged chemical pollution of New York's Love Canal typical of the "cancerous, toxic cesspools left by callous corporations." Thanks largely to horror stories circulated about possible chromosome damage to persons living adjacent to the canal, many families left their homes.

The Love Canal controversy inspired at least one TV movie and countless essays and stories. Once again, critics charged, American business had displayed its venal worst.

Soberer minds declined to rush to judgment. Although initial Environmental Protection Agency studies suggested the toxic wastes dumped in the canal had adversely affected residents, subsequent reports have disagreed. In 1980 a board of eminent scientists commissioned by New York State reviewed the original Love Canal reports. The commission uncovered numerous flaws and found that the EPA had "damaged the credibility of science." It said another study linking toxic wastes to health problems "cannot be taken seriously as a piece of sound epidemiological research."

It lamented that "many of the residents have come to believe that their health is in fact irreversibly damaged, that they are at future risk of cancer." In fact, a truly scientific study found "no evidence for higher cancer rates associated with residence near the Love Canal toxic-waste burial site in comparison with the entire state outside New York City."

The other day the Centers for Environmental Health of the National Centers for Disease Control cited another comprehensive study exposing the myths about the Love Canal. The new study "suggests that no specific relationship existed between exposure to chemical agents in the Love Canal area and increased chromosomal damage."

In other words, fears that pollution in the Love Canal had turned people into zombies and their unborn children into mental or physical defectives are proving unfounded. Yet never underestimate the emotional and psychcosomatic problems resulting from the fear hammered into residents by those seeking to exploit Love Canal for ideological purposes.

Indeed, doomsayers are the new ambulance-chasers. If an industrial mishap occurs, they rush out of the woodwork. They descended on the Love Canal. They flocked to Three Mile Island, which produced no "casualties." But when the first baby with birth defects is born in the area, look for the fear-mongers to blame nuclear power companies and to goad frightened parents into filing suits.

Of course, every society needs safeguards to protect it from deliberate or inadvertent health-threatening pollution. But too many of the self-anointed spokesmen for the "people" do not educate so much as they scare. The disposal of toxic waste in the Love Canal region occurred in the late '40s and early '50s. But the company responsible for the wastes conformed to standards set by the Carter administration decades later. Such behavior hardly justifies condemnation by Nader's flying squads.

The Love Canal controversy likely will persist for decades. Studies studying previous studies will update other studies. Each time a report challenges the accepted view, the Fear Network will cry, "Foul!" But the goal of scientific investigation — especially of something so emotional as toxic waste — ought not to be the amassing of evidence to buttress prior opinions. Rather, investigators should search for the truth — wherever it may lead. Regarding Love Canal, the evidence increasingly suggests the residents of the area have been lead to emotional slaughter by those supposedly speaking on their behalf.

Dioxins: Most Toxic Chemical Known to Man?

Dioxins are chemical substances formed as by-products in some chemical manufacturing processes. They can be fatal to animals in minute quantities, and have also been linked to cancer, birth defects and liver damage. The U.S. Centers for Disease Control considers environmental contamination by dioxins at a concentration of more than one part per billion to constitute a health hazard. Dioxins first became familiar to Americans as the result of lawsuits brought by Vietnam veterans, charging that health problems they were suffering could be traced to the use of Agent Orange during the war. Used to defoliate the dense jungles that provided cover for enemy forces, Agent Orange was composed of two herbicides, 2,4-D and 2,4,5-T. The component of the defoliant suspected of causing diseases, TCDD, is a contaminant of 2,4,5-T, or a substance inadvertently created in the manufacturing process. It is one of the most toxic of the approximately 75 chemical compounds known as dioxins. (The herbicide 2,4,5-T has also been sprayed domestically to clear pastures and other areas of unwanted vegetation, but in concentrations much weaker than those used in Vietnam.)

In 1983, dioxins again made headlines when the town of Times Beach, Missouri was declared a disaster area. The federal government offered in February to buy all the homes and businesses in Times Beach, after test results showed dioxin levels of more than 100 parts per billion in the town's soil. Russel D. Bliss, a waste hauler, had purchased 55 pounds of dioxin from a chemical plant in 1971, mixed it with waste oils and contracted with local governments to spread it on unpaved roads throughout Missouri as a dust-control measure. Dioxin contamination has since been discovered at inactive industrial sites, at municipal incinerators where dioxin has been produced as a byproduct, at toxic waste dumps and on roads, pastures and other areas sprayed with dioxin-contaminated materials. The Environmental Protection Agency, announcing a program to identify and clean up dioxin contamination sources in December, 1983, described the substance as "one of the most perplexing and potentially dangerous chemicals to pollute the environment."

It is nevertheless true, as the American Medical Association pointed out in June, 1983, that no human being is known to have died from dioxin contamination. In an attempt to lessen the aura of panic surrounding the issue and return the debate on dioxin to a more scientific level, the AMA adopted a resolution stating that the media's coverage of dioxins had caused "unjustified public fright". The AMA later, however, regretted its own intemperate language in the resolution, which accused the press of "hysterical malreporting" on the issue.

DESERET NEWS
Salt Lake City, Utah, January 20, 1982

The Environmental Protection Agency seems to be dragging its feet on cleaning up a chemical waste dump in Missouri so toxic it killed 90 horses and other animals in the early 1970s.

Worse, says Dr. Ellen Silbergeld, a toxics specialist, the EPA refuses to plug loopholes that would prevent a recurrence of that unhappy situation.

Dr. Silbergeld is chief toxics scientist for the non-profit Environmental Defense Fund. Though the fund is a self-appointed watchdog and frequent critic of the EPA, the charges raised by Dr. Silbergeld in House testimony this week still should raise plenty of eyebrows. Children who are still playing in the dump site face a cancer risk of up to one in 10. And the dioxin-contaminated soil, much of which was hauled away in the 1970s, is still much more dangerous than that left after the Love Canal cleanup in New York.

"Incredibly," says Dr. Silbergeld, "even when faced with their own risk assessments, EPA has not acted" to clean up such waste dumps.

The EPA has opposed amendments to the Resource Conservation Recovery Act to plug legal loopholes in current waste disposal methods. Unless that's done, there could be more lethal situations like the one in Missouri.

Deadly chemicals are so plentiful today in American industry that the most rigid health standards are needed to protect human life. It's often easier and less expensive to make sure those wastes are cleaned up than to incur lawsuits later.

St. Louis Review
St. Louis, Mo., November 26, 1982

For more than 10 years several sites quite near St. Louis have contained soil contaminated with the cancer-inducing agent dioxin. Other sites are spread across the state, but in less populated regions. This chemical dioxin was sprayed apparently unknowingly on earth which was later deposited in landfills at these sites. For some time authorities have been aware of the great danger which this chemical poses to the human organisms. It is now also apparent that such chemicals have a much longer period of potency than previously thought.

Now that the situation has become widely known, people living near these landfills are rightly worried and angry. Much of the population shares their feelings. Blame deservedly or undeservedly is being laid at many doors in Jefferson City and Washington, D.C. It seems that many must share the blame — if not of actually causing this situation at least of not doing enough to correct the danger to the population once it became known.

We are witnessing many frightening events these days. Danger comes in headache tablets, the overuse of saccharin and cigarettes, and in the pollution of the very air we breathe. Life in this technological age is a risky business for all citizens no matter where they live. To meet this challenge many people are now trained as experts in deciphering these dangers to the public health. Government agencies such as the Environmental Protection Agency have been established to protect our citizens as well as to coordinate the efforts necessary to forestall such dangers. Is it too much to expect that these trained experts and their governmental administrators do their job?

Situations such as this provide a field day for politicians both presently in office or hopeful about the next election. The air is now filled with charges and countercharges transferring blame and guilt hither and yon. While all of this political gain is being achieved little is being done about accurately assessing the danger and correcting the situation. This is what we deplore. We urge that immediate steps be taken to amend the situation.

A thorough investigation needs to be initiated immediately to assess both the danger to human life and the magnitude of the steps which need to be taken to remedy this public danger. Ignorance of danger and inaction is one thing, but continued inaction after an alarm has been sounded is idiocy. Let us stop wondering and start acting.

The Virginian-Pilot

Norfolk, Va., January 22, 1983

The continuing story of dioxin contamination in Missouri serves as a another warning about the dangers of the inadequate disposal of chemicals. Whether it's the headline-grabbing discoveries of Kepone in Virginia and the Love Canal in upstate New York, or merely the appearance of an abandoned dump site in a remote area, once-unsuspecting citizens are beginning to realize that their lives can be rearranged quickly because a toxic chemical has been disposed of improperly.

The dioxin tragedy is also a powerful reminder of how uninformed many in government and industry were about this danger until the past decade.

Dioxin, a chemical byproduct of several manufacturing processes, was deposited in landfills near residential sites, buried on farms and, at Times Beach, Mo., sprayed on dirt streets to hold down dust. Obviously, the people in charge of its disposal had no idea how dangerous it was: It is 150,000 times as toxic as cyanide.

Moreover, even when some government agencies figured out its toxic power (the Center for Disease Control discovered some of its effects back in 1975) other government agencies — those charged with enforcing environmental laws — took no action to clean it up.

Unfortunately, there may be dozens — or hundreds — of similar tragedies just waiting to be discovered. The Environmental Protection Agency figures that there are 14,000 dangerous toxic dumps in the nation. EPA has placed 418 on its "hit list," eligible for cleanup money from the $1.6-billion federal superfund. However, the EPA is not breaking any records in cleaning up this mess; it was 18 months late in announcing the worst-site list and it has not moved expeditiously on the sites it has targeted.

How can the government deal with the misuse of chemicals? First, strict and aggressive enforcement of existing statutes is a necessity. Violators must bear heavy costs, both as a punitive measure and to help pay for the huge damages that result from major chemical disposal problems.

Second, because there is a nationwide shortage of disposal sites, the federal government must persuade states to agree to create toxic waste sites, and the states must persuade localities to do the same — despite the complaints that will come from those most affected.

Finally, attitudes about waste disposal must be changed. Often, government does not take action until there is proof that something is amiss. But, on the subject of hazardous waste, it is prudent to assume that chemicals are going to be hazardous until otherwise proved, and their use should be severely restricted until they are deemed safe.

Detroit Free Press

Detroit, Mich., February 24, 1983

IN AGREEING to buy the town of Times Beach, Mo., the Environmental Protection Agency has shown reasonable efficiency and dispatch in closing the barn door after the horses are out. At least the people of Times Beach were subjected to a relatively short period of time of troubles before they got government help in getting out of their quandary, which was created by the detection of massive dioxin contamination in the soil of their town.

The role of buyer of last resort for people trapped in such environmental disasters is one only the federal government is equipped to play. But it is an expensive role — more than $33 million in the case of Times Beach alone — and it does little to undo the damage done to the lives of people by these disasters over which they have no control. The insidious effects of the contamination, which occurred over many years but became acute only after recent flooding brought concentrations of the dioxin to the surface, may never be erased.

If ever there was a case calculated to show the need for thoughtful environmental protection policies **ahead of the fact**, surely it is the Times Beach episode. For the last several years, the political support for environmental protection has supposedly deteriorated. Despite disasters such as Love Canal, the PBB case in Michigan and the Times Beach case, advocates of preventive action have found themselves hooted at and jeered at as opponents of progress.

The truth that prevention is easier and cheaper than trying to clean up the mess afterwards should not have needed proof. But the pressure to minimize costs and inconvenience in the short run is strong, and many individuals and businesses are still willing to risk long-term damage rather than confront potential problems now. And all too often, the costs of cleaning up the mess of environmental carelessness or ignorance are borne by the public.

In its management of the Environmental Protection Agency, the Reagan administration has put into governmental practice the laissez-faire approach to environmental regulation. Indeed, there are now substantial indications of possible favoritism and lax enforcement in the agency.

The government has a stewardship role to perform in trying, to the extent current technology and knowledge permit, to protect us from environmental disasters. So do individuals, and so, too, do corporations. The hidden liability can wind up destroying once-reputable firms and requiring huge outlays of public funds.

No amount of federal money after the fact can undo the full damage done by a dioxin disaster or a PBB tragedy or a Love Canal case. The sight of the EPA's administrator, Anne Burford (formerly Gorsuch), trotting out to Missouri to try to fix things up ought to strengthen our resolve to demand of her and her agency some stewardship **before the fact**.

Post-Tribune

Gary, Ind., February 25, 1983

Strange, the government buying up a whole town. But what does a government do when a community is devastated by toxic waste?

The offer to buy Times Beach, Mo., from its 2,500 residents is an act of compassion. It also is good realism. It also is good politics, but we'd like to think politics wasn't the motive.

Our opinions

The town has been sitting on a time bomb for a decade, since its streets were sprayed with a mixture of waste oil and dioxin. Now that the bomb is about to explode, the people want out — most of them. They will be paid for their property, but that won't compensate for the fright, the uncertainty, the breaking up of families.

The Environmental Protection Agency is an easy target, but it probably does not deserve most of the blame it is getting on this one. It may have moved with inexcusable slowness in assessing the danger, but EPA Chief Anne Gorsuch Burford says she acted the day after receiving the test results.

The more ominous thing about this sad incident is the EPA chief's matter-of-fact statement that "Cleaning up America will not be easy." Can it be done at all? The government can't go around buying up communities. Yet America's countryside may be like a minefield, with toxic wastes waiting to explode their deadliness. We hope that's an exaggeration, but that general fear is what has lifted the problem to its urgent state. The Reagan administration has not, until recently, acted with urgency.

After Times Beach's people leave, the government will take years to restore the area. This first buying-up of a community since Superfund was created in 1980 to clean up hazardous waste sites, will cost an estimated $33 million. The price of America's apparent determination to poison itself probably can't be measured.

CHARLESTON EVENING POST
Charleston, S.C., February 24, 1983

It's great for the EPA to say it is prepared to spend $33 million to buy a Missouri town and move residents and businesses to protect them from further contamination by the toxic chemical dioxin. Buying an entire town is an unprecedented move by the federal government. It will, apparently, provide an answer to the prayers of the townspeople of Times Beach, beset by floods and the dioxin danger — or, more precisely, a combination of both. There is, however, a dark side to the Times Beach story.

EPA Administrator Anne Burford is correct, up to a point, when she says flooding accentuated the Times Beach situation. When the waters receded, soil samples showed dioxin — mixed with oil and sprayed on streets to control dust a decade ago — at levels federal scientists considered a health risk. That happened three months ago, and Mrs. Burford says EPA could hardly have moved faster in reaching a decision on what to do about the plight of the people of Times Beach.

Perhaps not, but the sequence of events there raises questions. How many other Times Beaches might there be in this country, and what is the EPA doing to find the answer and to alert health authorities to possible risks from toxic contaminants? What has EPA done and what is it doing to reduce the chances of toxic chemicals getting from manufacturing plants into the hands of the public? Or into the soil?

The EPA, according to an internal audit, can't say for sure how it spent $53 million from a "superfund" earmarked for cleanup of hazardous waste. Congress is trying to find out how the money was spent, or if it was. Because the answer is not known, the EPA can't say with assurance that it has done and is doing all that should be done to clean up chemical waste dumps and possibly avert contamination threats in other communities. That's the dark side of the Times Beach story.

DAYTON DAILY NEWS
Dayton, Ohio, January 7, 1983

Times Beach, Mo., is a classic case of an environmental problem that got bigger because no one wanted to believe there was a problem. Today, it's too big to ignore, and still no one wants to take the correct measures.

It began innocently enough. In 1971, a private company used waste oil sludge to settle the dust on a three-horse farm, some roads in Times Beach and Fenton and maybe a dozen other places. Unfortunately, the oil contained 50 pounds of dioxin, a deadly chemical now directly connected to some types of cancers, liver damage and birth defects.

The horses started dying almost immediately so the contaminated soil was removed from the farms and buried.

In 1975, the Centers for Disease Control advised the Environmental Protection Agency that a health problem existed and the soil should be re-buried according to hazardous waste regulations. No one listened. Hardly worth the expense involved.

Last month, the Meramec River flooded the area and spread the contamination to the homes, the furnishings and other possessions.

Still, no one knows what to do. EPA has urged residents to locate elsewhere. Some will. But those who choose to begin again face with a pile of dioxin-contaminated trash that is drawing rats and breeding diseases. No one wants to take responsibility for dealing with it. Because the area qualifies for federal disaster funds, the Federal Emergency Management Agency plans to bury the stuff in ordinary landfill now and worry about dioxin later, so the contamination is still there.

That puts the Times Beach folks back at square one — waiting for direction on what to do and how to pay for it.

The EPA should step in and make some firm decisions about how to clean up the area, then start cleaning, even if it takes money.

The Washington Post
Washington, D.C., February 25, 1983

THE FEDERAL government is offering $33 million to buy the whole town of Times Beach, Mo. That's a decent and effective response, but it will work only in a case like this one, of intense and highly local pollution. It's possible to buy out a village with a population of 2,400. But dioxin was also sprayed, less heavily, over much wider areas of the state.

The Times Beach episode began 12 years ago when horses suddenly began to die on a breeding farm in the area. The cause was traced to the oil that had been thrown on the dirt to keep the dust down. The same mixture had been sprayed elsewhere—on dirt roads, for example—by a man whose business was collection of waste oil from service stations and, occasionally, other wastes from industrial plants. Soil samples were sent to the federal Center for Disease Control. It was three years before the CDC was able to identify the active agent as dioxin, a rare but extremely powerful poison. That led to the source, a small company manufacturing a chemical disinfectant. The dioxin was a waste product.

Before the issue of liability went much further, the company went out of business. There the matter rested for some years, apparently because state and federal authorities were under the impression that the dioxin would rapidly decompose in the earth. But retesting last year showed that, unfortunately, it was not decomposing.

The events at Times Beach illustrate clearly both the importance of controlling chemical wastes and the technical difficulty. It is work that sometimes, as in this case, makes demands that even excellent laboratories cannot easily meet and requires answers that even highly trained chemists cannot quickly provide. While it is true that larger budgets do not necessarily guarantee better scientific work, it's pretty safe to assume that steadily smaller budgets aren't going to improve things. The Reagan administration has cut the Environmental Protection Agency's funds for hazardous waste programs by 20 percent over the past two years. Those cuts are altogether unwise. As Congress looks into the internal affairs of the EPA, it needs to go beyond the accusations of scandal and look more carefully at the resources of the agency and whether they are commensurate with the public responsibilities that Congress itself has established.

For Times Beach, the buy-out provides compensation to people at least for their property losses. But, like the asbestos cases, it is another reminder that the country does not have an orderly and reliable system of compensation for the damage that dangerous pollution can inflict.

THE SUN
Baltimore, Md., January 20, 1983

Dioxin, a byproduct of certain chemical processes, is one of the deadliest substances known to man. Yet nothing happened for years after it was discovered that various roads, hamlets and ranches in Missouri had been contaminated with it through careless disposal. Now, finally, Missouri's dioxin problem is coming to public light, and it threatens to become a hazardous waste *cause célebre* on the scale of New York's Love Canal. It should.

In 1970, when waste disposal companies in Missouri and Illinois prepared to mix certain industrial wastes with used crankcase oil so they could spray the mixture on dirt roads and horse tracks, *someone* should have called an immediate, alarmed halt. True, it was not until 1974 that the federal Center for Disease Control firmly established that the dioxin contamination existed. But the furor over military use of dioxin-contaminated "Agent Orange" in Vietnam had shown what kinds of wastes commonly contained dioxin, and someone should have suspected its presence in the Missouri materials. Indeed, there is evidence the chemical manufacturing company which hired one of the disposal firms was aware of a chemical hazard of some sort and communicated it to the disposal firm's workers.

State and federal agencies, with the exception of the CDC, look worse than the companies. Evidence of the poisoning of animals, and possibly of humans, came almost immediately after the spraying, from ranchers who reported both dead animals and sick children. The CDC began investigating in 1971, and in 1975 urged excavation and disposal of some of the contaminated soil. Federal environmental officials insist they never heard from CDC. Missouri state officials admit getting the CDC's report, but say they thought that dioxin would soon break down chemically and thus posed no problems. A simple call to the CDC could have disabused them of this notion immediately.

The problem finally came to wide public attention in 1982, when flooding in one of the contaminated Missouri towns worsened the threat of human exposure to the contaminated soil. Evasion is no longer possible. Responsible state and federal agencies must prevent further human and animal exposure and make Missouri safe once again.

St. Louis Globe-Democrat
St. Louis, Mo., March 7, 1983

It appears that most legislators in Jefferson City favor appropriating $3.5 billion as the state's share of the buy-out of dioxin-plagued Times Beach.

But now that it has become apparent that the state also will face a sizable cost of the buy-out, one-tenth of the total, there is the realization that this is not just a case of hurrying up to vote the money and go on to other legislative business. It suddenly is dawning on some lawmakers that Times Beach might be just the first in a series of costly buy-outs the state could face.

A discovery that fish in 10 Michigan rivers are contaminated with dioxin and indication that similar pollution may exist in the waterways of other heavily industrialized areas gives some idea of how widespread the problem may be.

If an increasing number of sites are found to contain levels of dioxin considered potentially dangerous to humans, the question inevitably will be: how many areas can the federal government and states buy out? The sheer size of the problem could mandate that methods be found for removing dioxin contamination rather than buying out entire areas.

At this point there still has not been enough research to definitely say how great a risk high levels of dioxin pose to humans. So far it has only been linked to a severe skin disorder called chloracne. But researchers fear it could eventually be proven very dangerous to humans because certain types of dioxin have been proven to cause cancer, birth defects and liver damage in laboratory animals.

Those who are denouncing state legislators for not hurrying to take action on the buy-out should remember that it wasn't the state or the federal government that sprayed the dioxin at Times Beach and many other sites in Missouri. The state, as the guardian of public health, is attempting to meet the problem. But it should do so in a careful, well considered manner, remembering that what it does in Times Beach could set a precedent for a long time to come.

Until there has been more research definitely linking dioxin to serious human disorders, it is hard to make firm judgments on how great the effort should be to deal with dioxin contamination. How far can the state commit itself to deal with a threat that isn't fully known?

THE ATLANTA CONSTITUTION
Atlanta, Ga., February 28, 1983

The Environmental Protection Agency's decision to spend $33 million to buy the small Missouri town of Times Beach — all homes and all businesses — is the right answer to a situation that never should have been allowed to happen.

The 2,500 residents of Times Beach, located on the Meramec River south of St. Louis, learned late last year that their streets were contaminated with dioxin, a highly toxic chemical. The dioxin unknowingly had been deposited a decade ago, when unpaved streets were sprayed with used oil containing the chemical, a waste product from germicide and herbicide manufacture.

Minute amounts of dioxin have killed test animals. Its effects on humans remain under study, but scientists agree that dioxin is extremely toxic and almost surely contributes to various human illnesses and diseases. It is present in Times Beach in amounts up to 300 times the levels EPA considers safe.

After weeks of controversy about placing responsibility for Times Beach's predicament, EPA Administrator Anne McGill Burford announced Tuesday that her agency will buy the contaminated town with funds from the agency's $1 billion-plus Superfund. Congress established the fund in 1980 to clean up toxic-waste deposits and dumps, and currently is investigating charges that EPA, under Burford's direction, has been sluggish in implementing cleanups.

For the residents of Times Beach, the $33 million appears to offer at least an approximation of justice, though no amount of money can buy them out of possible long-term health damage. For the EPA, and for the Reagan administration, which has reduced the agency's research into, and enforcement of, environmental standards, the $33-million tab ought to offer a clear lesson: It is cheaper to prevent such events with firm environmental standards than it is to clean them up afterward.

The Wichita Eagle-Beacon
Wichita, Kans., February 24, 1983

The proposed "buy-out" of the dioxin-soaked town of Times Beach, Mo., at an estimated $33 million, comes as an expensive, but instructive, environmental lesson. The Environmental Protection Agency has proposed purchasing all the homes and businesses in the little Missouri town in the wake of findings that a recent flood had made the place virtually unlivable. Oil heavily laced with dioxin, a known cancer-causing agent, had been used to spray dusty unpaved streets. It ended up in the ditches and, last December, a flood that otherwise devastated the town also spread the poison, in amounts hundreds of times stronger than is deemed safe, throughout the town.

The only adequate answer to the Times Beach dilemma, now, is the extreme solution of the government buying the contaminated area, and trying to rid it of its lingering dioxin taint. That's a legitimate, if tragic, use for the environmental Superfund established by Congress to address just such dangerous cases of pollution. But the EPA shouldn't seek to divert attention from the current Superfund controversy in which it's embroiled by means of a "magnanimous" gesture in relieving the residents of Times Beach of their burden. Not many environmental hazards will yield to the "buying out" approach.

Possibly a hundred other communities in Missouri alone also may have had their streets sprayed with dioxin-poisoned oil. The sad lesson symbolized by the literal death of Times Beach is that there may be countless instances of careless pollution from the past that will come back to haunt the present.

Chicago Tribune
Chicago, Ill., March 2, 1983

Now that it has created an environmental panic in Times Beach, Mo., the federal government has offered to buy out all the home and business owners there because of the possibility residents are endangered by small amounts of dioxin in the soil. The $33 million for the buyout will come from the Environmental Protection Agency's superfund for cleaning up toxic wastes.

The amounts of dioxin involved are infinitesimal. But they are higher in some drainage ditches, roadways and under the macadam of some streets than the one-part-per-billion level that the Centers for Disease Control says is cause for concern. This level of contamination was found only in three residential yards, the highest of which had 1.5 parts per billion.

The dioxin has been in the soil for a decade and is not known to have harmed anyone in Times Beach, including residents who have lived there all during the past decade. There are reports that about 100 horses and other animals did die and a few people said they became ill elsewhere in Missouri when they were directly exposed to dioxin contained in waste oil when it was sprayed around their farms and stables to control dust a decade ago. Almost nothing was done about the residues of this dioxin in dozens of Missouri sites until EPA superfund money became available—and until a flood in Times Beach this winter raised fears that residual dioxin there might spread.

Despite all the fears, it is not at all clear that any danger exists in Times Beach. Dioxin is poisonous to some animals. It does cause a serious skin rash called chloracne in humans. But accidental exposure of people to dioxin in much greater concentrations than exist in Missouri now have not resulted in an increase in birth defects, cancer or other ills, as expected.

It is, of course, impossible to prove that the dioxin in Times Beach is harmless, or that exposure to small amounts of it over the last decade could not possibly cause cancer or unexpected damage of some kind in decades to come. The EPA has no guidelines as to what level of contamination should require evacuation of people or mandate clean-up efforts.

But the federal government has created an environmental panic at Times Beach and done more to destroy property values and upset people's lives than did the disastrous flooding. The EPA is under heavy fire in Washington for several unrelated reasons. EPA administrator Anne Burford (who was Anne Gorsuch until her marriage last week) needs some favorable publicity.

So without even waiting until tests were completed or health risks fully assessed, the EPA plunged into the $33 million decision to buy out the whole town. But the basis for the decision is political, not scientific. The real damage that has been done to Times Beach residents is psychological and financial, not physical.

It is easy to whip up hysteria over possible environmental hazards, especially when dealing with substances, like dioxin and other toxic chemicals, for which safe levels of exposure are not accurately known—and when there is concern about adverse effects like cancer or genetic damage that may not become apparent for decades or generations. Even when evidence accumulates that dangers are minimal or nonexistent, the panic that has already been created blights lives and destroys property values.

EPA has a responsibility to use its superfund to clean up the worst of the nation's toxic wastes. But it must find a way to do so without creating so much panic that it must buy out homes and evacuate a town because of fear alone—regardless of whether there is convincing evidence of actual hazard to health.

THE INDIANAPOLIS NEWS
Indianapolis, Ind., May 22, 1983

The federal government was busy buying out all the homeowners and businesses of dioxin-contaminated Times Beach, Mo., when, whoops, it found out that a Gray Summit, Mo., trailer park, into which it had moved some Times Beach families, had dioxin levels four times higher.

Now government officials have offered six months of free housing to all 28 families residing in the trailer park. The Federal Emergency Management Agency told the families they could be moved immediately into a hotel at government expense.

Hold it right there, men. Before you offer to buy out any more homeowners or pick up any more moving costs and rent, at least one taxpayer wants some reasonable assurance that the drain on the treasury is necessary.

The Missouri case makes a person wonder if indeed the reported dioxin levels are all that unusual. Maybe they're quite common in many regions of the country. Who knows? Evidently nobody at this point. Maybe somebody ought to find out before we begin shuffling people here and there and paying them to resettle.

Let the government scatter a platoon of those space-age suited soil samplers willy-nilly across the land. Let them plug away to their hearts' content for, say, two weeks. Send them in different directions, have them sample the soil in areas in which there have been no contamination scares. Just make sure they get a good cross-section of plugs from all regions of the country.

Maybe there's hardly a state or county in this country that's free of dioxin contamination. Maybe people are disrupting their lives for nothing. And maybe the taxpayers who are buying out homeowners are themselves living on land that is more highly contaminated.

Let's make sure the EPA and the FEMA and all the other acronyms in the environmental book know what they're doing with people's lives and people's money.

The Detroit News
Detroit, Mich., April 3, 1983

The Chicago office of the Environmental Protection Agency (EPA), in a new report, identifies Dow Chemical Co. as the source of dioxin and other chemical pollutants, including PCB, in the Tittabawassee River and Saginaw River and Bay.

The report also accuses the company of blocking access by government researchers to effluent pipes inside the plant property.

EPA found dioxin levels of 50 parts per quadrillion in the wastewater outside Dow's fence, and 100 parts per trillion in caged fish exposed to the effluent. The Food and Drug Administration (FDA) warns against eating fish with 50 parts per trillion content.

Dow says the contamination is infinitesimal, that a person would have to eat 25 tons of carp a year to be harmed. Carp, most of which goes outstate, is the only species mentioned.

Nevertheless, out-of-state sportsmen are discouraged by any report of pollution. Such stories are bound to affect tourism, which is Michigan's third largest industry.

Thus, the state may suffer more harm from perceptions than from pollutants.

This factor ought to spur Atty. Gen. Frank Kelley and the state Department of Natural Resources to do more than warn against eating fish from the two rivers and Saginaw Bay — their only substantive action in five long years.

Chief Assistant Atty. Gen. Stanley Steinborn certainly isn't leaping into the fray. He says he isn't sure whether the EPA data is current enough. Dr. Winifred Owen, Midland County health director, found "nothing new" in the latest EPA report, and said if the findings are based on 1981 data "I don't see a need for action."

The problem is that collection and analysis of samples often takes a year or two. In the Dow case, the data has never been current enough for law-enforcement action, according to Dr. Owen's criteria.

We aren't sure whether a dioxin peril exists, but EPA says it does. State officials ought to assure the public that they are seeking current data, and that they plan to take whatever further steps are necessary to protect the public health.

THE WALL STREET JOURNAL

New York, N.Y., May 31, 1983

Hardly a day goes by without some frightening headline about dioxin—or more precisely, 2,3,7,8-tetrachlorodi-benzo-p-dioxin, one of 75 dioxins. This is reported as the most toxic chemical made by man. Its presence in drainage ditches was one of the principal reasons families near Love Canal abandoned their homes. The federal government bought out the town of Times Beach, Mo., after the Center for Disease Control recommended evacuation at exposures above one part per billion in soil samples. Traces of the toxin have been found in fish. And more than 16,000 Vietnam veterans have filed disability claims attributing skin disease, baldness, impotence, cancer and their children's birth defects to the dioxin in the defoliant Agent Orange. The conventional wisdom was aptly summed up in the title of a recent New York news show—"Dioxin: In Search of a Killer."

Clearly, there are reasons to be worried about the stuff. High concentrations in the soil—30,000 times greater than the CDC evacuation standard—killed horses at Times Beach. In the laboratory, low doses in the daily diet have led to increases in liver cancer among rats, to birth defects such as cleft palates among mice and to abnormally high numbers of miscarriages, as well as anemia and hair loss among monkeys. The lab results vary enormously by species: Dioxin appears to be about 5,000 times more toxic for guinea pigs than hamsters. Since it would be unthinkable to conduct laboratory experiments on people, researchers really don't know at what dose it becomes dangerous for humans.

What is known, however, suggests that most of the scare stories are exaggerated. Human exposure to dioxin has so far been scientifically linked to only one health problem—a skin disease called chloracne, which tends to go away fairly rapidly. This is true even at fairly high exposure levels.

We have some real experience with dioxin, resulting from major accidents, in particular the infamous explosion of a trichlorophenol plant in Seveso, Italy, in 1976. More than 700 people were evacuated from Seveso, and 181 contracted chloracne. So far, however, no serious medium-term health problems have been discovered, according to a report last year by the Council on Scientific Affairs of the American Medical Association. "The mild chloracne, which occurred mainly in a small group of children, healed quickly. . . . Except for the skin, no organs or body functions were impaired. No derangement of gestation, no fetal lethality and loss, no gross malformations, no gross retardation at term and no cytogenetic abnormalities have yet occurred."

It's too early to tell whether the residents of Seveso face a higher risk of cancer as a result of their exposure, but studies of another dioxin accident more than 30 years ago suggest that this threat may be overblown. In 1949, there was an explosion at another trichlorophenol plant, this one owned by Monsanto in Nitro, W.Va. Soon afterward, 121 workers contracted chloracne. In 1980, the Journal of Occupational Medicine published a study by a Monsanto scientist and by the director of the University of Cincinnati Institute of Environmental Health, tracking the medical histories of those 121 workers through 1978. The authors discovered no abnormal incidence of cancer, and in fact the workers were on average *living longer* than would be expected from their age, race and sex distribution.

The Air Force has been conducting an epidemiological study of the 1,200 "Ranch Hands" who were directly involved in spraying Agent Orange in Vietnam; of all Vietnam veterans, they were presumably the most exposed to dioxin. Of the 67 deaths among the Ranch Hands so far, only three are attributable to cancer, a lower proportion of cancer-caused deaths than in the control population.

The Environmental Protection Agency now bans the spraying of the dioxin-contaminated herbicide, 2,4,5-T, on forests, pastures and rights of way. It based its suspension decision on a study attributing to 2,4,5-T an abnormally high number of miscarriages in the area around Alsea, Ore., in 1976. The EPA Alsea study has since been discredited by 18 separate investigations. In particular, a study by the Environmental Health Sciences Center of Oregon State University concluded in 1979 that the "rate of spontaneous abortions in the study area does not appear to be related to the use of 2,4,5-T."

None of this means that dioxin is harmless. Certainly toxic chemicals should be carefully controlled, and accidental releases should be carefully monitored. But there are a lot of other threats to worry more about, such as tobacco, marijuana, drunk driving or street crime, where the evidence of threats to health is clear. The notion that dioxin is a doomsday menace is based less on medical evidence than on some kind of psychological phenomenon.

THE SAGINAW NEWS

Saginaw, Mich., March 17, 1983

As usual when things get complicated, the questions about dioxin are as easy as they are grave. It's the answers that come hard.

What's the true danger, if any, to residents near the Pine, Chippewa, Tittabawassee and Saginaw rivers, as well as six others in Michigan where a Michigan State University study found fish contaminated with dioxin?

Is this the start of an environmental disaster at least on the scale of PBB? Is Michigan, like Times Beach, to become a poisoned wasteland, to be condemned and evacuated?

Or is the dioxin a natural phenomenon, produced by normal chemical processes, and not the result of careless, even villainous, disposal or discharge of industrial wastes?

Certainly we don't know, not when the MSU researchers don't, and state health officials don't. Unraveling the answers could take at least the three years it took to complete the study.

In that light, and despite the lesson of the PBB disaster, it's hard to quarrel with the state's cautious response to the findings. We simply don't know how bad it is, or even if it's that bad at all.

Caution, however, should not be confused with nonchalance. A lot more is known about dioxins than was known about PBB when that episode began, and what is known is ugly. Dioxins are a group of highly toxic chemicals. The type of dioxin found in the fish is the worst of all. One ounce of that is enough to kill 675,000 people. The U.S. Food and Drug Administration recommends limiting fish consumption at a dioxin level of 25 parts per trillion, and avoiding it at 50 parts per trillion. MSU found levels in Saginaw-area river fish ranging from 17 to 332 parts, and from 123 to 586 parts in the other rivers.

That doesn't make Michigan a Times Beach, where the contamination was a thousand times worse. But it's bad enough to warrant deep concern.

Actually, dioxin-based warnings against eating Tittabawassee and Saginaw river fish have been in effect since 1978. The renewed attention comes from the fresh controversy over an Environmental Protection Agency report pinning blame for the dioxin on Dow Chemical Co., and the EPA letting Dow "edit" that condemnation out of the final report.

That procedure may have been ill-advised. Certainly it gives the appearance of cover-up even though Dow scientists originated much of the data themselves, hardly the usual behavior of a polluter seeking to hide its dirty work. But it is a separate issue that should not be allowed to confuse the central one of public health that the MSU study raised. In that respect, it could be ironically reassuring. Here's why.

Dow declined to take the rap because its own research suggests that dioxin is a natural as well as industrial byproduct. In a couple of instances, the MSU study backs up the Dow argument. It found worse contamination of some fish taken from the Tittabawassee *above* Midland. Although a new state study muddies that point, MSU also detected no dioxin in fish from the industrialized Raisin River, but high levels in fish from the "pure" upper Muskegon River.

Dioxin is dioxin is poison, no matter how produced. But so are other natural elements. If it's been with us all along, as a natural material, then human tolerance for it may be higher, the risks lower, Dow an unfair target and Michigan no less attractive as a place to live and to play — specifically, to fish.

Admittedly, that's a lot of ifs, buts and maybes — too many. They have to be eliminated, and not only because PBB has warned Michigan for good against environmental complacency.

Despite the national news reports that have focused on Midland and Saginaw, this is a matter of national public health.

The MSU study found contamination throughout most of industrial lower Michigan. The implication is that waters in most industrial areas of the nation may contain dioxin beyond levels accepted as safe. It's just that no similar comprehensive study has been done elsewhere.

As a matter of national urgency, the EPA should give Michigan all the help it needs. If dioxin in the Tittabawassee and Saginaw is something for us to worry about, a lot of other folks should worry too, because this country's urban environment may be a lot worse off than anyone imagined. If it isn't, we'll be glad to share our relief as well as our natural bounty.

In any event, let's find out — and try not to take three years doing it.

The Hartford Courant
Hartford, Conn., July 2, 1983

The American Medical Association ought to find something more useful to spend its money on than a publicity campaign to convince people that the chemical dioxin may not be as bad for humans as it is cracked up to be.

But the AMA is convinced that the news media is inaccurately reporting the dangers of dioxin, causing unnecessary upset to "the lives and well being" of people in contaminated areas. It wants to set the record straight by pointing out in its campaign that no one has yet been known to die from dioxin, and that the evidence about its effects on humans is inconclusive.

AMA officials say they now regret some of the "imprudent" language in the resolution. If all the AMA wants to do is prevent people from going into a panic over dioxin, commonly described as one of the most toxic substances known to man, then no one will object. But the association shouldn't put out the all-clear sign yet.

Inconclusive evidence is not the same as no evidence. It took decades to fully establish the hazards of smoking, and, even now, against the overwhelming accumulation of scientific study, the Tobacco Institute still insists that it is an open question whether smoking causes cancer.

More than 20 years ago, when Rachel Carson made a case against the indiscriminate use of pesticides in "Silent Spring," her credibility was viciously attacked, and not just by the chemical industry.

Many eminent scientists, Time magazine observed after the book's publication, "objected that the book's exaggerations and emotional tone played on the vague fears of city dwellers." She was accused of "putting literary skill second to the task of frightening and arousing her readers." Of course, it turned out that Miss Carson was essentially correct in her assessment of the dangers of persistent pesticides.

This is not to say that some of the dangers of dioxin have not, in some cases, been overplayed. But clearly, there is enough smoke billowing around dioxin to justify genuine and serious concern about fire.

The links between dioxin, a contaminant produced in some chemical and combustion processes, and several human skin diseases are quite firmly established. Evidence has also been found linking the chemical to other adverse health effects in people exposed to it on their jobs, ranging from headaches to liver damage to deterioration of immune systems.

Cancer, reproductive abnormalities and many other serious problems have been caused in laboratory animals exposed to the chemical. Humans, of course, do not have all the same reactions as guinea pigs, but experiments are conducted on guinea pigs and other animals precisely to provide suggestive evidence about humans.

The evidence is not all conclusive, but it is convincing enough to justify the warnings and the evacuations that have resulted when sites with heavy concentrations of dioxin have been identified.

THE DENVER POST
Denver, Colo., May 23, 1983

NO ONE is known to have died as a result of exposure to dioxin at Times Beach and other Missouri communities. But there has been much anxiety and agony resulting, not from the effects of the toxic chemical, but from inexpert communication.

Some of the details appear in the June issue of Discover magazine. After the presence of dioxin was confirmed, physical examinations were given 112 adults and children. In time, some of them were notified by mail that the tests had shown "abnormal blood chemistry," "abnormal blood count" and "abnormal urinalysis." Some were told they had "elevated threshold for 256 frequency." A letter accompanying these reports said "three highly technical and important" tests of liver function had not yet been completed.

What the reports did not say was that few of the abnormalities are likely to be related to dioxin. Discover magazine points out that doctors would expect to find abnormal results in an intensive screening of any 112 persons, and that one-third of those tested probably had never been exposed to dioxin.

Were health officials trying intentionally to frighten the public? Not at all. They realized the test findings were complicated and should be discussed individually with each person. But there wasn't enough money for that. So they mailed out the raw laboratory reports without explanation and urged recipients to talk to their doctors.

Understandably, there was great alarm. "It was chaos," said one of the residents. "People were crying. We all thought we were at death's door."

Obviously, Missouri health officials had erred in their handling of an extremely sensitive situation. With more adequate training, it is likely they could have avoided the mistake.

There is irony in this unfortunate episode. The University of Missouri at Columbia is dropping its community health education program because of lack of funds. The program trains health workers who can bridge the often awesome communications gap between the public and the medical profession. The dioxin flap demonstrated the crying need for this program just at the time it is being killed.

The Oregonian
Portland, Ore., July 11, 1983

Intemperate language that accused the news media of "hysterical malreporting" of rumors about the dangers of dioxin has been regretted by a spokesman for the American Medical Association. The unfortunate language, introduced by the Missouri delegation with a resolution adopted by the AMA's national House of Delegates, put an excessive emotional charge on an issue already suffering from an excess of hysteria. It was beneath the AMA, and a congressional committee has been so informed.

If misunderstood, the AMA's position could do great damage by seeming to undervalue press reports of the dangers of chlorinated hydrocarbons, when uncovered in places like Times Beach, Mo., and in scores of other chemical dumps around the nation.

In contrast to the AMA's resolution, which in its best aspects advocated a public campaign to "prevent irrational reaction and unjustified public fright," the editors of Science, published by American Association for the Advancement of Science, in the same month wrote a balanced, concise editorial on the chlorinated dioxins and their risks.

The editorial explained the chemistry of TCDD, the most notorious of some 75 dioxins containing chlorine atoms, and pointed out that it is definitely lethal when given orally to animals. It is not only found in some chemicals, but in trace amounts when wood and other products are burned.

The most famous dioxin case, and the one that offers the most scientific guidance, is found in Seveso, Italy, where a chemical plant explosion in 1976 drenched a whole village with TCDD. No one died as a result, and follow-up studies have failed to uncover any significant change in spontaneous abortions, congenital malformations or postnatal problems. Further, the studies indicate little chance that a higher rate of cancer among 736 persons who lived in the village will occur. But thousands of animals that ate contaminated feed and grass died, and 176 people got a bothersome case of chloracne, a skin disease like common acne.

What seems clear from this and other studies, the editorial pointed out, is that the dioxin TCDD, when administered orally, is highly toxic, but when bonded to the soil, while slow to decompose, does not pose much of a hazard and can be destroyed in a few days if exposed to sunlight in the presence of foliage.

What the Science editorial did that the resolution failed to do was to issue a clear warning about TCDD. "In the matter of risk versus benefit, the balance is completely one-sided: all risk and no benefit," the Science editorial said.

This means that TCDD formations in the burning of industrial wastes must be considered along with efforts to reduce its quantities to safe amounts in industrial chemicals, such as herbicides like 2,4,5-T, which once contained much larger amounts of TCDD when sprayed as an ingredient of Agent Orange, the substance used in Vietnam that exploded the dioxin controversy.

Those most responsible for raising public fears are companies that filled dump sites, sprayed roads with dioxins and either ignored or failed to learn anything about the dangers of the chemical that is a minute part of many useful substances.

The Providence Journal

Providence, R.I., June 16, 1983

A noxious chemical compound by the name of dioxin has acquired a growing reputation as a public health menace. In minute quantities, it is lethal to laboratory animals. Its detection in Times Beach, Mo., prompted Washington to buy up dozens of homes as dwellers fled in panic. Recently there was alarm when dioxin was discovered at an abandoned herbicide plant in Newark.

The scare is understandable, since dioxin is clearly deadly in small amounts to some small animals. On the basis of these tests, it has been authoritatively described as one of the most lethal of man-made poisons. But before panic carries people to extremes, a few modest reappraisals are in order.

The most telling fact about dioxin, just now, is that no one really knows what lasting impact (if any) it has on the human organism. In 1949, after an accident at a West Virginia chemical plant released a small quantity of dioxin, 121 workers came down with a skin disorder known as chloracne. But this cleared up, and long-running medical follow-ups have found that these workers have a death rate lower than average and a rate of cancer no greater than normal.

One such test may prove nothing, and clearly dioxin is nothing you would want to pour on your breakfast cereal. But it remains at least possible that the stuff is not so dangerous to humans as all the hue and cry would suggest.

Until more is known, the most sensible course for public officials is to steer away from panicky overreactions and seek out more facts. New Jersey's Gov. Thomas Kean did just this with regard to the dioxin found in Newark. Although he ordered the temporary closing of a food distribution center near the dioxin site, he stressed that the step was merely "precautionary." He later took it on himself to brief city residents on what is known — and not known — about the compound.

This is straightfoward and welcome. Just as welcome is a new research program set up by the Dow Chemical Co., a major manufacturer of the herbicides of which dioxin is a byproduct. Dow has repeatedly discounted the dangers of dioxin, but it now seems to recognize that the public fears must be directly addressed. Accordingly, it plans to spend $3 million and bring in independent scientists for a thorough evaluation of the dioxin hazard in Michigan.

The Dow study, if honestly pursued, should help clarify the peril that dioxin poses for humans. Then, and only then, will it be possible to make prudent decisions on what levels of the chemical may constitute a hazard worth worrying about.

The Star-Ledger

Newark, N.J., July 5, 1983

The discovery in several places in New Jersey, as well as in industrial sites in 33 other states, of the toxic chemical dioxin has stirred up a national debate. The question being debated, which has enormous health and policy ramifications for the future, is: Just how harmful is dioxin?

One point of view that has been expressed in a number of scientific journals is that dioxin is less harmful than first was supposed. The interpretation is that, if swallowed, dioxin is poisonous but when confined to the ground, as it has been at most of these industrial sites, it is a good deal less of a hazard than the public has been led to believe.

The American Medical Association has gone so far as to state that some reports about dioxin have generated "hysteria." On the other hand, a number of environmental scientists continue to insist there is genuine cause for alarm. Samuel Epstein, an expert on hazardous waste and cancer at the University of Illinois Medical Center in Chicago, claims that the evidence about the toxic nature of dioxin warrants this extreme concern. He calls the medical association stance "a travesty."

Partly in response to criticism along these lines, the medical association shifted course and repudiated some of the extreme statements attributed to it, stating they did not represent its policy

During the midst of this controversy, a 1980 report by the U.S. Environmental Protection Agency surfaced. It listed hundreds of industrial sites in 33 states where dioxin might be found. The report had long been available to the states but received little or no attention until recently.

The controversy about the degree of toxicity of dioxin will continue, and so will the uncertainty about its effects on human beings. Until these questions are resolved, it is important that government develop an even-handed approach. There should be concern about the dangers of dioxin and adequate precautions taken to preserve public health. But the peril of over-reaction should also be avoided.

In this respect, New Jersey's response to the discovery of dioxin at a number of sites in the state is commendable. Testing has been thorough and there has been no attempt to downplay the possible dangers. But there also has been an unwillingness to panic and to adopt extreme solutions without sufficient thought.

The dangers of dioxin have been sufficiently emphasized. What is most needed at this time is information to determine just how great is the danger. Until this information is more widely available, the proper government response is to try to abate the problem without causing undue public concern.

THE PLAIN DEALER

Cleveland, Ohio, July 2, 1983

The American Medical Association has backed off its less-than-inspired resolution that, among other things, charged the media with witch-hunting and "hysterical malreporting [wonderful word!]" on dioxin contamination. The wording, sponsored by the understandably agitated Missouri delegation at the AMA's annual House of Delegates meeting in Chicago, has been wrongly interpreted — by President Reagan, among others — as saying that dioxin, one of the most toxic poisons known to man, is not harmful to humans.

In disclaiming the accusatory clauses, an AMA representative said: "We regret some of the imprudent language of our Missouri colleagues. The AMA does not pooh-pooh dioxin." We should think not.

Dioxin is a term used to refer to a potent group of waste products of herbicide manufacturing and other industrial processes. Although little is known about its effects on humans, it has been found to cause a number of disorders in animals. Death may result from such small doses that health officials are said to be acutely concerned. (So great is that concern that scientists meeting Thursday at the Center for Disease Control (CDC) in Atlanta established the safe limit of dioxin contamination at one part per billion.)

There are several ways of dealing with dioxin at the legislative level, and among them is closing your eyes and hoping that nothing happens. That is the strategy that has been employed, with limited success, by industry and government over a number of recent controversies. But, needless to say, ostrich-like environmental policies will simply not suffice, and we are never less than frightened by an administration that employs them.

That the Reagan administration is willing to play fast and loose with the environment is evident by its continual bungling of environmental issues. To this degree we have witnessed the near-destruction of the EPA and the diminution of the office of Secretary of Interior.

President Reagan, suddenly so responsive to the needs of education, is wrong for not being equally responsive to the needs of the environment. His stand on acid rain has been effectively undermined by both the National Academy of Sciences and his own science advisor. And there is no reason, in light of the AMA's retraction and the CDC's contamination limit, for him to remain equally silent on the threat posed by dioxin.

We fear, however, that he will come around too slowly; that we will have to wait until the environment is as bad as the education system before Reagan recognizes the problem and moves to remedy it. Will he?

The Idaho STATESMAN

Boise, Idaho, July 5, 1983

When the American Medical Association ripped into the media for their reporting on the chemical by-product dioxin, we were reluctant to comment. Sometimes it seems a newspaper's interest in an issue makes it hard to speak with credibility, and the occasion of the AMA's criticism seemed like one of those times.

The AMA's statements were criticized in turn by Michael Epstein, professor of occupational and environmental medicine at the University of Illinois Medical Center in Chicago.

Epstein said that dioxin "is the most potent carcinogenic . . . known to science." He added that the AMA historically has shown "a remarkable indifference to legislation to protect workers" and "a lack of concern for protective medicine."

Another authority disagreed with Epstein, but a representative of the AMA's Council on Scientific Affairs conceded the organization was "imprudent" in charging the press with conducting a "witch hunt" in its reporting on dioxin. Suddenly, a comment seemed very much in order.

It's hard to understand how the AMA's 351-member House of Delegates could have passed its witch-hunt resolution. The resolution conflicted with the organization's own research.

That research needs updating, the House of Delegates decided, but without waiting to see what its updated research would show, it went ahead and endorsed an extreme view on the dangers dioxin poses. In essence, the doctors made their diagnosis without waiting for the results of their tests.

Since experts disagree over the dangers posed by dioxin contamination, you'd think an organization of scientists would proceed with some hesitation in its public pronouncements on the subject. The very words "witch hunt" and "hysterical mal-reporting" indicate that the AMA acted with little caution, and the organization now offers the lame excuse that those statements, even though they are part of the resolution's verbiage, aren't official policy.

Rep. James Scheuer, D-N.Y., said he was "dumbstruck" as to how the House of Delegates could have accepted such language. So were we — for a while.

The Philadelphia Inquirer

Philadelphia, Pa., December 24, 1983

Suddenly, dioxin is back in the news. It seems almost passe. Wasn't something done about it in June or July? The woebegone town of Times Beach, Mo., was bought, fenced and turned into a federal dioxin lab of sorts. Then New Jersey had a dioxin anxiety attack; with the turning leaves, it dropped out of sight.

Now dioxin again has been declared worthy of official concern after being cold-shouldered and minimized by the pre-Ruckelshaus Environmental Protection Agency. This month the EPA pronounced dioxin "one of the most perplexing and potentially dangerous chemicals ever to pollute the environment."

Perplexing, for sure. Only days before the EPA announced plans to investigate hundreds of chemical plants and waste dumps at a possible cost of $250 million, a gathering of scientists convened at Michigan State University to pooh-pooh the threat. Dioxin dangers, they reported, aren't severe enough to justify crash programs.

While the final word on dioxin — whether its links to cancer and birth defects in animals portend human danger — is awaited, the EPA has set out on an appropriate if belated campaign of surveillance, relocation and cleanup. Playing it safe, at this juncture, means acting pre-emptively, not idling until the next scare.

EPA officials called the plans for the Diamond Shamrock plant site in Newark "emblematic of the kind of action we'd take" in the new campaign. The initial reaction to high dioxin levels seems melodramatic in retrospect — troopers sealing off roads, blitzes by Gov. Kean — but the panic seems to have matured into constructive action: Sampling of residential areas has been completed (much lower levels were found), the site has been fenced off and the tainted soil will be covered to prevent seepage.

Whether that is a perfect solution may be debatable. But it is, at least, an effort. Compared, for instance, with the EPA's lollygagging over the years about whether a massive asbestos waste dump in Ambler posed a hazard to residents (the agency approved an emergency $625,000 grant to cover the stuff last week), EPA's dioxin strategy is refreshingly welcome.

It confronts the Reagan administration with a new dilemma, though. Having concluded that the substance is "a very potent carcinogen" the question of how to treat or compensate Vietnam veterans sprayed with dioxin-laced Agent Orange takes on new urgency. At the moment, two critical issues remain unresolved. In the courts, manufacturers such as Dow Chemical are being sued by veterans who contend the health dangers of dioxin were covered up, and, in the Congress, a bill that would give veteran's benefits to GIs suffering from skin, liver and cancer conditions linked to Agent Orange has been reported out of committee.

Unless studies find dioxin innocent, it would be unseemly indeed for the administration to impede *any* efforts to aid veterans exposed to a compound Mr. Reagan's recast EPA has declared is a "potent carcinogen."

Rockford Register Star

Rockford, Ill., December 13, 1983

Last week's Time Beach, Mo., anniversary is one we cannot afford to ignore. It's as close to us as our next glass of water from a city well.

One year ago the town of Times Beach started to die in a river flood. Months later it was dead — its 2,000 residents forced out by the cancer-threatening chemical dioxin which washed out of roadways and permeated the town.

When 200 ex-residents came back to mourn last week, it was more than a sad anniversary. It should renew for all of us the flashing-red warning of our limitless capacity to endanger ourselves and our environment. It is a warning we dare never forget.

We don't dare forget it because the Evironmental Protection Agency (EPA) has not fulfilled its name. That clearly-vital governmental agency has failed to consistently implement and enforce procedures for safe disposal of hazardous and toxic materials.

The heavy dioxin dose that killed Times Beach had been mixed with oil used to cut down dust on Times Beach roads. Instead of the proper but costly disposal of dioxin, it was secretly and illegally distributed to firms like the one that spread death at Times Beach.

The Times Beach tragedy is a momument to both the EPA's inability to protect and its capacity to cost taxpayers heavily. The EPA's "cleanup superfund," formerly overseen by the now convicted Rita Lavelle, will spend $33 million to buy-out and relocate the former residents of Times Beach.

Indeed, Lavelle's conviction on felony counts of perjury and obstructing congressional investigation seems timed as a perversion of fate to coincide with the Times Beach anniversary. While the debacle of a community destroyed by toxic pollutants is commemorated, the leadership needed to circumvent that is proven to be not only unreliable but punishable.

Nor is there any end in sight for the costs of Times Beach. The EPA last week said the soundest and cheapest way to dispose of dioxin-tainted soil at six priority sites in Missouri would be to bury it in a concrete bunker at Times Beach. That project's cost is estimated at $15.8 million, again billed largely to the federal "superfund."

Obviously there is a limit to the "superfund." Taxpayers cannot go on indefinitely buying-out or cleaning up a contaminated Times Beach or Love Canal. Many dioxin-poisoned areas have been identified since the Times Beach shock. Dioxins alone are byproducts of many chemical manufacturing processes, including those for herbicides.

But that's just a drop in the bucket. It's estimated America produces 85 billion pounds of hazardous waste annually — and we don't know where all of it is being dumped.

Obviously we need to know and control that. We need to prevent a Times Beach and a Love Canal. We desperately need an EPA — but an EPA we can trust.

When do we get one? Times Beach started to die one year ago.

ST. LOUIS POST-DISPATCH

St. Louis, Mo., November 18, 1983

The clear lesson that emerges from the 48-page special section on dioxin, published by the *Post-Dispatch* earlier this week, is that preventive action against chemical poisoning could avoid much human misery and be far less costly than action after the fact. The dioxin report suggests that, although warnings of danger were there, government and industry were moved only by tragedy and public alarm to take strong corrective action.

In its comprehensive account of the dioxin contamination of Missouri, the special report reveals understandable errors of judgment on the part of government officials and private entrepreneurs; but it also reveals cases of government apathy and corporate deception. The federal government, for example, ignored warning signs of dioxin contamination from Agent Orange, the defoliant sprayed by U.S. forces in Vietnam. Had the government acted in response to scientists' proposals for tests, the domestic hazards of dioxin poisoning might have been realized earlier.

And, as the special report notes, manufacturers of herbicides containing dioxin — including the Monsanto Co. of St. Louis and Dow Chemical Co. — "placed competitive instincts ahead of public-health concerns." They withheld information about dioxin risks and exposed people to dioxin in medical experiments. Monsanto tends to minimize the dangers of dioxin, its spokesman saying, among other things, that there "is no acceptable evidence that dioxin has caused any type of cancer in humans." There is evidence, however, that it causes cancer in laboratory animals and, although the evidence is not conclusive, some reputable scientists believe it causes cancer in humans and suspect it causes other disorders far more serious than chloracne, the problem on which the industry focuses.

The difficulty is that the effects of a carcinogen may not show up in humans until many years after their exposure. Therefore, ordinary prudence demands caution in the handling of a suspected carcinogen before conclusive evidence is in. Otherwise the harm may be irreparable. The instruments for a cautious approach are provided in federal and state laws passed in the last decade in response to the evident growing menace of hazardous chemicals, including such federal laws as the Toxic Substances Control Act and the Resource Conservation and Recovery Act, which is complemented in Missouri by a 1977 hazardous waste law.

Still other laws are being considered, such as a bill now in Congress that would extend controls to small waste generators and require better methods of waste disposal. But even the wisest and most exacting laws will not do any good if they are not enforced. Frequently both federal and state laws offer only paper protection because required reports of conditions go unprocessed — sometimes because of willful neglect but more often because federal and state agencies are understaffed. Industry often delays enforcement through extended court challenges.

The *Post-Dispatch* special section reported that estimates of the dioxin cleanup cost in Missouri range up to a billion dollars. The obvious message is that strengthened agency staffs followed by stringent law enforcement could avert astronomical cleanup costs and avoid future hazardous waste disasters such as the dioxin problem now plaguing Missouri.

Minneapolis Star and Tribune

Minneapolis, Minn., December 26, 1983

The Environmental Protection Agency's new plan to clean up dioxin contamination is a rational response to an environmental disaster. If instituted quickly and pursued aggressively, the plan could be a powerful tool to safeguard human health.

Dioxin, a byproduct of the herbicide 2,4,5-T, is remarkably toxic. Tests on animals show that, over time, even minute amounts of the chemical can cause severe harm — from spontaneous abortions and birth defects to liver damage and inability to resist infection. But the biggest long-term threat is cancer: Rats fed one ten-millionth of their weight of dioxin develop tumors of the liver, mouth, lungs and thyroid. Research on humans shows the same cancer link: A 1978 study of Swedish forest workers who used dioxin-contaminated herbicides found that exposure produced nearly a sixfold increase in soft-tissue sarcomas.

To avoid the risk of cancer, the Centers for Disease Control now recommend that a contaminated area be evacuated if the proportion of dioxin in the soil exceeds one part per billion. That was the situation in Times Beach, Mo., where waste oil containing dioxin was sprayed more than a decade ago. Tests in 1975 showed dioxin concentrations there of about 30 parts per *million*. But the EPA acted slowly, evacuating inhabitants and buying homes only when a flood spread the poison all over town.

The EPA deserves credit for finally moving to clean up Times Beach. But one EPA buyout does not constitute a policy. Hundreds of other sites around the country are believed to be just as seriously contaminated with dioxin; residents of those areas face continuing health risks. Consistency and compassion demand a clear plan for cleaning up all the sites and helping the victims.

The EPA's aggressive new approach could fill the bill. The agency plans first to identify sites where dioxin-tainted chemicals have been produced or dumped and to assess risks to human health. Then it promises to do what is necessary to limit human exposure to dioxin. The agency says it will adhere to the Centers for Disease Control's one-part-per-billion guideline as the threshold of significant risk.

Such promises are encouraging, but they also point to the plan's chief shortcoming: It contains no specific plans for cleaning up sites after they are discovered or for helping those who are exposed. And, except for $4 million for a national study, the plan includes no additional money to combat the dioxin problem. The EPA says it will use the $1.6-billion superfund, established by Congress for cleaning up hazardous waste sites, to accomplish its task. But without clear procedures and earmarked funds, the new plan is no guarantee of action. Study is only a start. The EPA also has an obligation to move quickly to clean up the sites and help the victims.

The EPA "Superfund": Cleaning Up Our Chemical Legacy

The 96th Congress, in its last days, established a $1.6 billion fund to finance the cleanup of toxic-waste spills and abandoned chemical dumps. Much of the impetus for the 1980 legislation, commonly referred to as the "Superfund" law, came from political pressure generated by the Love Canal emergency. (See pp. 198-207.) The bill, supported by President Carter, gave the Environmental Protection Agency the authority to clean up leaks into the environment of hazardous substances, and to sue those responsible for the spills in order to recover cleanup expenses. The fund was to be established over five years, with about 87% of the amount coming from excise taxes levied on the chemical industry and oil companies that produced chemicals, and the remainder from government appropriations. A controversial provision that was not included in the final legislation would have allotted funds for the compensation of victims of a chemical spill.

The 1983 political scandal at the EPA that eventually resulted in the conviction of Assistant Administrator Rita Lavelle and the resignation of Administrator Anne McGill Burford grew out of an investigation into implementation of the Superfund law. A House subcommittee, concerned about a marked increase in the illegal dumping of toxic wastes, reported in 1982 that the deregulatory emphasis of the Reagan Administration had encouraged lax enforcement by the EPA of the Superfund and other hazardous waste laws. Later, multiple congressional investigations resulted in allegations of conflict of interest, political manipulation of Superfund grants and obstruction of congressional inquiry. The charges were levelled at Burford and a handful of other officials. Lavelle, the only official eventually convicted, had been head of the toxic-waste program until she was fired by President Reagan in February, 1983.

To date, the EPA has compiled a "national priority list" of 542 dumps that should be cleaned up as quickly as possible, and estimates that three or four times that number will eventually wind up on the priority list. The total number of abandoned toxic waste dumps in the U.S. runs into the tens of thousands. Although cleanup work has been completed at only six of the priority sites, the EPA under Burford had signalled that the Reagan Administration would not seek renewal of the Superfund when it expired in October, 1985. In part as a result of the political furor caused by the 1983 investigations, the House in August, 1984 overwhelmingly passed a robust expansion of the Superfund program. The legislation, which had not yet gone to the Senate for action, would create a $10.2 billion Superfund program over five years, and would require the EPA to speed up its cleanup efforts, initiating work on 150 new sites a year.

Minneapolis Tribune

Minneapolis, Minn., November 26, 1980

The hazardous materials superfund, one of President Carter's top environmental priorities, has been saved from lame-duck oblivion — ironically, by Republicans. After more than three years of work and debate, the legislation — which would provide an emergency mechanism to deal with spills and improper hazardous material dumps — seemed likely to die, a victim of the chemical industry's divide-and-conquer strategy. Without substantial Republican support, it would have.

Sen. Robert T. Stafford, Republican of Vermont, persisted, with support from the Republican leadership, in fashioning a compromise that undercut the chemical industry's righteous posturing and was acceptable to a Senate majority.

Sen. Dave Durenberger of Minnesota quickly signed to cosponsor Stafford's bill. Durenberger's action is commendably in the best interests of Minnesotans. This state knows firsthand the frustration of dealing with improperly discarded toxic substances. Liability often is so confused or the guilty party so recalcitrant that the state must choose between financing the clean-up itself, with funds diverted from other important goals, or delaying action until a court decides who should pay. When Minnesotans face a potentially serious health threat, a wait for court action is neither morally nor politically acceptable.

Superfund would solve that problem by providing a reserve fund to finance quick cleanups, with questions of liability decided as the courts are able. Those assigned liability would reimburse the fund. The reserve would be created primarily by fees on chemicals at their source, making the cost of clean-up part of the price paid for their use. Given the questions of liability, that's the only fair way to address the toxic materials problem. It's also a solution that only the federal government can impose.

The Senate received, debated and passed a revised Stafford bill Monday — an impressive achievement. The House should now show similar alacrity and accept the same legislation. The country would be grateful to responsible Republicans and Democrats in both houses who defied expectations for this lame-duck session by addressing a national problem with pragmatism and good sense.

THE TENNESSEAN

Nashville, Tenn., December 1, 1980

SENATE approval of "superfund" legislation means that some action will be taken to meet the problem of toxic waste dump sites which threaten the public safety. It is only one step forward, however.

Senators and representatives, who have worked on the legislation, are trying to work out a compromise between the Senate and House versions of the bill. With only a week remaining in the session, it is important to avoid a conference and the possibility of no final action. To get a bill passed at all has meant making concessions.

The dangers of toxic waste dump sites were made vivid with Love Canal where families had to leave their homes after reports of illnesses possibly related to the hazardous materials buried nearby. The Environmental Protection Agency has estimated there are thousands of similar dump sites and that many of them may pose dangers to the public health. The superfund legislation is aimed at cleaning up those sites.

The important fact is that money is being provided to clean up these dumps. The Senate approved a fund of $1.6 billion that would be used for cleaning up dumps and spills. The House had provided for almost $2 billion in three separate funds, with $1.2 billion aimed at cleaning up the dump sites. That means the amount targeted for the clean-up of toxic waste dumps is about the same.

The House and Senate versions are also close in the levels of contributions made by the chemical industry to the fund. The Senate bill calls for industry to contribute 87.5% of the fund with the remainder coming from federal revenues. The House had provided for the industry to provide 75% of the $1.2 billion earmarked for dump clean-up.

Where important concessions have been made is in the area of liability, an area in which Rep. Albert Gore Jr. of Tennessee has worked especially hard.

The Senate ruled out the concept of joint and several liability, and has left it up to the courts to decide how responsible individual companies may be in a situation where a number are involved. The Senate also left the definition of liability in muddy waters.

But the Senate version, in effect, left intact limits on the use of a third-party defense, and that is important. A federal cause of action was taken out of the Senate bill, as happened in the House, and that means individuals will still have to confront the maze of state laws.

The concessions on liability are important, but they probably do not cripple the legislation, and it is significant that the Senate decided to act. It should be possible for a quick compromise to be worked out and for superfund legislation to pass the Congress in this session.

The Register

Santa Anna, Cal., December 6, 1980

The toxic waste "superfund" legislation passed last week was the kind of horror that gives critics of "lame duck" sessions of Congress nightmares — a hasty, ill-considered, ill-advised and expensive program that will probably retard progress on the problems it is intended to solve.

The bill Congress passed lumped several problems — toxic spills, hazardous substance spills and inactive hazardous waste sites — into one category, and some congressmen wanted to throw in oil spills for good measure. It created a superfund — 87.5% of which is to be funded by a new tax on chemical companies — of $1.6 billion to handle problems and liabilities arising from disposal problems. Some congressmen tried to pretend that since most of the money will come from a special new tax, it will represent only a minimal increase in federal spending, and no loss to taxpayers. Anyone who has gotten past kindergarten economics could have told them better. The legislation won't solve anywhere near as many problems as it will create.

For starters, the technological and environmental problems associated with different kinds of disposal are different. Oil spills aren't the same as chemical spills. Relatively inactive chemicals differ from highly toxic ones. The technology for safe disposal of each of them will be different. Lumping them all together will create inflexibility, discourage innovation, and perhaps keep disposal technology mired in the practices of the past, which are widely conceded to be inadequate.

The "superfund" will come from chemical companies because they happen to be in the business — whether they have a history of scrupulous care in waste disposal or a record of wanton disregard for safety and environmental consequences. Polluters and non-polluters alike will pay into it. That's not only

unfair, it encourages practices few of us would approve. It provides incentives for companies to act as if once they pay into the fund, their responsibility for trying to improve their practices and come up with safer disposal methods is at an end. If you're going to pay whether you act responsibly or not, the inclination is to do the minimum the law requires.

The creation of a government "superfund" would emphasize collective responsibility rather than individual responsibility. Lack of foresight and individual responsibility are the main contributors to the problem we now face. To have government rather than individual or corporate polluters assume liability for disposal problems would slow down the process of finding ways to clean up old sites and make new ones safer.

It has just been in the past few years that most of us have become aware of the problem of hazardous waste disposal. We still know very little even about the magnitude of the problem. The EPA has estimated that there are 32,000 to 50,000 sites in the country. That's a big margin of error. But the House Commerce Subcommittee investigating the problem identified 3,383 sites while a chemical industry survey found 4,196. Not all of them can be correct.

It was certainly premature to pass legislation when so little is known about the scope and nature of the problem. Individuals and groups damaged by irresponsible dumping should have some recourse for damage and liability claims through the courts now, without new laws. And the existence of a "superfund" of whatever magnitude will bring the temptation to use that money to study the problem to death (at which government has proven adept) while doing little to clean up actual toxic wastes.

A few professional government contractors will be delighted at the passage of this legislation. The rest of us may be forgiven if we restrain our enthusiasm.

The Wichita Eagle-Beacon

Wichita, Kans., December 5, 1980

History can remember that the 96th Congress used some of the final hours of its lame duck session to do something very positive for the American people and their environment: It passed the "superfund" bill, making possible an immediate start on the cleanup of toxic wastes, such as those that made New York's infamous Love Canal area virtually uninhabitable.

Now, the nation's drinking water will be safer. Dumps or chemical spills that endanger groundwater will be one target of the program to be financed largely from taxes on industry. Provisions of the bill passed do not include oil spills, nor a way for persons whose health or property have been harmed to recover damages easily. There wasn't time to compromise

all the differences between earlier House and Senate bills.

But now, President Carter's signature on the bill for which he actively lobbied is sufficient to get the war on toxic garbage started. Spokesmen for the chemical industry have agreed the bill is a fair one. Now something can be done to neutralize the health and environmental problems caused by improper disposal of hazardous wastes in the past.

Some lame-duck Congresses and presidents in the past have bowed out petulantly. This outgoing Congress and president have given the nation a really practical Christmas gift, even though it is less than what many, including ourselves, had wanted.

THE ATLANTA CONSTITUTION

Atlanta, Ga., December 5, 1980

Not too many years ago, so-called "environmental issues" were usually considered the sole province of the bird-and-bunny crowd. No one else seemed to care what was pumped into the atmosphere, poured into the water and buried in the ground. After all, a few fish might die or a few animals might keel over, but nothing else of much significance would happen. Or so the public thought.

After a time, the people got wise. The polluters' propaganda was mostly untrue, they discovered. Toxic waste not only killed fish and rabbits; it killed people. It caused birth defects and strange sicknesses among human beings. Public opinion turned. The politicians responded. Tough laws concerning air and water pollution were enacted. And, yes, it should be emphasized that industry responded, too. It decided that the best public relations efforts could not match the amount of good will generated by a factory that worked at keeping the environment around it clean.

Environmental problems, however, extend beyond merely cleansing lakes and streams and the out-pourings of smokestacks. In recent months, the public has come to understand that toxic waste dumps and accidental spills can be just as deadly to the human environment as poison dumped into a river. Love Canal in New York served as a dramatic and tragic example of the ill effects on humankind of unguarded dumping of poisonous materials.

Once again, the politicians have responded to public pressure to clean up the poisonous dump sites. The House this week passed and sent to President Carter a $1.6 billion "superfund" bill to help clean up waste dump sites and toxic spills across the country over the next five years.

Most of the funding for the bill will come from polluting industries. President Carter, who was identified with the bird-and-bunny crowd long before fighting for the environment became a popular cause, telephoned dozens of congressmen to gain support for the superfund bill.

We are pleased that he plans to sign it into law. There are, however, some flaws in the legislation. It does not, for instance, cover oil spills which are a growing problem in many coastal areas of the country. Nevertheless, the measure does constitute a beginning and it serves notice on polluters to clean up their act.

The News Journal

Wilmington, De., December 5, 1980

The passage of the so-called Superfund bill to confront the cleaning up of wastes, principally involving the chemical industry, is a reassurance to the public that serious attention is being focused on a serious problem.

Perhaps some members of the House are justified in protesting that delays in the Senate ruled out serious efforts to reach further compromises on the measure. However, getting some legislation on the books in this lame duck session of the Congress was worthwhile for at last one reason: to send the aforesaid message.

The dimensions — $1.6 billion — and the sources of the Superfund may need amending later. There might be cause to revise the ways in which the fund can be applied.

The time to get the law enacted, however, was yesterday. Since that was impossible, today is the next best thing.

THE WALL STREET JOURNAL.

New York, N.Y., December 15, 1980

Now that the House and Senate have hastily agreed to a compromise toxic wastes superfund bill, the critical decisions about cleaning up abandoned dumpsites and chemical spills will be up to the Reagan administration. Shortly after he takes office, Mr. Reagan will be required to submit a contingency plan for administering the superfund, including criteria for choosing cleanup sites. He has the chance, therefore, to ensure that the Environmental Protection Agency doesn't abuse the open-ended authority it is given by the new legislation.

To begin with, it will be important to focus EPA attention on dumpsites and spills that threaten public health in a very real way. Seepages of toxic chemicals, for example, that risk contaminating drinking water. Or potentially explosive storage dumps, such as the Chemical Control depot which recently caught fire in Elizabeth, N.J. Under the sweeping language of the new law, the EPA is empowered to clean up any chemical release possibly endangering "public health or welfare or the environment"—blanket authority which could conceivably require that every dumpsite in the country be restored to conditions of pristine wilderness. Remedial work belongs where it is most needed, namely threats to human health, not eyesores or seepages that kill a few bushes.

The second task for the new administration will be to add some guidelines for keeping down costs. The superfund law specifies that cleanup and remedial work be "cost-effective" but fails to spell out what this means. All that is needed are some common sense standards for economizing, of the sort that are so uncommon at the EPA. Don't aim for zero discharge, for example, if the goal isn't required for public health reasons, and cleaning up the last 2% of a spill costs as much as the first 98%.

Cost control standards are especially important, because the superfund is unrestricted by any budgetary limitations. Much of the debate in the Congress focused on the size of the su-

perfund—proposals ranged from $600 million to $4.1 billion, with the compromise legislation settling on $1.6 billion, to be raised mostly by an excise tax on chemical feedstocks. But that $1.6 billion is just for starters; the EPA can constantly replenish the superfund by reaching into the pockets of chemical manufacturers, transporters and dumpsite operators that are found liable. Without clear cost guidelines, there will be a great temptation for the EPA to spend freely—at the expense of liable chemical companies.

The basic idea of the superfund, of course, is eminently sensible and consistent with market principles. Prices ought to internalize external costs, and if the manufacture, transport and use of hazardous chemicals occasionally leads to dangerous spills and oozings, consumers of the product ought to pay for cleaning up the mess. But how high are the external costs? The EPA's behavior so far suggests that the agency will grossly exaggerate them.

Throughout the whole toxic waste controversy, the EPA has been peddling slipshod studies that overstate the dangers involved. In May this year, it needlessly scared hundreds of Love Canal families out of their wits when it released a chromosome-breakdown study that has subsequently been discredited by three separate panels of scientific experts. On the basis of extrapolations from 232 actual cases, the EPA publicized a wholly unsubstantiated estimate of between 30,000 and 50,000 hazardous dumpsites requiring remedial action. Its rhetoric has screamed of "toxic time-bombs ticking away."

In truth, some dumpsites do threaten public health. Now that the superfund bill has passed, we can hope that after years of financial limbo—during which local governments have often been waiting for someone else to pick up the tab—genuine health hazards will finally be cleaned up. But the Reagan administration should make sure that the EPA doesn't use the superfund as an opportunity to go on a massive spending binge and destroy the chemical industry.

The Des Moines Register

Des Moines, Iowa, December 1, 1980

The U.S. Senate took an important step when it voted to create a $1.6-billion "superfund" to help clean-up hazardous chemical spills and waste dumps.

Iowa's Senator John Culver deserves much of the credit for the best features of this bill. Unfortunately, the legislation that Culver originally worked for became politically infeasible following the November elections. The bill was stripped of provisions that would have helped victims of hazardous waste dumps and spills recover their medical expenses and lost wages. The size of the fund was

whittled back severely, from $4.2 billion to $1.6 billion.

The Environmental Protection Agency has estimated that there may be more than 2,000 hazardous dump sites around the country that pose a threat to public health. Each year, there are about 3,500 chemical spills capable of inflicting harm on the environment.

Chemical spills and hazardous waste dumps have demonstrated their power — at Love Canal and elsewhere — to cause serious and sometimes fatal diseases, pollute water supplies, destroy farm products, ruin fisheries and damage other natural resources.

The Senate bill would establish a $1.6-billion fund — financed through a tax on the chemical industry and public tax revenues — to pay for the quick clean-up and containment of such spills and dumps over the next five years. If the responsible company or other party were later identified, it could be held liable for the costs of the clean-up. Senate aides believe most of the money won't be

recovered, due to difficulties in establishing responsibility. They are convinced that the $1.6-billion fund isn't adequate to deal with the problem.

An equally serious flaw with the Senate bill is that it won't help the victims of tragedies like Love Canal recover their out-of-pocket medical expenses and wages. It was the anguish of the Love Canal victims that originally helped spur Congress to action on this issue.

Thanks to Culver, the EPA is already working on containing the wastes at the LaBounty dump in Charles City, Ia., as part of a special demonstration project. Should this effort prove unsuccessful, the $1.6-billion fund could be used to pay for hauling the wastes from the dump. The fund also could be used to test the safety of drinking water that might be affected by the LaBounty dump, provide alternative drinking water supplies to people whose water was contaminated, and conduct tests to determine if contaminated water had caused certain diseases — such as bladder cancer.

The key word here is "could." There probably won't be enough money in the $1.6-billion fund to pay for all of these measures, even if they became necessary.

The Senate bill now goes to the House, which has passed its own, weaker version. A key flaw in the House bill is that it fails to cover all releases of hazardous chemicals. Had it been in force, the House bill would not have covered such incidents as the release of PCB's in a feed-plant accident in Montana. This contaminated feed eventually spread to 19 states and two foreign countries.

Time is running out on the lame-duck Congress. If a disagreement between the two chambers develops on this legislation, the problem may be left for the next Congress.

The Senate version is better than nothing. The country has waited long enough for action on this massive environmental problem.

The Senate hazardous-waste bill is better than nothing. The country has waited long enough for action on this massive environmental problem.

DAYTON DAILY NEWS
Dayton, Ohio, December 23, 1982

It's nothing really for Ohio to boast about to have within its borders 19 of the worst hazardous waste dumps in the country, but at least by receiving such mention earlier this week by the U.S. Environment Protection Agency, it will be easier for this state to get these dangerous messes cleaned up.

Congress in 1980 passed a $1.6 billion Superfund law to clean up dangerous dump sites. The EPA hasn't been too swift about naming the country's worst offenders to take full advantage of this special kitty, but last year four Ohio sites, including Chem-Dyne Corp.'s dump in Hamilton, were named on an interim list of 115 sites. That has helped the Ohio EPA with its cleanup campaign.

The Superfund provides money to pay for cleanup operations if the site owners won't do it themselves. The state EPA then can sue the owner to collect triple the amount spent. Such leverage often convinces owners to clean up their own dumps.

The U.S. EPA's latest list names 418 sites in the country, 115 of them, including the Arcanum Iron and Metal Co. and Chem-Dyne sites, considered more hazardous than America's most infamous dump, the Love Canal in New York, where hundreds of families were forced to leave their homes because of chemical pollution.

Ohio already has received $7 million from 13 Superfund grants, some of which were issued on an emergency basis for immediate cleanups. Now with additional money available, the Ohio EPA will be able to take action against obvious environmental spoilers and conduct studies at those sites suspected of causing pollution.

Chemical dumps pose substantial risks to public and private water supplies, soil health and air quality. The Ohio EPA diligently has been attempting to cut the damage in this state. It's good the U.S. EPA, whose concern for the environment at times seems lacking, now is lending a greater hand.

The Times-Picayune
The States-Item
New Orleans, La., October 28, 1981

At first glance, there might appear to be some solace in the fact that not one of Louisiana's hazardous chemical waste sites was included on a list of 114 targeted for federal "Superfund" cleanup money. That could mean that none of the Louisiana dumps is as dangerous to human health and the environment as any of the 114 that made the list.

But no one at the state level, including Gov. David C. Treen, seems to know exactly how the U.S. Environmental Protection Agency came up with its list of "winners." So it is not reassuring that not one of six potential Louisiana entries made the EPA's initial "priority" list.

As recently as last summer, some 23 chemical dumps in Louisiana were sufficiently hazardous to warrant the attention of the state Environmental Control Commission and the attorney general's office. Perhaps the two worst sites are the Cleve Reber dump near Sor-

rento in Ascension Parish and the Tate Cove dump near Ville Platte in Evangeline Parish.

Numerous news stories have documented the seriousness of the hazardous chemical waste disposal problem in Louisiana, particularly as it pertains to abandoned dumps and the dangers they pose to water supplies. During the last regular session of the Legislature, Attorney General William J. Guste Jr. felt the state needed at least $30 million to clean up the dumps.

The EPA reportedly will expand its list of eligible dumps in December. Since it is clearly the judgment of responsible state experts that the worst Louisiana sites fully qualify, state officials should do what they can to get them on that ignominious list. As the governor says, if federal funds are going to be dispensed, Louisiana ought to be given its share for what, by any measure, is an enormously costly cleanup job.

New Orleans, La., December 24, 1982

Louisianians undoubtedly have mixed emotions about their state qualifying once again for the United States Environmental Protection Agency's "Superfund" list. This time the state qualified by having two hazardous waste disposal sites ranked by the EPA as among the 418 most dangerous in the nation. Two other sites qualified earlier.

The bad side of this is that the EPA ranking underscores the fact that Louisiana has serious hazardous waste problems — a fact that all interested residents are aware of by now. The good side is that the state will be receiving federal funds to help clean up the sites.

The wonder to some close observers is that at least one other hazardous chemical dump was not included on the latest EPA list. We refer, of course, to the Petro Processors dump (actually, two locations are involved) north of Baton Rouge. Last summer state Attorney General William J. Guste Jr. caused a furor among other state officials when he said this dump, which sits atop the aquifer from which East Baton Rouge draws its public water supply, is potentially worse than New York's infamous

Love Canal. Mr. Guste's assertion was based on allegations contained in a suit filed by the U.S. Justice Department in behalf of none other than the EPA.

While the abandoned creosote plant on Bayou Bonfuca in St. Tammany Parish undoubtedly is a serious environmental hazard, it is questionable that it should rank above the Petro Processors site for federal cleanup funds.

In any event, Gov. David C. Treen has pledged "to see that all possible efforts are made to ensure cleanup of these and other dangerous sites as quickly and efficiently as possible."

The dreadful Cleve Reber chemical dump in Ascension Parish was the other site to qualify this time, and it was an appropriate choice. The 20-acre dump about two miles southwest of Sorrento contains an estimated 500 to 600 barrels of contaminated liquids in rusted barrels. It long has been a major source of complaint from nearby residents, as have the Bayou Sorrel dump in Iberville Parish and the Old Inger Oil Refinery site in Ascension Parish, the two sites chosen earlier for Superfund help.

The State
Columbia, S.C., January 4, 1982

IT IS gratifying that the U.S. Environmental Protection Agency is giving priority attention to the large chemical waste dump on Bluff Road south of Columbia.

This site, which contains 7,500 to 10,000 drums of toxic, flammable or volatile substances, is among 141 dumps in the nation identified in October as being eligible for aid from the $1.6 billion clean-up superfund authorized by Congress in 1980.

A spokesman for EPA's regional office in Atlanta said the Bluff Road site is considered the most environmentally harmful dump in the South. He estimated that it will take $4 million to $5 million to clean it up. The state will have to ante up 10 percent of the cost, and the law requires EPA to try to recoup as much as it can from companies that produced the dangerous wastes. The trouble is the generators cannot often be found.

This dangerous eyesore on Bluff Road was formerly operated by the S.C. Recycling and Disposal Co., which received the used chemicals from various sources. This company is said to be virtually bankrupt.

A smaller dump run by this outfit near Capital City Park in Columbia was cleared earlier with the use of emergency grants from the superfund. It was a hazardous undertaking. Some of the deteriorating drums had to be exploded.

While the federal law requires the chemical industry to contribute a substantial portion of the superfund with the U.S. government putting up the rest, we still consider this chemical waste situation to be outrageous. It is corporate irresponsibility at its worst. And we wonder why the state and local governments permitted problems like this to build up in the first place.

Surely the companies responsible for generating and disposing of these wastes and the governmental watchdog agencies employ people who understood the threats posed by these dumps. We continue to believe that something approaching criminal negligence is involved. But, as is too often the case, taxpayers will have to share the cost of others' wrongdoing.

THE MILWAUKEE JOURNAL
Milwaukee, Wisc., December 21, 1982

Anne Gorsuch, head of the Environmental Protection Agency, has a reputation for brusqueness, if not arrogance. She also has seemed less than enthusiastic about carrying out the EPA's enforcement responsibilities. (Her announcement Monday about the nation's worst waste-dumping sites is a notable exception.)

It was only natural, then, for the House Public Works Committee to become suspicious when she refused to surrender documents the committee had subpoenaed in its investigation of charges that the EPA is not requiring chemical companies to pay their full share of the cost of cleaning up toxic-waste abuses. And it was perhaps inevitable that the House would cite Gorsuch for contempt.

Nevertheless, that does not *necessarily* mean she is wrong when she says the documents are protected by executive privilege. The issue may have to be resolved by the courts.

Executive privilege is a right that presidents have asserted, in varying degree, to withhold from Congress information that the presidents preferred to keep confidential. Sometimes a president is justified, as when disclosure would compromise genuine military or diplomatic secrets. At other times, as when President Nixon invoked the doctrine as an excuse for withholding the incriminating Watergate tapes, executive privilege is grossly abused. The US Supreme Court properly rejected Nixon's claim.

Between the extremes, there is a broad, vaguely defined middle ground, a zone in which presidents and Congresses have engaged in tug-of-war. Lawmakers argue that they need particular items of information in order to determine whether a given agency is carrying out the laws properly; presidents contend that the congressmen are trying to usurp presidential powers. Is the Gorsuch case in that nebulous category? The House committee contends that it must have the documents to ascertain whether Gorsuch is breaking EPA laws by refusing to enforce them. The Reagan administration contends that release of the materials would jeopardize pending enforcement actions.

Given the Reagan administration's poor record on environmental matters, we are wary of the claim. Our skepticism is reinforced by the fact that 55 Republicans (including Wisconsin's James Sensenbrenner, Thomas Petri and Steve Gunderson) voted with the Democrats to cite Gorsuch for contempt.

If the administration persists in its defiance, it will have to justify its actions to the federal courts. It should not be hard for a federal judge to ascertain — in the secrecy of chambers, if necessary — whether the administration has an overriding need to withhold the information.

ALBUQUERQUE JOURNAL
Albuquerque, N.M., December 19, 1982

Anne Gorsuch is clearly in contempt of Congress for refusing to help the nation's leaders determine how well federal environmental laws are being enforced. President Reagan's reasons for ordering Mrs. Gorsuch to refuse to help simply do not wash.

Congress recently voted a contempt citation against Mrs. Gorsuch, the controversial head of the Environmental Protection Agency. One minute later the Reagan administration filed a federal court suit seeking to void the citation. Experts are calling the dispute the biggest separation-of-powers fight since Watergate.

Administration members can cite "reasons of national security" for refusing to give data to pertinent congressional committees. This argument was often used in the Watergate cases. Interior Secretary James Watt used the argument a few months ago in an administration-Congress dispute. But that was not the reason given by Mrs. Gorsuch.

At Reagan's order, she refused to produce documents relating to the government's efforts to clean up hazardous waste sites under "Superfund" legislation. She said that divulging the information, even to the relevant House committee, would "jeopardize" pending suits. Apparently Mrs. Gorsuch believes that disclosing the information would cause some companies — who may be called upon to help pay for cleanups of sites they helped pollute — to declare bankruptcy. This, according to her reasoning, would excuse them from participating in any future cleanups.

Congress created the Superfund in 1980 to help pay for cleaning up hazardous waste dump sites around the country. Many companies have volunteered to help pay cleanup costs since then. Such voluntarism has enabled those companies to fend off possible legal suits and has helped to stretch the $1.6 billion fund.

Congressional committees charged with accounting for taxpayer money spent in cleanups have an obligation to make sure the money is spent wisely. That explains the need for the documents held by the EPA, which is responsible for carrying out environmental law.

Mrs. Gorsuch's refusal to comply shows the administration will go to great lengths to prevent Congress from fulfilling that responsibility — almost as if the administration has something to hide. But Reagan can ward off those suspicions, help Congress do its duty, and possibly get the contempt citation dropped by ordering Mrs. Gorsuch to provide the information.

EVENING EXPRESS
Portland, Maine, December 17, 1982

The continued refusal of the Reagan administration to cooperate with congressional committees examining toxic waste cleanups does little to bolster an already dubious public perception of the administration's commitment to environmental health and safety.

President Reagan has shown no signs of weakening in his decision to order Environmental Protection Agency chief Anne M. Gorsuch to withhold documents subpoenaed by panels looking into the agency's handling of toxic waste cleanups. Mrs. Gorsuch has been cited for contempt by the House of Representatives in a decisive 259-105 vote Thursday night. Still she continues to withhold the documents.

Reagan is claiming executive privilege to withhold the material, and Mrs. Gorsuch says disclosure could put several legal cases in jeopardy.

But surely members of investigatory congressional committees are fully aware of the sensitivity of information involving pending litigation. It is likely true that public disclosure of some of the information would damage the government's position in court. Without examination of the subpoenaed documents, however, there is no way to know whether the claim of executive privilege is valid or whether their release would actually harm efforts to prosecute violators of environmental laws.

The committee chairmen who requested the documents say they are needed to ascertain whether EPA has acted properly in pursuing toxic waste cleanup cases.

Reagan and Mrs. Gorsuch ought to be eager to persuade both Congress and the public that the agency has been carrying out its mandate. Refusing to assist the committees in their examination of that question only lends fuel to suggestions that they have something to hide.

Newsday

Long Island, N.Y., December 31, 1982

Five years have passed since chemical hazards were discovered at Love Canal and the nation was awakened to the threat posed by industrial waste material. More than two years have passed since Congress adopted the Superfund legislation to deal with that threat.

After all that time, the Environmental Protection Agency appears to have progressed little beyond belatedly releasing its current list of the 418 most hazardous waste dumps in the country.

But making lists of dangerous sites is a long way from the more urgent task of cleaning them up — and there is scant evidence thus far of any major cleanup effort.

The new list is not the first issued by the EPA. It had previously identified 160 dump sites — eight of them in New York. But as Sen. Daniel Patrick Moynihan (D-N.Y.), a member of the Environment and Public Works Committee, points out, six of those — including the Old Bethpage landfill — have yet to receive a nickel from the Superfund for actual cleanup work.

Two other Long Island sites made the new, enlarged list: the landfills in Syosset and Port Washington. Altogether, 26 of the sites on the new list are in this state.

No one outside of the administration — including the members of Congress — seems to have a clear idea how much progress, if any, has been made toward cleaning up the identified dump sites. That's because the EPA has adamantly refused to tell anyone. But judging by how much money has been spent — $8 million out of $75 million appropriated in 1981 and $117.7 million out of $190 million allocated in 1982 — the cleanup campaign is far behind schedule.

At the very least, EPA should provide a report on the status of all the sites on its list, along with a schedule of planned cleanup work, complete with staffing and budget requirements.

Some of the 418 dumps are in established communities; others are close to populated areas. The people who live near these sites have a right to know just how long it will take to make their neighborhoods safe again.

EPA Administrator Anne Gorsuch was found in contempt of Congress earlier this month for refusing to provide documentation on the Superfund. She should understand there's more than a contest with Congress involved here; thousands of people who fear for their health need the EPA's information too. More than that, they need to have the chemicals that menace them removed.

The Oregonian

Portland, Ore., December 25, 1982

Publication of the federal Environmental Protection Agency's list of 418 hazardous waste sites is a positive starting point toward mandatory cleanup. The list offers affected industries a carrot — the chance of financial help from the environmental superfund. And the possibility of fines for dilatory cleanup provides the EPA with an enforcement stick.

Companies that produce hazardous waste as part of their process must modernize the way they handle waste. Standards for disposal of such materials have grown stricter as more has been learned about the penetration of heavy metals and chemical compounds into the air, food chain and aquifers.

Oregon has only two industries on the list — and they are far from the worst. Both already are moving toward handling their waste in a satisfactory manner in cooperation with the state Department of Environmental Quality.

One of the cases in Washington, a chromeplating company that discharges contaminated waste water, clearly shows the need to apply requirements for safe handling of hazardous waste evenly throughout any industry. The manager, after describing corrrective steps already taken by the company, pointed out that it would cease to be competitive if it is the only one of its kind required to give special handling to its waste. The listing of the one company should serve notice to others in the field that their number will come up.

Publication of the federal list reveals sites of hazardous waste in Idaho that otherwise would not be sniffed out because Idaho got rid of its environmental watchdog agency.

The listing of potentially dangerous waste sites is one of the most constructive environmental actions taken by the Reagan administration. No state seeks to be ranked on a hit list like this, so national exposure for the worst dumps in the nation might shame local governments and companies into cleaning up messes they have made.

Compilation of another Foul 400 should follow EPA and state action on the the present list of 418 sites. But let the public look at something more than soiled sites. EPA also ought to publish an honor roll of companies that have laundered their dirty linen.

"A POISONED APPLE. THE SLEEPING DEATH FOR THE ENVIRONMENTAL PROTECTION AGENCY! PERFECT!"

Arkansas Gazette.
Little Rock, Ark., March 22, 1983

The dimensions of the Reagan administration's failure in environmental protection seem to be expanding daily in each revelation about the way the Environmental Protection Agency has been shirking its responsibilities to the American people.

Especially unsettling is the thought that the EPA has presided over a hazardous waste program that actually has made the problem worse. The result flows from an attitude that President Reagan brought with him into office based on the assumption that government regulation is to be avoided, or voided, at all costs. In this case, the cost is measured in toxic wastes that are accumulating and posing increasing threats to the public health.

Part of this danger was brought into focus in an appearance that Dr. Samuel Epstein of the University of Illinois Medical Center made on CBS' "Face the Nation" the other day. Dr. Epstein holds the altogether sensible view that the basic trouble is to be found not in the person who runs the EPA but in the person who runs the White House: "Until there can be a major change in policies at the White House, and until the president can become * * * more in tune with national concerns, my basis for pessimism is of a high order." The priority at the White House now is on the costs of doing business, not on the costs of living in a fouled nest.

Hazardous waste dumping poses a far greater menace, in Dr. Epstein's expert opinion, than is generally recognized. He scoffs at the idea that the $1.6 billion superfund can pay for a cleanup of some of the worst abandoned toxic waste sites, and even dismisses the estimate of $40 billion that Congress' Office of Technology Assessment has placed on the cleanup. Cleaning up the hazardous waste, he figures, would cost a minimum of around $200 billion, a sum almost equal to a year's defense budget.

The EPA in the Reagan administration — it differs dramatically from the EPA in previous administrations — simply doesn't want to regulate those who don't want to bother with or pay for safe disposal of toxic wastes. The pity is that there are workable ways of recycling or disposing of hazardous wastes — some of them profitable — and the EPA could be using its time and effort to turn industry into the correct direction. Every year American industries generate 400 pounds of hazardous waste for every person in the United States. Only a tiny part of this is recycled, leaving massive amounts to be dumped (far too much of it illegally) or burned.

A survey of the possibilities produced recently for Worldwatch Institute by researcher Margee Ensign concludes that government does have a proper role in promoting recycling and proper disposal: "Although governments may be reluctant to provide waste treatment subsidies, recent environmental problems caused by hazardous waste demonstrate that governments can get involved now in solving the problem, or they will be dragged in later."

One of the most interesting ideas — converting hazardous wastes from one process into the raw material for another — offers considerable promise. Waste exchanges provide a way to match hazardous waste producers with businesses that can put the material to good use. Worldwatch Institute's study shows that there is great interest in American industry: in 1975 the United States had only one such exchange; today it has perhaps 30.

The point here is that protecting the American environment and helping to create a healthy business climate are not mutually exclusive. However, the point cannot be discerned from the Reagan administration's EPA, which obviously is more interested in protecting the short-term balance sheet of industries producing hazardous wastes than in making sure such industries do not poison the land and water of America.

The Boston Herald
Boston, Mass., July 31, 1983

THERE IS a new spirit over at the U.S. Environmental Protection Agency and Massachusetts is among the first places to benefit mightily from that spirit.

The EPA announced last week it will spend $1.9 million at five sites in the state to study and contain, if possible, chemicals dumped in those areas. If those efforts fail, the agency will then consider more expensive clean-up operations.

The point is, however, that the EPA acted swiftly and generously to protect residents of Holbrook, Groveland, Westborough, Bridgewater and Plymouth from whatever hazardous chemicals are infesting sections of those towns. And in Norwood as well, the EPA move in quickly to remove PCB-contaminated soil from a 16-acre site that has troubled local residents.

The once scandal-ridden EPA has obviously bounced back under the new leadership of William D. Ruckelshaus, who told reporters last week he believed the Reagan administration had initially misread the public's views on environmental laws. He said that while there was public support for deregulation of trucking and railroads and airlines, there was also broad public support for continued regulation of matters related to public health and the environment.

"We cannot deregulate in this area," Ruckelshaus said.

He added that while some laws need to be changed to make them "more reasonable, rational and less burdensome" that will not mean any change in the ultimate goal of protecting the public.

That is the kind of good common sense at the top that is leading to sound decision-making at the regional level. The result has been a restoration of faith in their government by the people in places like Holbrook and Norwood. And that's not a bad fringe benefit for the Reagan administration.

The Boston Globe
Boston, Mass., March 18, 1983

The Massachusetts Legislature, after faltering when the clock ran out on the 1982 legislative session, has now passed – and Gov. Dukakis has signed – legislation creating this state's $25 million "superfund." That should unlock $1.6 billion in federal superfunds and trigger cleanup action at New Bedford, Lowell, Woburn and other hazardous-waste sites; unfortunately, it doesn't.

Somewhat overlooked as the political scandal unfolds at the Environmental Protection Agency is the bureaucratic inertia which seems to define the agency's response to hazardous waste.

Action on the toxic PCB pollution of New Bedford harbor is apparently still at least two years away, and EPA has given Massachusetts environmental officials a timetable of 65 weeks just to complete studies on the Siresim site at Lowell.

Massachusetts, of course, has already completed extensive studies of these sites, but were it to go ahead on its own and spend some of the newly appropriated $25 million to clean up some of the worst hazards, state environmental officials fear they would forfeit any chance to obtain the federal matching funds. "It's a 'Catch-22' situation," explained one state official. "To get federal superfund money, the site must be 'immediately hazardous,' but it the state does anything by way of remedial action, EPA will try to take that site off the superfund list."

In attempting to explain that the failure of the EPA's hazardous-waste cleanup program involves something more than conflict-of-interest politics, agency apologists have blamed the lack of state 10 percent matching funds.

"The word 'superfund' conjures up the image of a massive pool of federal dollars sitting out there waiting to be spent," claimed superfund director William Hederman a few weeks ago. "But we can't spend a dime if the states can't make their contributions."

Massachusetts now has $25 million ready to contribute; unreasonable delays by the EPA at this point will make it clear that the agency's present bad reputation is deserved.

the Charleston Gazette

Charleston, W. Va., April 1, 1983

SUSPICION that industrial polluters got "sweetheart deals" from the Environmental Protection Agency is bolstered by a mess at Fairmont.

During the 1970s, severe pollution poured from a Fairmont coke plant owned by Sharon Steel Co. (headed by notorious Miami financier Victor Posner, indicted last year on tax charges).

The state Air Pollution Control Commission said the plant's dilapidated coke ovens leaked fumes and smoke that stained Fairmont and left workers coughing up black dust. Commissioner Carl Beard said Sharon had "a long and sad history" of false promises about repairs. The state Water Resouces Board said Sharon was caught dumping a seven-year accumulation of chemical muck into the river. Labor Department inspectors cited the plant in 1978 for 70 health and safety violations. The coke plant closed in 1979.

An EPA lawsuit filed against Sharon during the Carter administration demanded nearly $300 million in fines for the Fairmont atrocity.

Later, during the Reagan administration, EPA offered to drop the fines in return for a Sharon promise to spend $2.5 million cleaning up toxic wastes at the plant site. The city of Fairmont entered the suit to fight the proposed settlement. The case is pending.

West Virginia University law professor Patrick McGinley, who represents Fairmont, says the EPA settlement offer is "preposterous" because Sharon is required by law to remove the wastes, anyway, regardless of whether the fines are dropped.

The West Virginia Citizen Action Group calls the EPA deal with Sharon "an improper agreement." CAG's current newsletter asks: "Can you imagine the Internal Revenue Service being owed $300 by a taxpayer and agreeing to settle for $2.50?" People of Fairmont are being cheated, CAG says, and so are honest companies which have spent millions to comply with pollution laws.

The aroma of EPA "sweetheart" deals isn't limited to congressional hearing rooms. Part of the smell can be detected here in West Virginia.

THE DENVER POST

Denver, Colo., May 18, 1983

ONLY DAYS after an internal report sharply criticized the Environmental Protection Agency's handling of the $1.6 billion Superfund program to clean up toxic-waste dumps, EPA's acting deputy administrator has indicated there will be important changes in administering the program.

It's about time.

Serious public health threats are posed by the more than 419 top-priority waste sites on the Superfund cleanup list. But EPA under Anne Burford was snail-like in moving to clean up those sites.

The agency then was more concerned with saving Superfund monies than with spending them. And it required that states contribute 10 percent of the remedial costs, beginning at the planning phase.

States beset by budget problems rightly complained that EPA was using that requirement to slow Superfund's implementation.

Now, after two years of seemingly subverting the program, it appears finally that EPA will try to use Superfund for what it was intended — cleaning up some of the nation's most dangerous toxic-waste sites.

EPA's acting deputy administrator, Lee Thomas, said some cleanup planning will be done without waiting for the required cost-share contributions from the states.

At several dump sites, rapid cleanup of surface pollution will be initiated without making that contingent on the development of long-range cleanup plans.

It also has been decided that regional EPA officials, who are closest to the problem, will play a key role in deciding what cleanup actions are necesary.

These steps were among several that the internal EPA report recommended to carry out Superfund as Congress intended. The time for that to happen is long overdue.

Los Angeles Times

Los Angeles, Calif., August 4, 1983

California is finally receiving money from the federal Superfund to start cleaning up the Stringfellow Acid Pits in Riverside County. But the chemical legacy of the dump will last far longer than the former Environmental Protection Agency officials who held up the funds for political reasons. They were fired; you can't fire acids and heavy metals.

Tuesday's initial grant of $2.7 million in Superfund dollars—to grow to $10 million by October—could be called too little, too late. Even so, it is an important step toward knowing what must be done to rid the area of health-threatening chemicals, some of which have already seeped into the groundwater.

It also is an important and expensive lesson in the need for vigorous enforcement of toxic-waste laws to prevent new Stringfellows and the agonies of uncertainty that toxic wastes have meant for residents in the Stringfellow area.

Joel Moskowitz, new director of the toxic substances division of the state Department of Health Services, took the occasion of the grant announcement to promise just that. With the example of Stringfellow, California can settle for no less.

To a casual visitor the acid pits, north of the Pomona Freeway and west of the city of Riverside, look deceptively benign these days. Between 1954 and 1972 they were raw craters in a canyon of the Jurupa Mountains into which workers poured DDT, chlorinated biphenyls (PCBs), tetrachloroethylene, arsenic, cadmium, lead and hydrochloric, nitric and sulfuric acids. Now a layer of clay covers the pits. Daisies bloom in the dry earth.

The dump stopped accepting toxic wastes in 1969 after heavy rains caused an overflow south into nearby Pyrite Creek and through the community of Glen Avon. Five years after the dump was closed to all wastes in 1972, more heavy rains threatened to break the Stringfellow dam and pour 4 million gallons of acid water into the neighborhood. To prevent that, the state deliberately released 800,000 gallons into runoff channels, one of which passes immediately behind Glen Avon Elementary School

and neighboring homes. The neighbors did not receive adequate notice.

Last fall the state Regional Water Quality Control Board ordered a clay cap placed over the empty pits to keep rains from brewing new batches of acidic water. But the cap already is eroding in places, and there still is seepage below the pits.

Most of the initial Superfund grant will finance a study to find a solution more permanent than a clay cap. One possibility is to neutralize Stringfellow's chemicals with other chemicals. Another is to dig down to bedrock and haul the dirt away. But even the bedrock is so heavily contaminated that flowing water would pick up its chemicals.

While the search for long-range solutions goes on, the first priority is to clean up and halt the movement of contaminated groundwater, which now has seeped about 2,000 feet from the south end of the pits. In addition, water that enters the canyon area above the pits must be diverted so that it won't continue to flow through or under the pits.

Many local residents insist on additional steps. They do not believe that studies of the potential health hazards from the dump and its runoff have been broad enough, and they want their own independent analysts to assess information on cleanup efforts. Given the strong feeling among residents of the community that they have not always been told the whole truth, the request for at least a monitoring role in the study is reasonable.

No matter what is done, the process will be costly. Officials have already seen estimates of cleanup costs skyrocket from $370,000 in 1976 to $5.5 million for the state's existing interim program to a possible $40 million. Whatever the eventual cost to the Superfund, the law requires that it be borne by those who dumped chemicals at Stringfellow.

The people around Stringfellow cannot put the hazards of toxic wastes out of their minds. Neither should any other Californian—not until industries that create the wastes and governments that must supervise their disposal can assure the state that half-measures have been translated into full-fledged solutions.

Roanoke Times & World-News

Roanoke, Va., March 26, 1984

CONGRESS and the president now agree that Superfund, the law providing money for cleaning up hazardous waste dumps, should be extended beyond its expiration date late next year. Still a long way off is accord on how strict the new law should be.

When Superfund was enacted in 1980, legislators thought that its $1.6 billion would make a good start toward cleanup. They were mistaken. The problem was even bigger than they had thought. And they could not have anticipated the reluctant enforcement by the Environmental Protection Agency under Anne Gorsuch Burford, who thought that Superfund should die a natural death in 1985.

EPA produced a national priority list of 546 dumps that should be cleaned up. But that work has been completed at only six sites. Meantime, several hundred more sites have been put on the urgent list. There are thousands of such dumps across the country, and the total cleanup cost could exceed $40 billion.

Reversing his prior stand, Reagan said in his State of the Union message this January that he would ask William Ruckelshaus, his new EPA head, to develop a proposal for extending Superfund. Since then, Ruckelshaus has said that EPA hasn't finished studies on how big the fund should be, and that the administration proposal won't be sent to Capitol Hill before December.

Key members of Congress don't want to wait that long to pass their own tough legislation. Overriding objections from the Office of Management and Budget, they pried out of EPA preliminary figures that put the number of potential Superfund cleanup sites between 17,000 and 22,000. The original $1.6 billion Superfund might take care of fewer than 200.

"There isn't a congressional district in this country that doesn't have this kind of problem," says Rep. Guy V. Molinari, R-N.Y. The president would be under election-year pressure to sign such a bill.

The issue is not only the size of Superfund. Oil and chemical companies feel that the fund's chief source, a tax on raw materials used to mmake industrial chemicals, is unfair. Rep. James J. Florio, D-N.J., chairman of the Energy and Commerce subcommittee that has jurisdiction over hazardous waste, wants to increase that tax and add a levy on the hazardous wastes at disposal time. He also would fix strict timetables for site cleanup and would compensate victims of toxic wastes.

There are many practical problems in a task of such scope, and no ideal approach. Finding and/or penalizing those who created the worst of the existing dumps is sometimes impossible. Oil and chemical companies may feel it is wrong to hold them responsible for others' acts by taxing their "feedstocks." But as an alternative, it would be just as unfair to tax, say, the incomes of ordinary consumers.

As an added revenue source, the "waste-end" tax seems reasonable. It could also encourage companies to reduce the amount of their waste. The other side of the coin is that it may lead to more night-time dumping at unauthorized sites. Preventing that will be a continuing problem whether the waste is taxed or not.

Cleanup timetables? These can introduce an element of inflexibility. But something obviously is needed to break the EPA out of its snail's pace. At the current rate, regulators will always have more to do at sunset than they've accomplished during the day.

Alarm bells should ring and caution lights flash at proposals to compensate victims of toxic wastes. A degree of help, especially in emergencies, may be warranted. And certainly, people should be able to seek damages in court from known polluters.

But it is beyond Uncle Sam's capacity to right all past wrongs to individuals. The most he can do in this case — and that will be plenty in itself — is to clean up the most dangerous dumps and curb further poisoning of the environment. To hold out hope of more is a delusion.

The Houston Post

Houston, Texas, July 3, 1984

In 1980 Congress created the $1.6 billion Superfund program to begin cleaning up the worst of 20,000 hazardous waste dump sites in this country. But so far only six of more than 500 sites designated among the most dangerous have been cleaned up, and the Environmental Protection Agency is eventually expected to add more than 1,000 sites to that list. Ironically, the Superfund is due to expire next year, just as its attack on these concentrations of discarded industrial poisons seems ready to pick up speed.

Congress is now working on legislation to reauthorize the Superfund. The bill that appears to be making the swiftest progress through the legislative maze is sponsored by Rep. James Florio, D-N.J. The House Energy and Commerce Committee last week approved an amended version of the bill that would extend the cleanup fund for another five years and increase it nearly fivefold.

The Chemical Manufacturers Association opposes the Florio bill. If that doesn't surprise you, try this: The CMA says it supports both the Superfund and its reauthorization. It also favors a substantial increase in funding — but not to the level proposed in the Florio measure.

The chemical industry recommends that the annual funding be raised to $850 million, less than half that proposed in the Florio bill but more than twice the amount now available. The chemical makers, who pay most of the Superfund cost through a tax levied on their feedstocks (raw materials), raise two main arguments against the Florio bill: 1) It would provide more money annually than can be spent effectively on hazardous dump site cleanup. 2) It would saddle the petrochemical industry with an unfair share of cleanup costs.

The CMA cites EPA testimony before Congress to support the first argument. As for the second, it wants other industries that generate hazardous wastes to help pay for the cleanup program through a new "waste-end" tax that would be collected at the dump site. The CMA proposes that the feedstock tax on the chemical industry be continued at its present level.

The chemical industry has brought many of its public-image problems on itself by past stonewalling on toxic waste eradicaton. But it now contends that, in its own self-interest, it wants the dump sites cleaned up. But industry members, including major Houston-area petrochemical producers such as Exxon Chemical Co. and Dow Chemical Co., make a convincing case against the Florio bill. It's the wrong solution to a pressing national problem.

THE LOUISVILLE TIMES

Louisville, Ky., June 26, 1984

The uproar over Smith's Farm, the second huge toxic waste dump to draw national attention to Bullitt County, is a tribute to the shortcomings in the federal Superfund cleanup program and to the dereliction of federal officials who reluctantly administered it.

At the same time, the dump's rise to prominence makes it a splendid advertisement for the tougher Superfund bill now starting to move in Congress. Although the program expires in 1985, Rep. James Florio of New Jersey and others want to renew it now for fear President Reagan, if re-elected, will try to reduce rather than enlarge it.

Why Smith's Farm was overlooked for so long is a mystery. It is reputed to contain more barrels of poison than any dump in the country.

Federal officials got to know Bullitt County well back when the Valley of the Drums came to light. The Environmental Protection Agency has used aerial photography to locate dump sites across the state. The 10,000 to 20,000 barrels on the surface of Smith's Farm cannot have gone unnoticed both by human eyes and by the camera's.

The best explanation is that in the early days of the Reagan administration, the agency did not want fresh evidence of environmental problems. In spite of its size and potential for harming land, water and people, Smith's Farm never even made EPA's priority cleanup list.

To his credit, William Ruckelshaus, the current EPA boss, has accelerated the effort to get rid of the worst dumps. Smith's Farm is now officially deemed a health threat. About $1 million in Superfund money will be invested in clearing the drums from the surface.

But it is clear Mr. Ruckelshaus and his successors can't seriously attack the problem with the resources now available. The $1.6 billion derived mostly from a tax on raw chemicals has paid for cleaning up only six sites, although Mr. Ruckelshaus says the total will reach 100. Emergency work has been done at nearly 200 others.

That's at best a start at dealing with 1,400 to 1,600 dumps eligible for the priority list. EPA reckons the cleanup cost could reach $16 billion, a sum that doesn't include what is often the most important part of the job — restoration of contaminated ground water.

Throwing money at problems is never a good answer. But given the immensity of the mess, Mr. Florio's plan to enlarge the Superfund to $9 billion is reasonable. In the interest of fairness, he would shift some of the cost from chemical makers to all companies that produce toxic waste. And he would set a timetable requiring cleanup of hundreds of the worst sites in five years.

Anyone who doubts the need for speed and money might contemplate the leukemia attributed to poisoned wells in Massachusetts or the higher than average levels of PCBs in the blood of Bloomington, Ind., residents. This, plainly, is a job that can't wait.

The Star-Ledger

Newark, N.J., July 12, 1984

Approval of the superfund reauthorization by a key House committee had two significant legislative aspects besides its passage: The one-sided confirmation—a 38-3 vote—and the rejection of efforts by some panel members to critically weaken the extension of the national toxic waste cleanup effort.

The superfund extender still has to clear two other House committees, but its passage by the Energy and Commerce Committee removes major obstacles that could have stalled consideration by the full House. And the broad consensus evident in the House committee action is reassuring in that it could be indicative of strong support on Capitol Hill concerning this important environmental legislation.

The superfund extender is a more comprehensive, realistically financed version than the one created by Congress four years ago to begin attacking the massive hazardous waste cleanup. The funding formula in the reauthorization bill will be five times larger than the present financing mechanism. Under the new proposal, $9 billion will be available to the states over a five-year period.

The accelerated proliferation of abandoned toxic waste sites has far outpaced the capability of the Environmental Protection Agency (EPA) to deal with this enormous pollution problem. The $1.6 billion raised by a tax on the chemical industry is rapidly being exhausted. In four years, the agency has designated 133 new sites for cleanup.

New Jersey is at the top among states with the most serious hazardous sites. While some cleanup progress has been made, there is still a long way to go before the state can even get a handle on its massive environmental problem.

Rep. James Florio, the New Jersey Democrat who is one of the initial sponsors of the superfund law, has been urging the EPA to move with greater speed on the current 546 priority sites in the next five years. He would put the agency on a strict cleanup schedule, utilizing the substantially greater financing that would be raised under the reauthorization measure.

The need for speed in toxic waste cleanup is imperative. There has been a disturbing time loss in this initiative because of mismanagement during the Reagan Administration's first two years. Another year has been lost undoing the damage inherited by EPA Administrator William Ruckelshaus.

With the proposed sizable increase in funding and a strict timetable in place, the Environmental Protection Agency would be firmly positioned to move expeditiously on the enormous task of cleaning up hazardous waste sites.

St. Petersburg Times

St. Petersburg, Fla., April 7, 1984

Suppose you lived near a hazardous waste dump that was leaking poisonous chemicals into your drinking water, the air and soil? You would want someone to do something about the dump endangering your health. So, how would you feel if the money in the federal government's Superfund ran out before the dangerous dump in your neighborhood was cleaned up?

Angry, frightened, cheated — those are just some of the emotions you might feel.

NO ONE REALLY knows how much it will cost to clean up all the hazardous waste sites that pose a serious threat to the public health and the environment. An internal Environmental Protection Agency report estimates that the federal government will have to spend at least $8.4-billion and as much as $16-billion. Some environmental groups contend that $20-billion is a more realistic figure.

But there is only $1.6-billion in the Superfund authorized by Congress in 1980 for cleaning up the most dangerous dumps. The so-called Superfund is not nearly super enough to deal with a health hazard much bigger than anyone imagined four years ago. If the fund is not fattened, the money will run out before hundreds of hazardous waste sites are cleaned up.

Sensible members of Congress realize that. Alarmed by the updated figures and projections, a bipartisan group of 12 House members, led by Rep. James Florio, D-N.J., introduced a bill that would permanently reauthorize the Superfund program, which is scheduled to expire next year. The legislation would replenish the fund at a rate of $2.4-billion annually for any year in which its total dropped below $3-billion. The additional revenue would come from an expanded tax on chemicals and would be supplemented by a tax on hazardous waste generated by industry.

But this sensible solution suffered a sneak attack last week. In a surprise manuever that caught Florio off-guard, the House subcommittee he chairs approved legislation that would reauthorize the Superfund for the next five years at its current $1.6-billion funding level.

Rep. Barbara Mikulski, D-Md., voted against the bill and aptly called it "an act of taxidermy on the Superfund legislation — it disemboweled it, cleaned it all out and stuffed it with illusions that . . . the bird is still alive."

FLORIO AND REP. Dennis Eckart, D-Ohio, said they believed the Reagan administration, specifically budget director David Stockman, was behind the "crafty" subcommittee manuevering. The Democrats charged Republicans were attempting to avoid election-year pressure to beef up the fund.

If that is true, the administration is acting against its own advice. Administration officials have acknowledged that the $1.6-billion Superfund is not adequate to clean up the nation's estimated 22,000 toxic waste dumps.

The full House Energy and Commerce Committee should overturn the subcommittee's vote and approve Florio's bill instead. Millions of American people already have gotten or will get the bad news that they live near a dangerous dump. They deserve the reassurance that the federal government will have enough money to protect them from such a health hazard.

The Philadelphia Inquirer

Philadelphia, Pa., April 29, 1984

Thus far, it has cost about $800 million to clean up just six hazardous waste sites around the country, or half the $1.6 billion earmarked for the federal Superfund program. That's an average of about $133 million per site. An additional 540 sites have been given cleanup priority. The final number of dangerous sites may reach 1,600. The Environmental Protection Agency estimates that the final cost may be $16 billion, or about $10 million per site.

The Commerce Department, however, has done some calculating of its own. It figures that the cost of additional cleanup projects will plummet to an average of about $1.437 million per site. Or at least, that's the way the numbers work out in a recent report prepared by Commerce analysts. The report maintains that only an additional $1.5 billion will be needed to rid the United States of the threats posed by hazardous waste sites.

The Commerce document evidently was prepared to support a developing Reagan administration position on reauthorization of Superfund before it expires next year. Mr. Reagan has promised to support reauthorization but has declined to specify to what extent, particularly a dollar figure.

The report was released by Rep. James J. Florio (D., N.J.), who noted that the Commerce data on the cleanup costs "fly in the face of every estimate" provided by knowledgeable authorities. Commerce officials, including Secretary Malcolm Baldrige, quickly repudiated their report in response to Mr. Florio's comments, calling it "wrong" and saying that its release was the result of a "comedy of errors."

That excuse might carry some weight if it were the only signal on the Superfund coming from the administration. But time and again, the Reagan administration has demonstrated its disregard for the threats of hazardous wastes and its casual attitude toward ridding the public of those threats.

Someone at Commerce ought to get out the calculator again and do some more figuring if the agency intends seriously to advise Mr. Reagan on an effective federal commitment to the Superfund. If that report represents the thinking inside the administration about the necessary federal response to the life-threatening problem posed by hazardous waste sites, then the whole Reagan pledge of support for Superfund is a cruel sham.

DESERET NEWS
Salt Lake City, Utah, June 23, 1984

Progress in cleaning up more than 1,500 of the worst toxic waste dumps around the nation has been agonizingly slow. A bill now in Congress to speed up the work ought to be approved.

A $1.6 billion "superfund" previously set aside for the Environmental Protection Agency to carry out the cleanup has not been handled well, leading to the resignation in early 1983 of EPA chief Anne M. Burford and the firing of several top officials amid charges of business favoritism.

Detroit Free Press
Detroit, Mich., February 11, 1984

NOBODY KNOWS exactly how many hazardous waste sites, each one a ticking time bomb, threaten public health and the environment in the United States. But unless the federal government extends the Superfund program created to clean up our toxic mistakes, we may find out in the worst possible way — waking up to discover ever more instances of harmful chemicals oozing out of the ground and leaching into water supplies.

An internal Environmental Protection Agency report estimates that the federal government will have to spend at least $8.4 billion and as much as $16 billion to clean up the worst of the country's hazardous waste sites. The agency wants Congress to extend the four-year-old, $1.6 billion Superfund program, which uses funds from a tax on chemical and petrochemical stocks to cleanse toxic waste sites. The program is set to expire in 1985.

The EPA report estimates there are about 22,000 hazardous waste sites around the country. Of these, the study group concluded from 1,400 to 2,200 require a federal cleanup because they constitute a serious threat to public health and the environment. Though President Reagan recently called extending the Superfund program an "important task," he is said to oppose introducing legislation extending and enlarging the program until 1985.

But congressional supporters of Superfund fear the president may not support reauthorizing the program at all should he win re-election this fall and no longer have to worry about toxic wastes becoming a campaign issue. Rep. James J. Florio, D-N.J., has introduced legislation making Superfund permanent. Mr. Florio is right to press for action now, particularly given the environmental agency's own staff findings on the necessary scope and range of an extended toxic waste program.

The cost of Superfund is large, but so is the need for it. We are a technological society, dependent on chemical compounds for the ordinary comforts of life, as well as for agriculture and industry. We can't live without chemicals, but we also can't live with them if they are allowed to spill unguarded and uncontrolled into our soil and water. We should neither delay nor fail to enact the continuation of a program to clean up our toxic wastes before they contaminate our lakes and our lives.

The superfund program is due to expire in 1985. In any case, the money in the fund is only enough to take care of the first 170 toxic dumps on the "worst" list.

The new measure now before Congress not only would extend the life of the superfund program beyond 1985, but it would boost financing to $9 billion through 1990.

That kind of spending, particularly at a time when a struggle is being waged to cut the budget, is enough to make taxpayers swallow hard, but it may be more palatable when spread over five years.

In any case, the threat to the environment and to human health is great enough that the cleanup must be done. Utahns can hardly demand the removal of the Vitro uranium tailings and at the same time deny similar relief to people with toxic dump problems that may be even worse.

Along with the money, the proposed legislation contains some excellent features, including establishment of a mandatory schedule for the EPA to follow in tackling the worst dumps, and tougher rules to make those responsible for the dumping help pay for the cleanup.

President Reagan has opposed reauthorization of the superfund program this year pending an EPA study. But the agency has acquired — deserved or not — a reputation for foot-dragging in this matter.

No real reason exists for continued delay. The toxic dumps are there, they are dangerous, and they must be dealt with — the sooner the better.

THE CHRISTIAN SCIENCE MONITOR
Boston, Mass., July 2, 1984

CONGRESS needs to get its act together on an issue that remains largely overlooked by a considerable segment of the American people:

The need for an expanded "Superfund" to clean up hazardous waste.

Unfortunately, Congress as a whole has dragged its feet on the issue this year, despite the fact that the current Superfund legislation expires in September 1985. That may seem a long way off, but considering the time it takes to put new laws in place, to staff regulatory agencies, and to bring industry up to date on changes in existing regulations, September 1985 can be said to be just around the corner.

Besides, there is legitimate concern among environmentalists that if Congress waits to enact new legislation until next year — after the November presidential election — there will be insufficient public pressure on lawmakers to enact a truly meaningful bill.

Environmental legislation could well be lost in the shuffle of the deficit-reduction and tax-reform measures that are expected to dominate the 1985 legislative agenda.

Such concerns ought not be taken lightly.

Just how important is this business of cleaning up waste dumps throughout the United States? Very important. The Environmental Protection Agency, for example, has identified more than 540 hazardous waste dumps. Moreover, there are some 17,000 sites that are considered dangerous.

Unfortunately, as pointed out by US Rep. Dennis E. Eckart in an article on today's Opinion page, since Congress enacted the Superfund program more than three years ago, the EPA has cleaned up only six hazardous waste sites.

At that rate, Mr. Eckart wryly notes, it would take the EPA 273 years to clean up all those sites.

The House Energy and Commerce Committee, to its credit, has cleared a $9 billion Superfund bill that, among other considerations, sets stringent timetables for cleaning up chemical dumps.

Many legislative hurdles remain — both on the House side and, particularly, on the Senate side. In the Senate, a measure will first have to clear the committee level. It is not expected that a bill will come before the full Senate until later this year — assuming strong support emerges in that chamber for such a measure.

Such a snail's-pace effort need not be accepted by the American people.

It is time to clean up America's far too many hazardous waste sites. That means that Congress must move ahead with greater dispatch on Superfund legislation.

THE ARIZONA REPUBLIC
Phoenix, Ariz., April 9, 1984

ONE of the most politically sensitive issues before Congress is whether it will strengthen the Superfund law to clean up hazardous waste sites this year.

Less sensitive but also an issue with important political implications is a possible congressional impasse over federal clean air standards.

Both issues affect all states. They also cross party lines. Republicans are joining Democrats to oppose President Reagan's postponement of the renewal of the Superfund until next year.

They also are expressing concern about effective federal air standards.

The $1.6 billion Superfund will not expire until Sept. 30, 1985. However, its supporters believe the president would be under great political pressure to increase funding and make the law stronger if Congress renewed it this year. Many are convinced the administration would favor much weaker legislation next year.

They are probably correct since Reagan had earlier opposed renewing the cleanup.

Also, the administration, in cutting many programs, might include the Superfund in 1985.

The Environmental Protection Agency admits only six of the nearly 550 most dangerous waste sites have been cleaned up. Such sites are increasing more quickly than they are contained.

Congress seems willing to boost Superfund financing to speed the cleanup of more sites. Many members also believe victims of such waste should be compensated by polluters and, in some cases, by the federal government.

Congress also fears that EPA may weaken public health standards by continued bickering over what amount of particles in the air constitutes a general health hazard.

Those up for re-election are concerned that, unless EPA decides on new standards, the old ones would not be considered effective. Better health standards would be left in limbo while the government fiddled. No politician wishes to be attacked for not promoting higher standards.

So, every politico in the country now wants to be Mr. Clean. Support for the Superfund and clean air will resound in Congress.

Voters, regardless of political ideology, support higher public health standards.

Any politician who has not figured that out doesn't deserve to be re-elected.

Index

A

ACID Rain
Fossil fuels—102–121
AEROSOL Spray Cans
Fluorocarbons—88–92
AGENT Orange
Burning toxic wastes at sea—163–168
Dioxins—208–217
AGRICULTURE
Pesticides—2–9
EDB uses curbed—10–21
California's "Medfly" spraying—22–31
US spraying of marijuana—32–35
Food irradiation—42–44
"Greenhouse effect"—93–101
Great Lakes pollution—136–145
AIRPLANES
Passive smoking—78–83
AIR Pollution
Asbestos—46–56
Lead—57–65
Auto emissions—66–77
Cigarettes—78–83
Arsenic (copper smelting)—84–87
Fluorocarbons—88–92
"Greenhouse effect"—93–101
Acid rain—102–121
Clean Air Act—122–133
Toxic waste burning at sea—163–168
AMERICAN Medical Association (AMA)
Resolution on dioxins—208–217
AQUIFERS
Ground water pollution—146–153
ARSENIC
Tacoma, WA copper smelter—84–87
ASBESTOS
Health hazards—46–56
AUTOMOBILES
Leaded gasoline—57–65
Emissions, vehicle inspection programs—66–77
Clean Air Act—122–133

B

BIRDS—see WILDLIFE
BLISS, Russel D.
Times Beach, MO dioxin contamination—208–217
BLOOD Disorders
Lead—57–65
BROWN, Gov. Jerry (Calif.)
"Medfly" spraying—22–31
BURFORD, Anne McGill
EPA "Superfund" investigation—218–228

C

CALIFORNIA
EDB uses curbed—10–21
"Medfly" spraying—22–31
CANADA
Acid rain—93–101
Great Lakes pollution—136–145
CANCER
EDB uses curbed—10–11
Food irradiation—42–44
Asbestos—46–56
Passive smoking—78–83
Arsenic—84–87
CARBAMATES
Pesticides—2–9
CARBON Dioxide (CO2)
"Greenhouse effect"—93–101
CARBON Monoxide (CO)
Auto emissions—66–77
Clean Air Act—122–133
CARSON, Rachel
Pesticides—2–9
CARTER, President Jimmy
EPA's "Superfund"—218–228
CHEMICAL Waste—See HAZARDOUS Waste
CHESAPEAKE Bay
Surface water pollution—136–145
Clean Water Act—169–175
CHILDREN
Lead and brain damage—57–65
CHLORINATED Hydrocarbons
Pesticides—2–9
PCBs—36–41
CIGARETTES
Passive smoking—78–83
CIVIL Aeronautics Board (CAB)
Passive smoking—78–83
CLEAN Air Act
Auto emissions—66–77
Revision stymied—122–133
CLEAN Water Act
Sewage—154–162
House revision passed—169–175
CLIMATE
CO2 and "greenhouse effect"—93–101
COAL—see FOSSIL Fuels
COLOMBIA
US spraying of marijuana—32–35
CONGRESS
Clean Air Act—122–133
Ocean Dumping Act—163–168
Clean Water Act—169–175
Toxic waste law (RCRA) revision—178–197
EPA's "Superfund"—218–228
CONSTRUCTION
Asbestos—46–56
COPPER
Tacoma, WA smelter (arsenic)—84–87
COURT Cases
Manville bankruptcy (asbestos)—46–56
Clean Air Act "bubble policy"—122–133
Agent Orange—208–217

D

DEFOLIANTS—see PESTICIDES
DIOXINS
Health hazards; Times Beach, MO—208–217
DRINKING WATER—see WATER Pollution
DRUG Enforcement Agency, U.S. (DEA)
Marijuana spraying—32–35

E

EDB—see ETHYLENE Dibromide
ENVIRONMENTAL Protection Agency (EPA)
EDB uses curbed—10–21
Leaded gasoline—57–65
Tacoma, WA copper smelter (arsenic)—84–87
"Greenhouse effect"—93–101
Ocean Dumping Act—163–168
Hazardous waste and RCRA—178–197
Love Canal—198–207
Toxic waste "Superfund"—218–228
ETHYLENE Dibromide (EDB)
Uses curbed by EPA—10–21

F

FARMERS—see AGRICULTURE
FISH—see WILDLIFE
FLORIDA
EDB uses curbed—10–21
FLUOROCARBONS
Ozone layer depletion—88–92
FOOD and Drug Administration (FDA)
Food irradiation—42–44
FOODS—see also AGRICULTURE
Pesticides—2–9
EDB uses curbed—10–21
PCB contamination—36–41
Irradiation—42–44
FORESTS—see VEGETATION
FOSSIL FUELS
"Greenhouse effect"—93–101
Acid rain—102–121
Clean Air Act—122–133
FRUIT—see FOODS

G

GASOLINE
Lead levels—57–65
GEORGIA
US spraying of marijuana—32–35
GORSUCH, Anne—see BURFORD, Anne McGill
GRAIN—see FOODS
GREAT Lakes
Surface water pollution—136–145
Clean Water Act—169–175